D0897771

FLORAL BIOLOGY

FLORAL BIOLOGY

Studies on Floral Evolution in Animal-Pollinated Plants

EDITED BY David G. Lloyd
University of Canterbury, New Zealand
Spencer C. H. Barrett
University of Toronto, Canada

CHAPMAN & HALL

New York • Albany • Bonn • Boston • Cincinnati • Detroit • London • Madrid • Melbourne
Mexico City · Pacific Grove • Paris • San Francisco • Singapore • Tokyo • Toronto • Washington

Printed in the United States of America

For more information, contact:

Chapman & Hall
115 Fifth Avenue
New York, NY 10003

Chapman & Hall
2-6 Boundary Row
London SE1 8HN
England

Thomas Nelson Australia
102 Dodds Street
South Melbourne, 3205
Victoria, Australia

Chapman & Hall GmbH
Postfach 100 263
D-69442 Weinheim
Germany

Nelson Canada
1120 Birchmount Road
Scarborough, Ontario
Canada M1K 5G4

International Thomson Publishing Asia
221 Henderson Road #05-10
Henderson Building
Singapore 0315

International Thomson Editores
Campos Eliseos 385, Piso 7
Col. Polanco
11560 Mexico D. F.
Mexico

International Thomson Publishing - Japan
Hirakawacho-cho Kyowa Building, 3F
1-2-1 Hirakawacho-cho
Chiyoda-ku, 102 Tokyo
Japan

1 2 3 4 5 6 7 8 9 10 XXX 01 00 99 98 97 96

Library of Congress Cataloging-in-Publication Data

Floral biology: studies on floral evolution in animal-pollinated
plants / edited by David G. Lloyd, Spencer C.H. Barrett.
p. cm.
Includes bibliographical references and index.
ISBN 0-412-04341-6
1. Flowers. 2. Pollination by animals. 3. Pollination by
insects. 4. Angiosperms-Evolution. I. Lloyd, David G.
II. Barrett, Spencer Charles Hilton.
QK653.F58 1995
582.13'04463—dc20 94-42927
CIP

To order this or any other Chapman & Hall book, please contact **International Thomson Publishing, 7625 Empire Drive, Florence, KY 41042.** Phone: (606) 525-6600 or 1-800-842-3636.
Fax: (606) 525-7778. e-mail: order@chaphall.com.

For a complete listing of Chapman & Hall's titles, send your request to **Chapman & Hall, Dept. BC, 115 Fifth Avenue, New York, NY 10003.**

Contents

Preface

Studies in floral biology are largely concerned with how flowers function to promote pollination and mating. The role of pollination in governing mating patterns in plant populations inextricably links the evolution of pollination and mating systems. Despite the close functional link between pollination and mating, research conducted for most of this century on these two fundamental aspects of plant reproduction has taken quite separate courses. This has resulted in suprisingly little cross-fertilization between the fields of pollination biology on the one hand and plant mating-system studies on the other. The separation of the two areas has largely resulted from the different backgrounds and approaches adopted by workers in these fields. Most pollination studies have been ecological in nature with a strong emphasis on field research and until recently few workers considered how the mechanics of pollen dispersal might influence mating patterns and individual plant fitness. In contrast, work on plant mating patterns has often been conducted in an ecological vacuum largely devoid of information on the environmental and demographic context in which mating occurs. Mating-system research has been dominated by population genetic and theoretical perspectives with surprisingly little consideration given to the proximate ecological factors responsible for causing a particular pattern of mating to occur. Over the past decade, however, this situation has begun to change as a new generation of floral biologists are trained with backgrounds in both ecology and genetics, as well as an appreciation of the natural history tradition of pollination biology and the need for theoretical approaches to guide problem-solving. This book is an attempt to highlight this new synthesis that is rejuvenating the 200-year-old discipline of floral biology and turning it into one of the most challenging and fast-moving areas of ecology and evolutionary biology.

This volume was initially conceived as a way of celebrating the bicentenary of the publication of Christian Konrad Sprengel's book *Das entdeckte Geheimnifs der Natur im Bau in der Befruchtung der Blumen (The Secret of Nature in the*

Form and Fertilization of Flowers Discovered) in 1793. Sprengel's volume stands out as one of the most significant works in plant reproductive biology because it effectively began the systematic study of the relations between flowers and insects. In his book, Sprengel described in fine detail structural adaptations of hundreds of species and provided for the first time functional interpretations of many floral mechanisms that are now the focus of much contemporary research in floral biology. It has been said that the study of pollination before and after the appearance of Sprengel's work were two different things.

To commemorate the publication of Sprengel's book, we organized the Sprengel Bicentenary Symposium on Floral Biology that was held at the XVth International Botanical Congress on August 29, 1993 in Yokohama, Japan. Using the eight presentations made at Yokohama as the nucleus of this work, we invited additional authors to write chapters on other areas of floral biology. The main goal in producing this volume was to provide an account of current issues and research opportunities in floral biology. Because of the enormous scope of contemporary research in plant reproduction, we have made no attempt to be comprehensive in our coverage of topics represented in the book. Instead, the chapters in this volume focus almost exclusively on research conducted on the floral biology of animal-pollinated plants, with a primary emphasis on understanding the evolution and functional significance of the spectacular range of floral variation displayed by flowering plants. The authors, all of whom are internationally recognized leaders in their field, were encouraged to present new research results and ideas. This they have done, giving the volume an up-to-date flavor that enables the reader to judge how far floral biology has come since its birth 200 years ago. This book should be of interest to a wide audience including advanced undergraduate and graduate students and research workers in the fields of plant reproductive biology, plant ecology and evolution, plant systematics, plant genetics and entomology. We also hope that because of the strong natural history tradition found in floral biology, this book may be of value to anyone fascinated by plants and their pollinators and who wishes to know more about how scientists investigate floral evolution.

This volume is divided into three parts. The first, "Historical Perspective," commences with an English translation by Peter Haase of the introduction of Sprengel's book on structural adaptations found in flowering plants that encourage cross-pollination. This is of considerable historical importance since it is the first time that Sprengel's writings have been translated from the original German into English. The translation will enable biologists to appreciate the full significance of the many novel discoveries concerning the functional significance of floral adaptations that were made by Sprengel following his careful observations of flowers and their pollinators. This is followed by Stefan Vogel's account of the historical importance of Sprengel's work and some of the reasons why his ideas were not fully appreciated by naturalists of the time. For it was not until Charles Darwin and other nineteenth century biologists began to interpret the living world in a radically different way that the true significance of Sprengel's work became evident.

The second section of this volume, "Conceptual Issues," covers current conceptual problems in floral biology. Foremost in the minds of most biologists when they contemplate floral evolution is the question of how pollinators have driven the diversification of floral form. In Chapters 3 and 4 by Carlos Herrera, and Paul Wilson and James Thomson, respectively, this issue is addressed from a critical and somewhat unorthodox perspective. In both chapters the authors question the common assumption that current selection pressures exerted by pollinators provide a universal explanation for the maintenance of variation in floral traits in populations of animal-pollinated plants and provide experimental data to support this view. Although in neither chapter do the authors doubt that floral syndromes in animal-pollinated flowers arise through pollinator-mediated selection, they challenge students of floral biology to provide more convincing evidence of this process and present several novel ways of approaching the problem.

Why do some flowers last for only a few hours, whereas others may be functional for several weeks? This question has often intrigued floral biologists, but until now there has been little attempt to address the problem in a rigorous manner. In Chapter 5, Tia-Lyn Ashman and Daniel Schoen provide a conceptual framework for approaching the question of the adaptive significance of floral longevity. The authors develop and test a model that predicts floral longevities are optimized by natural selection acting on the interaction between fitness returns through male and female reproductive function and floral construction and maintenance costs. Their chapter serves as an excellent illustration of how the development of phenotypic selection models can be applied to addressing long-standing issues in floral biology.

The next two chapters by Lawrence Harder and Spencer Barrett (Chapter 6) and Allison Snow and colleagues (Chapter 7) are both concerned with the general problem of the functional significance of inflorescence displays and the linkage between pollination and mating patterns. Harder and Barrett focus on how the design and display of flowers influence pollen dispersal. Based on theoretical analysis and empirical work using genetic markers, they propose novel adaptive explanations involving the promotion of fitness through male reproductive function for a range of traits (e.g., dicliny, dichogamy, and heterostyly) traditionally interpreted as antiselfing mechanisms. They argue that such floral mechanisms decouple the benefits of large inflorescences through increased pollinator attraction, from the costs associated with lost mating opportunities through male function. Under this view, the inflorescence rather than the individual flower is the more relevant unit of male function for most animal-pollinated plants. In Chapter 7, Snow and her collaborators also examine mating costs associated with large inflorescence displays, but in this case largely from the perspective of female reproductive function. The authors examine ecological factors that affect the frequency of self-fertilization as a result of geitonogamous pollinations in plants that simultaneously expose many flowers to pollinators. Geitonogamy is one of the least studied aspects of plant mating despite the fact that it is probably near ubiquitous in mass flowering species. Snow and colleagues argue for more attention to be paid to the ecology of geitonogamy and review experimental methods that can be used to study the phenomenom.

The last chapter on conceptual issues is by Lynda Delph and deals with the problem of flower-size dimorphism in species with unisexual flowers. The view that staminate flowers are generally larger than pistillate flowers has a long tradition beginning with Darwin's observations on dioecious and gynodioecious plants. More recently, this pattern has been used as evidence that the evolution of many floral traits is driven by sexual selection operating on male reproductive function. Delph reports on the results of a comprehensive survey of flower size in over 900 species of monoecious and dioecious species. In the survey female flowers are larger than male flowers in nearly half the species where a dimorphism exists and strong geographical patterns are evident in the data indicating that a reexamination of functional explanations for flower-size dimorphism in plants is in order. Delph proposes several adaptive and nonadaptive hypotheses that could account for the patterns revealed by her studies.

Many of the conceptual advances in floral biology have come from detailed work using model systems. These involve plant groups with attributes that make them suitable for addressing particular questions using observational, experimental, and/or comparative approaches. In the last part of the book, "Model Systems," a selection of research projects involving different model systems is presented. The section begins with Chapter 9 by Scott Armbruster who reviews his extensive research on *Dalechampia* (Euphorbiaceae), a genus of tropical vines, the flowers of which produce resins that are collected by bees for nest building. By combining phylogenetic reconstruction with field studies of the floral biology of populations, Armbruster has been able to take a truly integrative approach to the study of floral evolution in the genus. In our view, the research conducted by Armbruster represents the "way forward" in floral biology since it abandons the largely ahistorical approach taken by most workers in the field and considers both micro- and macroevolutionary time scales. In his chapter Armbruster focuses on the role of pollinator-mediated selection, genetic constraints, nonadaptive divergence, and adaptive compromise in generating the patterns of floral variation found within *Dalechampia*.

Long-term experimental field studies are often necessary to provide convincing evidence of the evolution of floral traits in response to pollinator-mediated selection. In Chapter 10, Candace Galen reviews her own studies in this regard on the long-lived alpine herb *Polemonium viscosum* of the Polemoniaceae. Galen argues that two basic requirements are necessary for thorough evolutionary analysis of floral form: the demonstration of heritable variation in floral morphology under field conditions and measurements of selection under similar circumstances. Few workers in the field of floral biology to date have attempted to satisfy these requirements. Galen reviews two sets of field experiments she has conducted on these topics. The results provide strong evidence for the evolution of clinal variation in flower size in *Polemonium* in response to selection exerted by bumble bees. Experimental field studies also involving bumble bees are employed by Douglas Schemske, Jon Ågren, and Josianne Le Corff in their examination of deceit pollination in *Begonia oaxacana,* a neotropical monoecious herb reviewed in Chapter 11. Borrowing manipulative approaches more commonly employed by behavioral ecologists, Schemske and colleagues constructed artificial flowers to investigate whether preferences by foraging bumble bees were evident for flowers of varying size. Their studies indicate that deceit pollination in *B. oaxacana* is highly effective. Although bumble bees display a striking preference for male flowers, rewardless female flowers receive sufficient pollen to achieve high reproductive success.

Deceit-pollinated plants are also used as model systems by Anna-Lena Fritz and L. Anders Nilsson to investigate the influence of variation in inflorescence size on male and female reproductive success. In their studies, presented in Chapter 12, populations of three nectarless orchid species (*Anacampsis pyramidalis, Orchis palustris,* and *O. spitzelii*) are used. The authors demonstrate that the gender of plants varies with the size, timing, and duration of the floral display and this is directly related to changes in pollinator visitation. Plants with small displays reproduced relatively more as male parents, whereas large-display plants were more likely to be functionally cosexual. Fritz and Nilsson argue that these changes in gender with display size are likely to be universal in plant species in which reproductive success is commonly pollinator-limited.

Despite the fact that the genus *Narcissus* (Amaryllidaceae) is one of the most important groups used in ornamental horticulture, the floral biology of wild species is virtually unknown. In Chapter 13, Spencer Barrett, David Lloyd, and Juan Arroyo investigate the patterns of sex-organ polymorphism in wild species from the Iberian peninsula. Their results suggest that this familiar genus may also have a bright future as a model system for floral biologists. In their chapter they address functional questions concerning the optimal positioning of sex organs in flowers experiencing contrasting pollinator regimes. Barrett and colleagues also settle a long-standing controversy by confirming the occurrence of the genetic polymorphism heterostyly in *Narcissus.*

The evolution of self-pollination has fascinated floral biologists for over a century. Shifts from outcrossing to selfing are particularly likely following island colonization, and comparisons of the floral biology of mainland and island populations can provide valuable model systems for investigating the selective mechanisms responsible for this shift in mating pattern. In the final chapter of the volume, Ken Inoue, Masayuki Maki, and Michiko Masuda provide evidence that floral traits and mating patterns in island populations of *Campanula microdonta* (Campanulaceae) are differentiated from those of mainland populations. The authors provide evidence that these changes are the result of adaptation to contrasting pollinator faunas. The studies by Inoue and colleagues illustrate a recurrent theme throughout this volume concerning the close functional link between the pollination biology of populations and their mating systems.

Floral Biology has benefited from the assistance of numerous colleagues. We offer particular thanks to the authors for their patience and goodwill during the editorial phase and to Jon Ågren, Tia-Lynn Ashman, Robert I. Bertin, Steven B. Broyles, Martin Burd, Deborah Charlesworth, Mitchell B. Cruzan, Rivka Dulberger, Christopher G. Eckert, Sean W. Graham, Lawrence D. Harder, David Houle, Brian C. Husband, David W. Inouye, Mark O. Johnston, Joshua R. Kohn, Susan J. Mazer, Martin T. Morgan, Randall J. Mitchell, Heather C. Proctor, Jennifer H. Richards, Kermit Ritland, Douglas W. Schemske, Daniel J. Schoen, Joel S. Shore, James D. Thomson, and Paul Wilson for careful reviews of individual chapters. We also thank Shoichi Kawano, Chairman of the Program Committe for Systematics and Evolution, for enabling the Sprengel Bicentenary Symposium on Floral Biology to be included as a special symposium at the XVth International Botanical Congress. We are indebted to Oswald Wyss for making available reproductions of original plates from Sprengel's book, and Taline Sarkissian for typing the index to *Floral Biology*. Christine M. Kampny for help in interpretation of material from that book. Finally, we offer particular thanks to Gregory W. Payne and the staff at Chapman & Hall for editorial assistance and Suzanne Barrett, William W. Cole, Lynda Delph, Lawrence D. Harder, Linda Newstrom, Diane M. Smith, and Colin J. Webb for numerous favors as well as support and encouragement over the three-year period that has seen this work evolve from an idea to a reality.

David G. Lloyd
Christchurch
Spencer C.H. Barrett
Toronto
October 1994

Contributors

Jon Ågren
Department of Ecological Botany
University of Umeå
S-901 87, Umeå
Sweden

W. Scott Armbruster
Department of Biology and Wildlife
and Institute of Arctic Biology
University of Alaska Fairbanks
Fairbanks, AK 99775-0180
U.S.A.

Juan Arroyo
Departamento de Biología Vegetal y
Ecología
Universidad de Sevilla
41080 Sevilla
Spain

Tia-Lynn Ashman
Department of Biological Sciences
University of Pittsburgh
Pittsburgh, PA 15260
U.S.A.

Spencer C.H. Barrett
Department of Botany
University of Toronto
Toronto, Ontario M5S 3B2
Canada

Lynda F. Delph
Department of Biology
Indiana University
Bloomington, IN 47405
U.S.A.

Anna-Lena Fritz
Department of Systematic Botany
Villavägen 6
S-752 36, Uppsala
Sweden

Candace Galen
Division of Biological Sciences
105 Tucker Hall
University of Missouri
Columbia, MO 65211
U.S.A.

Lawrence D. Harder
Department of Biological Sciences
University of Calgary
Calgary, Alberta T2N 1N4
Canada

Carlos M. Herrera
Estación Biológica de Doñana, CSIC
Apartado 1056, E-41080
Sevilla, Spain

Ken Inoue
Biological Institute and Herbarium

Faculty of Liberal Arts
Shinshu University
3-1-1 Asahi
Matsumoto 390
Japan

Robert A. Klips
Department of Plant Biology
Ohio State University
1735 Neil Ave.
Columbus, OH 43210-1293
U.S.A.

Josiane Le Corff
Department of Botany
University of Washington
Seattle, WA 98195
U.S.A.

David G. Lloyd
Department of Plant & Microbial Sciences
University of Canterbury
Christchurch 1
New Zealand

Masayuki Maki
Department of Biology
College of Arts and Science
University of Tokyo
Tokyo, Japan

Michiko Masuda
Department of Biology
College of Arts and Science
University of Tokyo
Tokyo, Japan

L. Anders Nilsson
Department of Systemic Botany
Villavägen 6
S-752 36, Uppsala
Sweden

Douglas W. Schemske
Department of Botany
University of Washington
Seattle, WA 98195
U.S.A.

Daniel J. Schoen
Biology Department
McGill University
1205 Avenue Docteur Penfield
Montreal PQ, H3A 1B1
Canada

Rachel Simpson
Department of Biology
University of Michigan
Ann Arbor, MI 48109
U.S.A.

Allison A. Snow
Department of Plant Biology
Ohio State University
1735 Neil Ave.
Columbus, OH 43210-1293
U.S.A.

Timothy P. Spira
Department of Biological Sciences
Clemson University
Clemson, SC 29634-1903
U.S.A.

James D. Thomson
Department of Ecology and Evolution
State University of New York
Stony Brook, NY 11794-5245
U.S.A.

Stephan Vogel
Institut für Spezielle Botanik
der Johannes Gutenberg Universität
Mainz, Germany

Paul Wilson
Department of Ecology and Evolution
State University of New York
Stony Brook, NY 11794-5245
U.S.A.

PART 1

Historical Perspective

1

Discovery of the Secret of Nature in the Structure and Fertilization of Flowers*

Christian Konrad Sprengel†

Introduction

When I carefully examined the flower of the wood cranesbill (*Geranium sylvaticum*) in the summer of 1787, I discovered that the lower part of its corolla was furnished with fine, soft hairs on the inside and on both margins. Convinced that the wise creator of nature had not created even a single tiny hair without definite purpose, I wondered what purpose these hairs might serve. And it soon came to my mind that if one assumes that the five nectar droplets which are secreted by the same number of glands are intended as food for certain insects, one would at the same time not think it unlikely that provision had been made for this nectar not to be spoiled by rain and that these hairs had been fitted to achieve this purpose.

The first four figures of the 18th copperplate (see Fig. 1.1) may serve as an illustration of what I am saying. They show the swamp cranesbill (*Geranium palustre*) which is very similar to the wood cranesbill. Each nectar droplet is located at its nectary immediately beneath the hairs, which are on the margins of the two nearest petals. Since the flower stands erect and is rather large, raindrops must fall into it when it rains. But none of the raindrops can reach the nectar droplet and mix with it because they will be stopped by the hairs which are located above the droplet, just as a bead of sweat which has run down a man's forehead is caught by his eyebrow and eyelashes and is prevented from running into his eye. On the other hand, an insect is not prevented by the hairs from getting to the nectar droplets. Now I examined other flowers and noted that various of these had something in their morphology which appeared to serve

*Translated by Peter Haase, University of Canterbury, Christchurch, New Zealand.
†Vieweq, Berlin, 1793.

Das
entdeckte Geheimniſs
der
NATUR
im Bau und in der Befruchtung
der
Blumen
von
CHRISTIAN KONRAD SPRENGEL,
Mit 25 Kupfertafeln.
Berlin, 1793.
bei Friedrich Vieweg dem ælteren.

Figure 1.1. The title page to C.K. Sprengel's book "Discovery of the secret of nature in the structure and fertilization of flowers" published in 1793.

the same purpose. The longer I continued this investigation, the more I realized that those flowers which contain nectar are equipped in a way that rain cannot spoil the nectar although insects can easily reach it. From this I deduced that, at least in the first place, the nectar of these flowers is secreted for the insects' sake, so that they can enjoy it pure and unspoiled, protected from rain.

In the following summer I investigated the forget-me-not (*Myosotis palustris*). Not only did I find that this flower possesses nectar but also that this nectar is completely protected from rain. At the same time I noticed the yellow ring which

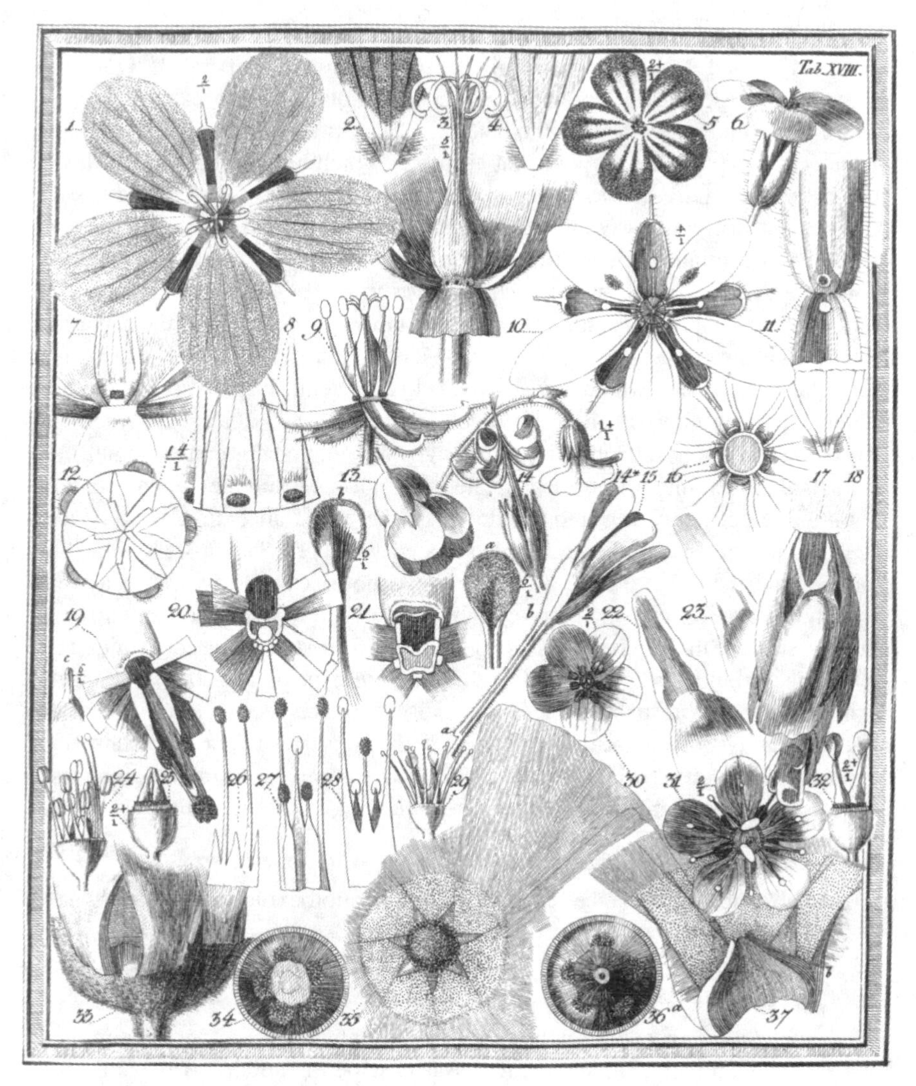

Figure 1.2. The 18th copperplate from Sprengel's book. The first four figures illustrate *Geranium palustre* (Geraniaceae) with hairs on the petal margins that protect nectar from rain (see text for further discussion).

surrounds the opening of the corolla tube and contrasts well against the sky-blue color of the corolla lobes. I thought this feature might be related to insects too. Should nature have particularly colored this ring for the purpose of showing the insects the way to the nectar container? With this hypothesis in mind, I examined other flowers and found it confirmed by most of them. I saw that those flowers whose corolla is differently colored in one place than it is elsewhere always have these spots, figures, lines, or dots of particular color where the entrance to the nectary is located. Now I deduced from the particular to the general. If, I thought, the corolla is specifically colored in a particular place for the sake of insects, then it is colored altogether for the insects. And if that particular color of a part of the corolla serves the purpose that an insect which has perched on the flower can easily find the right way to the nectar, then the color of the corolla has the purpose that flowers possessing such a corolla already from afar catch the eyes of insects swarming in the air in search of food.

When I examined some species of *Iris* in the summer of 1789, I soon discovered that Linné was erroneous in his view of the stigma as well as that of the nectaries. I found that the nectar is completely protected against rain, and that there is a particular colored spot which guides insects to the nectar. I discovered still more, namely, that this flower can be fertilized in no other way than by insects, and only by insects of considerable size. Although I did not find this concept confirmed in practice at this time (this only happened the following summer when I actually saw bumble bees crawling into the flowers), the appearance already convinced me of its correctness. I therefore investigated whether other flowers were also designed in a way that their fertilization could not occur in any other way than through insects. My investigations convinced me even further that many flowers, perhaps even all those which possess nectar, are fertilized by insects which feed on the nectar and that consequently this nourishment of insects is in fact the final purpose with respect to themselves. But with respect to the flowers, the nourishment is only a means, and indeed the only means, to a certain final purpose which consists of their fertilization. The complete morphology of such flowers can be explained if one bears the following points in mind during their examination:

1. These flowers are to be fertilized by one or another or by several species of insects.
2. This should occur in such a manner that, while seeking the nectar of the flowers, and for this reason either moving on those flowers in no specific manner, or crawling into them in a certain way, or running in circles on them, the insects necessarily wipe the pollen off the anthers with their mostly hairy bodies or only with a part thereof, and bring it onto the stigma, which is covered at its top either with short and fine hairs or with a particular, often sticky, liquid.

In the spring of 1790 I noticed that *Orchis latifolia* and *Orchis morio* exhibit the complete morphology of a nectar flower but do not contain any nectar. This observation, I initially thought, ought to completely upset my previous discoveries or at least render them very dubious. These flowers have, for example, a nectar guide (so I call the differently colored spot of the corolla), although this cannot be a guide for insects to the nectar because no nectar is present. From this it appeared to follow that the nectar guide, here and also in those flowers which indeed contain nectar, would not be there for this final purpose and consequently the idea was a mere fantasy. I must confess that this discovery was not at all a pleasant one. But this encouraged me to investigate these flowers more closely and observe them in the field. And there at last, I finally discovered that these flowers are fertilized by certain flies which, deceived by their appearance, assume there is nectar in the funnel and therefore crawl into it. While doing this, they pull the anthers from their compartments and bring them onto the sticky stigma. Such flowers, which have the complete appearance of nectar flowers without containing nectar, I call false nectar flowers. That such flowers exist I verified in the same year with the common birthwort (*Aristolochia clematitis*). I found that this flower, which does not contain nectar, is also completely designed like a nectar flower, and because of this many small flies crawl into it. In the following summer I fully realized that this flower is a true miracle of nature in the sense that these flies are tempted by the flower's appearance to crawl into it and thereby fertilize it, and that they are trapped inside until they have fertilized it. As soon as this is accomplished, however, they are again released from their prison.

In the summer of the above mentioned year, I discovered something about *Epilobium angustifolium* which I would never have stumbled on by myself. This hermaphrodite flower is fertilized by bumble bees and honey bees, but each flower does not do so through its own pollen. Instead, the older flowers are fertilized by means of pollen which the insects transfer from younger flowers. This discovery shed a bright light onto many of my previous discoveries. I took particular pleasure when I discovered just this method of fertilization in the fennel-flower (*Nigella arvensis*). In the summer of 1788 I had discovered the excellent organization of the nectar mechanisms of this flower. In the following summer I learned by observation that it is fertilized by honey bees. I then believed that I had fully understood how this occurs. But now I found that I had been wrong about the last point, because I still believed then that all hermaphrodite flowers would have to be fertilized by their own pollen.

When I finally investigated the common spurge (*Euphorbia cyparissius*) last summer, I found that it possesses a mechanism which is exactly the opposite of that just described, namely, that this flower is fertilized by insects but in a way that they carry the pollen of the older flowers onto the stigmas of the younger.

On these six main discoveries made in the course of 5 years, I founded my theory of flowers.

Two Common Misconceptions

Before I present it, I cannot leave untouched two common misconceptions which have hitherto been held regarding the final purpose of the sweet nectar of flowers. As well as being opposed to each other, both also contradict my theory.

Various botanists have believed that nectar would be directly and first of all useful to the flowers themselves, either by promoting the fertilization of the carpel by keeping it moist and flexible, or by maintaining the seed, which it permeates, by its ability to germinate. According to this concept, the fact that insects seek out the nectar would have to be viewed not only as something accidental and quite unimportant but even as something detrimental to the flower.

In many flowers the nectar is certainly close enough to the carpel; in some it is even produced and secreted by the carpel. But from this it still does not follow that nectar is of direct benefit to the carpel. If the carpel is to be kept flexible by the nectar or the enclosed seeds are to be permeated by it, then it would be more practical if the carpel would keep the nectar instead of secreting it. In many flowers, however, the nectar is distant and separated from the carpel in such a way that one cannot conceive how it could get to the carpel. The author of the dissertation *De nectario florum,* which is included in Linné's *Amoenitatibus academicis,* has already realized this. He says that in contradiction to this hypothesis male flowers, which are often far away from the females, possess a nectary. Roth has published his remarks on this subject in the Magazin für die Botanik (1787, Vol. 2, p. 31). In order to verify this hypothesis, he says, among other things, that in the African cranesbills the nectar is located in a long corolla tube and ascends upward to the carpel. This carpel, however, is surrounded by the basally fused filaments, and consequently, cannot come into direct contact with the nectar. *Antirrhinum linaria,* which he also mentions, should have suggested another concept to him. He correctly notices that the nectar of this flower is not secreted by the spur in which it is contained, but by a nectary located at the bottom of the carpel, and that it flows down into the spur from there. How can it now ascend again from the spur to the carpel? Even if this happened, what a useless complicated process it would be. How can the nectar which is enclosed in one or more special containers away from the carpel get to the carpel in *Passiflora, Helleborus, Nigella,* and *Aconitum?* Perhaps by means of insects. But what have the insects, after they have consumed the nectar, to do with the carpel?

The other hypothesis was introduced by Krünitz in his *Oekonomische Enzyclopädie* (Part 4, p. 773). He says that honey bees provide a threefold service to the flowers. First: "The nectar which the flowers secrete is going to be detrimental to them if it is not collected by the honey bees. It is initially liquid, but changes without evaporating, very soon accumulates, finally thickens, and where it remains blocks and covers the finest exits and impedes and destroys the subsequent full development and growth of the most delicate fruits." This hypothesis is exactly opposed to the previous one. According to the first hypothesis, the nectar is useful to the carpel; according to the second hypothesis, it is detrimental. According to the first hypothesis, the fact that the nectar is consumed by the insects is something accidental and detrimental to the flowers; according to the second hypothesis, it is useful to flowers and appears to be an arrangement of nature.

In order to verify that this hypothesis too is unfounded, I do not have to look for some flower which is in any way advantageous for this purpose, since I can use those which I have already mentioned. For the same reason as I deduced that the nectar could not be of benefit to the carpel, it also follows that it could not be detrimental, because it is always located at some distance from the carpel. The nectar may change as it likes, but this has no effect on the carpel. And if the nectar is close to the carpel in other flowers, from this it follows just as little that the nectar is detrimental to the carpel as that it is beneficial. Concerning those flowers where the carpel itself secretes the nectar, it appears from the fact that it secretes the nectar that the nectar is detrimental to it. Meanwhile one can in part already deduce the opposite from analogy. In part it will also be possible to adequately demonstrate in the following that the carpel of these flowers secretes the nectar not as something detrimental but with a certain purpose, and consequently that the insects become beneficial to the carpel by collecting the nectar. The benefit comes not directly from the collection itself but from the fertilization of the carpel that necessarily thus results.

Second, Krünitz says that honey bees collecting pollen carry it onto the stigma in flowers with partly or fully separated sexes as well as in hermaphrodite flowers. With regard to the latter he says: "How frequently is the natural effect of these sexual organs on each other diminished, impaired, or thwarted by very common and usual circumstances, that on one hand the pollen of one flower is in poor condition but the style is still in good condition, or the other way round. Besides honey bees, this benefit is also achieved by other nectar sucking insects which do not go after the pollen but carry it away accidentally. . . ." Here truth and misconception are mixed up. That honey bees and other insects carry the pollen onto the stigma is certain, but that the former only do so when collecting pollen is wrong because even if they would only pursue the nectar without caring for the pollen, they must, whether they want to or not, necessarily carry the pollen to the stigma. I will subsequently verify this in the most obvious manner. It is also incorrect that honey bees and other insects only promote the fertilization of hermaphrodite flowers in so far as these have certain accidentally developed defects which prevent fertilization, which is often the case. (It would follow from this that in an unspoiled state these flowers are fertilized without the intervention of insects.) First, this concept would not pay much credit to nature. According to it, nature gives rise to hermaphrodite flowers with the intention that they will fertilize themselves, but does not ensure that they can always do so, and lets it happen that often or even routinely changes occur in them that would prevent this important final purpose unless by pure luck the insects visit and fertilize the flowers. If this is not occurring accidentally but with the intention and by arrangement of nature which thereby tries to correct those defects, then, according to this idea, nature proceeds here just as a human being who because he is incapable of thinking of a single reliable method to achieve some aim, chooses two methods so that if the first should not lead him to his aim, he can use the other. A flower with a deteriorated stigma cannot be fertilized at all, not even by an insect. So only half of the desired final purpose would be achieved by this method. Second, the deteriorated state of the sexual parts of the flowers is anything but normal, but rather something as rare as it is in animals. One can see this for oneself from daily experience. And if this deteriorated state of the sexual parts was something that occurs more commonly, it should occur in those flowers which do not possess nectar and are pollinated by wind as well as in nectar flowers. According to that idea, it would follow that fertilization of the wind-pollinated flowers would fail more often than that of the insect-pollinated flowers because the former, unlike the latter, are not visited by insects. In this matter observation tells just the opposite. In nectarless flowers, or at least in several species of them, fertilization occurs if not more assuredly then more commonly than in nectar flowers. The reason for this is easy to understand. If, for example the wind carries the pollen of male aspens to neighboring female trees, it can easily happen, considering the large quantity of pollen which often falls as a cloud on the female trees, that a considerable number of carpels receive some of the pollen dust and are fertilized in this way. Many a March violet can wither, however, without having received a visit from a honey bee or a similar insect. Then it cannot set any fruit because it can neither fertilize itself nor be fertilized by wind. Small flies crawl into most flowers of the common birthwort and fertilize them, but not into every one. These cannot be fertilized in any other way. Fertilization of flowers by wind takes place on a large scale, by insects on an individual scale. A single gust of wind directed from the male tree to the female can fertilize many thousands of flowers in one instant; a honey bee, on the other hand, can fertilize only one flower at a time. Third, most hermaphrodite flowers possess a structure such that they cannot by any means, even with their sexual parts in perfect condition, be fertilized in any other way than by honey bees and other insects. In the following I will verify this with so many examples and in such a manner that even the most stubborn skeptic will not question it any further.

Third and finally he says that honey bees suck the harmful wax and honey exudates from the flowers of the meadows and pastures, because it has been noticed in various countries that in places where many honey bees are kept, the pastures are much more healthy and nutritious for the livestock, particularly for sheep; also the hay of such places is more fragrant, richer, and healthier. Here a benefit is attributed to the honey bees which is as little due to them as to other insects. They promote the fertilization of many species of flowers which would remain absolutely unfertilized without their assistance, thus assuring that these plant species can reproduce and that none of them perishes. They cannot contribute anything whatsoever toward the correction and improvement of plants. If honey bees improve pastures, particularly sheep pastures, then this can only happen in a way that they especially visit and fertilize the flowers of such plants which are beneficial to the livestock. And this is very probable, particularly concerning sheep pastures. For among those plants mentioned by Gleditsch (Vermischte Abhandl., Part I, p. 284, etc.) as being particularly sought after by sheep, most bear flowers which cannot possibly fertilize themselves or be fertilized by wind, but have to be fertilized by honey bees and other insects. I know from experience that various of these plants are in fact visited by honey bees.

Structures Associated with Nectar

In all those flowers that actually secrete nectar, the following five parts should be noticed:

1. *The nectary*. The nectary is that part of a flower which produces and secretes the nectar. Its shape and location are highly diverse and variable. Sometimes it instantly strikes the eye if one looks at the flower, but sometimes it is rather hidden so that it takes some effort to find it, particularly if it is very small. It is sometimes the carpel itself or a part of it, sometimes separate and removed. It is fleshy or of a certain thickness because if it was to be as thin as, for example, the petals of most flowers, it could not produce a sufficient, albeit very small quantity of nectar. If the distal part of a funnel or spur is fleshy, then that is the nectary, but if it is as thin as the remaining part, one has to look elsewhere for the nectary. It is also bare and glossy. There is no reason why it should be covered with hairs or wool, as are other parts of many nectar flowers. It has to be smooth for the only reason that it is usually a part of the nectar container, or often is the nectar container itself, which I will subsequently demonstrate is invariably smooth. If therefore the carpel is covered with hairs, it cannot be the nectary. But if the upper part of it is hairy and the lower smooth, or the other way round, then the smooth part is the nectary, particularly if it is distinguished by a protruding shape and by a particular color. Finally the nectary is usually colored, seldom green. The most common color is yellow, less frequently white, orange-yellow, cherry-red, etc. The particular color presumably arises usually only from the different properties and combination of its components; but at times it appears that a certain purpose is to be achieved by it, namely, that the nectary strikes the eyes of the insects.

2. *The nectar container*. The nectar container is that part of a nectar flower which receives and stores the nectar secreted by the nectar gland. Its inner surface is always smooth, for two reasons. First, just as the inner surface of containers in which one wants to store liquids has to be smooth, particulary if the liquids are precious and valuable, so that nothing remains on emptying them, which would happen if their inner surface is rough, so the nectar container must be internally smooth so that insects can fully suck out the nectar or lick it. Second,

a body with a smooth surface attracts a liquid substance better than one which is rough or covered with hairs or wool, because the latter has more points of contact. Now the nectar is to remain in the nectar container until it is collected by insects, but on no occasion must it fall out by itself or be thrown out by the wind shaking the flower to and fro. The nectar container must therefore strongly attract nectar and consequently be smooth. The shapes and locations of the nectar container are numerous and diverse. In most cases it is found directly at the nectar gland, but it is sometimes separate; often the nectar gland itself also serves as the nectar container.

3. *Protection of the nectar from rain. The nectar cover.* Nectar flowers are furnished in such a way that although insects can easily reach their nectar, raindrops which have fallen on or into the flower always remain some distance apart from the nectar and consequently can neither mix with it nor spoil it. Just as human beings plug the openings of those containers in which they store delicious liquids so that these liquids neither evaporate nor are mixed with dust, rain, and other foreign substances, so has the kind and wise creator of nature, not satisfied with having prepared in the flowers a delicious nectar for the insects, made appropriate and excellent provisions so that this nectar is secured from any waste by rain. I do not believe that the first mentioned purpose, namely, that the nectar does not evaporate, also takes place. The author of the above-mentioned dissertation states this for *Campanula* and some other genera. At least concerning *Campanula,* I shall verify that he has been erroneous. Hence on the one hand, the second purpose inevitably occurs in these genera. On the other hand, there are many genera for which one cannot possibly think of the first purpose because their nectar is completely exposed to the air so that its evaporation is not prevented by any structure, and at the same time completely secured from mixing with a raindrop, even one that is very close. This purpose is already sufficiently achieved either by the structure and position of a flower or by some other special feature present somewhere in the flower which only serves to achieve the purpose. This I call the nectar cover.

The wisdom of a human being presents itself in its brightest light if he knows how to achieve two goals at the same time, one of which appears to hinder the achievement of the other or even render it impossible. Equally one can already imagine *a priori* that the arrangement in the flowers by which two purposes that apparently neutralize each other, namely, that access to the nectar be open to insects but closed to raindrops, is fully achieved at the same time, has to show the wisdom of the creator of the flowers in the most evident way. The wisdom must have many aspects, particulary if one considers the highly diverse development of the flowers.

To initially note some frequent methods which serve this final purpose: These include above all that the corolla is usually very thin and consequently because it has only a small body mass, also possesses little attraction to raindrops; that its inner surface, and at times also the outer, is covered with fine hairs or wool or powder; if this surface is smooth, the corolla appears to exude a finely dispersed oil. In all these cases, the parts of a raindrop which has fallen onto the corolla are little attracted by it, but show their attraction mainly toward themselves. The raindrop takes on a spheroid shape so that its area of contact with the corolla is smaller than that which is oriented parallel through its center. In such a manner it cannot adhere to the corolla for long but must fall off as soon as the flower is shaken by the wind. But even if it stays, it still cannot get to the nectar. While running downward, it encounters a row of hairs which are attached above the nectar container and directed upward, mostly forming an acute angle with the surface of the corolla. Consequently, the hairs point their tips toward the raindrop and keep it away from the nectar container. Alternatively, the raindrop gets to a neck and has to stay before it. At times a raindrop touches some anthers; because these are thicker than the filaments, they also attract the raindrop more strongly. Then it remains between the anthers and the corolla and cannot get to the nectar droplet attached to the bottom of the filaments. Often the filaments are thicker at the top than at the bottom. If a raindrop falls on their uppermost part, it stops here for the same reason. A similar effect can be noticed on pine needles after a rainfall. Looking at those needles which point their tip toward the ground, one finds a raindrop not at the bottom, at the tip, but somewhat above it. For if a raindrop has fallen onto such a needle, then because of its weight it must flow down; the needle cannot prevent this because it has the same thickness throughout its length. But when it has flowed down to the point where the needle begins to end in a tapering tip, then it must stop there because it is more strongly attracted by this part of the needle than by its tip.

Many tubular flowers have a rather wide opening, but because it is divided by five or more filaments into just as many smaller openings, no raindrop can enter the tube through them. Alternatively, five or more anthers are located at the opening and almost fill its space; here too no raindrop can enter. In both cases, however, smaller insects can easily crawl in and larger ones can insert their proboscises. Often nature has employed flexibility in order to achieve this double aim. It has attached certain lids which can easily be lifted up or pushed down by an insect in order to reach the nectar, but which close again when the insect returns so that no raindrop can penetrate. The flexibility does certainly not occur in flowers to the extent to which some capsules possess it. In part this is not possible because a flower is of a much softer substance than a capsule; in part it is also not necessary because here it is only intended that a lid lifted by an insect drops down again but on no account that certain bodies are thrown a long distance in the manner that capsules hurl their seeds far away. Finally, the feature which many flowers possess, opening only in fine weather but remaining closed in rainy or cloudy weather, is also related to this purpose.

Most flowers occupy a distinct position. If the nectar contained in them has to be secured from rain, then their structure must vary depending on their different positions because of the perpendicular direction of falling raindrops.

First, there are erect flowers. These are regular because nature always prefers regularity to irregularity, and at least with regard to rain, there is no reason why it should diverge from this rule in the case of these flowers. Since their inner side faces the falling raindrops, and the raindrops which have fallen into it gravitate toward the nectar located at the bottom of the flowers because of their weight, these flowers must be guarded the most by particular measures against the penetration of raindrops. Their petals are often split into narrow parts. Because every corolla has to be as large as possible, as will be proven below, the corolla of these flowers, if it was large and at the same time entire, would receive and keep too many raindrops which could easily mix with the nectar. One can particularly expect that these flowers will not open in rainy weather.

Second, there are pendulous flowers. These are regular too, for just the same reason as the previous. They present their outer side toward the falling raindrops; the inner side is exposed to little or no rain at all, particularly if they are bell-shaped or have a cylindrical or spherical shape. The nectar is located above (at the base of the flowers), where the raindrops cannot ascend because of their weight. One may least expect special arrangements for keeping out raindrops in these flowers. Their petals must be entire so that raindrops are kept on their outer side; if they were divided into narrow parts, the raindrops could easily get to their inner side and into the nectar container. These flowers do not need to close in rainy weather.

Rain is mostly associated with wind; this fact is certainly advantageous to all flowers, even to those which do not secrete nectar. Since the wind shakes the flowers vigorously, it causes most of the raindrops that fall on it to fall off again, so they can spoil neither the nectar nor the pollen. For the erect or pendulous flowers, however, this circumstance is also advantageous in another way. I have shown this in the 25th copper-plate (Fig. 1.3). In Fig. 4, *Ranunculus acris* is shown in its natural upright position. The five dotted lines indicate the direction of the same number of raindrops which they possess in calm conditions. With this direction of the raindrops, this position of the flower is the most disadvantageous one because all five raindrops fall into the flower. In Fig. 5, one can see the position of the flower and the direction of the raindrops in a moderate wind. With this direction of rain, the flower position is more advantageous, since not more than two raindrops fall into the flower. Finally, the position of the flower and the direction of rain in an intense wind are shown in Fig. 6. Here not one of the five raindrops falls into the flower; they all fall on its outer side. This is the most advantageous position that the flower can have with this direction of rain. In Fig. 9 *Campanula rotundifolia* is shown in its natural position. This is the most advantageous position that this flower can have in calm conditions with regard to perpendicular falling raindrops. If a wind blows, however, be it light or strong, then the spike of the flower and the line in which the raindrops fall lie in approximately the same direction. The flower always keeps the most advantageous position with regard to the rain. The flower is shown in Fig. 6 in that position which it has in a moderate wind, and in Fig. 10 in that into which it is moved in the strongest wind. Consequently, wind which accompanies rain is useful in this manner to the erect nectar flowers by bringing them from a disadvantageous position into a less disadvantageous or advantageous position; and to the pendulous flowers wind renders the service that it continues to keep them in the most advantageous position which they can have.

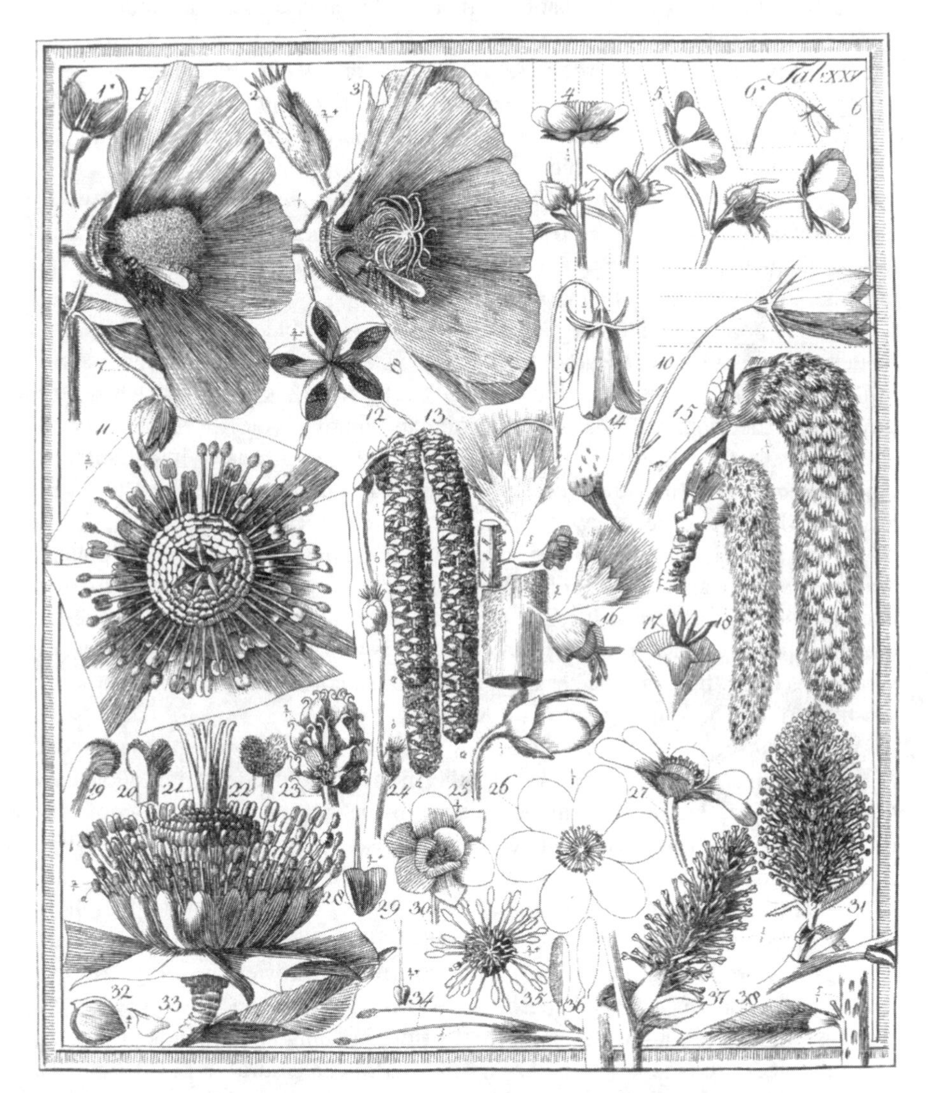

Figure 1.3. The 25th copperplate from Sprengel's book. Figures 4, 5 and 6 illustrate different floral orientations in *Ranunculus acris* (Ranunculaceae) and Figures 9 and 10 in *Campanula rotundifolia* (Campanulaceae). See text for further discussion.

Third and finally, there are horizontal flowers. The opening of their corolla is turned toward the horizon; their tube may be either also horizontal or more or less approaching the perpendicular. The flowers are mostly irregular and have two lips. If their nectar is to be secured against rain, the upper lip of the corolla has to be shaped and designed quite differently from the lower one, because the raindrops fall on the outer side of the former but on the inner side of the latter. Consequently, the upper lip must be similar to the corolla of the pendulous, the lower lip to the corolla of the erect flowers. The former is therefore curved and entire and has no hairs inwardly; the latter is flat and often divided and hairy at the opening of the tube. These flowers are either permanently closed, as in the Scrophulariaceae, or their nectar is completely guarded against rain by another method so that they do not need to close in rainy weather.

4. *Arrangements that insects can easily find the nectar in nectar flowers. Corolla. Scent. Nectar guide*. The facts that most flowers secrete nectar and that this nectar is secured against rain would not help the insects unless at the same time there is provision for them to easily find this food intended for them. Nature, which does nothing in halves, has made the most suitable arrangements in this matter also. First, it has ensured that the insects already notice the flowers from a distance, either by sight or by scent or by both senses together. All nectar flowers are therefore adorned with a corolla and many give off a scent which to humans is usually pleasant, sometimes unpleasant, occasionally intolerable; to those insects for which their nectar is intended, the scent is always pleasant however. Except in very few species, the corolla is colored, that is, colored other than green in order to contrast strongly against the green color of the plants. Sometimes, the calyx is also colored; if a complete corolla is present, the calyx is colored differently, or if it forms an entity with the corolla, it is the same on the inner side as the corolla. If the corolla is missing, however, the calyx substitutes for it. In many species the bracts are also colored for this final purpose, mostly differently from the corolla however.

When an insect, attracted by the beauty of the corolla or by the pleasant scent of a flower, has perched on it, then it either notices the nectar instantly or does not because it is located in a hidden place. In the latter case, nature comes to its aid by means of the nectar guide. This consists of one or more spots, lines, dots, or shapes of a color other than that the corolla in general possesses, and consequently, it contrasts more or less against the color of the corolla. It is always located where the insects must crawl in if they want to get to the nectar. Regular flowers have a regular nectar guide, irregular flowers an irregular one. If the nectar container is separated from the opening through which the insects crawl in, the nectar guide begins in front of the opening and extends through it up to the nectar container, thus serving the insects as a sure guide. If a flower has several entrances to the nectar container, then it also possesses the same number of nectar guides. If a flower has several nectar containers which surround the carpel, or only one which encircles the carpel in the form of a ring, and the nectar of which cannot be consumed by the insect in any other way than by running in a circle around it and inserting its proboscis into it several times, then the nectar guide has a circular shape and guides the insect in a circle.

On the topic of the nectar guide, I have to talk of differences among the nectar flowers which depend on the time of day in which they flower. Just as there are insects which only swarm by day and those which only seek their food at night, there are also day flowers and night flowers.

Day flowers open in the morning. Many of them close in the evening, or incline themselves if they were erect during the day, or another change occurs in them from which one can deduce that they are only intended for day insects. Some close on the first evening and do not open again the following day, thus flowering only for a single day; most flower several days.

Most, although not all, day flowers are adorned with a nectar guide.

Night flowers open in the evening. By day most of them are closed or withered and unsightly; hence, it appears that they are not intended for day insects. Some flower for several nights; the evening primrose (*Oenothera biennis*) flowers 2 nights.

Night flowers have a large and light-colored corolla so that they catch the eyes of insects in the darkness of the night. If their corolla is inconspicuous, then this shortcoming is substituted by a strong scent. On the other hand, a nectar guide does not occur because if, for example, the white corolla of a night flower had a nectar guide of another but also light color, then it would not contrast against the color of the corolla in the darkness of the night, and would consequently be of no use. If it had a dark-colored nectar guide, however, then this would not catch the eye and consequently would be as useless as the former color.

5. *Fertilization of nectar flowers by insects. Dichogamy*. I have already mentioned above that all these arrangements are first and directly related to insects, but through their intervention the final purpose to the flowers themselves is that the flowers are fertilized by insects.

That insects contribute their share toward fertilization of the flowers has in principle already been noticed by others. To my knowledge, Kölreuter, who has discovered and very well verified it in *Iris* and some other genera, has progressed farthest in this matter. But so far nobody has demonstrated that the complete structure of nectar flowers is aimed at this final purpose and can be fully explained by it, because nobody has recognized what I call nectar covers and nectar guides for what they are, although everyone has seen them. No one has as yet explained the structure of either this or any other flower in such a complete, satisfactory, and unquestionable manner as I have explained the structure of the wild fennel-flower, for example.

The mechanism of very many hermaphrodite flowers which I first discovered and by means of which each flower cannot be fertilized by its own pollen but only by the pollen of another flower is an undeniable proof of this fertilization of flowers by insects. This is so because if those flowers had to be fertilized in a mechanical way, that is, either the anthers would directly touch the stigma and transfer its pollen to it, or the pollen of the former would fall onto the stigma or be brought onto it by wind, then this arrangement would in the first case absolutely foil the achievement of this intention, and in the two latter cases it would at least render it much more difficult. Consequently, the argument would in the first case be absurd and in the latter at least inappropriate.

This arrangement I call the nonsimultaneous flowering of the sexual parts or actually of the anthers and the stigma, or more concisely dichogamy, which means the same. After the flower has opened, the filaments, either all at the same time or one after another, occupy a distinct position in which their anthers open and present their pollen for fertilization. In the meantime, however, the stigma is located at a place removed from the anthers and is still small and tightly closed. Thus, the pollen of the anthers cannot be brought onto the stigma, either in a mechanical way or by an insect, because it is still undeveloped. This condition lasts for a specified time. After this time has passed, and if the anthers have no pollen left, various changes occur in the filament with the result that the anthers do not occupy the same place which they held so far. In the meantime, the pistil has changed in a way that the stigma is now located at the position where the anthers formerly were. Since it now also opens or unfolds its parts, it often also occupies approximately the same space which was formerly occupied by the anthers. At that time it cannot receive pollen from the anthers because they do not have any left. The particular place where at first the flowering anthers and later the flowering stigma are located, however, is selected in each flower in such a way that the insect, for which the flower is intended, cannot get to the nectar without at the same time touching the anthers in the younger flower and the stigma in the older with a part of its body, wiping the pollen off the former and bringing it onto the latter. In this manner, the older flowers are fertilized with the pollen of the younger.

With respect to fertilization, the dichogamous hermaphrodite flowers are therefore similar to the flowers with semiseparated gender. At first they are male and later female flowers.

That this arrangement is very useful is easily demonstrated. If the anthers and the stigma flowered at the same time, the former would prevent the insects from touching the latter, and the latter would prevent them from touching the former. But with this arrangement, the insects find in the younger flower only the anthers obstructing their way and must consequently wipe off the pollen, and in the older flower only the stigma over which they consequently must fully brush the pollen adhering to their body.

I had discovered this arrangement in *Epilobium angustifolium* in July 1790. From this time onward until May of the following year, I noticed the same arrangement in different genera, even in whole families such as the umbellifers, so easily and so clearly that I was surprised that it had not already been discovered long ago by others and much earlier by myself. During that complete period, however, the thought never came to my mind that the opposite of this arrangement would also have been chosen by nature, in which there were flowers whose stigma functions first, the anthers of which only start to flower after fertilization of the carpel is completed. As natural as it was to come to this idea by oneself, it still remained unknown to me until nature herself suggested it to me. And this occurred when I investigated *Euphorbia cyparissias* in May of the following year. That is to say, I saw that as soon as the flower has burst open, the stigmas emerge, standing erect and unfolded. After several days, the complete pistil, which is attached on its own stalk, emerges from the flower, gradually looses its upright position and finally turns the stigmas toward the ground. Only then do the anthers appear one by one from the flower and occupy just that place which was formerly occupied by the stigmas. Since I had already discovered long before that this flower is a nectar flower, I thus realized that because of this arrangement, it could not be pollinated in any other way than by insects, and hence, must be pollinated by them. Because if they visit an older flower, insects must necessarily wipe the pollen off the anthers. So that they can do this without hindrance, the pistil has left its former place and turned toward the ground. If they now visit a younger flower, then they must again necessarily touch the stigmas with their dusted body, fertilize them, and in this manner fertilize the younger flower with the pollen of the older one.

Since there are thus two kinds of dichogamy, they must be distinguished from each other by different adjectives. The first-discovered I call male-female dichogamy and the last-discovered female-male dichogamy (*Dichogamia androgyna* and *Dichogamia gynandra*). The opposite of dichogamy is called homogamy.

Since the last flowers of a dichogamous plant of the first kind communicate their pollen to the next preceding flowers and their stigma remains unfertilized, they cannot set any fruit. Similarly, the first flowers of a female-male dichogamous plant communicate their pollen to the next following flowers, and their stigma also remains unfertilized; thus, they also can set no fruit. I shall later verify by various examples that observation confirms this.

Insects That Fertilize Flowers

It is certain that many flowers are fertilized by several species of insects, for example, the umbellifers, the euphorbias. These are visited by all kinds of insects because their nectar catches their eyes as soon as they approach the flowers, so that even the most clumsy fly can easily find it. While these insects are running about on these flowers in an imprecise way and rob now the older now the younger flowers of an umbel of their nectar, they must thus touch now anthers now a stigma and bring the pollen of the former onto the latter, and in a very unspecific way. But it is also certain that many flowers are fertilized by only one species of insect and in a very specific way, because either other species are too clumsy to know where the nectar is hidden and how they can get to it, or if they know it, are either too large to be able to crawl into the flowers or too small to touch the anthers and the stigma when crawling in. In this way, as I shall verify, *Nigella arvensis* is only fertilized by honey bees and *Iris xiphium* only by bumble bees, both in a very specific way, however. For the latter flower, the honey bees are too small and too weak and cannot work their way into it. In specific ways, *Antirrhinum majus* is fertilized by a large bumble bee and *Antirrhinum linaria* by a small bumble bee. The large bumble bee cannot fertilize the latter flower because this is too small for it to crawl into. Therefore, the bumble bee uses force, bites a hole in the spur which contains the nectar, pushes its proboscis through the hole, and consumes the nectar.

Those insects which I can verify from observation fertilize flowers are mainly honey bees and bumble bees. The skill of these little animals in finding nectar, even when it is well hidden, often filled me with astonishment. How tiny are the nectar machines of the wild fennel-flower? And how much smaller is that part of it, formed like a small box and furnished with a flexible lid, that contains the nectar? The honey bee, guided by the annular nectar guide, runs in a circle, opens each box, and takes out the nectar. If someone who has no knowledge of the flowers sees *Antirrhinum majus* for the first time, he may perhaps believe that the lower lip of it forms a single unit with the upper lip, because both are tightly sealed to each other. From the yellow spot on the lower lip he will be even much less able to conclude the opposite, since its final purpose has hitherto not been known to a single botanist. If a bumble bee has approached the flower, however, it will not first try whether it could enter, and how. Because it knows very well what the yellow spot signifies, it immediately perches on the lower lip, separates it from the upper lip, and crawls between both into the flower. In order that these animals can fertilize the flowers, their body is hairy all over because they have to wipe the pollen off the anthers with one part of their body in this flower and bring it onto the stigma of a different flower with the same part of their body. That these animals hold a superior status among insects appears not only from their skillfullness, but also from the provision which nature has contributed in this affair toward the maintenance of their life. Flies which visit and fertilize several species of *Asclepias* are often caught in a certain part of these flowers as in an iron trap and must either die a miserable death or at least leave behind a leg in order to keep their life. Small flies which fertilize several orchid flowers become stuck on the sticky stigmas like birds on lime-twigs and must die. But I have never observed a bumble bee meet an accident during a visit to a flower, and have seen honey bees do so only a few times.

That these and other insects, while pursuing their food in the flowers, at the same time fertilize them without intending and knowing it and thereby lay the foundation for their own and their offspring's future preservation, appears to me to be one of the most admirable arrangements of nature.

* * *

Procedures to Investigate Flowers

Because the fertilization of the carpel by insects is the final purpose to which the complete structure of most, presumably even all, true nectar flowers furnished with a corolla is related, this structure is fully explained as soon as one has demonstrated that and how all parts of it contribute toward achievement of this final purpose.

The first question which has to be answered on investigating any flower is whether it is a nectar flower or not because if one regards a nectar flower as lacking nectar, one will not quite possibly be in a position to give any reason why it has this particular structure and not another. Someone who considers, for example, the disk-florets in *Viburnum opulus* or in the many species of *Centaurea* as devoid of nectar will never explain the purpose of the sterile marginal florets.

If one has satisfied oneself that a flower is a nectar flower, then the second question is whether it is visited and fertilized by insects. Anybody who does not try to answer this question properly, but believes that the flower is fertilized in a mechanical way, will make the most serious errors when he tries to explain its structure and the changes which he notices in it with this preconceived idea. This happened to Linné and other great botanists. They noticed that various changes occurred in the sexual parts of various flowers during their time of flowering. Quite correctly, they judged that this was nothing accidental but an arrangement of nature by which it wants to achieve a certain final purpose, namely, the fertilization of the flowers. They failed in directing their attention merely to the flowers. They took the circumstance that they are visited by insects, which they must have noticed often enough, as something accidental and not worth their attention. Thus, they always observed those changes from an incorrect perspective because they believed that the flowers are fertilized in a mechanical way. Their explanation naturally had to be prone to many doubts and objections, never giving the impression of a natural explanation for a natural phenomenon, one which completely satisfies the reader who is only interested in investigating the truth because he notices with pleasure the unforced nature of it. Their method of explanation, however, made it quite impossible for them to even make the attempt to answer the following questions: What purpose does the nectar of this or that flower serve? What is the particularly colored spot on it meant for? What connection do all parts of the flower have, and what relationships do they have to the fruit which should originate from the flower? And how does everything which we see and notice in a flower throughout its complete flowering period unite into a single beautiful whole?

Someone who therefore has carried flowers from gardens and fields and examines them in his study will never discover the intention of nature in their structure. One rather has to investigate the flowers in their natural habitat and pay particular attention to whether they are visited by insects, and by which insects, how these behave while they crawl into the flowers and consume their nectar, whether they touch the anthers and the stigma, whether they bring about any change in appearance of any part of the flowers, etc. In short, one has to try to catch nature in action. I would never have been able to discover the splendid structure of *Nigella arvensis* and the secret of its fertilization if I had not observed it in the field. The honey bees which I found on it put me on the right track. The small fly which I found caught in a spider web and covered with pollen grains of *Serapias longifolia* completely convinced me of the soundness of the concept which I had of the fertilization of that species. This concept was also based on other previous observations made in the field, however. One must not be discouraged from staying at a flowering plant for a long time and repeating such observations of one species of flower many times, because it is not always the first visit by an insect which is intended for its fertilization.

One must observe and examine the flowers at different times of the day in order to learn whether they are day or night flowers, and in different weather conditions, for example, during and following rain, in order to realize how their nectar is secured against rain. In particular, however, the noon hours, when the sun, standing high in the cloudless sky, shines warm or even hot, is the time when one must make diligent observations. Then the day flowers appear in their greatest beauty and compete for the visit of insects with all their appeal; then their fertilization can proceed much more easily because the pollen, even of those anthers which are exposed to the open air, is completely dry. The insects which most like the extreme heat are now most active around and on the flowers in their greatest activity. It is their intention to feast on their nectar, but it is the intention of nature that they fertilize the flowers at the same time. In the plant kingdom, the wisdom of which is not less admirable than its beauty, wonderful things happen of which the armchair botanist, who at this time is occupied with satisfying the demands of his stomach, does not have even a clue.

In searching for the nectary of a flower, one has to think of its above-mentioned properties all the more, namely, that it is fleshy, smooth, and usually colored, because it is often very small and hard to see with the naked eye. If one supposes a certain part to be the nectary because of its location, shape, or other features, and these three properties are present in it, then it certainly is the nectary. Occasionally, if it is near the carpel or even a part of it, it is green but a lighter or darker green than the carpel or the remainder of it, so that it can be easily recognized in this case too.

Further, if one searches for the nectary of a flower, one must start at the carpel as its center, and if one does not find it here, continue toward its outer parts. Someone who proceeds in the opposite direction and starts at the periphery of the flower and proceeds from there toward the center may easily consider a part as the nectary when it is something completely different.

If the lower part of a flower is a tube, or if the corolla is multipetalled and tubelike, one always has to look for the nectary at the bottom of the tube and never at its opening. This is because nature either gave the flower this shape in order to protect the nectar located at the bottom of the tube from rain. Or if nature has done this for another reason, it had to make use of this shape which is advantageous for the protection of the nectar against rain, and consequently attach the nectary at the bottom of the tube rather than at the opening of it where the nectar would be exposed to the rain.

Without a very thorough previous examination, one must not consider a very small flower to be devoid of nectar merely because it is very small. Although the nectar droplet which it is able to secrete must be extremely small, it can still provide food for some insect. The smaller the flowers of a plant are, the larger the numbers in which they are present. Although each one thus contains only a very small nectar droplet, the nectar droplets of all flowers together constitute a considerable quantity. The flowers of umbellifers are very small but they still possess a nectary and nectar. How large is the number of flowers of such a plant, however? The nectar which they contain altogether provides an abundant meal for a fly. Since I have found nectar in many larger flowers of the Compositae, I conclude by analogy that all, even the smallest flowers of this class, such as those of *Achillea* and *Artemisia,* contain nectar. I only exclude the marginal florets in *Syngenesia frustranea* and in *Syngenesia superflua,* since those are present for another final purpose. Someone who is surprised that such small flowers should have a nectary also has to wonder that they have fertile parts. Just as the latter belong to the most essential parts of these flowers, so the former also belongs to them, and just as the latter are extremely small, so are the former. And just as the nectar droplet is extremely small, so it is intended for extremely small insects. One has only to think of the size of the thrips which are present on almost all flowers, or of the even much smaller insects which one occasionally finds in the flowers, and then try to determine exactly how large a flower must be if it has to secrete and store as much nectar as is necessary for the maintenance of such small animals.

One should not dispute whether a flower has a nectary, particularly if one actually finds nectar in the flower, merely because it is not a special organ distinct from other organs. The author of the above-mentioned dissertation *de nectario florum* therefore judges quite incorrectly when he says that although one finds nectar at the bottom of the corolla tubes in *Lamium, Anchusa, Galeopsis,* and several foreign genera and in those flowers whose *receptaculum* or calyx contains nectar, one cannot assign a true nectary because they would not possess a particular nectar gland. First, the former three genera do not belong here at all because they have in fact special structures which are exclusively intended for preparation and secretion of the nectar, which he did not see because of their minute size. Second, if some flowers do not possess a true nectary, then they have a false nectary. A false nectary, however, is an expression which makes one think of nothing. Third, someone who judges in this manner appears to completely ignore the noble simplicity and great economy of nature. According to this kind of conclusion, one would also have to say that nature had armed the ox by giving it horns but not the horse, which has not received particular weapons although it is able to defend itself with its hind legs. If nature can prepare nectar in a flower without a specially shaped nectar gland distinct from other parts, then it would be a useless length to give them such a nectar gland. In this case, therefore, that part of the flower which secretes the nectar, be it either the carpel or a part of it, or the base, or a part of the corolla or the filaments, at the same time holds the nectar.

When the flower withers, the nectary either falls off together with the corolla or it remains. In the latter case, it may be segregated from the carpel and dries, shrinks, and becomes unsightly, or be part of the carpel and enlarge together with the carpel but it is still always distinguishable by its external appearance, smoothness, etc. In the latter case, one can become more certain of its presence and get a better idea of its former shape if it has been very small and hardly noticeable during the flowering time. Thus in the completely matured rye grain, one can see the former nectary very clearly which one can hardly see with the naked eye during flowering because the carpel itself is then very small. In capsules of *Silene* enclosed in the calyx, one can still tell from the outside where the former nectary was located.

Since the nectar container is always smooth, this is a good aid to finding it. In flowers which are furnished with a tube, one will mostly find that the upper longer part of the tube is internally covered with hairs or wool, but the lower shorter part is smooth. In this case, the latter is always the nectar container.

If one has found nectar in a flower, one has at the same time located the nectar container and will also encounter the nectar gland not far from it. One only has to be sure that the liquid observed is indeed nectar and not a raindrop. One will usually see that this liquid is located in a place where a raindrop cannot get to or cannot get to easily. Often, however, one will find liquid on a part which is detached and exposed to the air, and then one will sometimes not know whether it is nectar or a raindrop. One cannot always decide this by taste. Although nectar always tastes sweet, who has such a fine taste that he could sense the sweetness of a droplet which is much smaller than a pin head? If one finds that several droplets sit regularly on a flower, or that all flowers are furnished with one or several droplets in precisely the same place, or if one finds such droplets in dry weather, then one can reasonably conclude that this is likely to be nectar. One will achieve complete certainty, however, if one takes such flowers home and puts those which have not yet opened into water. If they are nectar flowers, as soon as they have opened, they will begin to secrete nectar. In such a way, I have convinced myself that the droplets which, for example, I found in the heath on *Anthericum ramosum* were in fact nectar drops. They were sitting on the carpel in a manner that one could easily believe they would be raindrops, for which a botanist to whom I showed the heathland mistook them.

Occasionally, one does not find any nectar in flowers which are in fact nectar flowers, either because it has already been consumed by insects, which is much more probable if one has only a few specimens to examine, or because the lateness of the season is responsible. Some plants bring forth flowers late in the season but do not appear to have sufficient vigor to be able to produce nectar in them. The same applies to flowers which one gets in winter from a greenhouse or hothouse. The artificial warmth does not appear to be able to bring some flowers to such perfection that they actually secrete nectar. Nevertheless, in these cases someone who has some knowledge of the structure of flowers will often be able to convince himself that these flowers are nectar flowers. I examined *Jasione montana* in late autumn. I did not find nectar in it, but I still concluded that it was a nectar flower from a feature of its structure which I noticed. I learned by observation in the following summer that I had deduced correctly. I did not find nectar in *Coronilla emerus,* which I had received from a greenhouse in winter. From its overall structure, however, I realized that it was a nectar flower. When I later examined the flower in summer, I indeed found nectar in it.

The insects can be very helpful in such an investigation. A flower which is frequently visited by one or more species of insects probably has nectar. One has to exempt only honey bees, because these also visit nectarless flowers for the pollen, and at least one species of bumble bees which also collects pollen. One can easily be misguided by single insects, however, if one fails to conduct a proper investigation because they occasionally search in nectarless flowers or in those parts of nectar flowers which do not contain the nectar. I give *Lychnis dioica* as one example among others. This, however, applies only to flies, aphids, rose beetles, and other lower insects, but certainly not to honey bees and bumble bees which know how to find the nectar of each flower very easily.

If a flower has a structure such that raindrops are kept out of its interior, one can expect that it contains nectar. Tubular flowers belong to this category, as do those which are pendulous, particularly if they are also bell-shaped or even cylindrical. Flowers which have a tube contain nectar so frequently that only the false nectar flowers are an exception. But this is exactly the reason why the false nectar flowers have a tube or tubular part, at least as far as the four false nectar flowers known to me are concerned. Because if nature wanted to achieve her intention, which is to deceive the insects and to induce them to crawl into these flowers, then it had to give them such a shape that the insects must regard it as a nectar flower. Consequently, nature had to furnish them with a tube because the insects know from experience that a tube contains nectar.

Flowers with a special nectar cover also have to be nectar flowers. Consequently, if one finds hairs in a flower, then one can regard them as a nectar cover and one will soon find the nectar below. Someone who does not know this will search long and perhaps in vain for the nectar glands in many malvaceous flowers because they are located in a rather hidden place. Someone to whom this is known, however, will instantly conclude from the hairs which he sees at the bottom of the corolla that the nectar has to be located below them, and soon find the nectar and the nectar glands. If a tubular flower has certain appendages around the opening of the tube, then one should not consider them as nectar glands, as Linné has occasionally done, but as the nectar cover. If one concludes from their presence that the flower must have nectar, and looks for it in the bottom of the tubular part, then one will easily find it there.

Flowers which possess a nectar guide are mostly nectar flowers. And just as this is helpful to the insects to find the nectar, so we also can make use of it for the same purpose.

Not every flower furnished with a corolla has nectar. Disregarding the false nectar flowers, there are also others which have a sizable corolla but no nectar. The corolla of these flowers is either something completely inexplicable or it serves to catch the eyes of pollen-collecting honey bees from a distance. If this is correct, it follows that these flowers are also fertilized by honey bees, which I will verify from experience by various examples. Otherwise, if they were fertilized in a mechanical way so that honey bees collecting their pollen were not advantageous for the flowers, but detrimental, their fertilization would be rendered the more difficult the more their pollen supply is diminished. Consequently, their corolla would cause honey bees, attracted by it, to complicate their fertilization and it would only cause them the greatest damage uncompensated by any advantage, which is absurd.

Are all flowers which have a scent nectar flowers? I do not dare to answer this question in the affirmative. The flowers of the elder (*Sambucus nigra*), for example, have a strong scent but so far I have encountered neither nectar nor insects on them, except for May beetles and a rare fly the size of a large bumble bee, which, however, consumed the pollen as I observed thoroughly.

Wind-Pollinated Flowers

All flowers which have neither a true corolla, nor in its place a handsome and colored calyx, nor scent, and which are usually called flowers, are devoid of nectar and are not fertilized by insects. They are fertilized in a mechanical way, namely, by wind which either blows the pollen off the anthers and onto the stigmas, or by shaking the plant or flower, causes the pollen to fall from the anthers onto the stigmas. That mechanical fertilization, however, could also occur even in nectar flowers by the anthers directly touching the stigma and transferring their pollen to it, I would not believe at all if the fertilization of *Lilium martagon* could be explained in any other way. In the meantime I will prove by several examples that the observations from which one wanted to conclude this type of fertilization with regard to many other nectar flowers are incorrect.

For the first part of this statement, the flowers of the grasses make an exception, however, because they do not have a true, sizable, colored, and eye-catching corolla and yet have nectar. I will prove that they are not fertilized by insects but by wind, although they contain nectar. But first I will indicate the difference between flowers which are fertilized by wind and those which are fertilized by insects.

The flowers of the former kind are distinguished from flowers of the latter kind first by the larger amount of pollen. If the flowers of the female poplar, for example, are to be fertilized by wind with the pollen of a neighboring male tree, then the male tree must prepare much more pollen than is necessary just for the fertilization of all flowers of the female tree. This is because the wind does not blow the pollen only toward the female tree, and does not bring each pollen grain onto a flower which is still unfertilized. Likewise, the rain not only washes much pollen from the anthers, which are very exposed to it in such flowers, but also precipitates the pollen already released and present in the air. And if the female spikes of a sedge are to be fertilized by pollen falling from the male spikes located above them, then the largest part of the pollen falls past them. Consequently, in this case also much more pollen must be present than is necessary for fertilization. This is confirmed by observation. The two mentioned genera produce abundant pollen. The pine (*Pinus sylvestris*) has so much pollen and disperses it into the air in such quantities that during its flowering period it sometimes rains sulphur, as the common people say. Are not the male catkins of the hazel and alder much larger than the female flowers and catkins? For the flowers of the other kind, it is a very different story. Suppose a plant has male and female flowers and the latter should be fertilized by honey bees with the pollen of the former. Suppose further that by crawling into the male flowers, honey bees wipe the pollen off the anthers with their back, and if after this they crawl into a female flower, with their dusted backs they touch the stigma, which for this purpose is located exactly where the anthers are in the male flower. One realizes that not much pollen is necessary in this case. If one hits on the flowering branch of a pine, a hazelbush, or an alder with a stick, then one will produce a large cloud of pollen. If one hits a flowering currant or gooseberry bush, however, no such pollen cloud will appear. The two-lipped flowers do not have more than four anthers, some only two, and therefore can produce only a little pollen; this is completely sufficient for fertilization, however, because this does not occur by wind but by insects.

It should be noted, however, that an attempt to convince oneself of the amount of pollen in the flowers of the first kind must take place only in calm weather. If a wind blows, little or no pollen will appear, because the wind has already blown it away. Hence, these flowers are also distinguished from the flowers of the other kind in that their pollen is very light and easily carried off by the gentlest breeze, whereas the pollen of the latter is more firmly attached. In spring, one has to break off branches of a hazel, or aspen, or alder which are furnished with male catkins that are not yet flowering but are not very far from flowering and consequently have not yet lost any of their pollen. If one puts these in a water-filled container at a window through which the midday sun shines, one will find that after a few days the catkins have extended and the anthers have opened. If one now blows on these branches, then a large cloud of pollen will appear. If, however, one waits a few days longer without having made this attempt, until all anthers have opened, and then blows, one will blow all the pollen clean off. If one repeats this experiment after a few days, one will not notice any more pollen. One will notice the same result if the branches are shaken. In the anthers of a flower of the other kind, however, one cannot blow off the pollen as easily. It is attached more firmly and is more similar to flour which is slightly moistend, and is therefore somewhat sticky, than to dry pollen which the slightest breeze carries off. If one performs this experiment with a branch of the male goat willow (*Salix caprea*), one will find that one can cause such a pollen cloud neither by blowing nor by shaking. If one blows on the anthers of *Crocus, Tussilago farfara, Cornus mas,* or *Ornithogalum luteum,* then one will only be able to blow off single grains but not the whole supply of pollen in the form of an actual pollen cloud. Even in *Hepatica nobilis* and *Papaver dubium,* which do not have nectar but possess a corolla, the same result will appear. From this and from other facts which I will note in support, I conclude that these and similar flowers are fertilized by honey bees. That this different property of the pollen is very appropriate anyone can realize for himself. The opposite of this mechanism would completely foil the intentions of nature. If the pollen of the flowers of the first kind was firmly attached, it could not be carried by the wind onto the often distant stigmas, and if the pollen of the flowers of the other kind was easily blown away by the wind, then the insects would wipe off little or no pollen when they visit the flowers and consequently could not fertilize them.

Finally, in the flowers of the first kind, the anthers as well as the stigmas must be freely exposed to the air so that the wind can carry the pollen from the former onto the latter, and the stigmas have to be of considerable size, because if they were small, they would only rarely receive pollen. In the flowers of the other kind, however, neither the former nor the latter is necessary. For them it only depends on whether the anthers and stigmas are located so that they must be touched by the insect intended for their fertilization while it crawls in, and even if the stigma is very small, it will still always be pollinated by the insect in this case.

To return to the flowers of the grasses, I demonstrate that they are not fertilized by insects but by wind. This follows first from the amount of pollen which they produce and second from its lightness. If, for example, one taps or blows at the flowering panicle of *Dactylis glomerata* in fine, calm weather, one produces a pollen cloud which disperses in the air. Third, the filaments are very long and thin so that the anthers hang at some distance below the flowers. This obviously serves the purpose that the wind can better shake the anthers and release their pollen. Fourth, the considerable size and shape of the stigmas enable them to receive many pollen grains blown toward them by the wind. Fifth and finally, I have not encountered any insects on these flowers. The flowers of the grasses thus hold an intermediate position between the flowers of the sedges and similar plants and the nectar flowers. They are similar to the former in that they are fertilized by wind but are dissimilar in that they have nectar. They agree with nectar flowers in the latter aspect but are distinguished from them with respect to the former. But what purpose does their nectar serve? I am not in a position to answer this question.

The reason that Linné has already noticed, namely, that many flowers appear earlier than the leaves so that the wind is not prevented from carrying the pollen away, only applies to flowers of the first kind, for example, those of the elm, poplars, hazel, etc. The leaves of the spruce species cannot effectively prevent fertilization by wind because they are very narrow and smooth. In the European lime, on the other hand, the leaves would certainly do this. From this fact alone it can be assumed that its flowers are nectar flowers and fertilized by insects. To these remarks of Linné, I add that such trees must not only flower before they themselves have leaves, but also earlier than the trees in general have leaves. If, for instance, the aspens which stand in a heathland start flowering only when other trees which stand among them are already in leaf, these would prevent the wind from carrying the pollen of the male aspen to the female trees.

Miscellaneous Topics

Now there are also nectar flowers which make their appearance earlier than the leaves. These include, among others, the cornel (*Cornus mas*), the common daphne (*Daphne mezereum*), the coltsfoot (*Tussilago farfara*), the pestilence wort (*Petasites albus*), and the meadow saffron (*Colchicum autumnale*). The reason given by Linné cannot apply to these flowers because they are not fertilized by wind but by insects. The proper reason for this mechanism appears to lie in the flowering time. The meadow saffron is one of the last to flower; the following species belong to the earliest flowering nectar flowers. Since they all flower in a season when there are few flowers among them, it was necessary to arrange that honey bees and other insects can find them all the easier since they are the only flowers which can provide food for them, or almost the only ones. And for the achievement of this purpose, it was suitable to let the flowers bloom before the leaves appear, so that the flowers are not covered by the leaves and already catch the eyes of the insects from afar.

Because the final purpose of the corolla, which always applies, is that the flower catches the eyes of insects from afar, it must always be as large as possible. This possibility rests mainly on its shape, however. If it is flat, then it can be very large, indeed as large as its thickness allows. We find this, for example, in the mallow, poppy, and carnation families, and in the marginal florets of *Viburnum opulus* and the composites. But if it has a globular shape, as in the blueberry (*Vaccinium myrtillus*), it cannot possibly be larger than it is, because otherwise the flower itself would have to be larger. But since its thickness does not directly contribute anything to the achievement of this final purpose, it is always very thin if only this single purpose has to be achieved by it. Thus, *Convolvulus tricolor,* for example, has a very thin corolla although it serves three additional purposes as well, namely, that it spreads into a conical shape by day, which serves the achievement of the first purpose, that it closes by night, and that a larger insect can presumably stay on it to get to the nectar. Since these purposes require a certain although still very small thickness of the corolla, the corolla would presumably be thinner still if they were absent. Thus, whenever the corolla is thick or fleshy, another additional purpose has to be connected with it. So it is usually fleshy in those flowers which do not have a calyx, both because it substitutes for the calyx in the bud phase and has to protect the still tender sexual parts, and also because after these parts have burst open, they are not supported by any calyx but have to support themselves in their positions.

Each flower must always have a shape such that in the position which it occupies it can most easily catch the eyes of the insects. This position itself, however, has to be inferred from the fruit, because the flower exists not for itself but for the fruit. One must not argue that the fruit has this or that position because the flower has to have the same position, but the other way round—the flower has this or that position so that the fruit can have the same position.

The filaments and style are only present so that the anthers and stigma are located in just that place where they must necessarily be touched by the insects intended for the fertilization of the flower while they crawl into it. If therefore the location of the stigma is directly above the carpel and that of the anthers is directly above the bottom of the flower, then in the former case the flower has no style, and in the latter no filaments. For almost all flowers mentioned in the dissertation, one can easily recognize that the filaments and style are in fact present for this purpose, without me reminding him. The previously mentioned *Orchis* flowers result from the filaments being missing for this purpose. Finally, that the style is missing for this purpose can be recognized in *Parnassia palustris*. One therefore has to be very attentive about the absence or presence of these parts when examining flowers and particularly how, if they are indeed present, they behave throughout the complete flowering period, how they gradually elongate, bend, straighten out, etc. As all this can be easily observed, one will often soon come onto the right track if one investigates the purpose for which this behavior occurs.

One cannot deny that nature has protected the anthers and stigma of many flowers well against rain because it is detrimental to both organs. The harm to the former comes from sticking their pollen together, and perhaps also making it unfit for fertilization. To the latter, whether it is covered with hairs or with a certain moisture, the harm comes from preventing it from receiving pollen. There also not a few flowers in which one does not find such a mechanism although their nectar is completely secured from rain. There are even flowers whose stigmas and stamens are exposed to the rain so that they catch the raindrops and prevent them from penetrating to the nectar located behind or below them. These include, for example, various malvaceous flowers. The reason why nature has provided more protection from rain for the nectar than for the anthers and stigma is easily discovered. The nectar is for the flowers what a spring is for a clock. If one takes the nectar from the flowers, one renders all their remaining parts useless; one thereby destroys their final purpose, namely, the production of the fruit. The same ensues if rainwater mixes with the nectar and spoils it because the insects, which are excellent tasters, scorn the diluted food and leave the flowers unvisited and consequently unfertilized. If therefore the anthers and stigmas of flowers have been rendered incapable of fertilization by the rain in this way, this disadvantage still affects only them; the insect, which has found an unspoiled meal in them, continues with pleasure the fertilization function given to it and promotes this, at least for those flowers which did not suffer from the rain. If the nectar had been spoiled by the rain, however, then the insect might easily develop a dislike for the whole species, and consequently turn to another and leave it unfertilized.

Nature has determined a specific life span for each flower, a shorter time for this one, a longer for another. Some, such as *Hemerocallis fulva,* flower only 1 day, others for several days. The one which flowers the longest according to my observations so far is *Vaccinium oxycoccos,* which flowers for 18 days. One has to pay particular attention to the duration of the flowering time. I should have already learned that my first concept of how *Nigella arvensis* is fertilized by honey bees was wrong from the condition that this flower, after the presumed fertilization has been accomplished, still continues flowering for a fairly long time and only then loses the petals, stamens, and nectar mechanisms. At that time I overlooked this circumstance, however. I still did not realize how nature, always having only her main purpose in mind, namely, the production of the fruit, lets each flower merely exist for so long as is required for the fertilization of the carpel, and how, as soon as the carpel is fertilized, it strips the flower of all its adornment which it had until then displayed so marvellously because it would now be a completely useless state. The corolla therefore soon drops off, or if it stays, it becomes withered, unsightly, and completely unrecognizable. If the calyx was previously colored, then it now turns green because the young fruit until its maturity should be less eye-catching so that it can continue to grow and ripen, unnoticed and undamaged by any animal.

Kölreuter and Medikus claim to have observed in various species of (*Scrophularia*) that the stamens which initially bend toward the base of the corolla and subsequently, one after the other, straighten and put their now mature anthers on the stigma. They claim that the flowers are consequently fertilized in that mechanical way which I have attributed above to arising from erroneous observations. If these men had paid attention to the duration of the flowering time of these flowers, they could not only have found that they had been wrong in this observation, but they might also have easily discovered dichogamy. They would then have noticed that these flowers blossom for approximately 2 days before an anther appears. Had they now reasoned that nature cannot possibly let the flowers blossom in vain throughout this time, they might easily have noticed that the stigma is receptive in the first 2 days, and that consequently these hermaphrodite flowers are female flowers during this time. They would further have found that the first anther does not appear before the upper part of the style with the stigma has withered and bent downward, and that consequently these hermaphrodite flowers are male flowers as long as the anthers flower, which also lasts approximately 2 days. They would also have found that pollination cannot possibly occur while the anthers are gradually appearing during this time, but has already occurred earlier when the anthers are still tucked in the bottom of the flower. From all this the conclusion would have automatically followed that these flowers are not able to fertilize themselves by their own pollen; consequently, their fertilization cannot occur in any way other than by insects carrying the pollen of older flowers onto the stigma of younger ones. Had they therefore frequently observed the flowers in fine weather in order to see the soundness of this conclusion confirmed by observation, they would have found that wasps and other insects visit them, and that these cannot partake of the nectar without at the same time firmly touching the anthers in the older flowers and the stigma in the younger ones with their body. Consequently, they rob the former of their pollen and supply it to the latter.

If the flowers are to be visited and fertilized by insects, then they must be easily noticed by them, and that already from a distance. Consequently, they must stand exposed and be covered neither by the leaves of their own plant nor by other neighboring plants. If this cannot be done for other important reasons, however, then they must have an even stronger scent. I will verify by several examples in the dissertation that this is confirmed by observation.

Regular and Irregular Flowers

There are three circumstances through which one can explain, among many other structural features of flowers, why they are regular or irregular. The first is the inflorescence, or the way in which the flowers are attached to the stalk or branches of a plant. The second, which I have already mentioned above, is that the raindrops fall perpendicularly onto the flowers at least in calm conditions. The third is the intention of nature that the insects shall fertilize the flowers, if at the same time one considers the natural position of the insects which is always upright in flight and is usually so while they are walking and standing. Although insects can walk and stand in an inverted position, they will not do so without urgent reason, because it causes them more effort by having to cling in order not to fall off. As examples, I choose an erect flower, *Dianthus superbus,* plate XIV. 15.18. (Fig. 1.4); a pendulous one, *Leucojum vernum,* plate X.42.47. (Fig. 1.5); a horizontal one, *Lamium album,* plate XVI.8.9. (Fig. 1.6); and a downwardly inclining one, *Digitalis purpurea,* plate XVII.22.25.33. (Fig. 1.7)

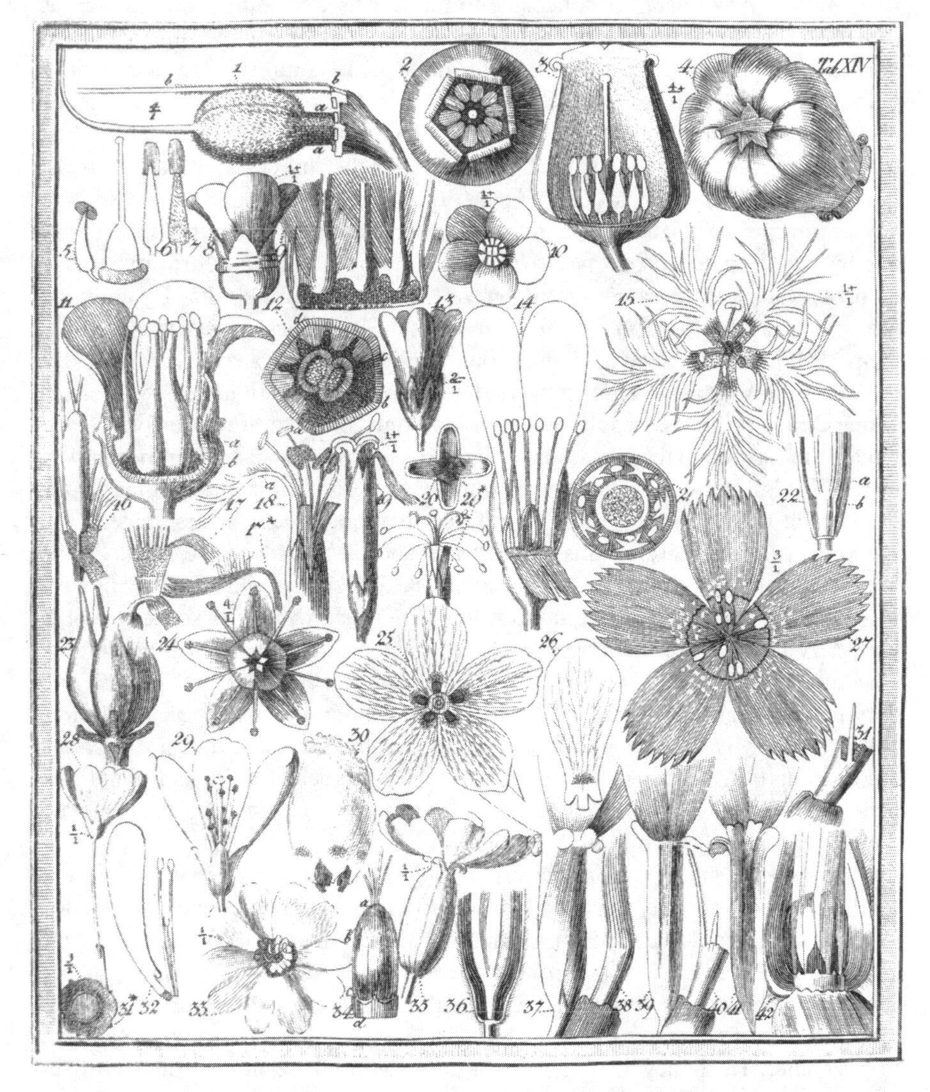

Figure 1.4. The 14th copperplate from Sprengel's book. Figures 15 and 18 illustrate the erect flowers of *Dianthus superbus* (Caryophyllaceae). See text for further discussion.

With regard to the first flower, one easily realizes that there is not the least reason why it should not be regular, neither because of the rain nor because of the insects. It is attached to the end of a branch and is single and upright, and is thus not prevented by anything from spreading its corolla in all directions as far as is necessary in order to catch the eyes of the insects from afar. It will therefore spread the corolla in the same way toward all directions because an insect is sometimes on one side, sometimes on another. There is no reason why it should not make itself as noticeable to the insects from either side as well as the other. The upper part of the corolla projecting from the calyx, the platform, now also serves for the purpose that the insects can comfortably stand on it to get to the nectar, whichever direction it flies in from. For both reasons, the petals must be equal with respect to this platform, and in particular because of the first reason they must be of considerable size. After the insect has perched on the flower, a spot of particular color in the form of the nectar guide should show it the way to the nectar located at the bottom of the calyx. Since the insect now has by chance perched on one or the other platform, each platform must have its nectar guide at the same distance from the opening of the tube. By crawling into the tube, the insect has to fertilize the flower in such a way that it wipes the pollen off the dehisced anthers in younger flowers and again brushes it against the receptive stigmas in older flowers. Consequently, the stigmas as well as the anthers must possess a regular position, not only in relation to the five petals, but also in relation to the axis of the flower. Because of this particular mechanism of fertilization, however, the stigmas must occupy approximately the same position which the anthers occupy. Therefore, they both stand in the center. The raindrops fall vertically on this erect flower and do not get to the nectar in the bottom of the calyx. Since raindrops cannot easily penetrate into the narrow tube in which the stamens and stigmas are located and partly fill its space, it is not necessary to prevent a raindrop which has fallen onto the corolla from coming near the opening of the tube. For this purpose, the petals are first incised into very narrow segments in order to catch as few raindrops as possible, and there is no reason why they should not be incised in this way. Second, not far from the opening of the tube, just in that place where the nectar guide is located, the petals have hairs which are turned outward, and one can think of no reason why they should not all have these hairs and in that place.

In some aspects, the second flower (*Leucojum vernum*) agrees with the first, but in others it is exactly opposite, particularly in being pendulous. It is attached to the end of the downwardly pointing stalk, spreads evenly in all directions, and can catch the eyes of the insects from all directions. The spreading is not at all, and the eye-catching only very little, prevented by the larger upright-standing part of the stalk. Consequently, its six petals must be completely equal to each other. The honey bee which visits the flower and presumably also fertilizes it perches on the outer surface of the corolla on whichever side it wants. It then has to find a nectar guide, by which it is directed to crawl into the flower.

Consequently, the nectar guide has to be displayed on the outer surface of the corolla, and in a regular manner. Each petal must therefore have a spot of a different color at its end. After the insect has crawled into the flower, it has to fertilize the flower while consuming the nectar. This probably happens by the insects brushing against several anthers at the same time as it licks the nectar located on the style. Hence, the pollen contained in the anthers falls out of the openings located at their ends and onto the body of the honey bee, from which

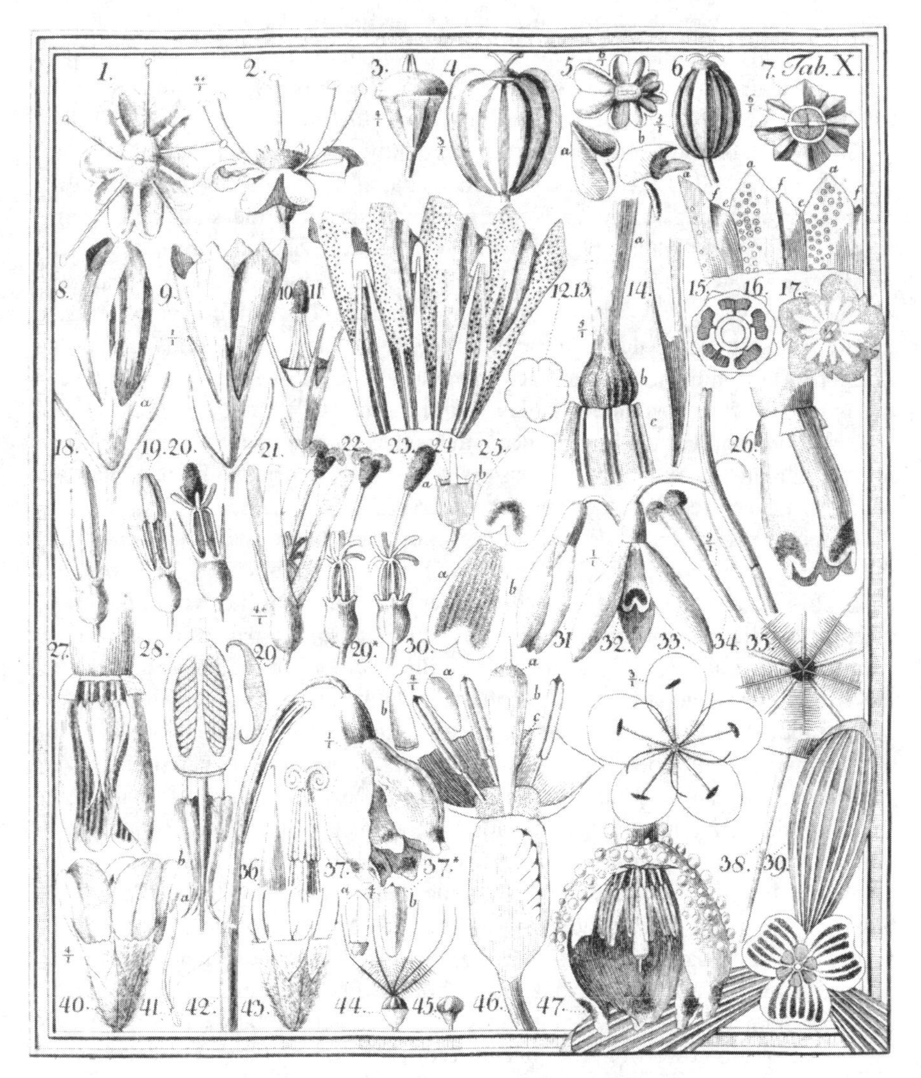

Figure 1.5. The 10th copperplate from Sprengel's book. Figures 42 and 47 illustrate the pendulous flowers of *Leucojum vernum* (Amaryllidaceae). See text for further discussion.

a portion gets onto the stigma which the honey bee must touch with the dusted part of its body. Because the fertilization has to occur in this manner, from whichever direction the honey bee may have crawled into the flower, there is no reason why an irregularity should occur with respect to the anthers and the style with its stigma. With regard to the rain though, the flower structure has to take into account another consideration, since the raindrops fall on the outer but not inner surface of its corolla as they did in the first flower (*Dianthus superbus*). The petals, for example, had to be entire and not incised into narrow segments as in the first flower. And since in this way the internal structures of the flower are sufficiently secured against the rain, it was in this case unnecessary to attach a particular nectar cover as in the first flower. There is also with regard to the rain no reason why the flower should not be regular.

The matter is quite different in the third flower (*Lamium album*). It is not attached at the end of the stalk or a branch, but at the side of the stalk and then not singly but alongside several others, which surround the stalk and form a whorl. It can only spread properly toward the front and then because of the neighboring flowers more in length than in breadth, but not at all toward the rear because of the stalk. It also attracts the bumble bees which have to fertilize it only from the front, although the complete whorl does so from all directions. Similarly, the bumble bees can reach the whorl from all directions, but approach a flower only from one side, namely, from the front. While there is no reason from the viewpoint of the bumble bees, as well as the rain, why the whorl should not be regular, there are several reasons why the flower has to be irregular and has the shape which it in fact possesses.

If a bumble bee is attracted by all corollas of the whorl and approaches it, it then perches on that part of a flower which is most comfortable for it because of its upright stance. This part is the lower lip of the corolla, which has to be of considerable size for this reason, and also because as a part of the corolla, it contributes its share to the just-mentioned final purpose of the corolla, to attract the insect. The nectar guide, located on the lower lip and extending to the opening of the tube, shows the bumble bee the way to the nectar container, which is the bottom part of the tube. If one compares this flower with the first (*Dianthus superbus*), one can imagine it with respect to its lower lip, as a fifth part of the former. The former has five petals and as many nectar guides and nectar covers, which stand regularly around its axis; the latter (*Lamium album*) has only one petal (the lower lip) and one nectar guide. With respect to the nectar cover, however, although other related flowers such as *Nepeta cataria* and *Glechoma hederacea* have hairs on the lower lip, in this flower these hairs have been attached not to its lower lip, but at the bottom of the tube directly above the nectar container. This is one irregularity. Now the bumble bee will not consume the nectar of the flower for free, but in return fertilize it and that presumably in exactly the way which has been demonstrated for the first flower. For this purpose, it is necessary that the flowering anthers occur in exactly that place in the younger flowers which the flowering stigma occupies in the older, in order that the bumble bee touches the former as well as the latter with the same part of its hairy body. But because the bumble bee does not crawl into the flower from different directions, but only from one direction, and each time in the mentioned way, it is not necessary that the anthers and stigma have a regular position with respect to the axis of the tube; instead, they have the one which is the most suitable. The filaments and style therefore bend away from the axis outside the tube and toward the front side. This is the second irregularity. Finally, the anthers and stigma, as well as the tube which contains the nectar, have to be protected against the rain by the upper lip. Consequently, the upper lip also has to have a quite different structure from the lower lip because of the different final purpose. It has to be curved, whereas the former is flat, and furnished with hairs at the margin which the former does not have; it does not require the nectar guide, which was necessary in the former, and has to be entire, whereas the former is divided into several sections. This is a third irregularity. In this irregularity, however, the flower also possesses a regularity. One can mentally divide it into two completely equal parts by a perpendicular plane. Because although it had to be irregular from the top to bottom, there is no reason why it could not be regular from one side to the other, neither with respect to the rain nor to the bumble bee, the body of which, regardless of its irregularity, is also built regularly to the extent that it can be divided into two completely equal parts by a perpendicular plane. The flower is then similar to the first with respect to the lower lip, except for the nectar cover, and similar to the second with repect to the upper lip, except for the nectar guide.

Finally, we will compare the fourth flower (*Digitalis purpurea*) with the second (*Leucojum vernum*). This is not attached at the end of a stalk like the latter, or at the end of a branch like the first (*Dianthus superbus*), but by a pedicel at the side of a branch, and many flowers form a onesided raceme (*racemus fecundus*), which strikes the eyes mostly from the front. Like the complete raceme, each flower also attracts the attention of the bumble bees and honey bees intended for its fertilization mainly from the front, and so it is to be counted with the horizontal flowers. Consequently, it also had to receive an irregular shape. Nature deemed it suitable to give it a position which puts it approximately halfway between completely horizontal and pendulous flowers. In this respect, it is similar to the second flower (*Leucojum vernum*), with which it therefore also agrees in that it mainly attracts attention with the outer surface of its corolla, but by no means with its inner surface like the first and third. But it deviates from the regularity of the latter in the following points:

1. At the margin, the corolla is divided into four segments of which the ones on both sides are equal to each other, but the lower is wider and longer than the upper. The final purpose of this irregularity is that the opening of the corolla is

Figure 1.6. The 16th copperplate from Sprengel's book. Figures 8 and 9 illustrate the horizontal flowers of *Lamium album* (Lamiaceae). See text for further discussion.

better presented to the insect after it has approached the flower and the part of the corolla on which it can perch comfortably because of its upright position, like the nectar guide located on the corolla, immediately catches its eyes.

2. The nectar guide could not be attached to the outer surface of the corolla as in the second flower. Neither could the nectar guide be attached in such a way that all four segments would have received a spot of a different color, since these spots would scarcely be noticed on the two marginal segments and not at

all on the rear or lower ones. Nor could it be attached in such a way that only the frontal or upper segment received such a spot, because then, according to the instructions of this nectar guide, the insect would have to sit on the upper part of the corolla, then turn around and crawl into the flower in an inverted position. The insect would not go to so much trouble, but rather it would have crawled in on the lower side of the corolla without caring about the nectar guide, and so would not accomplish the fertilization which had been intended with the first way of crawling in. Consequently, the insect naturally crawls in on the lower side, and so the nectar guide also had to be attached on the inner surface of the lower side.

3. When the insect crawls in to obtain the nectar located in the bottom of the corolla tube, it has to fertilize the flower in exactly the same manner as has been reported for the first flower (*Dianthus superbus*). For this reason, the filaments do not stand regularly around the axis of the corolla, nor is the style located in the axis as in the second flower; instead, both nestle closely against the upper side of the corolla as soon as they have left the short tube, so that in the younger flower the insect wipes the pollen off the anthers with its hairy back and in the older flower brings it onto the stigma.

4. Finally, for the exclusion of raindrops from the nectar, the corolla is furnished with hairs, though on the lower segment. In this respect, the flower is dissimilar to the second flower, but not to the remaining three which it resembles in this respect.

From the comparison of all four flowers with each other comes the general principle that erect and pendulous flowers have to be regular, since a lower and upper side is not present in them but all sides are of equal height so that the insect, whichever side it perches on, can fertilize them. In contrast, horizontal flowers must be irregular because they have an upper and lower part, and each time the insect perches on the lower part and crawls in on one of the two parts (in the March violet the honey bee perches on the lower part but then turns around and crawls in on the upper part). The manner in which fertilization by the insect should occur has to be determined only according to this fact.

Conclusion

There are various circumstances from which one can determine whether an insect which visits a flower is intended to fertilize it or not. That honey bees are intended to fertilize the common sage (*Salvia officinalis*) one recognizes from the fact that they visit this flower extremely frequently (consequently, its nectar is very nourishing to them) and that they are just as large as they must be so that when crawling in, they necessarily touch the anthers of the younger flowers and the stigma of the older flowers. The same applies to the lime, although its fertilization occurs in another way. On the other hand, honey bees should not fertilize *Iris*

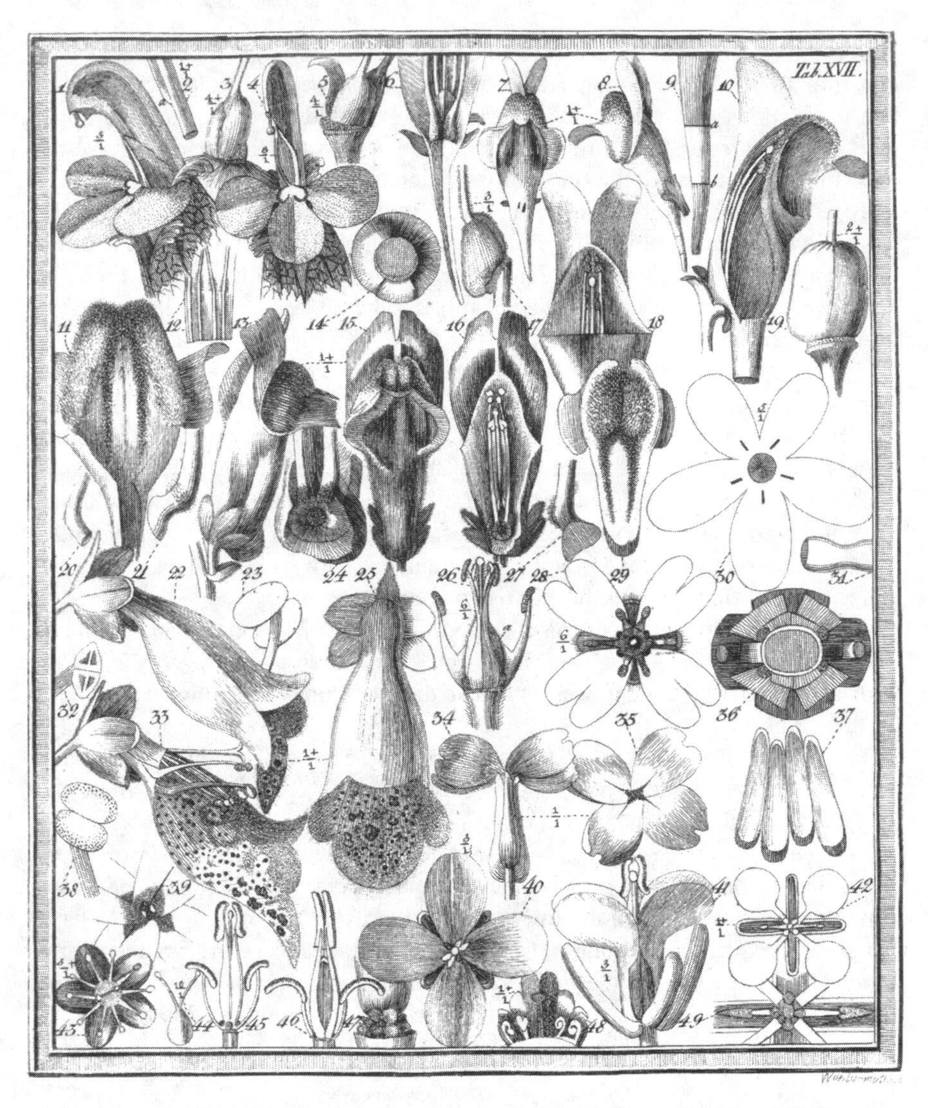

Figure 1.7. The 17th copperplate from Sprengel's book. Figures 22, 25 and 33 illustrate the downwardly inclining flowers of *Digitalis purpurea* (Scrophulariaceae). See text for further discussion.

germanica because they die from its nectar, nor *Parnassia palustris* because they become unconscious when visiting it. *Iris xiphium* has to be fertilized by a large bumble bee because this knows how to find the nectar easily and can crawl into the flower, for which the honey bee is too weak. An insect which commits an outrage against a flower is not intended to fertilize it. The small May beetles which gnaw the flowers of *Viburnum opulus* will not fertilize them.

The large bumble bee which, as I have said above, forcibly seizes the nectar of *Antirrhinum linaria* from the outside, because the natural entrance is too small for it, is not intended to fertilize this flower. The ear-wigs are not at all intended to fertilize flowers because they do not seek their nectar but consume their tender fertile parts, thus rendering their fertilization impossible.

Because many flowers have separate sexes and probably just as many hermaphrodite flowers are dichogamous, nature does not appear to intend any flower to be fertilized by its own pollen. I can mention one single experiment which confirms this assertion with respect to homogamous flowers. This is that last summer a plant of *Hemerocallis fulva* flowered in my garden. I tried to artificially fertilize some of its flowers with their own pollen (since only one flowered at a time), but not a single one set a seed-capsule.

Because the nectar flowers are intended either for several species of insects or for only one species, the fertilization of the carpel and the production of the fruit must also proceed more easily in the first case than in the latter. Observation confirms this. The umbellifers and euphorbias which are visited by many kinds of insects produce abundant seed. On the other hand, various *Iris* species which are visited only by bumble bees often have imperfect capsules with no seed in them. If rain, by washing the pollen off the anthers, were the only reason for the infertility of the flowers, then observation should show just the opposite. This is because in the umbellifers and euphorbias, the anthers are completely exposed to the rain, whereas in *Iris* they are protected from the rain. In *Iris xiphium,* for example, a raindrop cannot reach the anthers by any means. I have also noticed in flowers attached to the stalk in spikes that occasionally after the complete spike had long withered, some had set fruit but others had not. These flowers were partly of such a structure, however, that their anthers and stigma were completely safe from rain, as in *Hyacinthus comosus,* for example. This cannot be explained in any other way than that only one species of insects is intended for the fertilization of such flowers, because one spike does not flower all at once, but the lower flowers start to flower first and then progressively the upper ones. The flowering spikes had thus by chance been visited by the insect intended for the fertilization of the flowers at one particular time and not at another. Nevertheless, although for another reason, the fertilization of the false nectar flowers often has to remain undone; this is observed in the common birthwort and in those *Orchis* species which are false nectar flowers, in which few flowers set fruit. I will verify this in the dissertation.

Exotic flowers can remain unfertilized in our gardens for two reasons. First, if they flower only in winter, they consequently do so in greenhouses or hothouses and thus cannot be visited by insects. This applies to those plants which have been brought to Europe from the southern hemisphere, and which after this migration continue to flower in the summer there, or our winter. Second, they may be fertilized in their country of origin by an insect which does not occur in our regions.

* * *

It appears that certain species of spiders know how to distinguish nectar flowers from those devoid of nectar and that the need of the insects to visit the former is very well known to them. Hence, they stay in the neighborhood of such flowers or crawl into them and wait for the insects at the bottom of the flowers.

* * *

Someone who investigates the purposes of nature in the organization of fruits will probably find before him a similarly extensive field, as rich in potential discoveries as that in which the flower investigator roams. The former is still largely unknown to me, but since both are related, I was sometimes naturally induced to go from the latter over to the former. The few observations which I gathered there and which relate only to one kind of fruit, I will leave to the further examination of the reader.

Just as the flowers are fertilized either in a mechanical way or by insects, so also are the plant embryos which are contained in the fruits and are called seeds brought into the womb of mother earth either in a mechanical way or by animals. And just as those flowers which are fertilized by insects mostly contain something edible, namely, the nectar, by which the insects are attracted to carry out those duties, so also the fruits whose seeds are sown onto the ground by animals have something edible for this final purpose, namely, the flesh. Various species of birds consume various kinds of berries and digest their flesh but not their seeds, which they pass undigested and undamaged. Because considerable time has meanwhile passed and they are consequently in a place distant from the plant which provided them with the berries, they thereby promote the purpose of nature which is to the effect that the seeds should be sown a large distance from the parent plant. In the droppings in which the seeds are located, they find their first nourishment at the time that they germinate. Finally, just as most nectarless flowers are very unattractive and the nectar flowers in contrast make themselves conspicuous by means of their colored corolla, so those fruits whose seeds are brought onto the ground in a mechanical way are unattractive and uncolored. On the other hand, the fruits whose seeds shall move through the body of animals in order to be scattered on the ground are sizable and colored so that the animals notice them from afar and, attracted by their inviting appearance, consume them. My remarks in the dissertation relate only to the first kind of fruits. I prove that these are equipped in such a way that the seeds enclosed in them are scattered on the ground as far as possible from the parent plant, whereby nature achieves its greatest final purpose, namely, the preservation of species and the propagation of individuals of each species.

Nature has made use of various methods for this purpose. She has furnished some seeds with a hairy fringe, others with a wing which is much larger but at the same time much lighter than the seed, and by means of which the seed can be carried for many miles. Others are covered everywhere with hooks, and are caught in the wool of passing animals and in the clothing of humans and are carried everywhere by both. Some capsules are flexible, and when dried in the heat of the sun, burst open and throw out the seeds contained in them with great strength. Others are equipped so that when they have opened, the seed cannot drop out by itself but can only be thrown out by the wind and is consequently scattered far away.

In order to achieve just this final purpose, nature has bestowed on these plants such fertility with respect to the amount of the seeds which they produce that it appears to be wasteful in this respect, which it by no means is. To what Busch (*Encyclopädie,* p. 95) says about this, I add the following: As he correctly notices, a particular Providence by no means watches over the plant embryo so that it does not perish, and the Creator does not by individual arrangement bring each embryo to a spot on the ground which is suitable for it, but leaves the sowing of the seeds to the wind, for example. This, however, carries few seeds exactly to a place where they can germinate and turn into plants. *Chondrilla juncea* may serve as an example. This plant only occurs on infertile sandy and at the same time somewhat shallow and dry soils. Its seeds are furnished with a hairy fringe and can be carried far away by the wind. Will the wind sow them all onto a soil they require? Or will it carry most of them into water, swamps, meadows, fertile soil, forests, or gardens, where not a single one will germinate? And even of those which the wind carries onto a soil suitable for them, most still fail. Many fall onto the small turfs of the sand-grasses and other sand plants, thus not even reaching the soil. Many germinate but are smothered by neighboring plants. Few fall onto such a spot where they can in fact turn into plants. Thus of a hundred, perhaps even a thousand seeds, a single one succeeds, and such a large amount of seed has to be produced annually so that the species does not become extinct at some time or another.

2

Christian Konrad Sprengel's Theory of the Flower: The Cradle of Floral Ecology*

Stefan Vogel[†]

Introduction

As the plant taxonomist Karl Suessenguth once wrote, there are two kinds of discoveries: Detecting a thing nobody has seen before, or thinking what no one has thought before about something that everybody sees. A discovery of the second kind, and one of great moments in our science, was Sprengel's theory of the flower. This theory clearly expressed for the first time the notion that flowers are designed for the transmission of pollen by foreign vectors, that is, animals or wind, and they can be understood only from this perspective.

The Birth of Floral Ecology 200 Years Ago

The book in which Sprengel conveyed this perception was published in 1793 by Friedrich Vieweg in Berlin under the title:

Das entdeckte Geheimniss der Natur im Bau und in der Befruchtung der Blumen

(*The Secret of Nature in the Form and Fertilization of Flowers Discovered*)

This work initiated the discipline of floral ecology two centuries ago. This chapter commemorates that event, the importance of which was recognized by few of Sprengel's contemporaries and fully appreciated only 70 years later. Since then,

---

*The Sprengel Bicentenary Symposium on Floral Biology, organized by Drs. Shoichi Kawano, David Lloyd, and Spencer Barrett, was held as part of the XV International Botanical Congress in Yokohama, Japan, on August 29, 1993 (S1.12.2). The present chapter is a revised and final version of a paper read by the author.

†Institut für Spezielle Botanik der Johannes Gutenberg Universität, Mainz, Germany.

as far as I know, no fewer than 45 essays or chapters have been written on the life and work of this author, chiefly in German (for references, see the key articles of Wunschmann, 1893; Mittmann, 1893; Meyer, 1953, 1967; Wichler, 1936; Mayr, 1986; and the additional references on p. 61–62). Little can be added to the subject. However, there are two reasons why another account seems justified: The work of Sprengel, printed in Gothic, has apparently never been translated into English or other languages, and it is by no means a mere historical document but a treasure of anthecological data, still useful and worth citing today.

Nowadays it seems self-evident that flowers attract insects by color and fragrance, and are pollinated by them. One may ask, were flowers ever interpreted differently? Oh, yes! That recognition, to use a statement of Proctor and Yeo (1973), is hardly older than the invention of the steam engine.

State of Knowledge in Sprengel's Age

Since the dawn of history, mankind had been aware of some relationship existing between flowers and insects. However, that it had something to do with fertilization was generally unknown, although the Babylonians recognized that minute animals were necessary for the reproduction of fig trees. Until the seventeenth century, the beauty of flowers was regarded merely as a display of nature, created for its own sake or the delight of man. Except in a few cases—for example, ancient Egypt date palms were manually fertilized by using male spikes—plants were long regarded as nonsexual. It was mainly by the experiments of Camerarius (1694) with dioecious plants and by those of Kölreuter (1761–1766) who produced artificial hybrids, both in Germany, that the sexuality of flowers was uncovered. Independently, the observations of Grew (1671), Bradley (1717), and Miller (1724) in England provided further evidence. Nevertheless, it remained a moot point for long, being denied or called into question by notable botanists such as deTournefort (1700). Even by Sprengel's time, 100 years after Camerarius, it was not generally accepted. Only the experimental results of Gärtner (1844) were taken as definite proof.

Linnaeus, who was early convinced of the sexual role of flowers and based his system largely on the number of sexual parts, recognized nectaries and realized that flowers are visited by insects. He had little inclination, however, to interpret the functional significance of certain flower organs or even to "explain" them this way. In his opinion, the chief goal of botany was to describe and classify.

Sprengel also took the sexuality of flowers for granted. That insects were occasionally involved in the process of fertilization was already supposed in his time. Conclusive observations on tulips and honeybees, published in England by Miller (1724) and Dobbs (1750), were unknown to Sprengel, but he was acquainted with Kölreuter's work, to which he mainly referred. Kölreuter's emphasis had been on demonstrating the sexuality of flowering plants. He was less interested in exploring how the natural process of fertilization worked. Yet in the course of his experiments, he had made a number of casual observations leading him to recognize clearly the necessary interaction of insects in certain cases, especially in dioecious plants. Referring to the pollination of sword flag (*Iris*) and a few other plants, he made a comment that sounds surprisingly modern: "The decline of the insect (species) would inevitably be followed by the decline of the plant species" (1761, p. 31).

According to Kölreuter, and commonly believed in Sprengel's day, the visits of insects to flowers were by mere chance, and the dusting of the stigma with pollen was an occasional side-effect, something like contamination. What mattered was simply that some of the *farina fecundens* (the "fertilizing flour" as the pollen was called) had to come in contact with the stigma, usually the adjacent one of the same flower. This was believed to happen either spontaneously or by slight vibration, be it by gusts of air or animals. The display of color and fragrance had no other function in this process than to "manifest the solemnity of marriage." It was known, at least by beekeepers, that insects take nectar. Yet it was thought the nectar was not meant for insects. There were strange and controversial suppositions about this liquid. Partly it was held to be a substance that nourished the ovules, or was even involved in fertilization. In these cases, insects seemed detrimental because they robbed a substance indispensable for seed formation. By other authors, the nectar was thought be an excretion of matter harmful to the ovary; hence, insect visitors were useful because they liberated the pistil from it.

In order to appreciate Sprengel's discoveries, this state of confusion should be kept in mind. We shall return to these traditional ideas when we try to understand why his theory was ignored, rejected, or ridiculed at the time. Before discussing Sprengel's book, let me give a brief account of the author's life and personality.

It should be noted in advance that Christian Konrad is sometimes confused with Kurt Sprengel who is better known among plant taxonomists. This gentleman, a nephew and contemporary of the former, was Professor of Botany at the University of Halle, Saxony. He wrote taxonomical (Umbelliferae) as well as pharmacological and historiographic works. His treatise "Elements of the Philosophy of Plants" appeared in 1820 in English translation. Unlike his uncle, he was much celebrated. He died in 1833.

Short Summary of Sprengel's Life

Christian Konrad, son of a preacher and the last of 15 children, was born in Brandenburg, Prussia, on September 22, 1750. A student of theology and philology at the University of Halle, he later became a teacher at two secondary schools in Berlin. In 1780 he was appointed director of the Great Lutheran Town School of Spandau near Berlin. It was only at the age of 30, when already a professional in classical philology, that he became interested in botany. Reportedly, Sprengel's pursuit of botany was due to the advice of his surgeon. Prone to hypochondria and struck with a serious irritation of his eyes, Sprengel was advised to avoid indoor work and spend as much time as possible in the open. The surgeon, Ernst Ludwig Heim of Berlin, later a physician to the Prussian queen Louise, was himself a stimulating amateur mycologist and botanist (the lythraceous genus *Heimia* was named in his honor). He also won the young Alexander von Humboldt's heart for the *scientia amabilis*. Sprengel at first was busy in floristic field work around Spandau and he became a helpful informant for Willdenow (*Florae Berolinensis Prodromus,* 1787). Soon he was fascinated by the world of flowers. The discoveries he made so completely occupied his mind that this kind of work became the mission of his life, without compromise. He neglected his duties as a school teacher and refused to give private lessons, which led to permanent conflicts with his clerical supervisor and the pupils' parents. A chronicler of the school even wrote in 1853: ". . . During the directorship of Sprengel . . . , an irascible and obstinate man, the school began to relapse into ruin" (anonymous, quoted by Wichler, 1936). This verdict was certainly exaggerated. Nevertheless, the struggle of a haughty, insubordinate character with a narrow-minded, unappreciative environment ended with Sprengel's compulsory dismissal. He left the school the same year that his famous book was published.

After his dismissal, Sprengel spent his life as a private scholar, supported by a small pension and housed in an attic in Berlin. On occasions when he offered botanical excursions to laymen for a small fee, he showed, as several participants attest, his encyclopedic knowledge, but also his sense of humor. He continued his studies, planning a second volume of his book. Nevertheless, after failing to receive the attention and support he had hoped for, he abandoned the project, although he was never diverted from his ideas. The only further treatise he wrote on floral ecology was entitled, *"The Usefulness of Bees and the Necessity of Bee-Keeping, Viewed from a New Perspective"* (1811). He recommended that bee hives should be placed in clover and alfalfa fields in order to increase seed set, another topic that anticipated modern practices. At the end of his life, Sprengel broke with botany, returning to studies of classical literature. As a fruit of this labor, a book entitled *New Criticism of the Classic Roman Poets, with Comments on Ovid, Virgilius, and Tibullus* appeared in 1815, apparently a work of minor significance. Sprengel, unmarried, died on April 7, 1816 in Berlin at the age of 65. His grave is not known, and no portrait exists of him. Only a memorial stone, erected in the Botanical Garden of Berlin-Dahlem by Adolf Engler on the occasion of the 100th anniversary of Sprengel's death in 1917, reminds one of this great naturalist (see Engler, 1917).

The Flower Book

The book in which Sprengel unveiled the "secret of nature" was not a publication of the genre customary in scholarly literature of the time. Quite comparable to Goethe's contemporary essay entitled *"Attempt to Explain the Metamorphosis of Plants"* (1790), it addressed a wide-educated public. It was written to communicate a truth of which the author was convinced, but also intended to delight and entertain the reader. In this respect, Sprengel's book paralleled the much more successful *Insect Amusements* of Rösel von Rosenhof (1746–1761).

In the announcement of his book (1790), Sprengel called himself a "philosophical botanist." By approaching the flower philosophically, Sprengel aimed to explain the construction, conformation, and purpose of the floral organs. This was not a Linnean view. Indeed, it was the kind of perspective that, in the opinion of the great Swede, a scientific botanist ought to eshew. Today we are conscious of the fact that a flower as well as an entire organism can be "explained" in different contexts or at different levels: as a historical product of evolution or result of ontogenesis, as an expression of general and inherent developmental laws. The latter view was the starting point of Goethe's morphology. Sprengel's way of explanation, likewise revolutionary, was a functional or, as we would say, an ecological one.

Teleology and Utilitarism

It was a sudden inspiration, Sprengel admitted, that had let him pose the question "What for?" when he examined the delicate cilia lining the edge of the petal stalks in the flower of stork's bill, *Geranium sylvaticum*. Because they were exactly positioned above the five nectary glands, the answer seemed obvious: They serve as protection for the "honey" against its spoilage by admixture with raindrops. Sprengel was amazed at the ingenuity of this neat little arrangement that could cope simultaneously with two contradictory purposes: It prevented rainwater from entering, while allowing an insect's tongue to reach the nectar. In terms of creationism, an invention like this was for him a sign of the Lord's boundless wisdom, who ". . . has created not a single hair without a definite intention." In other words, a transcendental *causa finalis* had been at work; a purposeful finalistic concept governed every detail. The fact of expediency itself was proof of God's existence in the sense of Thomas Aquinus and Spinoza. Consequently, Sprengel's interpretation was teleological, and so was his diction: The flower of *Stapelia hirsuta,* for example, ". . . stinks like carrion only in order to lure bottle and carrion flies to which the stench is highly agreeable, and seduce them to fertilize the flower." The corolla of night candle, *Oenothera biennis,* ". . . has to be light-colored, because if dark-colored it would not catch the insect's eye. The flower had to be deprived of a nectar guide, because the latter would remain unnoticed in the darkness of the night," etc. Wording like this, we know, is taboo in the age of selectionism, but it gives no more or less information, indeed, than if we omit the intention ascribed to the plant by this statement.

Only after beginning his own investigations did Sprengel become acquainted with Kölreuter's work, published 30 years earlier, and it seems he discovered insect pollination independently. Kölreuter was still alive at this time, but there was no personal contact between the two. The aspects Kölreuter had annotated almost marginally occupied the center of Sprengel's interest; they were related to his own findings and systematized by clear definitions.

Sprengel's Discoveries and Novel Insights

Kölreuter (1761, p. 21) had been wondering about the circumstance that ". . . nature leaves so important a thing as reproduction to mere coincidence, to pure chance." As already pointed out, Kölreuter and his followers considered the process of pollen transfer a casual or, as Sprengel said, a "mechanical" event, chiefly taking place within the same flower. This alleged fortuity served as a starting point for Sprengel who, step by step, provided evidence that pollination is in reality a strictly organized and often sophisticated process, in most cases mediated by insects that act as "living brushes." The symbiotic nature of this relationship, the provision of food in return for pollination services, constituted for Sprengel the "discovered secret" to which the title of his book refers.

It has been suggested (Wichler, 1936) that Sprengel, being a professional in classical literature, probably would not have arrived at his discoveries if he had studied botany academically. The pedantic schematism of school botany taught in his age would have burdened him with a prejudice that might have prevented him from discovering a new paradigm. As an autodidact and self-educated person, he was in the position to arrive at his own ideas intuitively.

In his 43 double-column pages of introduction, the author describes how he arrived at his discoveries; he explains the principal tenets of his theory by means of selected case histories; he gives instructions on how to scrutinize a flower and which features deserve special attention. He specifies the details by which a nectary can be recognized in a flower. He stresses the necessity to observe a flower in its natural surroundings, at more than one location, at the right moment, and with due patience, ". . . because not always the first observed visit is by the insect which is intended for its fertilization" (p. 23).* In the main part of his book, spanning some 396 pages, the author analyzes, in the sequence of Linnean taxonomy, the flower structures of 461 species of flowering plants,

---

*Page numbers refer to those of Sprengel's original work of 1793. The quotations of Sprengel in the text were translated by the author.

mainly of his home country. In every case, he asks and discusses the following questions:

> What kind of insects transmit the pollen, or alternatively, how is the job done by the wind?
>
> What are the means of attraction?
>
> How do the floral organs fit together functionally?

The author experiences the surprising fruitfulness of his functional concept. One discovery follows the other. All of them reveal a fundamental realization: The construction and arrangement of flowers can only be understood given the pollination service of insects or wind.

Working without any special equipment, simply using a pocket lens, Sprengel examined his flowers, and he did it principally in natural settings. Even the latter was something new among botanists. Rather than collecting plants for herbaria, Sprengel observed their way of life in their natural habitats. ". . . One must try to catch nature in action," he said, apparently despising those armchair botanists who in breaking for lunch at noon miss the right hour to witness the life of flowers.

Camerarius and Kölreuter gained their insights by systematically experimenting with only a few species and therefore had not been in a position to generalize their circumstantial observations on floral ecology. Sprengel, on the other hand, though occasionally conducting experiments as well, drew his conclusions principally by the comparative method. How conscious he was of his synthetic and deductive method is shown in the following (p. 96):

> . . . Because if nature had applied in every part of the arrangement of any flower some particularity not occurring elsewhere, the knowledge about flowers would perhaps be a science for supernatural beings but not for mankind. For then in each case, when examining a flower, we would have to start from scratch and nothing of what we had learned before from 99 kinds of flowers would help us in the hundredth one, and there would be no analogy whatsoever. Yet, how can human intelligence operate without analogy?

Sprengel presented his results in a lively manner, but in clear and concise formulations. His own drawings, of almost unsurpassed accuracy, served as originals of the 25 black-and-white copperplates that accompany his work. Dictated by economy of space, the plates are crowded with figures totalling 1117 in number. They show nearly all flowers treated in the text in different projections and details, including their respective pollinators.

As space does not permit us to go into more detail, Sprengel's tenets may be summarized as follows:

1. The majority of flowering plants possess floral organs designed to attract the insects that pollinate them. Insect visitation and pollination are not fortuitous, but rather they are recurrent processes. Dimensions, positions, and movements of the sexual parts correspond to the kinds and sizes of the respective visitors, and are aimed at manipulating them and dusting their bodies with pollen at the appropriate place.
2. The nectar, called sap (*Saft*) by the author, serves to attract pollinators and this is its only function. Nectaries (called sap engines, or *Saftmaschinen*) are clearly defined and previous misinterpretations clarified (pp. 5 and 9).
3. In many cases, the nectar is concealed within the corolla in special containers termed *Safthalter* (meaning nectar holders), such as tubes or spurs, furnished with smooth inner surfaces (pp. 10, 25, 68).
4. The nectar is protected from spoilage by rainwater by means of nectar covers (*Saftdecken*), such as scales, hair fringes, etc. (p. 10).
5. The purpose of the coloration and fragrances of the corolla is to attract pollinators.
6. Nectar guides (*Saftmale*), local color patterns contrasting with the ground color of the corolla, signal the presence of nectar and indicate where insects can gain access to it (p. 15).
7. Bisexual flowers may be homogamous, that is, male and female functions are simultaneous; or they may be dichogamous, with noncoincident maturation of anthers and stigma (p. 17). There is protogynous dichogamy ["dichogamia gynandra," *Euphorbia* (pp. 4, and 266), *Scrophularia* (p. 222)] and protandrous dichogamy "dichogamia androgyna," *Epilobium* (p. 223), *Oenothera* (p. 217), *Aesculus* (p. 209), *Tropaeolum* (p. 213), *Ruta* (p. 236), Umbelliferae (p. 153). (Dichogamy in *Epilobium* had been discovered independently by Kölreuter).
8. Insect flowers may be generalists (pp. 17 and 32) or specialists (p. 20). The latter are adapted to bees, bumble bees, butterflies, or flies, respectively.
9. There are also flowers pollinated by wind. These are nectarless and inconspicuous; their stigmas and stamens are freely exposed to air currents, and they produce much more pollen than animal-pollinated flowers. Whereas insect pollen is sticky, wind pollen is dry and dusty (pp. 30 and 432).

Apart from these basic findings for which Sprengel coined terms, he implicitly observed and documented additional important phenomena, although he did not realize how frequent they are among flowering plant taxa. Of these, the following may be mentioned, using modern terminology. A few of these insights date back to Kölreuter, but most originated with Sprengel:

10. *Gynodioecism*. Occurrence of hermaphroditic and purely pistillate morphs (individuals) in a species (*Thymus*, p. 310).
11. *Hetero(di-)styly*. Occurrence of two morphs with different complementary levels of style and anther lengths in a species (*Hottonia*, p. 103).
12. *Herkogamy*. Spatial separation of anthers and styles in a flower, preventing self-pollination (*Iris*, p. 69, and *Convolvulus*, p. 107).
13. Experimental evidence of self-incompatibility (*Hemerocallis*, p. 202).
14. *Diurnal and nocturnal flowers*. The latter inconspicuous outside, and often closed during daytime (p. 16; *Oenothera, Portlandia, Chiococca*, and *Mussaenda*, p. 118).
15. Flowers conducting visitors to convey pollination by circling around the sexual organs (*apparecchi perambulatori* of Delpino) (*Passiflora*, p. 160; *Nigella*, p. 280; *Aquilegia*, p. 279).
16. *Scatter cones (Streukegel)*. Central clusters of poricidal anthers, releasing dusty pollen when shaken (*Borago*, p. 94, and *Viola*, p. 386).
17. *Buzz pollination*. Anthers designed to release pollen by the bees' vibratory movements (*Leucojum*, p. 181).
18. *Secondary pollen presentation*. Pollen exposed to insects by deposition on floral organs outside the anthers, for example on the style (*Campanula*, p. 109, composites p. 365).
19. *Kettle traps*. Flowers that keep insects temporally arrested and release them after dusting with pollen (*Aristolochia*, p. 418).
20. Pollen packets (pollinia) attached to insects by means of adhesive tags or clamps (retinacles) in orchids (p. 401) and *Asclepias* (p. 139), respectively.
21. *Floral deception*. Rewardless flowers lure visitors by faking presence of nectar (*Orchis*, pp. 28 and 401).
22. *Flower larceny*. Short-tongued visitors bypass the sexual organs by biting holes to gain access to the nectar (*Rhinanthus*, p. 314; *Linaria*, p. 320) [for *Linaria* first described by Gleditsch, 1766].
23. *Illegitimate visitors*. Exploitation of nectar or pollen by visitors unfit to pollinate a flower (incapable of doing so) (p. 43).
24. *Resource allocation*. A monoecious plant with large fleshy fruits keeps the number of pistillate flowers low to secure sufficient nourishment of the fertilized ovaries (*Cucurbita*, p. 434).

Observations on Seed Dispersal

It is seldom realized that Sprengel, when discussing the pollination of a flower, often paid attention to the ways in which the resulting seeds were spread. Such inquiries represented a pioneering effort about a process largely neglected at the time; thus, Sprengel can be considered the founder of dispersal ecology as well. His functional reasoning was no less successful in this field than it had been in the case of pollination, and he anticipated much of what we know today. Above all, he characterized anemochory and zoochory; within the latter category, he distinguished the two modes now termed epi- and endozoochory.

Gaps and Errors, as Judged from the Modern Standpoint

In a number of cases, Sprengel's interpretations of floral mechanisms differ from current ones, they are either erroneous or exaggerated. Perhaps other phenomena had confused him and remained unexplained.

In Sprengel's opinion, floral spurs and tubes are little more than nectar containers preventing the admixture of rainwater. Their main function, to filter out ineffective visitors, is nowhere explicitly mentioned.

His finding that there were insect flowers lacking a nectar gland irritated him, because it was inconsistent with the key role he attributed to nectar. His almost correct interpretation of the *Orchis* flower (p. 3) as deceptive is finally abandoned (p. 404): ". . . It is inexplicable to me that the flower has no nectar." The case remained a dark spot in the light of his theory. The same holds true for pollen flowers, now a familiar category of nectarless bee flowers (first defined by Müller, 1881). Sprengel failed to recognize surplus pollen as an alternative attractant, although he was aware that honeybees and ". . . at least one species of bumble bee" visited certain flowers only for gathering pollen (p. 27). Lack of nectar in a flower seemed to him a malfunction, caused by poor growth conditions or drought. So he erroneously ascribed nectar, for example, to *Solanum* (p. 128), *Lupinus* (p. 353), *Chironia* (p. 130), and *Sisyrinchium* (p. 417). Dealing with mullein (*Verbascum*), he confines himself to stating the absence of nectar. Sprengel obviously refrained from considering pollen as an alternative reward because he did not realize it could have two different functions: fertilization and reward. However, he concludes that bees do fertilize this type of flower also, because otherwise, they "merely would deprive them of pollen" (p. 28).

Sprengel did not come to terms with grasses either, because he mistook their lodiculae for nectar glands. How can flowers, he asked, being so obviously designed for wind pollination, possess such organs? The puzzle seemed all the deeper because the similar sedges did not show this inconsistency.

His explanations of the pollinium mechanisms in Asclepiadaceae and orchids contained some mistakes, and Sprengel also erroneously believed that the cyathium of *Euphorbia* was a flower.

All these structural misinterpretations do not lessen our admiration for Sprengel's descriptions, which for the most part are correct and still useful today. Neither were they the reason for the lukewarm or negative reaction Sprengel's book experienced during his life and long thereafter. Various circumstances may have been responsible for this.

The Dogma of Predominant Self-Fertilization, an Obstacle to Accepting Sprengel's Theory

As almost always happens, novel thoughts first meet skepticism. This was the initial reaction to Sprengelian theory, not least from trained botanists. Probably Sprengel himself contributed to scholars' reserved attitude by his disregard for authority and arrogant certainty regarding his own correctness. Above all, the traditional concepts established botanists taught seemed to justify their rejection of Sprengel's theory: Seen from the conventional standpoint, it was indeed weakly founded.

The prevailing dogma that not even Sprengel questioned suggested that flowering plants, rooted and sessile as they are, for the most part achieve sexual intercourse by means of a simple and most elegant solution, association of both genders within the same flower. Therefore, it was natural and logical to assume that fertilization normally took place by self-pollination. Floral arrangements preventing autogamy hence seemed absurd. There was no stringent reason, either, to postulate the obligatory role of insects in this process or consider floral organs as adapted to them. The presence of insects was a mere side-effect.

So Sprengel's ideas seemed arbitrary and unnecessary, wanting in the salient point of justification. Only in an aside did Sprengel come close to an understanding of the significance of outcrossing; when discussing dichogamy, he stated, ". . . Nature does not seem to allow any flower to be fertilized by its own pollen" (p. 34).

As mentioned, Kölreuter's experiments had provided the proof of sexuality in plants. His interspecific crosses, "botanical mules," demonstrated that also in plants paternal and maternal contributions blend; the "oil" inside the pollen grain and the stigmatic secretion were taken to be the delicate "fertilizing fluids," which had to be mixed up. They were believed to contain parental "essences" that represented the indivisible entity of an organism. The concept of essences was a consequence of creationism (Mayr, 1986). In this deterministic view, the entity of a species was constant and invariable from the beginning. The sterility Kölreuter had found in his hybrids corroborated the common belief that nature did not permit the rise of new species. Therefore, the inheritance and segregation of single characters were not concerns at the time, nor was the possibility of spontaneous variation. By pairing, equal and equal were united. Thus, the significance of mixis remained obscure. Despite occasional evidence (as presented, e.g., by Knight, 1799), that offspring resulting from outbreeding were more vigorous than those resulting from selfing, crossing within the same flower, between flowers of the same plant, and between separate genets was considered equivalent.

That nature bewildered botanists for so long was also due to the frequent occurrence of autogamous mechanisms, such as the movement of stigmas toward the pollen masses in *Campanula,* and the existence of permanently self-pollinating taxa never visited by insects. One hundred and seventy years had to pass until these cases were recognized as facultative fail-safe mechanisms accompanying cross-fertilization, or replacing it for some time in evolution.

Rarity of Insect Visits

Another argument advanced against Sprengel's theory was the putative rarity of insect visitation in many kinds of flowers. Could so important an affair as fertilization really depend on an accidental occurrence like this? Everybody realized that certain flowers such as the umbels of Apiaceae were usually crowded with insects. On the other hand, in orchards abounding with flowers, sometimes not a single visitor could be observed. Sprengel himself never saw insect guests on some of the flowers he studied. As we know now, every outbreeding species has its appropriate pollen vectors, and patient watching at the right time and place is necessary to witness their action. Even today, insect approaches to flowers remain unnoticed by most people. Even field botanists not focused on pollination often fail to remember having seen any bee or butterfly on a flower. In short, one can imagine why, in an epoch less informed than ours, Sprengel's assertion of the necessity of insects for pollination was doubted.

Philosophical Reasons for the Skepticism About Sprengel's Discoveries

Principal tenets of the zeitgeist were also at variance with Sprengel's philosophy; his reasoning about function in some way foreshadowed materialism; some even thought that this was blasphemous. Flowers were believed to be, as was nature in general, the work of the creator, and were supposed to be perfect and harmonic throughout and as a whole. Interpreting floral organs as intended for a particular material purpose seemed to imply that other components were not thus created. Furthermore, Sprengel spoke of "stupid insects" being deceived or incapable of managing certain flower arrangements; he discussed flower larceny, spoiling of pollen, and the like. The existence of such disorder was not accepted; it would have conflicted with the belief of nature's preestablished harmony.

That Sprengel's utilitarian view received little sympathy is reflected in Goethe's (1790) disagreement with him, expressed in the contemporary essay cited above. Nature, Goethe says, behaves like an artist rather than a workman: Its creatures and their organs exist for their own sake (Meyer, 1953). Only narrow-minded people can assume they are like tools, intended for a particular purpose.

The Fate of Sprengel's Theory During the Nineteenth Century

After his book was published, Sprengel lived for another 23 years. He received—with few exceptions—little or no positive response. As mentioned, because of this disappointment he never wrote a second volume. One of his opponents, the Professor of Botany, A.W. Henschel in Breslau (1820), who in a treatise still called the sexuality of flowers into question, wrote about Sprengel's theory of the flower: ". . . One has the impression it is conceived to entertain a schoolboy by such an amusing fairy tale." Among those contemporaries who were aware of his genius and praised his masterpiece were Willdenow (1802) and apparently James Edward Smith. The latter, who was the founder and first president of the Linnean Society, in 1794 named in Sprengel's honor a new genus of Epacridaceae, *Sprengelia*.

According to various sources, Sprengel's discoveries remained almost forgotten during the subsequent seven decades, being neglected in the manuals and largely ignored by university lecturers. They were appreciated, or cited at least, in a few botanical (Wiegmann, 1828; Treviranus, 1838) and entomological treatises (Kirby and Spence, 1815–1826, Burmeister, 1832), but research on plants has been so dominated by microscopy during the past century that Sprengel's work generated little interest among botanists. A measure of this was the book's low commercial value: Only one shilling and sixpence was paid by Eduard Strasburger in the 1850s when he bought the volume—simply because of its strange German title—from a second-hand dealer (F. Müller, 1884). However, *Habent sua fata libelli:* When in 1873, Friedländer's catalogue offered the book at 10 times this price, there was a very good reason for such an increase. The work had been rescued from oblivion and suddenly found itself in the limelight due to another man of genius, Charles Darwin.

Rediscovery by Darwin

As is well known, Darwin's thoughts on descent and evolution were strongly influenced by livestock breeding. Species characters were not invariable as the essentialist doctrine had maintained. Following inbreeding, an array of individual sports made their appearance in the progeny. Because wild species usually did not behave that way but remained homogeneous, Darwin inferred that the process responsible for preventing dissolution of the species or its deterioration was natural cross-fertilization.

On the other hand, certain small advantageous variations could by crossing spread within a reproductive community, thus slowly altering its features or even giving rise to new species. It was this entirely new view of evolution that shed light on the meaning of sexuality, at least in animals. Should it not apply to plants that were believed to be predominantly inbreeders?

Darwin, irritated by this inconsistency with his concept, began to take a closer look at pollination. For reasons now understood, the first experiments he made with field crops (mainly Fabaceae) failed to show clearly that cross-fertilization occurs and yields more vigorous offspring than selfing. At that time, he happened to remember the book of Sprengel, which had been recommended to him long before by his friend Robert Brown, who held it in high esteem. What could have been more welcome to Darwin than the ample evidence in Sprengel's work of the apparent tendency of flowering plants to escape self-fertilization? Adopting the methods of Sprengel, Darwin started to make empirical observations of British orchids, the results of which provided strong support for his own theory and were presented in the wellknown orchid book of 1862. He then took up *Linaria* and heterostylous plants and the results again corroborated the expectation that autogamy was disadvantageous for the progeny (1876). "Nature . . . tells us in the most emphatic manner that she abhors perpetual self-fertilization. . . ." (1877, p. 293). What Sprengel's aside had anticipated was of central importance for Darwin; only the concept of evolution gave Sprengel's theory of the flower its full meaning. "He clearly proved by innumerable observations how essential a part insects play in the fertilization of many plants. But he was in advance of his age. . . ." (1876). Because of Darwin's praise, Sprengel's work finally received the publicity among botanists it had previously lacked. It soon was debated, for instance, by Hagen (1883, p. 29; 1884, p. 572) and Müller (1884, p. 334) in *Nature* how it happened that Sprengel had been forgotten or ridiculed.

The First Flourishing of Floral Ecology

Darwin's own botanical studies inaugurated the first period of pollination research during the 1860s. Sprengel's way of reasoning and method soon caught on, were easily modified for adaptive explanations, and they were zealously extended to exotic floras. Most of those who became reproductive ecologists in this period, among them—to name some of the most prominent—Hermann Müller, Severin Axell, and Friedrich Hildebrand in Europe, Asa Gray and Charles Robertson in North America, and Fritz Müller in Brazil, were fervent supporters of Darwin's theory of selection and descent. Some of them went so far as to proclaim flowers an exclusive creation of insects. The ingenious Federigo Delpino in Italy, on the other hand, while adopting the theory of descent, claimed the intervention of a psychovitalistic intelligence to be responsible for the ingenuity of floral mechanisms (1868–1874), thus approaching the teleology of Sprengel. Sprengel's ecological classification of flowers was elaborated by Delpino, who coined the terms ornithophily, chiropterophily, etc., that are still in use today. Hermann Müller thoroughly investigated insect behavior in flowers and the structural relationships of both (1873). These two naturalists clarified, among other things, the pollination mechanisms of the fabaceous flag blossoms, which Sprengel had failed to fully understand. The enormous increase of knowledge achieved in this period notwithstanding, floral ecology became discredited and lost much of its interest toward the end of the century, mainly because uncritical and unproven speculations about floral mechanisms had begun to abound.

However, from the 1920s onward, Sprengel's theory again enjoyed confirmation, stemming from the new science of experimental sensory physiology practiced by, among others, Frederic Clements, Frances Long, Karl von Frisch, Fritz Knoll, and Hans Kugler. What naively had been taken for granted by Sprengel was now critically questioned. Do insects really see colors, perceive scents and taste? All these capacities were demonstrated by countless experiments. In addition, unexpected facets became apparent when insect UV vision, red-blindness, and the language of bees were discovered.

Sprengel and Modern Anthecology

Modern floral ecology owes its second rebirth to its relevance for evolutionary studies. The different branches of floral ecology, still united in the person of Sprengel, have developed today into lines of research so separate that communication between them is problematic. Neo-Darwinism now calls for quantitative evidence of how selection works in molding flower structures and insect strategies, and how mating patterns influence population structure—problems far from Sprengel's thoughts. And a long way from the hand lens he used are also modern analytical techniques that allow the analysis of scent components, calculation of energy budgets, or measurement of the wavelengths of floral pigments. The discovery of new kinds of rewards, such as fatty oils and liquid perfumes, incompatibility systems, and the adaptive control of sex distribution, opened new fields of investigation. Now as before, however, we are working, consciously or not, with the tenets and many of the terms of Sprengel. Considering the enormous number of plant taxa unexplored or insufficiently known in terms of their floral life, Sprengel's prognosis still rings true when he stated that his book's value was not so much its actual content as much as the key it provides to ". . . magnificent discoveries philosophical botanists will make in the future."

The philosophical basis, however, for the interpretation of floral structure, a matter so dear to Sprengel's heart, has changed entirely. His teleological concept, resting on creationism, has been discarded and replaced by "teleonomy" in the Kantian sense (Mayr, 1979; Simon, 1991): Although the fitness and utility of organs appear, *as if* purposefully designed by some intelligence, they result *a posteriori* from unintentional, evolutionary processes. Generations of students from Darwin to the present have demonstrated the explanatory power of this approach. Yet it has its moot points. There are floral coadaptations so sophisticated they resist Darwinian interpretation unless one believes in the omnipotence of natural selection. Another problem is the existence or nonexistence of selectively neutral traits. In not a few current papers the exclusive utilitarian view of Sprengel continues. Can we really maintain that even the tiniest hair must have its particular purpose? Recent reasoning emphasizes that the criterion of utility has its limits. The diversity of strategies as well as constraints indicate that little-understood internal organizing processes based on molecular interactions and structures to a certain extent determine organic forms and faculties, including the microcosm of the flower. Inherent morphogenetic patterns precede and guide the evolution of exotropous adaptations (Bateson, 1972; Riedl, 1976; Gould and Lewontin, 1979; Haken, 1980; Vogel, 1991; Wesson, 1993). The basic features of floral construction are controlled, but not necessarily shaped, by ecology. This aspect deserves to be considered by modern floral ecologists.

With this in mind, we recall the thoughts of Goethe, like Sprengel a dilettante in botany, whose book on the metamorphosis of plants was published 3 years before Sprengel's volume (1790). This essay, marking the birth of comparative plant morphology, appears antipodal, as well as complementary, in juxtaposition to that of Sprengel. Although Goethe's ideas on fertilization and the destination of nectar are outdated, his concept of gestalt and bauplan to some extent anticipated modern thought.

Perhaps new inspirations as ingenious as those of Sprengel and Darwin will be necessary to one day unravel remaining secrets and help us to fully understand how flowers were shaped in the evolutionary interplay and antagonism of morphogenetic laws and adaptation.

Acknowledgment

Cordial thanks are due Professor Klaus Dobat, University of Tübingen, for generously providing bibliographic data, and Professor Susanne Renner, University of Mainz, who offered her kind help in improving the English of this contribution.

References

Bateson, G. 1972. *Steps to an Ecology of Mind. Collected Essays in Anthropology, Psychiatry, Evolution and Epistemology,* Intertext Books, London.

Bradley, R. 1732. *New Improvements of Planting and Gardening*, both philosophical and Practical. Knapton London.

H. Burmeister, H. 1832. *Handbuch der Entomologie,* Vol. I. G. Reimer, Berlin.

Camerarius, R.J. 1694. *Über das Geschlecht der Pflanzen* (De sexu plantarum epistula). Reprint of Ostwald's Klassiker No. 105, W. Engelmann, Leipzig, 1899.

Darwin, C. 1862. *On the Various Contrivances by Which British and Foreign Orchids Are Fertilized.* Murray, London.

Darwin, C. 1876. *The Effects of Cross and Self Fertilization in the Vegetable Kingdom*. Murray, London.

Darwin, C. 1877. *Die verschiedenen Einrichtungen, durch welche Orchideen von Insekten befruchtet werden,* 2nd ed. Schweizerbart, Stuttgart (German translation of Darwin, 1862 by V. Carus).

Delpino, F. 1868–1869, 1873–1874. Ulteriori osservazioni sulla dicogamia nel regno vegetale. I, II. G. Bernardoni, Milano

Dobbs, A. 1750. Concerning bees and their method of gathering wax and honey. *Phil. Trans. Roy. Soc.,* 46: 536.

Engler, A. 1917. Bericht über die Enthüllung des Denksteins für Christian Konrad Sprengel im Königl. Botan. Garten zu Dahlem. *Notizbl. Königl. Bot. Gart. Mus. Berlin-Dahlem,* 6: 417.

Gärtner, C.F. 1844. *Versuche und Beobachtungen über die Befruchtungsorgane der vollkommeneren Gewächse und über die natürliche und künstliche Befruchtung durch den eigenen Pollen*. E. Schweizerbart, Stuttgart.

Gleditsch, J.G. 1766. *Vermischte physikalisch-botanisch-ökonomische Abhandlungen* Teil II. Abraham Gerhard, Berlin.

Goethe, J.W.V. 1790. *Versuch, die Metamorphose der Pflanzen zu erklären,* C.W. Ettinger, Gotha. Reprint (D. Kuhn, ed.), Verlag Chemie, Weinheim, 1984.

Gould, S.J. and R.C. Lewontin. 1979. The spandrels of San Marco and the panglossian paradigma: A critique of the adaptionist programme. *Proc. R. Soc. Lond. B.,* 205: 581–598.

Grew, N. 1671. The anatomy of vegetables begun, with a general account of vegetation founded thereon. Spencer Hickman, London

Hagen, H.A. 1883–1884. Christian Konrad Sprengel. *Nature,* 29: 29, 572.

Haken, H. 1980. Synergetics. Are cooperative phenomena governed by universal principles? *Naturwissenschaften,* 67: 121–128.

Henschel, A.W. 1820. *Von der Sexualität der Pflanzen*. W.G. Korn, Breslau.

Kirby, W. and W. Spence. 1815–1826. *Introduction to Entomology*. Longman, London.

Knight, T. 1799. An account of some experiments on the fecundation of vegetables. *Phil. Trans. Roy. Soc.,* 89: 195–204.

Kölreuter, J.G. 1761. *Vorläufige Nachricht von einigen das Geschlecht der Pflanzen betreffenden Versuchen und Beobachtungen*. Gleditsch, Leipzig 1761; with supplements I, II, III; Leipzig 1763, 1764, 1766 Reprint ed. by W. Pfeffer, of Ostwald's Klassiker No. 41, W. Engelmann, Leipzig, 1893.

Mayr, E. 1986. Joseph Gottlieb Kölreuter's contributions to biology. *Osiris,* Ser. 2: 135–176.

Meyer, D.E. 1953. Biographisches und Bibliographisches über Christian Konrad Sprengel. *Willdenowia; Mitt. Bot. Gart. Mus. Berlin-Dahlem,* 1: 118–125.

Miller, P. 1724. *The Gardener's and Florist's Dictionary,* or a complete system of horticulture. London.

Müller, F. 1884. Christian Konrad Sprengel. *Nature,* 29: 334.

Müller, H. 1873. *Die Befruchtung der Blumen durch Insekten und die gegenseitigen Anpassungen beider.* Engelmann, Leipzig.

Müller, H. 1891. *Die Alpenblumen, ihre Befruchtung durch Insekten und ihre Anpassungen an dieselben.* Engelmann, Leipzig.

Proctor, M. and P. Yeo. 1973. *The Pollination of Flowers.* Collins, London.

Riedl, R. 1976. *Die Strategie der Genesis.* Piper, München, Zürich.

Rösel von Rosenhof, A.J. 1746–1761. *Insecten-Belustigung,* 4 parts. Bauer & Raspe, Nürnberg.

Simon, J. 1991. *Subjekt und Natur.* In *Die Struktur Lebendiger Systeme* (W.Marx, ed.), Klostermann, Frankfurt am Main.

Sprengel, C. 1790. Versuch die Construction der Blumen zu erklaren (Announcement). Mag. Bot. 4th piece, 160–164 Zürich.

Sprengel, C.K. 1793. *Das entdeckte Geheimniss der Natur im Bau und in der Befruchtung der Blumen* I. Vieweg sen., Berlin Reprint by J. Cramer and H.K. Swann, Lehre, and by Weldon and Wesley, Codicote, New York, 1972.

Sprengel, C.K. 1811. *Die Nützlichkeit der Bienen und die Nothwendigkeit der Bienenzucht, von einer neuen Seite dargestellt.* Reprint (A. Krause, ed.) by F. Pfenningsdorff, Berlin, 1918.

de Tournefort, J.P. 1700. *Institutiones rei botanicae.* Typ. Regia, Paris.

Treviranus, L.C. 1838. *Physiologie der Gewächse,* Vol. 2. Marcus, Bonn.

Vogel, S. 1991. *Struktur lebender Systeme: Grundzüge und Problematik.* In *Die Struktur Lebendiger Systeme.* (W. Marx, ed.), Klostermann, Frankfurt am Main.

Vollmer, G. (ed.) 1988. *Was können wir wissen?,* Vol. 1. *Die Natur der Erkenntnis.,* 2nd ed. Hirzel, Stuttgart.

Wesson, R. 1993. *"Die unberechenbare Ordnung," Chaos, Zufall und Auslese in der Natur.* Artemis & Winkler, München (translated by P. Gillhofer).

Wiegmann, A.F. 1828. *Über die Bastarderzeugung im Pflanzenreiche.* Vieweg, Braunschweig.

Willdenow, C.L. 1802. *Grundriss der Kräuterkunde, zu Vorlesungen entworfen,* 3rd ed. Haude & Spener, Berlin.

Additional Literature on the Work and Life of Christian Konrad Sprengel

Ascherson, P. 1894. Zur Erinnerung an Ch. K. Sprengel und sein vor 100 Jahren erschienenes Werk "Das entdeckte Geheimnis der Natur im Bau und in der Befruchtung der Blumen." *Verh. Bot. Ver. Provinz Brandenburg,* 35: 8–13.

Endress, P. 1992. Zu Konrad Christian Sprengel's Werk nach zweihundert Jahren. *Vierteljahresheft der Naturforsch. Ges. Zürich,* 137: 227–233.

Frisch, K. von. 1943. Christian Konrad Sprengel's Blumentheorie vor 150 Jahren und heute. *Naturwissenschaften,* 31: 223–229.

Harvey-Gibson, R.J. 1919. *Outlines of the History of Botany.* A. & C. Black, London.

Hoffmann, P. 1920. Urkundliches von und über Christian Conrad Sprengel. *Naturwiss. Wochenschr. N.F.,* 19: 692–695.

Junker, T. 1989. *Darwinismus und Botanik. Rezeption, Kritik und theoretische Alternativen im Deutschland des 19. Jahrhunderts.* Deutscher Apotheker-Verlag, Stuttgart.

Kirchner, O. von. 1893. Christian Konrad Sprengel, der Begründer der modernen Blumentheorie. *Naturwiss. Wochenschr.,* 8: 101–105, 111–112.

Knoll, F. 1932–1933. J.G. Kölreuters und Ch. K. Sprengel's Blütenforschungen. *Der Biologe,* 2: 156–161.

Knuth, P. 1898. *Handbuch der Blütenbiologie,* Vol. I. Engelmann, Leipzig.

Kugler, H. 1942. 150 Jahre "Blumentheorie"—Ch. K. Sprengel zum Gedächtnis. *Der Biologe,* 11: 326–331.

Mägdefrau, K. 1973. *Geschichte der Botanik.* Fischer, Stuttgart.

Meyer, D.E. 1967. Goethes botanische Arbeit in Beziehung zu Christian Konrad Sprengel (1750–1816) und Kurt Sprengel (1766–1833) auf Grund neuerer Nachforschungen in Briefen und Tagebüchern. *Ber. Deutsch. Bot. Ges.,* 80: 209–217.

Mittmann, R. 1893. Material zu einer Biographie Christian Konrad Sprengel's. *Naturwiss. Wochenschrift,* 8: 124–128, 138–140, 147–149.

Müller, H. 1879. Koelreuter und Sprengel. *Kosmos; Zeitschr. für einheitl. Weltanschauung (Leipzig),* 3: 402–404.

Sachs, J. 1860. *Geschichte der Botanik vom 16. Jahrhundert bis 1860.* Oldenbourg, Munich (English translation: *History of Botany,* Glarendon, Oxford, 1889).

Schmid, R. 1975. Two hundred years of pollination biology: An overview. *The Biologist,* 57: 26.

Wagenitz, G. 1993. Sprengel's "Entdecktes Geheimniss der Natur im Bau und in der Befruchtung der Blumen" aus dem Jahre 1793 und seine Wirkung. *Nachr. Akad. Wiss. Göttingen, II, math.-nat. Kl.,* Annual Vol. 1993, Nr. 1, 1–11.

Wichler, G. 1936. Kölreuter, Sprengel, Darwin und die Blütenbiologie. *Sitz. ber. Ges. naturforsch. Freunde Berlin,* Annual Vol. 1935 305–341.

Willdenow, C. 1787. Florae Berolinensis Prodromus. W. Vieweg, Berlin. Reprint (ed. R. Böcher) Koeltz Scientific Books, Vönigstein 1987

Wunschmann, E. 1893. Christian Konrad Sprengel. In *Allgemeine Deutsche Biographie,* Vol. 35, Duncker & Humblot, Berlin, p. 293.

PART 2

Conceptual Issues

3

Floral Traits and Plant Adaptation to Insect Pollinators: A Devil's Advocate Approach

*Carlos M. Herrera**

> When one considers that populations are so rich in genetic variation and that responses to artificial selection almost invariably occur, the remarkable fact is not that some populations rapidly adapt to changed conditions, but that so few do.
>
> —D.J. Futuyma (1979)

Introduction

Certain natural history phenomena may provide a vivid illustration of selection in action and its adaptive products, and nearly every evolutionary biologist would agree that the pollination of flowers by animals provides a most illustrative example. It was surely not by chance that the first of Darwin's books to be published after *The Origin of Species* was precisely his treatise on the "contrivances by which orchids are fertilised by insects" (Darwin, 1862), the first in a series of monographs aimed at providing detailed supporting evidence for the theory of natural selection. Darwin's book on orchids evoked a major revolution in botany and gave rise to an enormous literature on pollination ecology (Ghiselin, 1984). It also marked the starting point for a tradition in the practice of pollination biology.

The Darwinian research program in pollination biology has been characterized by the search for the adaptive value of floral traits in relation to pollinating agents and, particularly, in promoting cross-pollination [see Baker (1983) for a historical account]. This activity identified a number of spectacular plant adaptations to pollinators, later becoming the textbook examples in pollination biology (including, typically, such biological oddities as fig trees, yuccas, and aroid inflorescences). However, the success of this research program was not restricted to identification of such canonical examples in pollination biology, as most interactions between plants and pollinators certainly are much less spectacular. The elucidation of relationships between the major pollinators of a plant species and particular combinations of structural and functional floral features ("pollination syndromes") was a further important achievement. The convergence of the flow-

*Estación Biológica de Doñana, CSIC, Apartado 1056, E-41080 Sevilla, Spain.

ers of disparate plant lineages into relatively few distinct floral types, along with their rather predictable association with pollination by different higher taxa of animals, was taken as *prima facie* evidence of adaptation by plants to their animal pollinators (Baker, 1961; Percival, 1965; Faegri and van der Pijl, 1966; Baker and Hurd, 1968; Proctor and Yeo, 1973). Most progress in pollination biology resulted from interspecific comparisons, and it has only been in the last two decades that pollination studies included intraspecific analyses, involving natural or experimentally induced variation in floral traits (e.g., Waser, 1983*a;* Nilsson, 1988; Galen, 1989; C.M. Herrera, 1993; and references therein). This latter approach led to analyses of the reproductive consequences of intraspecific variation in the arrangement, color, scent, morphometry, and shape of animal-pollinated flowers. By adopting a "phenotypic selection" approach (Lande and Arnold, 1983; Arnold and Wade, 1984), some of these investigations quantified pollinator-mediated selection on floral traits (e.g., Galen and Newport, 1988; Nilsson, 1988; Campbell, 1989; Galen, 1989; Schemske and Horvitz 1989; Robertson and Wyatt, 1990; C.M. Herrera, 1993; Anderson and Widén, 1993).

The century-old, Darwinian approach to pollination biology has thus established two important *qualitative* conclusions, namely that certain floral traits of some animal-pollinated plants must be interpreted as adaptations to pollinating agents, and that animal pollinators can effectively exert selective pressures on floral traits. The array of proximate mechanisms by which pollinators exert their selective pressures on plants is also reasonably well understood. In contrast, we are still profoundly ignorant of some *quantitative* aspects that are equally relevant from an evolutionary perspective. Demonstration that pollinators exert selection on some floral traits and that plants can respond to such selection does not verify either that both phenomena occur universally or that most or all floral traits of every insect-pollinated plant have been shaped in evolutionary time by the selective action of their current pollinating agents. (Reciprocally, demonstration that adaptations of plants to their pollinators are not universal in nature would not detract from the validity of any well-documented case of floral adaptation.) Two questions immediately arise if one accepts this conceptual distinction between the certainty that a phenomenon occurs and its actual frequency of occurrence. First, how common are species whose floral traits were predominantly shaped by selection from their pollinators (i.e., have evolved in response to the latter's selective pressures)? Second, how much of the floral phenotype of an average animal-pollinated plant has arisen as a consequence of selection exerted by pollinators? These two questions are subsumed by the less formal question, how widespread are floral adaptations to pollinators in nature?

Rather than attempting to answer these questions conclusively, the main aim of this contribution is to call attention to the actual relevance, and potential implications, of the questions themselves. Although foreign to mainstream tradition in pollination biology, the issue of the relative frequency of floral adaptations to pollinators deserves more consideration than it has received so far [but see

Schemske (1983), Waser (1983*a*), Howe (1984)]. I will first briefly review selected evidence that justifies questioning the commonness of floral adaptation to pollinators. Then, I quantitatively analyze the patterns of intraspecific variation in corolla-tube depth of insect-pollinated plants in a region of southeastern Spain, to illustrate that the above questions are amenable to rigorous quantitative testing. Finally, I enumerate some ecological factors that may constrain either the strength of selection by pollinators on plants, or the latter's response to such selection.

Despite its central role in evolutionary biology, there are numerous, often conflicting, definitions of the concept of "adaptation," and no consensus seems yet to exist about the requisites needed to identify a given phenotypic trait as an adaptation (e.g., Williams, 1966; Endler, 1986; Reeve and Sherman, 1993). In some cases, my reasoning will be based on the distinction between "exaptation" and *sensu stricto* adaptation suggested by Gould and Vrba (1982), a "history-laden" definition of adaptation (Reeve and Sherman, 1993). Under this view, true floral adaptations to pollinators would involve features that promote fitness *and* were built by selection exerted by current pollinators. In contrast, floral exaptations are characters that, although contributing to fitness, did not evolve as a consequence of selection by current pollinators. In other cases, I will adhere to a nonhistorical definition of adaptation. Under this view, which underlies all studies adopting the phenotypic selection approach, an adaptation is a phenotypic variant that enjoys the highest fitness among a specified set of variants in a given environment, irrespective of its history. This inconsistency of usages is chosen deliberately to examine whether questions about the frequency of occurrence of floral adaptations to pollinators are definition-dependent or, on the contrary, make sense regardless of the particular definition of adaptation one adheres to.

Some Cautionary Tales

Questions about the frequency of occurrence of floral adaptation to pollinators make sense only given the assumption that floral traits do not always represent adaptations to a plant's pollinators. I am not aware of any radical claim that all floral traits of all species exemplify actual adaptations, and every pollination biologist would certainly deny such imputation. Nevertheless, and probably for reasons less related to biology than to a well-defined tradition that had its inception in the urgent need to gather support for the newly born evolutionary theory, positive evidence for floral adaptations has traditionally been sought, publicized, and given more weight than possible negative evidence. However, independent lines of evidence caution us that a plant's floral traits need not represent adaptations to its pollinators.

Pollinator Diversity

Indiscriminate application of the concept of pollination syndromes has tended to exaggerate artificially the degree of adaptation of plants to pollinators (Baker,

1963; Macior, 1971; Waser, 1983*b*). In practice, the floral traits that characterize syndromes (e.g., color, shape, symmetry, nectar production) are generally of little use in predicting the pollinators of a given plant species. Furthermore, there is now overwhelming evidence that syndromes are of little value in explaining interspecific variation in pollinator composition (e.g., J. Herrera, 1988; Olesen, 1988; McCall and Primack, 1992; Waser and Price, 1990). I further illustrate this lack of conformity between syndromes and pollinator composition by examining the actual pollinator diversity experienced by insect-pollinated plants in nature.

The diversity of insects pollinating a given plant species can be evaluated at different levels in the taxonomic hierarchy, but the diversity of insect orders is particularly informative, as these higher-level taxa have been generally associated with different suites of floral characters or syndromes. I compiled information on the number of insect orders visiting the flowers of individual plant species in seven habitats on different continents. I selected studies providing comprehensive information on the flower visitors of at least 20 locally or regionally coexisting plant species. In one case (site 1 in Fig. 3.1), no reliable information is available to assess the extent to which flower visitors were pollinators. In four cases (sites 2–4 and 7 in Fig. 3.1), information presented by the authors indicates that most or all taxa involved were pollinators. In the remaining two cases (sites 5 and 6 in Fig. 3.1), there was evidence that all flower visitors were pollinators. As

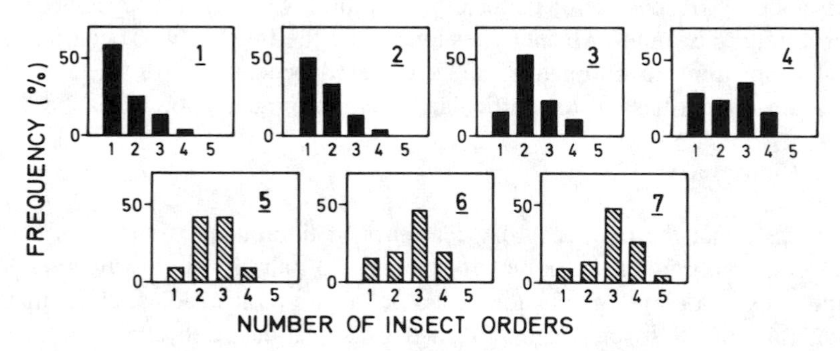

Figure 3.1. Ordinal diversity of insect floral visitors to plants in three Mediterranean (lower row, hatched bars) and four non-Mediterranean (upper row, filled bars) habitats. Each graph depicts the frequency distribution of the number of insect orders visiting the flowers of individual plant species in each habitat, coded as follows: 1, coastal tropical scrub, Jamaica (Percival, 1974) (N = 60 plant species); 2, montane temperate scrub, Chilean Andes (Kalin Arroyo et al., 1982) (N = 121 species); 3, alpine and montane habitats, Australia (Inouye and Pyke, 1988) (N = 40 species); 4, subalpine grassland and scrub, New Zealand (Primack, 1983) (N = 108 species); 5, montane scrub and woodlands, Sierra de Cazorla, southeastern Spain (Jordano, 1990; C. M. Herrera, unpublished) (N = 34 species); 6, coastal scrub, Doñana National Park, southwestern Spain (J. Herrera, 1985, 1988) (N = 26 species); 7, coastal scrub in Barcelona, northeastern Spain (Bosch, 1986, 1992; Retana et al., unpublished) (N = 23 species).

would be expected, insect orders were the same in all plant communities, namely, Hymenoptera, Coleoptera, Diptera, and Lepidoptera (a fifth order, Heteroptera, was recorded at a single site). Although sites differed somewhat in the frequency distributions of the number of insect orders visiting individual plant species, the prevailing picture from these studies is that, in most sites, a large fraction of plant species are pollinated by two or more insect orders (Fig. 3.1). This was particularly marked at the three Mediterranean sites, where plant species were pollinated by an average of 2.7 insect orders (the three sites combined).

Each of the major insect orders involved in the above pollination studies is generally associated with a distinct suite of floral characters or pollination syndrome (Percival, 1965; Faegri and van der Pijl, 1966; Baker and Hurd, 1968). Presumably each syndrome exemplifies alternative adaptive pathways to differences among higher insect taxa in sensory abilities, nutritional requirements, mobility, and morphology. The relatively loose association of plants to particular insect orders for pollination may indicate a prevailingly low degree of floral specialization to particular pollinators. Alternatively, the result might reflect some bias in the composition of the species considered with regard to floral morphology. For instance, this could occur if the species involved primarily produced "unspecialized" flowers with exposed floral rewards and an absence of structural or functional features that restrict the morphological or behavioral range of insect visitors. However, there is no evidence that this was the case with the species sampled. In the three Mediterranean sites, precisely where plants had the highest taxonomic diversity of pollinators, species with tubular corollas and concealed floral rewards accounted for between 50–76% of species in each site. Furthermore, in none of the three Mediterranean sites were species with exposed floral rewards (dish-shaped corollas) visited by more insect orders than those with concealed rewards (tubular corollas) (Table 3.1). McCall and Primack (1992), in an investigation of the pollination biology of three widely contrasting habitats, also found that the insect orders visiting flowers with open and tubular corollas were essentially identical.

Table 3.1 Mean (± 1 SD) number of insect orders visiting the flowers of species with open and tubular corollas at three Spanish sites (sites 5–7 in Fig. 3.1). Number of plant species in each category (sample size for means) in parentheses. Difference between floral types tested with single-factor ANOVAs (data square root-transformed for the analysis).

Site	Flower Type		Significance of Difference (P-value)
	Tubular	Open	
Cazorla	2.5 ± 0.7 (25)	2.6 ± 1.1 (8)	0.90
Doñana	2.7 ± 1.0 (13)	2.7 ± 1.0 (13)	0.98
Barcelona	3.0 ± 1.1 (15)	3.1 ± 0.7 (7)	0.81

Successful Pollination and Adaptation

The observation that the floral morphology and function of a given plant "fit nicely" the energetic, morphological, or behavioral traits of their pollinators has frequently prompted adaptive interpretations. However, that kind of evidence may convey little information about the origin of floral adaptations, because plants may be quite successfully pollinated even though the floral traits at work did not actually evolve in relation to their present pollinators. In other words, floral traits are often exaptations rather than *sensu stricto* adaptations.

Large-scale "experiments" brought about by human-induced, artificial expansions of plant geographical ranges provide unequivocal support for this claim. In contrast with the oft-quoted, popular examples of alfalfa and red clover (Faegri and van der Pijl, 1966; Proctor and Yeo, 1973), plants introduced into foreign continents for ornamental or commercial purposes provide many examples of successful pollination by completely new pollinator assemblages (e.g., Rick, 1950; Milton and Moll, 1982; Smith, 1988; Podoler et al., 1984; Kohn and Barrett, 1992). The phenomenon is quite familiar to keepers of botanical gardens, where artificial assortments of exotic plants are often successfully pollinated by pollinator assemblages whose taxonomic composition differs dramatically from those found by the plants in their native habitats. Heterostyly is a pollination mechanism where close reciprocal correspondence between the flower and particular pollinators seems most critical for successful pollination (Darwin, 1877; Ganders, 1979). Yet Kohn and Barrett (1992) found that the polymorphism in *Eichhornia paniculata* functioned in a similar manner in its native Northeastern Brazil and Canada, where it is an introduced plant and pollinated by bumble bees (which do not visit the plant in its native region).

When introduced plants are involved, we are aware that there is no history of interaction between the plants and their newly acquired pollinators (a prerequisite for the existence of adaptations if one adheres to a history-laden definition of the term), and no one would dare to propose adaptive explanations in these cases. But the success of pollination systems in the absence of a shared history of interactions between plants and pollinators is not restricted to cases affected by human intervention, and it may also occur in much less evident contexts. At least 12 species and six genera of typical "bird-flowers" (red-orange-yellow large corollas, plenty of nectar, little or no scent) occur on the Canary Islands and Madeira, yet no true flower-birds occur in these islands. That set of plants is successfully pollinated by several species of sylviid warblers that feed opportunistically on nectar (Vogel et al., 1984; Olesen, 1985). Although the plants involved are endemic relicts surviving from tertiary tropical or subtropical forests, their current avian pollinators are widely distributed in the western Palaearctic region and have originated more recently. The most likely interpretation in this case is that the modern avian visitors of Canarian bird flowers are newcomers to the scenario, and are not the original pollinators that interacted with these plants in the continental tertiary flora where they originated (Vogel et al., 1984; Olesen, 1985). Pollination of these plants is thus successfully accomplished by pollinators they certainly are not adapted to.

I close this section by describing an example from my own investigation (C.M. Herrera, 1993) that further documents the perils of inferring adaptation to present pollinators from simple observations of morphological matching. A nonhistorical definition of adaptation is adopted on this occasion. *Viola cazorlensis* is an endemic violet from southeastern Spain whose reproductive peculiarities make it a particularly favorable subject for investigating the adaptedness of floral morphology. The species is characterized by a closed corolla with a long, thin spur (mean = 25.0 mm, range = 8–42 mm), the longest of all European species in its genus. *Viola cazorlensis* also has the longest corolla tube of all regionally coexisting species with tubular corollas in its native range (see below). The shape of the corolla and size of flower parts (including spur length) vary widely among individuals. Nectar accumulates at the tip of the spur and, due to the highly restrictive floral morphology, pollination is accomplished by virtually a single species of day-flying hawkmoth (*Macroglossum stellatarum*), an unusual feature for the genus *Viola* (Beattie, 1974). Hawkmoth pollination systems generally involve sets of Sphingidae rather than single species, and documented cases of monophily are rare (Nilsson et al., 1985, 1987). Mean spur length of *V. cazorlensis* flowers matches nearly perfectly the mean proboscis length of *M. stellatarum* (26.4 mm). These observations might suggest that the unusually long (relative to congeneric species) spur of *V. cazorlensis* evolved in response to directional selection from its sole long-tongued pollinator. This hypothesis would be supported if spur length were significantly related to maternal fecundity, as found in other species with long tubular corollas (Nilsson, 1988). I tested this prediction in a 4-year study of seed production and, although flower size (peduncle length, size of petals) and shape (corolla outline) experienced significant phenotypic selection, spur length did not. In fact, the latter characteristic was precisely the only floral trait among those examined yielding negative evidence of phenotypic selection. Results thus did not support an interpretation of the characteristic long spur of *V. cazorlensis* as an adaptation to its current pollinator.

Pollinator Selection and Variability in Corolla-Tube Depth

Corolla-tube length has perhaps evoked adaptive interpretations more often than any other floral trait. Indeed, correlations have been found between the distributions of corolla-tube length and the length distribution of pollinators' mouthparts across regional floras (Inouye, 1979; Inouye and Pyke, 1988). For this reason, corolla-tube length seemed an appropriate character to quantitatively test the adaptedness of floral phenotypes to pollinators. The unanticipated results for *V. cazorlensis* mentioned above encouraged me to study patterns of corolla-tube length in a diverse regional species assemblage that included this species.

Hypotheses and Predictions

Darwin (1862) envisaged the evolution of flowers with deep corolla tubes as an "evolutionary race" between plants and pollinating agents resulting in directional selection for increased corolla length. Nilsson (1988) found evidence of directional phenotypic selection on corolla depth in two temperate orchids but, as noted earlier, I failed to find it for *V. cazorlensis*. This disagreement suggests that not all flowers with deep corolla tubes exemplify adaptation to current pollinators.

Fenster (1991) provided a quantitative tool for testing the generality of floral adaptedness to pollinators in plant species assemblages. Fenster's hypothesis relies on two demonstrable assumptions. First, increasing corolla-tube length is accompanied by reduced taxonomic diversity of pollinators and increased specificity of pollen placement on pollinators' bodies (Fenster, 1991). Second, pollinator taxa ordinarily differ in important components of pollinating effectiveness, and thus in the nature of their potential selective pressures on plants (e.g., Price and Waser, 1979; Motten et al., 1981; Schemske and Horvitz, 1984; C.M. Herrera 1987*a*). Given these relations, the strength of selection (either stabilizing or directional) on corolla-tube length will increase with increasing corolla-tube length. Based on these considerations, and the demonstration that *both* stabilizing and directional selection decrease the phenotypic variance of characters subject to selection (Lande and Arnold, 1983), Fenster predicted that, under selection by pollinators, the phenotypic variance of corolla depth should be negatively correlated with mean corolla depth. He found support for the prediction among 10 hummingbird-pollinated species.

I now test this hypothesis for a larger set of insect-pollinated plants from southeastern Spain. The following questions will be addressed: (1) Do most or all species exhibit significant individual variation in corolla depth? (2) Do species differ in the extent of individual variability in corolla depth? And if they do, (3) does variability vary inversely (across species) with mean depth? Questions 1 and 2, although not originally associated with Fenster's prediction (question 3), are directly relevant to the proper interpretation of results.

Methods

I measured corolla-tube depth for 58 insect-pollinated species from the Sierra de Cazorla region in southeastern Spain. Species were selected for study on the basis of the morphology of their corollas, and were sampled from a number of habitat types, including scrublands, mixed montane woodlands, forest edges, and disturbed places. I chose species with tubular corollas with fused petals, closed corollas with spurred petals, or functionally tubular corollas formed by fusion of sepals even though the petals themselves were not fused. Whenever possible, I measured at least 100 flowers for each species (5 flowers from each of 20 randomly chosen plants from a single locality; see the appendix to this chapter for species and sample sizes). Detailed data on the composition of pollinator faunas are available for only approximately 25% of species in the sample. Although limited, these data indicate that the assumption of Fenster's hypothesis of decreasing pollinator diversity with increasing corolla depth holds for my sample. In particular, species with corolla tubes >20 mm (near the upper extreme of the range for all species) are each pollinated by 1–11 insect species, whereas those with corolla depths <10 mm are pollinated by up to 90 insect taxa (J. Herrera, 1985; C.M. Herrera, 1988, 1993, and unpublished; Jordano, 1990).

To obtain comparable measurements (from a pollinator's perspective) of tube length from flowers of different morphologies, I used a measuring device that simulated an insect proboscis. This device consisted of a piece of fine (0.25-mm diameter), flexible nylon thread fitted inside a tubular plastic sheathing (1.5-mm diameter). The nylon thread was gently pushed into the corolla aperture, mimicking an insect pollinator probing the flower's interior for nectar. When the nylon thread could not be inserted further, I slid the plastic sheathing down the thread until it was impeded by some floral structure (generally the anthers or some perianth part). Sheath and thread were then held together, drawn from the flower, and the straightened length of thread protruding from the edge of the sheath was measured to the nearest 0.05 mm with a digital caliper. The nylon thread behaved much like a moderately rigid bumble bee or bee-fly proboscis when inserted into short, straight corollas, and it also served as an acceptable analogue of a hawkmoth or butterfly proboscis when applied to long, curved corollas or spurs. I am confident that this method produced truly comparable measurements of effective corolla depth.

Results

Mean corolla depth was computed separately for each plant and these "individual means" were then averaged to obtain species means (listed in the appendix). Species means ranged between 4.2–29.6 mm (Fig. 3.2A). Most species exhibited moderate individual variation in corolla depth, with coefficients of variation (CV) generally falling in the range 10–25% (Fig. 3.2B). Three of the five species with CV's >30% were gynodioecious. In these species, perfect-flowered individuals had significantly deeper corollas than male-sterile ones, and this dimorphism was responsible for the increased intraspecific variability. Individual variation in mean corolla depth was highly significant in all except one species (see the appendix), and the proportion of intraspecific variance accounted for by individual variation ranged between 13–94% (Fig. 3.2C).

I tested the significance of interspecific differences in variability with the variant of Levene's method proposed by Sokal and Braumann (1980). Individual means (Y_{ij}) were first log-transformed, and then the absolute deviates with respect

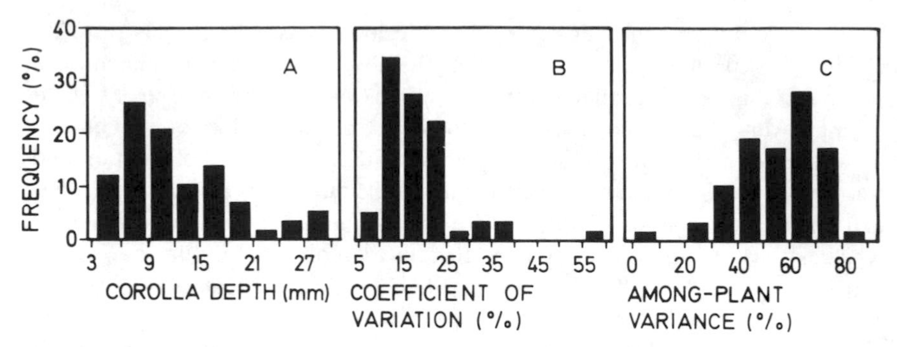

Figure 3.2. Summary of intra-and interspecific variation in corolla depth for 58 insect-pollinated species with tubular corollas from the Sierra de Cazorla region in southeastern Spain (see the appendix). Shown are frequency distributions of (A) species means, (B) coefficients of variation, and (C) proportion of total intraspecific variance accounted for by individual variation ("among-plant variance").

to the species mean of these transformed values were computed as $d_{ij} = |\log Y_{ij} - \overline{\log Y_i}|$, where d_{ij} and Y_{ij} are, respectively, the deviate and mean corolla depth of the jth individual in the ith species, and $\overline{\log Y_i}$ is the mean of the logarithms of the ith species. Absolute deviates d_{ij} were then compared across species using Kruskal–Wallis nonparametric analysis of variance (excluding gynodioecious species, which would have artificially inflated interspecific heterogeneity). Species differed significantly in individual variability of corolla depth ($\chi^2 = 136.5$, df = 54, $P <<0.001$).

The variance of log-transformed individual means [var(log Y_{ij})] was computed for each species, and the resulting figures plotted against species means (Fig. 3.3). Variance and mean were uncorrelated, regardless of whether the three gynodioecious species were included ($r_s = -0.070$) or not ($r_s = 0.045$) in the computations. Nonparametric regression analyses similarly failed to reveal any relationship between variance and mean of corolla depth (Fig. 3.3).

As species are related among themselves to varying degrees and belong to a common phylogeny, the data in Fig. 3.3 do not represent statistically independent observations. To account for the influence of possible phylogenetic correlations (Felsenstein, 1985; Harvey and Pagel, 1991; Gittleman and Luh, 1992), I applied a modification of the pairwise comparative method proposed by Møller and Birkhead (1992) (see also Lessios, 1990) to the 27 species belonging to genera represented by at least two species. All possible species pairs were constructed within each genus, and the relationship between variance and mean corolla depth was scored in each case as positive (increased variance with increased mean) or negative (decreased variance with increased mean). Of the 27 pairwise, intrageneric contrasts, 17 scored negative, and thus in the direction predicted by Fenster's hypothesis. The probability of obtaining by chance alone as many negative scores is 0.124 (from the binomial distribution); hence, the hypothesis of a negative

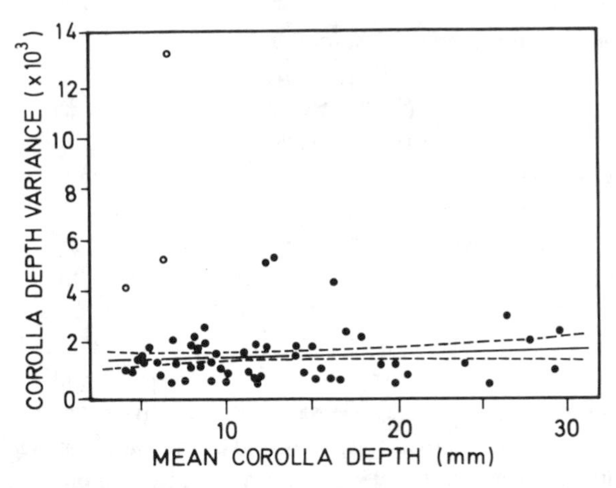

Figure 3.3. Relation of individual variance (variance of log-transformed individual means) to mean corolla depth for 58 southeastern Spanish species (see the appendix). Gynodioecious species (characterized by floral dimorphism and hence high variance in corolla depth) are shown as open dots. The illustrated regression line (continuous line) for all species, except the gynodioecious species, was fitted using a nonparametric procedure (cubic splines obtained by the generalized cross-validation method; Schluter, 1988). Dotted lines indicate ± 1 SE of mean predicted values from 1000 bootstrap regression replicates.

relationship between variance and mean is not supported when the statistical nonindependence of data points is controlled by restricting the analysis to pairs of closely related, congeneric species. It may therefore be concluded that, in the set of species studied, no significant relationship exists between variability and mean corolla-tube depth. This result is not consistent with the idea of generalized selection by pollinators on this floral trait in the study region, and confirms in a broad species assemblage the findings obtained for *Viola cazorlensis*.

Ecological Constraints on Floral Adaptation

The degree of adaptedness of plants to pollinators may be limited by factors intrinsic or extrinsic to the plants. Well-known intrinsic limitations include, for instance, genetic factors and life history or developmental constraints (Futuyma, 1979; Howe, 1984; Kochmer and Handel, 1986; Zimmerman et al., 1989), and I will not consider them here. This section will focus on extrinsic, or ecological, factors. For convenience, these can be classed into two nonexclusive categories depending on whether they reduce the probability of occurrence of selection by pollinators, or limit responses to selection when this effectively occurs. The two sections below examine some of these factors.

Constraints on the Occurrence of Selection

Spatio-temporal unpredictability in the composition of pollinator assemblages is probably one of the most important factors reducing the possibilities of selection on floral traits by pollinators. Whenever comparative studies have been undertaken, the pollinators of a given plant species have been shown to vary in their effects on plant reproductive success, an important fitness component. This variation may be due to differences in the frequency or amount of pollen removal or delivery, "quality" of delivered pollen from the viewpoint of the recipient plant, or some combination of these (e.g., Price and Waser, 1979; Motten et al., 1981; Schemske and Horvitz, 1984; C.M. Herrera, 1987*a;* Davis, 1987; Snow and Roubik, 1987; Ramsey, 1988; Wolfe and Barrett, 1989; Wilson and Thomson, 1991; Harder and Barrett, 1993). This variation provides, in theory, the opportunity for plants to specialize on (meaning to become adapted to) taxa providing the best pollinator services. Nevertheless, variation at various temporal and spatial scales in the composition and relative abundance of pollinators will most often limit seriously this possibility. Recent studies have shown that plants generally experience complex spatio-temporal mosaics in their pollinator assemblages, whose composition and abundance often vary tremendously between years and locations (C.M. Herrera, 1988; Waser and Price, 1990; Pettersson, 1991; Eckhart, 1992). Differences between plant populations in pollinator abundance and composition reflect the patterning in space of animal populations, whereas seasonal or annual variations reflect the population fluctuations of component species. For insects, annual fluctuations are universal even in the "stable" tropics (Wolda, 1983), and even affect the epitome of specialized pollinators, euglossine bees (Horvitz and Schemske, 1990). The stochastic nature of much of this spatio-temporal variation in pollinator composition, coupled with differences between pollinator taxa in pollinating effectiveness, will result in an inconstant pollination regime, as illustrated in Fig. 3.4 for a population of the insect-pollinated *Lavandula latifolia* in southeastern Spain. This species is pollinated by hymenopterans, lepidopterans, and, to lesser degree, dipterans. Compared to lepidopterans, hymenopterans transfer pollen to stigmas more often, and fly shorter distances between consecutive flower visits (thus promoting more geitonogamy) (C.M. Herrera, 1987*a*). The relative proportions of these major groups of pollinators vary annually, thus inducing fluctuations in average probability of pollen transfer per pollinator visit and mean flight distance between consecutive visits (Fig. 3.4). The variation in time and space of these and other important variables in the pollination environment generates a heterogeneous selective regime for plants, which weakens selection on floral traits (Schemske and Horvitz, 1989; C.M. Herrera, 1988; Horvitz and Schemske, 1990; Eckhart, 1991). For instance, in the bee-pollinated *Calathea ovandensis,* Schemske and Horvitz (1989) detected phenotypic selection on corolla-tube length during only one of three reproductive episodes because of marked annual variation in the composition of the pollinator fauna.

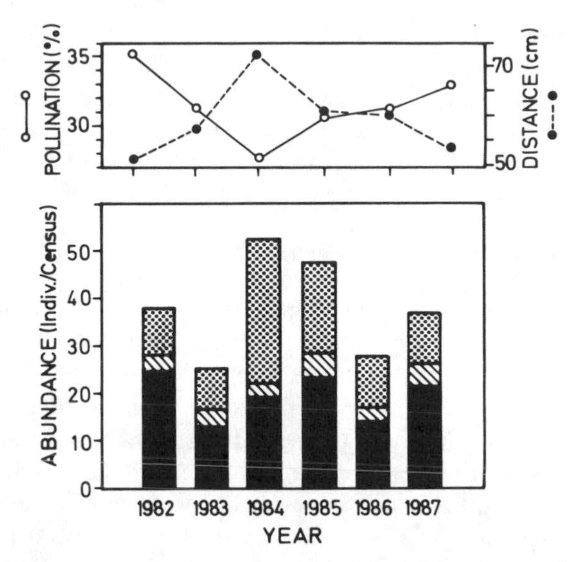

Figure 3.4. Annual variation in the composition of the insect pollinator assemblage of *Lavandula latifolia* at the Sierra de Cazorla, southeastern Spain (lower graph; solid, dashed, and dotted portions of bars correspond to hymenopterans, dipterans, and lepidopterans, respectively), and concomitant changes in the predicted mean distance flown by pollinators between consecutive floral visits ("distance") and the proportion of floral visits resulting in pollen delivery to the stigma ("% pollination"; upper graph). Data in the upper graph are averages weighted by the abundance of the different pollinators. Based on data in C. M. Herrera (1987*a, b,* 1988).

Expectations of floral adaptedness rely, to a considerable degree, on acceptance of Stebbins' (1970) "most effective pollinator principle," which states that "the characteristics of the flower will be molded by those pollinators that visit it most frequently *and* effectively in the region where it is evolving" (italics added). Nevertheless, pollinators providing pollination of higher quality are generally neither the most abundant nor the most predictable in time or space, thereby limiting the possibilities of selection on floral traits. For instance, although both diurnal and nocturnal visitors pollinate *Asclepias syriaca,* the latter provide higher-quality pollination (more pods produced per pollinator visit), but diurnal flower visitors are more abundant and eventually account for most effective pollinations (Bertin and Willson, 1980; Jennersten and Morse, 1991). An analogous situation was reported by Guitián et al. (1993) for the bee- (diurnal) and hawkmoth- (nocturnal) pollinated *Lonicera etrusca.* There have been relatively few investigations simultaneously studying pollinator effectiveness and abundance, but they suggest that decoupling of abundance and pollinating quality occurs frequently among insect-pollinated plants (Montalvo and Ackerman, 1986; Sugden, 1986; Schemske and Horvitz, 1989; Pettersson, 1991). *Lavandula latifolia* provides a further example. In a southeastern Spanish locality, pollinators of this species differ in several aspects of pollinating effectiveness, including the proportion of floral visits that result in effective pollen deposition to the stigma (C.M. Herrera, 1987*a*). They also differ broadly in mean abundance and extent of annual variation in abundance (C.M. Herrera, 1988). In this set of pollinators, pollinating effectiveness (measured as the proportion of visits resulting in effective pollen transfer) was unrelated to either mean abundance or population variability over a 6-year period (Fig. 3.5).

Constraints on the Response to Selection

Ecological factors may constrain adaptive responses of plants to selection by pollinators even when selection actually occurs. Recent studies have demonstrated phenotypic selection on floral traits (see references in the introduction); however, few of these studies evaluated the importance of variation in floral traits relative to other concurrent factors that can also influence fitness in natural populations. Even if variation in a given floral trait correlates significantly with differential reproductive success, the proportion of within-population variance in reproductive success explained by that attribute may vary widely, depending on the relative importance of other factors (Schemske and Horvitz, 1988; C.M. Herrera, 1993). Studies of phenotypic selection on floral traits have often shown that a statistically significant proportion of the "opportunity for selection" (the variance in fitness) depends on the existence of individual variation in floral traits, but the magnitude of this proportion has rarely been evaluated. In the *Calathea*

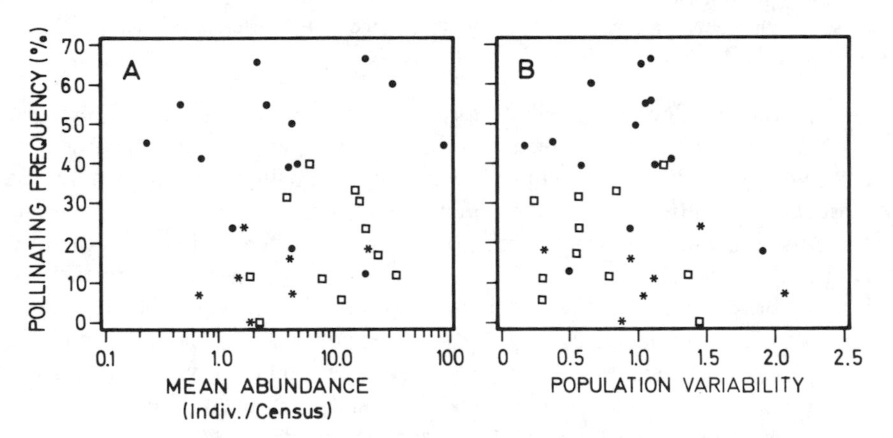

Figure 3.5. Relation between the frequency of pollen deposition on stigmas per visit ("pollinating frequency") and (A) mean abundance and (B) population variability of insect pollinators in a *Lavandula latifolia* population over a 6-year period (N = 32 insect taxa; hymenopterans, lepidopterans, and dipterans are coded, respectively, by dots, squares, and stars). Population variability is measured by the standard deviation of the natural logarithm of species abundance (Williamson, 1972). Pollinator effectiveness is unrelated to either abundance ($r_s = 0.060$, $P = 0.74$) or population variability ($r_s = -0.006$, $P = 0.97$) in this pollinator assemblage. Based on data in C. M. Herrera (1987*a*, 1988).

ovandensis study mentioned above, individual variation in corolla depth accounted for only 8% of total variance in fruit production in the single year when selection occurred (Schemske and Horvitz, 1988, 1989). In *Viola cazorlensis*, individual variation in floral morphology (corolla shape and petal and peduncle length) explained only 2% of total variation in fruit production (C.M. Herrera, 1993). Despite the statistical significance of selection on components of floral morphology, other factors (e.g., herbivory, microhabitat type, size differences) caused considerable variation in fecundity in these species, so that phenotypic selection on floral morphology probably has little impact on individual variation in maternal fitness. The opportunity for selection in these species was therefore only minimally attributable to phenotypic variance in floral morphology. More field studies on natural populations are needed to evaluate the generality of these results, but they suggest that selection on floral traits, even when it occurs, may be heavily "diluted" by the overwhelming influence of other factors (Schemske and Horvitz, 1988; C.M. Herrera, 1993). Maternal fecundity is only one possible component of fitness in hermaphroditic plants and it may be argued that selection on floral morphology via the male function might still occur in the cases mentioned above (Primack and Kang, 1989). However, studies on natural populations of hermaphroditic plants have documented close correlations (across individuals) between absolute measures of reproductive success via pollen (seeds sired) and ovules (seeds produced) (Broyles and Wyatt, 1990; Devlin and Ellstrand, 1990; Dudash, 1991), thus suggesting that maternal fecundity figures often represent satisfactory estimates of fitness.

Concluding Remarks

I have shown in the preceding sections that independent lines of evidence justify questions about the actual frequency of occurrence of floral adaptations to pollinators in nature. First, patterns of phenotypic variation in corolla depth in a regional species assemblage are not consistent with the idea of generalized selection by pollinators on this floral trait. Second, apart from possible intrinsic factors limiting adaptedness of plants to pollinators (e.g., life history or genetic constraints), ecological mechanisms often limit both the occurrence of pollinator selection on plants and the likelihood of adaptive responses. At the very least, these observations highlight the need to expand the prevailing tradition in pollination biology to include in its research program the additional set of questions outlined in the introduction.

Disagreements with the main tenet of this contribution will likely stem from discrepancies with my usage of the concept of "adaptation." As stressed in the introduction, there is not a common agreement about its meaning, and definitions may be classed into either "history-laden" or nonhistorical categories (Reeve and Sherman, 1993). I have indistinctly adhered to the two definitions, and this inconsistency has proven useful to illustrate that questions about the frequency of occurrence of floral adaptations to pollinators make sense, regardless of the particular definition of adaptation one adheres to.

We already understand how and why floral adaptations may come about, but we still need to know how frequent their occurrence is in nature. Plant interactions with animals for reproduction may successfully persist even in the absence of mutual adaptation and a shared history of interaction between counterparts. Consequently, animal-related plant traits, including floral ones, may persist unchanged for extended geological periods (C.M. Herrera, 1986, 1992). Some floral traits probably are exaptations rather than actual adaptations to present-day pollinators, whereas others may even be inconsequential in terms of differential fitness. Evolutionary pollination biologists should not let "the appeal of orderly pollination syndromes" (Waser and Price, 1990) obscure the richness and complexities involved in the relationship between plants and their floral visitors, and include in their research program an unprejudiced, unbiased scrutiny of the relative frequencies of occurrence of floral adaptations and exaptations in nature. After more than a century of successfully following Darwin's lead in search of support for evolutionary theory, there is no longer any question about the real occurrence of plant adaptations to pollinators: The time is now ripe to ask about their universality. Evolutionary theory would certainly not be challenged in any serious way by accepting that such adaptations may not occur. The question is, how often?

Appendix

Summary of corolla depth measurements for 58 insect-pollinated plants with tubular corollas from the Sierra de Cazorla region, southeastern Spain. Shown are the number of flowers (N_f) and individuals (N_i) sampled for each species, and the mean, standard deviation (SD), and coefficient of variation (CV) of individual means. Significance of individual variation was tested using single factor ANOVAs (P-values < 0.001 are shown using exponential notation).

						Individual Variation	
Species	N_f	N_i	Mean (mm)	SD	CV (%)	F-value	P
Acinos alpinus	100	20	11.94	1.58	13.3	7.76	2E-11
Anarrhinum bellidifolium	100	20	4.17	0.65	15.5	7.85	1E-11
Anchusa azurea	93	20	11.79	2.60	22.0	18.71	8E-21
Anchusa undulata	98	20	11.09	1.96	17.7	14.91	1E-18
Ballota hirsuta	100	20	9.19	1.64	17.8	33.79	2E-30
Calamintha sylvatica	100	20	11.41	1.62	14.3	12.70	6E-17
Cerinthe major	93	21	19.82	1.74	8.8	19.41	2E-21
Chaenorhinum serpyllifolium	90	20	16.14	1.73	10.7	5.64	41E-9
Cleonia lusitanica	79	25	16.33	3.61	22.1	5.90	33E-9
Coris monspelliensis	89	19	5.53	1.10	20.0	7.42	3E-10
Dactylorhiza elata	99	20	12.91	4.92	38.1	24.10	3E-25

Species	N_f	N_i	Mean (mm)	SD	CV (%)	Individual Variation	
						F-value	P
Daphne gnidium	100	20	4.48	0.65	14.5	1.72	0.05
Daphne laureola[G]	101	20	6.65	3.86	57.9	37.46	3E-32
Dipcadi serotinum	77	20	8.84	1.73	19.6	4.09	19E-6
Fedia cornucopiae	100	20	12.44	4.23	34.0	16.40	3E-20
Jasminum fruticans	70	14	9.48	1.99	21.0	10.82	5E-11
Lavandula latifolia	100	20	6.86	0.71	10.4	6.87	3E-10
Lavandula stoechas	100	20	6.19	0.86	13.9	13.85	5E-18
Linaria aeruginea	124	28	20.54	2.51	12.2	12.17	1E-20
Linaria anticaria	78	20	19.84	2.72	13.7	4.61	33E-7
Linaria viscosa	93	20	15.56	2.32	14.9	5.59	36E-9
Lithodora fruticosa	97	20	8.82	2.03	23.1	6.83	4E-10
Lonicera arborea	100	20	4.99	0.99	19.8	4.87	24E-8
Lonicera etrusca	100	20	26.46	7.12	26.9	13.08	3E-17
Lonicera implexa	138	27	27.83	6.42	23.1	18.73	2E-29
Lonicera splendida	101	17	29.30	4.89	16.7	17.45	3E-20
Marrubium supinum	100	20	6.97	1.71	24.5	11.30	1E-15
Mucizonia hispida	96	20	8.56	1.33	15.5	6.93	4E-10
Muscari comosum	112	20	8.04	1.39	17.3	93.90	1E-51
Narcissus cuatrecasasii	30	30	15.13	1.49	9.8	—	—
Origanum virens[G]	150	30	6.36	2.26	35.5	35.41	7E-46
Phlomis lychnitis	100	20	25.39	2.46	9.7	10.02	3E-14
Phlomis purpurea	110	22	14.61	2.11	14.5	6.77	5E-11
Pinguicula vallisneriifolia	44	20	17.01	2.90	17.0	7.44	5E-6
Pistorinia hispanica	79	20	12.42	2.63	21.2	10.18	2E-12
Polygala boissieri	83	24	16.65	1.72	10.4	10.44	2E-13
Primula vulgaris	78	20	14.08	2.32	16.4	7.94	4E-10
Prunella hyssopifolia	81	18	10.03	1.04	10.4	3.40	21E-5
Prunella laciniata	113	19	11.69	1.61	13.7	20.85	3E-25
Prunella vulgaris	83	20	9.71	1.27	13.1	7.00	2E-9
Rosmarinus officinalis	100	20	6.01	1.06	17.6	7.43	5E-11
Salvia argentea	100	20	8.35	1.87	22.4	23.46	4E-25
Salvia blancoana	86	20	23.94	3.89	16.2	8.46	3E-11
Salvia verbenaca	58	12	8.12	1.88	23.2	11.46	7E-10
Saponaria ocymoides	100	20	11.85	1.24	10.5	3.64	26E-6
Satureja intrincata	100	20	5.11	0.96	18.7	6.65	5E-10
Scabiosa turolensis	100	20	8.23	1.85	22.5	12.59	8E-17
Sideritis arborescens	65	13	7.11	1.24	17.4	20.16	2E-15
Sideritis incana	100	20	5.26	0.92	17.5	9.64	9E-14
Silene colorata	89	20	15.24	1.67	11.0	12.98	6E-16
Silene lasiostyla	53	20	18.98	2.62	13.8	14.34	7E-11
Silene legionensis	83	20	17.85	4.07	22.8	22.69	3E-21
Silene vulgaris	45	16	14.22	2.35	16.5	12.50	7E-9
Teucrium rotundifolium	100	20	10.14	1.47	14.5	5.29	54E-9
Teucrium webbianum	99	20	9.19	1.01	11.0	5.08	12E-8
Thymus orospedanus[G]	100	20	4.16	1.42	34.1	40.34	4E-33
Trachelium caeruleum	100	20	7.66	0.86	11.2	15.86	9E-20
Viola cazorlensis	80	20	29.60	7.34	24.8	13.28	6E-15

G = Gynodioecious, florally dimorphic species.

Acknowledgments

I am most grateful to J. Retana, J. Bosch, and X. Cerdá for generously supplying unpublished pollinator data, and S.C.H. Barrett, J. Herrera, P. Jordano, and two anonymous reviewers for constructive comments on earlier versions of the manuscript. J. Herrera and S. Talavera helped with some difficult plant identifications. My pollination studies in the Sierra de Cazorla were partly supported by grant PB87-0452 from the Dirección General de Investigación Científica y Técnica (DGICYT), and made possible by the facilities and permits provided by the Agencia de Medio Ambiente. While preparing this contribution, I was supported by grant PB91-0114 from DGICYT.

References

Andersson, S. and B. Widén. 1993. Pollinator-mediated selection on floral traits in a synthetic population of *Senecio integrifolius* (Asteraceae). *Oikos,* 66: 72–79.

Arnold, S.J. and M.J. Wade. 1984. On the measurement of natural and sexual selection: Theory. *Evolution,* 38: 709–719.

Baker, H.G. 1961. The adaptation of flowering plants to nocturnal and crepuscular pollinators. *Quart. Rev. Biol.,* 36: 64–73.

Baker, H.G. 1963. Evolutionary mechanisms in pollination biology. *Science,* 139: 877–883.

Baker, H.G. 1983. An outline of the history of anthecology, or pollination biology. In *Pollination biology* (L.A. Real ed.), Academic Press, Orlando, FL, pp. 7–28.

Baker, H.G. and P.D. Hurd. 1968. Intrafloral ecology. *Ann. Rev. Entomol.,* 13: 385–414.

Beattie, A. 1974. Floral evolution in *Viola. Ann. Missouri Bot. Gard.,* 61: 781–793.

Bertin, R.I. and M.F. Willson. 1980. Effectiveness of diurnal and nocturnal pollination of two milkweeds. *Can. J. Bot.,* 58: 1744–1746.

Bosch, J. 1986. *Insectos florícolas y polinización en un matorral de romero.* Unpubl. thesis, Univ. Barcelona, Barcelona.

Bosch, J. 1992. Floral biology and pollinators of three co-occurring *Cistus* species (Cistaceae). *Bot. J. Linn. Soc.,* 109: 39–55.

Broyles, S.B. and R. Wyatt. 1990. Paternity analysis in a natural population of *Asclepias exaltata*: Multiple paternity, functional gender, and the "pollen-donation hypothesis." *Evolution,* 44: 1454–1468.

Campbell, D.R. 1989. Measurements of selection in a hermaphroditic plant: Variation in male and female pollination success. *Evolution,* 43: 318–334.

Darwin, C. 1862. *On the Various Contrivances by Which British and Foreign Orchids Are Fertilised by Insects.* Murray, London.

Darwin, C. 1877. *The Different Forms of Flowers on Plants of the Same Species*. Murray, London.

Davis, M.A. 1987. The role of flower visitors in the explosive pollination of *Thalia geniculata* (Marantaceae), a Costa Rican marsh plant. *Bull. Torrey Bot. Club,* 114: 134–138.

Devlin B. and N.C. Ellstrand. 1990. Male and female fertility variation in wild radish, a hermaphrodite. *Am. Nat.,* 136: 87–107.

Dudash, M.R. 1991. Plant size effects on female and male function in hermaphroditic *Sabatia angularis* (Gentianaceae). *Ecology,* 72: 1004–1012.

Eckhart, V.M. 1991. The effects of floral display on pollinator visitation vary among populations of *Phacelia linearis* (Hydrophyllaceae). *Evol. Ecol.,* 5: 370–384.

Eckhart V.M. 1992. Spatio-temporal variation in abundance and variation in foraging behavior of the pollinators of gynodioecious *Phacelia linearis* (Hydrophyllaceae). *Oikos,* 64: 573–586.

Endler, J.A. 1986. *Natural Selection in the Wild*. Princeton Univ. Press, Princeton, NJ.

Faegri, K. and L. van der Pijl. 1966. *The Principles of Pollination Ecology*. Pergamon, Oxford.

Felsenstein, J. 1985. Phylogenies and the comparative method. *Am. Nat.,* 125: 1–15.

Fenster, C.B. 1991. Selection on floral morphology by hummingbirds. *Biotropica,* 23: 98–101.

Futuyma, D.J. 1979. *Evolutionary Biology*. Sinauer, Sunderland, MA.

Galen, C. 1989. Measuring pollinator-mediated selection on morphometric traits: Bumble bees and the alpine sky pilot, *Polemonium viscosum*. *Evolution,* 43: 882–890.

Galen, C. and M.E.A. Newport. 1988. Pollination quality, seed set, and flower traits in *Polemonium viscosum*: Complementary effects of variation in flower scent and size. *Am. J. Bot.,* 75: 900–905.

Ganders, F.R. 1979. The biology of heterostyly. *N. Zeal. J. Bot.,* 17: 607–635.

Ghiselin, M.T. 1984. Foreword to *The Various Contrivances by Which Orchids Are Fertilised by Insects,* 2nd ed., by C. Darwin. Univ. Chicago Press, Chicago, IL (reprint ed.).

Gittleman, J.L. and H.K. Luh. 1992. On comparing comparative methods. *Ann. Rev. Ecol. Syst.,* 23: 383–404.

Gould S.J. and E.S. Vrba. 1982. Exaptation—a missing term in the science of form. *Paleobiology,* 8: 4–15.

Guitián, P., J. Guitián, and L. Navarro. 1993. Pollen transfer and diurnal versus nocturnal pollination in *Lonicera etrusca*. *Acta Oecol.,* 14: 219–227.

Harder, L.D. and S.C.H. Barrett. 1993. Pollen removal from tristylous *Pontederia cordata*: Effects of anther position and pollinator specialization. *Ecology,* 74: 1059–1072.

Harvey, P.H. and M.D. Pagel. 1991. *The Comparative Method in Evolutionary Biology*. Oxford Univ. Press, Oxford.

Herrera, C.M. 1986. Vertebrate-dispersed plants: Why they don't behave the way they should. In *Frugivores and Seed Dispersal* (A. Estrada and T.H. Fleming, eds.), Junk, Dordrecht, pp. 5–18.

Herrera, C.M. 1987*a*. Components of pollinator "quality": Comparative analysis of a diverse insect assemblage. *Oikos,* 50: 79–90.

Herrera, C.M. 1988. Variation in mutualisms: The spatio-temporal mosaic of a pollinator assemblage. *Biol. J. Linn. Soc.,* 35: 95–125.

Herrera, C.M. 1992. Historical effects and sorting processes as explanations for contemporary ecological patterns: Character syndromes in Mediterranean woody plants. *Am. Nat.,* 140: 421–446.

Herrera, C.M. 1993. Selection on floral morphology and environmental determinants of fecundity in a hawk moth-pollinated violet. *Ecol. Monogr.,* 63: 251–275.

Herrera, J. 1985. *Biología reproductiva del matorral de Doñana*. Unpubl. thesis, Univ. Sevilla, Sevilla.

Herrera, J. 1988. Pollination relationships in southern Spanish Mediterranean shrublands. *J. Ecol.,* 76: 274–287.

Horvitz, C.C. and D. W. Schemske. 1990. Spatiotemporal variation in insect mutualists of a neotropical herb. *Ecology,* 71: 1085–1097.

Howe, H.F. 1984. Constraints on the evolution of mutualisms. *Am. Nat.,* 123: 764–777.

Inouye, D.W. 1979. Patterns of corolla tube length of bumble bee flowers from two continents. *Maryland Agric. Exp. Sta. Spec. Misc. Publ.,* 1: 461–463.

Inouye, D.W. and G.H. Pyke. 1988. Pollination biology in the Snowy Mountains of Australia: Comparisons with montane Colorado, USA. *Aust. J. Ecol.,* 13: 191–218.

Jennersten, O. and D.H. Morse. 1991. The quality of pollination by diurnal and nocturnal insects visiting common milkweed, *Asclepias syriaca*. *Am. Midl. Nat.,* 125: 18–28.

Jordano, P. 1990. Biología de la reproducción de tres especies del género *Lonicera* (Caprifoliaceae) en la Sierra de Cazorla. *Anal. Jard. Bot. Madrid,* 48: 31–52.

Kalin Arroyo, M.T., R. Primack, and J. Armesto. 1982. Community studies in pollination ecology in the high temperate Andes of central Chile. I. Pollination mechanisms and altitudinal variation. *Am. J. Bot.,* 69: 82–97.

Kochmer, J.P. and S.N. Handel. 1986. Constraints and competition in the evolution of flowering phenology. *Ecol. Monogr.,* 56: 303–325.

Kohn, J.R. and S.C.H. Barrett. 1992. Experimental studies on the functional significance of heterostyly. *Evolution,* 46: 43–55.

Lande, R. and S.J. Arnold. 1983. The measurement of selection on correlated characters. *Evolution,* 37: 1210–1226.

Lessios, H.A. 1990. Adaptation and phylogeny as determinants of egg size in echinoderms from the two sides of the Isthmus of Panama. *Am. Nat.,* 135: 1–13.

Macior, L.W. 1971. Co-evolution of plants and animals—systematic insights from plant-insect interactions. *Taxon,* 20: 17–28.

McCall, C. and R.B. Primack. 1992. Influence of flower characteristics, weather, time of day, and season on insect visitation rates in three plant communities. *Am. J. Bot.*, 79: 434–442.

Milton, S.J. and E.J. Moll. 1982. Phenology of Australian acacias in the S.W. Cape, South Africa, and its implications for management. *Bot. J. Linn. Soc.*, 84: 295–327.

Møller, A.P. and T.R. Birkhead. 1992. A pairwise comparative method as illustrated by copulation frequency in birds. *Am. Nat.*, 139: 644–656.

Montalvo, A.M. and J.D. Ackerman. 1986. Relative pollinator effectiveness and evolution of floral traits in *Spathiphyllum friedrichsthalii* (Araceae). *Am. J. Bot.*, 73: 1665–1676.

Motten, A.F., D.R. Campbell, D.E. Alexander, and L.H. Miller. 1981. Pollination effectiveness of specialist and generalist visitors to a North Carolina population of *Claytonia virginica*. *Ecology*, 62: 1278–1287.

Nilsson, L.A. 1988. The evolution of flowers with deep corolla tubes. *Nature*, 334: 147–149.

Nilsson, L.A., L. Jonsson, L. Rason and E. Randrianjohany. 1985. Monophily and pollination mechanisms in *Angraecum arachnites* Schltr. (Orchidaceae) in a guild of long-tongued hawk-moths (Sphingidae) in Madagascar. *Biol. J. Linn. Soc.*, 26: 1–19.

Nilsson, L.A., L. Jonsson, L. Ralison, and E. Randrianjohany. 1987. Angraecoid orchids and hawkmoths in central Madagascar: Specialized pollination systems and generalist foragers. *Biotropica*, 19: 310–318.

Olesen, J.M. 1985. The Macaronesian bird-flower element and its relation to bird and bee opportunists. *Bot. J. Linn. Soc.*, 91: 395–414.

Olesen, J.M. 1988. Floral biology of the Canarian *Echium wildpretii*: Bird-flower or a water resource to desert bees? *Acta Bot. Neerl.*, 37: 509–513.

Percival, M. 1965. *Floral Biology*. Pergamon, Oxford.

Percival, M. 1974. Floral ecology of coastal scrub in southeast Jamaica. *Biotropica*, 6: 104–129.

Pettersson, M.W. 1991. Pollination by a guild of fluctuating moth populations: Option for unspecialization in *Silene vulgaris*. *J. Ecol.*, 79: 591–604.

Podoler, H., I. Galon, and S. Gazit. 1984. The role of nitidulid beetles in natural pollination of annona in Israel. *Acta Oecol., Oecol. Appl.*, 5: 369–381.

Price, M.V. and N.M. Waser. 1979. Pollen dispersal and optimal outcrossing in *Delphinium nelsonii*. *Nature*, 277: 294–297.

Primack, R.B. 1983. Insect pollination in the New Zealand mountain flora. *N. Zeal. J. Bot.*, 21: 317–333.

Primack, R.B. and H. Kang. 1989. Measuring fitness and natural selection in wild plant populations. *Ann. Rev. Ecol. Syst.*, 20: 367–396.

Proctor, M. and P. Yeo. 1973. *The Pollination of Flowers*. Collins, London.

Ramsey, M.W. 1988. Differences in pollinator effectiveness of birds and insects visiting *Banksia menziesii* (Proteaceae). *Oecologia (Berl.)*, 76: 119–124.

Reeve, H.K. and P.W. Sherman. 1993. Adaptation and the goals of evolutionary research. *Quart. Rev. Biol.*, 68: 1–32.

Rick, C.M. 1950. Pollination relations of *Lycopersicon esculentum* in native and foreign regions. *Evolution,* 4: 110–122.

Robertson, J.L. and R. Wyatt. 1990. Evidence for pollination ecotypes in the yellow-fringed orchid, *Platanthera ciliaris*. *Evolution,* 44: 121–133.

Schemske, D.W. 1983. Limits to specialization and coevolution in plant-animal mutualisms. In *Coevolution* (M.H. Nitecki, ed.), Chicago Univ. Press, Chicago, IL. pp. 67–109.

Schemske, D.W. and C.C. Horvitz. 1984. Variation among floral visitors in pollination ability: A precondition for mutualism specialization. *Science,* 225: 519–521.

Schemske, D.W. and C.C. Horvitz. 1988. Plant-animal interactions and fruit production in a neotropical herb: A path analysis. *Ecology,* 69: 1128–1137.

Schemske, D.W. and C.C. Horvitz. 1989. Temporal variation in selection on a floral character. *Evolution,* 43: 461–465.

Schluter, D. 1988. Estimating the form of natural selection on a quantitative trait. *Evolution,* 42: 849–861.

Smith, J.M.B. 1988. *Prunus* (Amygdalaceae) in New South Wales. *Telopea,* 3: 145–157.

Snow, A.A. and D.W. Roubik. 1987. Pollen deposition and removal by bees visiting two tree species in Panama. *Biotropica,* 19: 57–63.

Sokal, R.R. and C.A. Braumann. 1980. Significance tests for coefficients of variation and variability profiles. *Syst. Zool.,* 29: 50–66.

Stebbins, G.L. 1970. Adaptive radiation of reproductive characteristics in angiosperms. I. Pollination mechanisms. *Ann. Rev. Ecol. Syst.,* 1: 307–326.

Sugden, E.A. 1986. Anthecology and pollinator efficacy of *Styrax officinale* subsp. *redivivum* (Styracaceae). *Am. J. Bot.,* 73: 919–930.

Vogel S., C. Westerkamp, B. Thiel, and K. Gessner. 1984. Ornithophilie auf den Canarischen Inseln. *Plant Syst. Evol.,* 146: 225–248.

Waser, N.M. 1983*a*. The adaptive nature of floral traits: Ideas and evidence. In *Pollination Biology* (L.A. Real, ed.), Academic Press, Orlando, FL, pp. 241–285.

Waser, N.M. 1983*b*. Competition for pollination and floral character differences among sympatric plant species: A review of evidence. In *Handbook of Experimental Pollination Biology* (C.E. Jones and R.J. Little, eds.), Van Nostrand Reinhold, New York, pp. 277–293.

Waser, N.M. and M.V. Price. 1990. Pollination efficiency and effectiveness of bumble bees and hummingbirds visiting *Delphinium nelsonii*. *Collect. Bot. (Barcelona),* 19: 9–20.

Williams, G.C. 1966. *Adaptation and Natural Selection*. Princeton Univ. Press, Princeton, NJ.

Williamson, M. 1972. *The Analysis of Biological Populations*. Arnold, London.

Wilson, P. and J.D. Thomson. 1991. Heterogeneity among floral visitors leads to discordance between removal and deposition of pollen. *Ecology,* 72: 1503–1507.

Wolda, H. 1983. "Long-term" stability of tropical insect populations. *Res. Pop. Ecol.*, Suppl. 3: 112–126.

Wolfe, L.M. and S.C.H. Barrett. 1989. Patterns of pollen removal and deposition in tristylous *Pontederia cordata* L. (Pontederiaceae). *Biol. J. Linn. Soc.*, 36: 317–329.

Zimmerman, J.K., D.W. Roubik, and J.D. Ackerman. 1989. Asynchronous phenologies of a neotropical orchid and its euglossine bee pollinator. *Ecology*, 70: 1192–1195.

4

How Do Flowers Diverge?

*Paul Wilson and James D. Thomson**

> Two populations of the same plant species living in different territories may both receive effective pollination visits from the same two classes of pollinators, say bees and beeflies, but receive the two types of visitations in different relative frequencies. If one local race receives a greater number of effective bee visits and the other population a greater number of effective beefly visits, natural selection will favor closer floral adaptations to the special characteristics of bees in the first race and closer coadaptations with beeflies in the second. The point of compromise in floral adaptations will shift in correspondence with the climate of pollinators in each given territory.
>
> —V. and K.A. Grant (1965)

Evolutionary Thought and Floral Biology

Darwin's orchid book (1862) has been cited as his first detailed example of how to study evolution (Ghiselin, 1969; Gould, 1986). The book starts as a presentation of observations showing that the morphology of orchids is, in most cases, wonderfully well suited to having insects remove and deposit pollinia. It ends by tracing how the enormous diversity of orchids can be seen as arising through modifications from ancestral forms. What Darwin did not do was to explain how orchid flowers come to be different. He probably thought that by showing how to study the origin of adaptation he had shown how to study the origin of diversity. Mayr has often pointed out that Darwin failed to see genetic isolation as a precondition for speciation, and thus for diversification (e.g., Mayr, 1959). Likewise, we contend that evolutionists have seldom clearly dissected the alternatives for how divergence occurs, given isolation. We shall concern ourselves here with how different environments—in our case, different pollinator regimes—do or do not provide heterogeneity in selection that might adaptively drive the divergence of flowers.

Before the neo-Darwinian synthesis, it was common for biologists to argue that the distinguishing characters of closely related species are often nonadaptive (Provine, 1986). Richards and Robson (1926) provided the most influential review, arguing forcefully that there are many contrasts for which stories about special adaptations had simply not been proven. Their perspective was thoughtfully adopted by Elton (1927). On the one hand, Elton imagined that the minor

---

*Department of Ecology and Evolution, State University of New York, Stony Brook, NY 11794-5245.

characters distinguishing geographical races, subspecies, and species are not due to selection in different environments, but rather spread through fluctuations in population size (Elton's Chap. XII). And, on the other hand, Elton believed that organisms are elegantly adapted to niches such that badgers and weasels end up filling different roles in the structure of their community (Elton's Chap. V). Our reading of Elton is that he was thinking of two scales of evolutionary differentiation, and that his interpretation of neutrality vs. selection is scale-dependent (Elton, p. 185). This is a point that we will return to at the end of our chapter when reviewing the many ways by which flowers might diverge.

At the beginning of the neo-Darwinian synthesis, Dobzhansky (1937) and the other masters of the synthesis did not insist on the adaptive nature of species differences (Gould, 1983). Stebbins (1950) had his adaptationist leanings, but still admitted he could not fully back them up for floral characters (pp. 118–121). But, as the synthesis hardened and MacArthurian ecology was born (MacArthur, 1958, 1972), the celebration of pervasive adaptation spread to an almost consensus view (e.g., Lack 1971). It seemed as if every species of animal claimed for itself its own niche unlike the niche of any other (coexisting) animal. A naive extension to plants would suggest that each flower is adapted to a pollination niche unique in its community.

In fact, floral biology did not walk lock step with zoological thought, but there was a loose connection. During the third quarter of this century, much work was done on flowers from the perspective of comparative functional morphology (Vogel, 1954; Baker, 1961; van der Pijl, 1961). The classic example was Grant and Grant's (1965) exploration of adaptive radiation in the phlox family. The data of that period generally consisted of observing what species of animals visit various kinds of flowers, how the visitors behave at the flowers, and whether or not they come into bodily contact with anthers and stigmas. The results were that different types of flowers attract different types of visitors, that flowers have many features used by their particular visitors, and that only some visitors are effective pollinators. Floral differences came to be treated as adaptations to diverse pollinator communities. This conclusion was based on broad-scale comparisons between flowers pollinated by butterflies vs. bees vs. bats vs. birds. Although there was an implication that differences between species pollinated by the same general type of animal are also adaptive, this now seems mostly like an extrapolation down from the major products of evolution. It assumes that floral divergence, whether fine or coarse in scale, is uniformly driven by one set of evolutionary mechanisms.

Floral syndromes as adaptations to fixed pollinator types have a certain typological ring about them that made the concept distasteful to a new generation of pollination biologists in the late 1970s. Trained in ecology rather than systematics, they emphasized community-level interactions such as competition and facilitation (Brown and Kodric-Brown, 1979). In these interactions, they saw not only rules for community assembly based on limiting similarity, but also a mechanism for the potent selection of floral attributes via niche partitioning and character displacement (Waser, 1978, 1983*a;* Pleasants, 1980, 1983; Thomson, 1978, 1982; Nilsson, 1983). The distinction between population-dynamic processes and character evolution was often blurred, and certain aspects of flowers, such as nectar production and phenological timing, were considered to be more responsive to community influences than floral morphology (see Zimmerman, 1988). These researchers' faith in their ability to comprehend floral evolution was soon shaken by a series of realizations—by the null-model revolution in community ecology (Rabinowitz et al., 1981; Strong et al., 1979, 1984), by the onerous burden of verifying character displacement (Grant, 1972), by criticisms of Panglossian adaptationism (Gould and Lewontin, 1979; Waser, 1983*b*), by the comprehension that the results of competition experiments can have many interpretations (Thomson, 1980; Bender et al., 1984), and by numerous complications that their own studies revealed. Plant-pollinator communities came to look less equilibrial, more dominated by spatio-temporal heterogeneity, historical influences, and convoluted species interactions (Feinsinger, 1987).

Many plant reproductive ecologists responded by turning toward processes that act within populations, trying to supply the details of genetics and the mechanisms of selection whose absence had crippled earlier arguments. With a few exceptions, such as Armbruster (e.g., 1988 and Chapter 9), they stopped working on diversification *per se*. Much of the recent work has focused on *how* natural selection acts on floral characters in a single population (Waser and Price, 1983; Nilsson, 1988; Campbell, 1989; Stanton et al., 1989; Galen, 1989; Schemske and Horvitz, 1989; Devlin et al., 1992; Mitchell, 1993; Chapters 10 and 11). This work tends to be more reductionistic and empirically demanding than previous work. A field that once freely indulged in ornate adaptive speculation now wishes for measurements of lifetime reproductive success through both male and female function.

One approach to estimating components of male and female function has been to focus on the rates of pollen removal from anthers and pollen delivery to stigmas. Pollen grains are small, numerous, and difficult to track, but new techniques have allowed for some quantitative study of their dispersal (Stanton et al., 1992). With simplifying assumptions, pollen-transfer parameters can be treated as surrogates of fitness (e.g., Thomson and Thomson, 1989), and some progress has now been made in understanding what influences how much pollen is dispersed by pollinators (Harder et al., 1985; Galen and Stanton, 1989; Harder and Thomson, 1989; Young and Stanton, 1990; Harder, 1990; Wilson and Thomson, 1991; Harder and Barrett, 1993; Chapter 6). Efficient new techniques now let us measure pollen removal and pollen deposition for large samples of flowers. By controlling or noting the species of pollinator and controlling or measuring the morphology of flowers, we can then use statistical procedures to examine the influence of pollinator-species variation, floral-morphological variation, and the interaction between the two in determining how much pollen is dispersed. We believe this interaction is of particular interest in understanding the origin of the sorts of floral differences that distinguish closely related plants pollinated by similar types of animals.

Why the Interaction Term Is Important

In 1984, Schemske and Horvitz entitled one of their papers, "Variation among floral visitors in pollination ability: A precondition for mutualism specialization." To provide an example of such variation, they presented data showing that bees are better than butterflies at tripping *Calathea* flowers, and they stressed that adaptation to one pollinator over another is driven by such differences in the average effectiveness of pollinators. A similar focus on overall (i.e., average) differences in effectiveness had been made previously by Primack and Silander (1975) and has been repeated subsequently by Herrera (1987). Schemske and Horvitz alluded to something else also: "In view of the variation in the pollination ability of different visitors, the reproductive success of individual plants is a function of plant characters that determine the number and kinds of visitors that a plant attracts. The extent to which variation in plant fitness is attributable to such characters determines the potential for selection of pollinators on plants." We extend this to mean that, in order for specialization to occur, visitors must differ not only in their overall quality, but also in the way they affect the relationship between floral traits and fitness.

We are interested in the visitor species × floral character interaction term, in other words, in the way different animals affect selection gradients with differing slopes (Wade and Kalisz, 1990). By selection gradient, we mean the relationship between some measure of fitness and a character. Strictly speaking, it is the standardized partial regression of fitness on a character, as it naturally varies, holding other characters constant, and it is considered a measure of the direct action of selection on a focal character (Lande and Arnold, 1983). Our usage, however, will be fairly loose in that we will treat pollen counts as fitness, and we will not present standarized coefficients. Our aim is simply to look for any evidence that different bees might affect selection on flowers differently.

Specialization can occur when there is variation in floral characters that can effectively limit the impact of inferior types of visitors while promoting pollination by superior visitors. Consider Fig. 4.1 and imagine we are at a site where Bees A and B occur. With either of the bees, flowers having shorter corollas work better than those with longer corollas. The two bee species differ in their means—Bee B is a relative parasite compared to Bee A—but still the flowers will evolve toward shorter corollas and still they will be visited by Bee B. Now, imagine we are at a site with Bees B and C. These bees affect selection gradients in opposite directions, and because Bee C is the better pollinator, the flowers will evolve to be specialized on Bee C by an increase in corolla length. In a

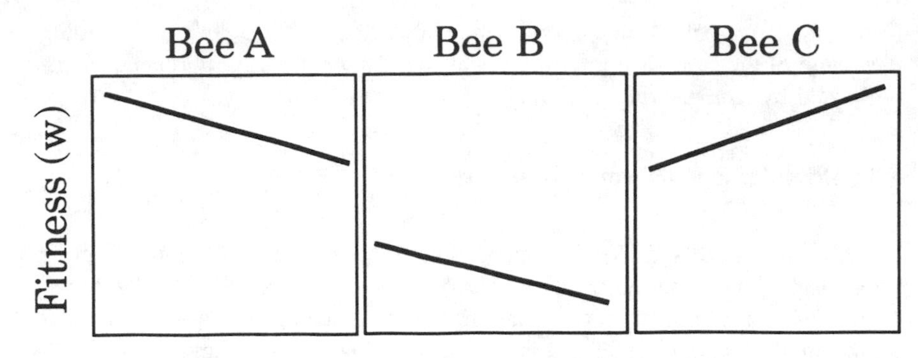

Figure 4.1. Various relationships between floral morphology and fitness, as determined by different hypothetical bee species. Bee A differs from bee B in its *average* effect on fitness. Bee A differs from bee C in the *slopes* of its effect.

single place inhabited by Bees A and C, Schemske and Horvitz are right in saying that there will be no specialization because A and C are effectively the same.

However, because we are concerned with the origin of divergence, we wish to consider a series of isolated plant populations visited by various admixtures of bees, but with one bee species more or less predominating at any one site. Where Bee A predominates, the corolla tubes will evolve to be shorter. Where Bee C predominates, the corollas will evolve to be longer. Where Bee B exists in the virtual absence of the other two, corollas will become shorter, but if Bee A is present in secondary abundance, the flowers will still evolve under A's influence (following "the most-important pollinator principle" of Stebbins, 1970). Thus, the contrast that most clearly leads to adaptive divergence is between sites dominated by A vs. C: The pollinators impose opposite selection gradients (slopes) but are actually comparable in their overall quality (means). It is this crucial interaction term—visitor species × floral attribute, or more generally, environment × character—that we wish to examine as a first step in reattacking the nature of divergence. Here we present two studies of wildflowers pollinated by bumble bees, both of which allowed us to evaluate such interaction terms. We did not try to locate actual populations served by different suites of pollinators, but we did examine pollen removal and deposition in single visits by different species of bumble bees to flowers that exhibited natural character variation.

Jewelweed Study

Impatiens pallida (Balsaminaceae) has gullet-shaped protandrous flowers (Fig. 4.2). The gender phases are absolutely distinct in that the androecium completely

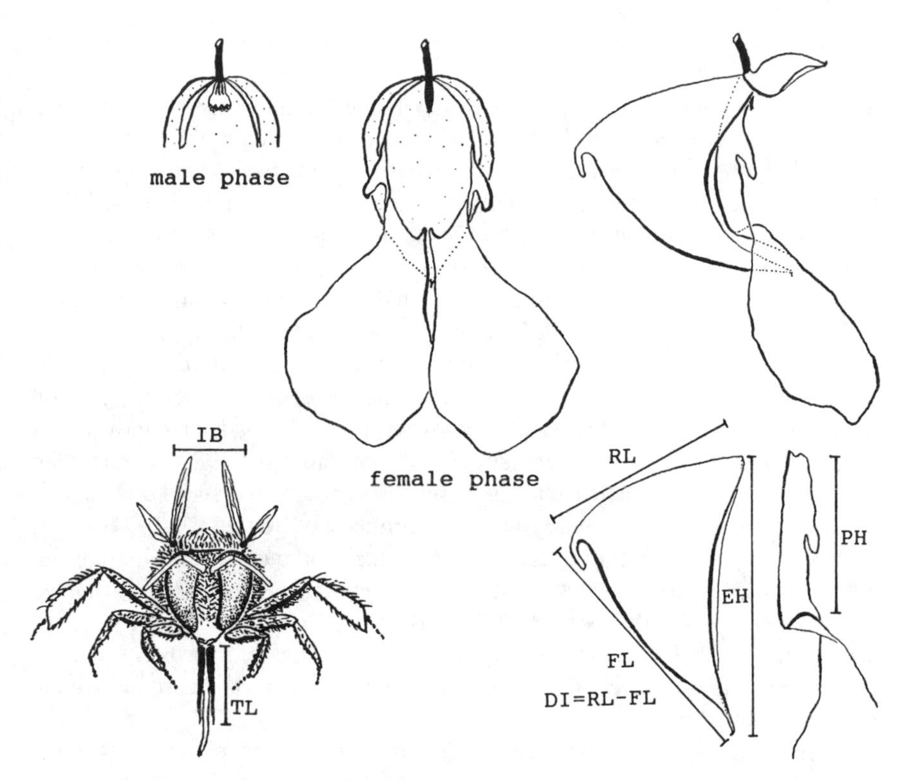

Figure 4.2. *Impatiens pallida* and *Bombus impatiens* showing how measurements were taken. Notice that the flowers fit around the bees quite snugly.

covers the gynoecium until the gynoecium swells and pushes the androecium away; thus, both the androecium and gynoecium are presented in the same place, pointing down from the roof of the vestibular sepal. The male phase is somewhat longer than the female phase (Schemske, 1978). Local populations are remarkably differentiated in morphology and in the plants' ability to grow and reproduce (Schemske, 1984), and in one population in Quebec, phenotypic selection on plant size and phenology was found to be strong and spatially heterogeneous (Stewart and Schoen, 1987). *Impatiens* flowers are visited by a variety of animals but are as clearly adapted to bumble bee pollination as any flower. Bee activity varies from nil (rarely) to hundreds of visits (often) during the life of a flower. Bees crawl into the vestibular sepal in order to drink nectar from a recurved spur, and in the process their backs brush against the androecium or the stigma. While foraging, bees develop an obvious stripe of white pollen that runs down their backs. The fit between bee and flower can be snug, and one would expect that bee size and flower size could strongly influence pollen removal and deposition.

Methods

We were interested in how much pollen is removed from anthers and deposited on stigmas in a visit. Since pollen removal is not directly measurable, we counted the number of grains *not* removed, which can be treated as the opposite of removal or used to calculate removal by subtracting from the amount of pollen produced in unvisited control flowers. Our two surrogates of fitness, then, were pollen remaining in anthers and pollen deposited on stigmas in a single visit. It is hard to know how these variables are related to reproductive success (Snow, 1989; Wilson and Thomson, 1991; Stanton et al., 1992). High pollen removal may be either negatively or positively correlated with male reproductive success, depending on how much of the removed pollen is taken to stigmas, and on whether or not the pollen that is not removed in one visit is later removed and subsequently deposited in another visit (Harder and Thomson, 1989). High pollen deposition may be correlated with high female reproductive success, but only when seed set or progeny quality is visitor-limited (Wilson et al., 1994). Still, pollen removal and deposition are necessary for reproduction and must be in some way related to reproductive success. Moreover, they have the advantage of being the variables that bees most directly affect, and so, of all surrogates of fitness, they should be the most deterministic and statistically tractable. If neither removal nor deposition were related to pollination variation, it is hard to imagine how fitness could be.

During August of 1990, *Impatiens pallida* flowers were covered by wax paper bags, in bud or in male phase. Overnight, the buds matured into virgin male-phase flowers, and the previously male-phase flowers matured into virgin female-phase flowers. Each flower, except for unvisited controls, was cut and put in a florist's cut-flower holder taped to the end of a stick, where it was presented to a naturally foraging *Bombus* worker for a single visit. The visit was timed with a stopwatch. *Bombus vagans, B. impatiens, B. fervidus,* and unvisited control treatments were administered in roughly rotating order (except that additional *B. vagans*-visited flowers, taken for other purposes, are here included in the data). For male-phase flowers, the androecium was carefully detached, placed in a microcentrifuge tube, and preserved in 70% ethanol. For female-phase flowers, the stigma was clipped, placed on a microscope slide with a small piece of fuchsin-tinted glycerin jelly, heated with an alcohol burner, and squashed under a cover slip. The rest of the flower was put in a plastic vial and preserved in ethanol.

Flowers were measured in random order under a dissecting microscope while being floated in alcohol over 1 mm graph paper. Measurements were precise to about 0.2 mm. Three floral characters will be reported on here and are illustrated in Fig. 4.2: the entrance height (EH) of the vestibule when gently flattened, the porch height (PH) of a flattened petal, and the difference (DI) between the length of the floor and the roof of the vestibule. We imagined that an increase in any of these characters might have led to less contact between the bee and the androecium or stigma and therefore less pollen moved. Such a relationship could also have been expected to itself vary with bees of differing sizes. The number of pollen grains left in androecia was estimated using an Elzone® electronic particle counter with a 76 μ aperture (Harder, 1990). Pollen deposited on stigmas was determined using a compound microscope with an ocular grid and mechanical stage.

Results

The three *Bombus* species differed considerably in overall size, tongue length, and the duration of their visits (Table 4.1). *Bombus vagans* was the smallest, *B. impatiens* intermediate, and *B. fervidus* the largest, as indexed by the distance between the bases of the wings. Tongue lengths did not correspond to body size in that the order was *B. impatiens* with the shortest tongue, followed by *B. vagans,* and *B. fervidus*. Probably because of its greater tongue length, *B. fervidus* did not crawl into the flowers as deeply as the smaller species and may have fit more loosely despite its greater size. *Bombus vagans* spent the most time at a flower, *B. impatiens* less, and *B. fervidus* the least time. Pollen removal and deposition were not significantly related to visit duration (statistics not shown).

Both pollen remaining in androecia (our measure of pollen removal) and the log of pollen deposited on stigmas have approximately normal distributions. Table 4.2 shows a series of progressively simpler statistical models by which we tried to explain variance in the amount of pollen remaining and the log of pollen deposition. In the full models, shown in the top panel, only 17 and 11%

Table 4.1. Comparisons of three bee species: morphometrics, visit length, and effects on pollen movement in one visit to flowers of Impatiens pallida. *Numbers are means ± standard errors (sample sizes).*

	B. vagans	*B. impatiens*	*B. fervidus*
Distance between wing bases (mm)	3.47±0.130(8)	3.74±0.082(8)	4.35±0.141(6)
Tongue length (mm)	4.46±0.161(8)	3.73±0.103(8)	5.89±0.090(6)
Duration of visit (sec)*	20 at ♂, 10 at ♀	16 at ♂, 8 at ♀	8 at ♂, 6 at ♀
Pollen remaining (grains)†	474,000±22,300(53)	511,000±20,500(39)	520,000±21,100(38)
Log (pollen deposited + 1)‡	2.517±0.0499(67)	2.680±0.0660(35)	2.576±0.0536(43)
= back transformed value	= 328 grains	= 478 grains	= 376 grains

*The distribution of durations was very skewed so the geometric mean was calculated; standard errors are not presented.

†There were on average 707,000±17,200(46) grains produced in unvisited flowers; the number removed could be calculated by subtraction.

‡There were a very few pollen grains on stigmas of unvisited flowers: 0.415±0.0540(41) = 1.6 grains.

Table 4.2. Analyses of variance (bottom), covariance (middle), and heterogeneity in slopes (top) for pollen remaining and pollen deposited in a single visit of one of three Bombus *species to flowers of* Impatiens pallida. *Unvisited control flowers are not included in the analysis. EH, PH, and DI are measurements of the flowers that might have affected how tightly they fit around bees and thus how much pollen was moved. Three successively simpler models are presented for each variable. The difference between the top model and middle model represents the effect of the bee-species × floral-morphology interaction, which was not significant for either removal or deposition (see text) despite the marginal significance of bee species × DI.*

	Pollen in Androecium			Log (Pollen Deposited + 1)		
	df	III-SS[1]	*F*	df	III-SS	*F*
Bee species	2	143		2	0.437	
EH	1	791		1	0.002	
PH	1	921		1	0.042	
DI	1	178		1	0.063	
Bee × EH	2	397	1.05 n.s.	2	0.559	1.90 n.s.
Bee × PH	2	43	<1 n.s.	2	0.044	<n.s.
Bee × DI	2	1175	3.12*	2	0.202	<1 n.s.
Model	11	4525	R^2=0.17	11	2.424	R^2=0.11
Error	118	22,244		133	19.596	
Bee species	2	618	1.61 n.s.	2	0.684	2.24 n.s.
EH	1	779	4.05*	1	0.019	<1 n.s.
PH	1	1009	5.25*	1	0.041	<0.1 n.s.
DI	1	104	<1 n.s.	1	0.084	<1 n.s.
Model	5	2921	R^2=0.11	5	0.783	R^2=0.04
Error	124	26,768		139	21.021	
Bee species	2	545	1.32 n.s.	2	0.612	2.03 n.s.
Error	127	26,223	R^2=0.02	142	21.408	R^2=0.03

*Type III sums of squares × 100,000,000.

n.s. $p>0.1$.

*$p<0.05$.

of the variances, respectively, were explained. The interaction between bee species and the three dimensions was not significant except for bee species × DI for pollen remaining ($p = 0.048$): DI was positively related to pollen remaining for *B. vagans* and *B. fervidus,* but negatively related for *B. impatiens*. If we drop the three interaction terms and calculate the decreases in the models' degrees of freedom and sums of squares, then we can evaluate the collective significance of those interaction terms. For pollen remaining, SS = 1604 and $F_{6,118} = 1.42$; for pollen deposited, SS = 1.641 and $F_{6,133} = 1.86$. Neither was significant.

As shown in the simplified models in the middle of Table 4.2, EH and PH had a significant effect on pollen removal (or possibly they were correlated with pollen production, although this was not evident among the unvisited control flowers; Wilson, in press). There did not seem to be any effect of EH, PH, or

Table 4.3. Comparisons of two bee species: size, visit length, and effects on pollen movement in one visit to flowers of Erythronium grandiflorum. *Numbers are means ± standard errors (samples sizes).*

	Bombus occidentalis	*Bombus bifarius*
Distance between wing bases (mm)	7.09±0.095(8)	5.92±0.081(8)
Duration of visit (sec)[1]	21	16
Pollen remaining (grains)	25,900±1730(76)	31,700±1640(79)
Pollen deposited (grains)	1791±104.5(76)	988±54.2(78)

[1]Visit durations were skewed so the geometric mean was calculated and standard errors are not presented.

DI on pollen deposited (Table 4.2, middle panel). The collective effect of floral morphology was likewise significant for pollen remaining (SS = 2376; $F_{3,124}$ = 3.67) and not significant for pollen deposited (SS = 0.171; $F_{3,139} < 1$).

Finally, as shown in the bottom panel of Table 4.2, bee species by itself had no affect on pollen remaining or pollen deposited. Bees did remove and deposit pollen, but the three *Bombus* species did so in comparable amounts (Table 4.1).

Glacier Lily Study

Erythronium grandiflorum (Liliaceae) has large open pendant flowers (Fig. 4.3). It is visited primarily by queen bumble bees with occasional visits by hummingbirds and small bees. Fair-weather visitation rates are a bit less than one visit per flower per day, and flowers last for about 4 to 5 days (Thomson and Thomson,

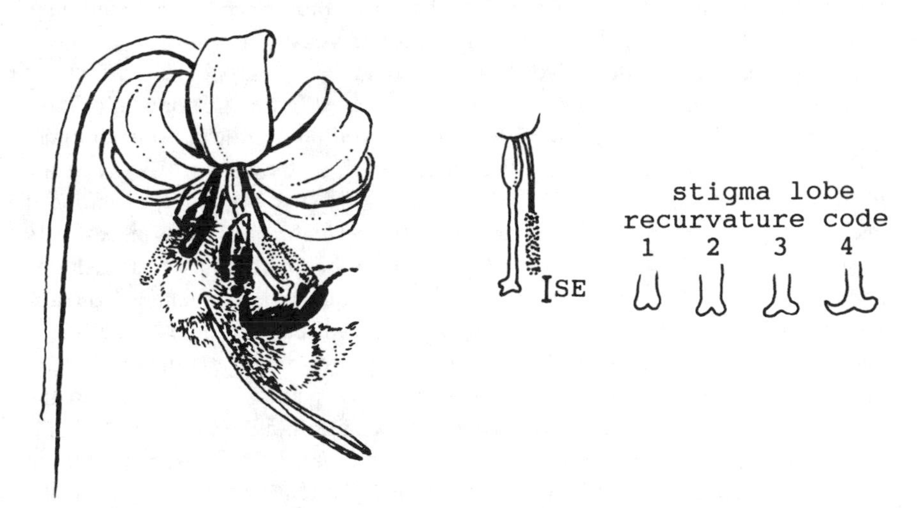

Figure 4.3. *Erythronium grandiflorum* and *Bombus occidentalis*. Stigma exsertion (SE) varies quite considerably and affects the amount of pollen that gets deposited on a visit.

1992). Bumble bees forage exclusively for nectar, which is produced at the base of the tepals. Pollen is incidentally removed from the six dangling anthers, and on flights between flowers, bees groom off pollen and discard it (Thomson, 1986). Stigmas vary from being exserted up to 5 mm beyond the anther tips to being included 2 mm behind them (Thomson and Stratton, 1985). Thomson (1986) previously suggested that small *Bombus* frequently fail to contact stigma. Here we were interested in how pollen transfer is related to stigma exsertion, bee species, and most importantly to the interaction between them.

Methods

We studied pollen removal and deposition in flowers of *Erythronium grandiflorum,* comparing queens of *Bombus occidentalis* and *B. bifarius,* the most important pollinators in the subalpine meadows of western Colorado where we did the work. We picked buds that had begun to open, placed them indoors in florists' cut-flower holders, and measured with calipers the length of the longest anther before any anthers had begun to dehisce. It was already known that anther length is linearly related to pollen production (Thomson and Thomson, 1989). Within 1 to 2 days, flowers typically had opened completely, all six anthers had dehisced, and stigmas were receptive with well-developed papillae.

We caught bees on *Erythronium grandiflorum* or *Taraxacum officinale*. In the latter case, we enclosed the bees in flight cages with large numbers of *E. grandiflorum* flowers for at least 2 days, so that they became proficient visitors before being tested. For testing, we would enclose a chilled bee in a flight cage with about 20 measured flowers. After the bee had warmed up, visited several flowers, and begun feeding at a normal rate, we began recording data. We timed each flower visit by speaking into a tape recorder. As soon as a flower visit was complete, we removed the flower from the arena very gently, being careful not to dislodge pollen. The details of technique followed those of Thomson (1986).

At the end of a run of about 12 flowers, we did the following to each once-visited flower. (1) We measured stigma exsertion as the distance between the tip of the longest anther and the end of the stigma after bending the stamen if necessary to lie parallel to the style (Fig. 4.3). This measure was necessarily approximate because we had to avoid any contact that might transfer or dislodge additional pollen. (2) Holding a microcentrifuge tube under the anthers to catch falling pollen, we removed the anthers with forceps, placing them in the tubes. We rinsed adhering grains off the forceps with a stream of 70% ethanol directly into the tube, thereby filling it and preserving the anthers. (3) We scored the degree of stigma-lobe recurvature on a subjective four-point scale (Fig. 4.3). (4) Again using forceps, we removed the entire stigma to a microscope slide on which had been melted a drop of fuchsin-tinted glycerin jelly. We remelted the jelly with an alcohol burner and firmly squashed the stigma under a cover slip.

The number of pollen grains that had remained in the anthers was estimated using the electronic particle counter with a 150 μ orifice. The number of pollen grains deposited on each stigma was counted using the compound microscope with ocular grid and mechanical stage.

Results

Queens of *Bombus occidentalis* are noticeably larger than those of *B. bifarius,* and they spend more time per visit (Table 4.3). For *B. occidentalis,* visit duration was related to pollen remaining at $r = -0.17$ and pollen deposited at $r = 0.32$. For *B. bifarius,* the correlations were $r = -0.22$ and $r = 0.07$ (sample sizes all >70).

Table 4.4 shows a series of general linear models by which we tried to explain variance in pollen remaining in anthers and in pollen deposited on stigmas. For analyses of pollen remaining, undehisced anther length was always included as a covariate to adjust for the number of grains the flower produced. Likewise, for analyses of pollen deposited, stigma lobe recurvature score was included as a discrete variable to adjust for the amount of receptive stigmatic surface area. These variables were not the focus of our study, but they are clearly significant and were included in the hope of increasing statistical resolution.

Our initial models were aimed at testing the effect of the bee-species × degree-of-exsertion interaction. For both pollen remaining and pollen deposited, the F statistic for this term was not significant (Table 4.4, top panel). Clearly, there was no suggestion of any interaction, and the term was dropped from subsequent models.

Stigma exsertion as a main effect showed no tendency to influence pollen remaining, but it did have a significant influence over pollen deposited (Table 4.4, middle panel). The correlation between stigma exsertion and pollen deposited after adjusting for bee species, bee individual, and stigma code was $r = -0.24$ ($N = 153$).

In the even more simplified analysis, the two *Bombus* species were found to differ in both the amount of pollen remaining and deposited (Table 4.4, bottom panel). *Bombus occidentalis* removed and deposited more pollen than *B. bifarius* (Table 4.3). Individual bees within bee species also seem to differ (although it must be admitted that bee identity was not interspersed or in any way randomized).

Discussion of Results

Plants can affect their own pollination success through influencing either visitation rates or the amount of pollen removed and deposited in a visit (Müller, 1883). Our studies of the latter revealed that strikingly little of the variance in pollen removal and deposition is explained by floral morphology and bee species. Furthermore, we found very little evidence for difference among *Bombus* species in the selection gradients they imposed on floral morphology.

Table 4.4. Analyses of variance (bottom), covariance (middle), and heterogeneity in slopes (top) for pollen remaining and pollen deposited by two Bombus *species in single visits to* Erythronium grandiflorum *flowers. Stigma exsertion is the distance between the stigma and the tip of the longest anther. Anther and stigma size are included to control for pollen production and stigmatic area, respectively. Successively simpler models are presented for each variable. In the top models, the interaction of bee species by stigma exsertion is evaluated, and found to be nonsignificant; for pollen remaining, exsertion itself is also found to be nonsignificant. In the middle model, exsertion is found to be significant for pollen deposited. The bottom models evaluate the importance of bee species ignoring morphology except for anther or stigma size.*

	Pollen Remaining in Anthers					
	df	III-SS[2]	*F*	df	III-SS	*F*
Anther or stigma size[1]	1	2790		3	11,922	
Bee species	1	2203		1	23,736	
Indiv. within bee species	13	5300		13	11,853	
Stigma exsertion	1	76	1 n.s.	1	5263	
Bee × exsertion	1	135	0.82 n.s.	1	112	0.26 n.s.
Error	137	22,533	R^2=0.35	133	56,967	R^2=0.46
Stigma size				3	11,811	9.24***
Bee species				1	26,710	29.54***
Indiv. within bee species				13	11,756	2.12*
Exsertion				1	5344	12.55**
Error				134	57,079	R^2=0.46
Anther or stigma size[1]	1	2893	17.67***	3	8361	6.03***
Bee species	1	3160	7.65*	1	27,120	34.60***
Indiv. within bee species	13	5373	2.52**	13	10,191	1.70 p=0.07
Error	139	22,764	R^2=0.35	135	62,423	R^2=0.41

[1]Anther length is a continuous covariate; stigma code is a categorical variable.

[2]Type III sums of squares for pollen remaining is ×100,000 and for pollen deposited is ×1000.

n.s. $p>0.1$:

*$p<0.05$.

**$p<0.01$.

***$p<0.001$.

For *Impatiens*, there was a suggestion that *Bombus impatiens* removed more pollen when the dimension DI was small, whereas *B. vagans* and *B. fervidus* removed more when that dimension was large. Although this is only weakly indicated, let us accept the significance of the interaction term for the sake of argument. It would exemplify our main concern—heterogeneity in selective regime that could promote adaptive divergence—but it would not fulfill Schemske and Horvitz's (1984) condition for floral specialization. The different bees would drive floral evolution in different directions even though they remove equivalent amounts of pollen. In terms of Fig. 4.1, the means are similar, but the slopes are different.

For *Erythronium,* we did find differences between bee species in the average amount of pollen moved, and pollen movement was related to stigma exsertion, but here the direction of the relationship was definitely not reversed between *Bombus occidentalis* and *B. bifarius*. Schemske and Horvitz would have found their condition met. Our interpretation is that, although there may be selection on stigma exsertion, it seems to act in the same direction regardless of bee species, an interpretation that Schemske agrees with (personal communication).

The characters we studied were, at the outset, good candidates for floral traits that might have influenced the amount of contact with pollinators' bodies, and hence the amount of pollen dispersed. It could easily be supposed that gullet-shaped flowers are molded by selection to fit their own pollinators like finely tailored gloves fit the hands for which they are made. Snapdragons (*Antirrhinum*) and turtleheads (*Chelone*) are vestibular flowers that can only be entered by large bees since they need to be pushed open. Beardstongue (*Penstemon*) species are known to differ in the size and shape of their gullet, and this difference has been attributed to adaptation to pollination by hummingbirds, wasps, and carpenter bees (Straw, 1956). Our data, however, portray jewelweed flowers as less like gloves than like mittens—one size fits all.

Among angiosperms, the spatial separation of anthers and stigma, or herkogamy as it is known, is achieved in many ways, and is generally thought to be involved in the avoidance of self-interference and self-fertilization (Webb and Lloyd, 1986; see Chapter 13). Thomson and Stratton (1985) showed that shorter-styled *Erythronium* flowers had a larger proportion of self-pollen deposited than longer-styled flowers. That stigma exsertion might (and did) have an effect on pollen deposition is not surprising. It is less clear that this character should influence the amount of pollen removed, but both the hypotheses of self-interference (Webb and Lloyd, 1986) and pollen-discounting (Holsinger and Thomson, 1994) assume a relationship between herkogamy and effective pollen removal, which we did not detect.

The interaction term pollinator species × floral morphology does not seem to have been previously of much interest to biologists working at the level of variation among individuals within a species. In order to study this, one needs to have both different species of pollinator and different floral morphologies in a factorial design. Other workers have studied how visitors differ in their effectiveness at pollination, and there have been numerous studies of how phenotypes differ in their propensity at being pollinated. A few researchers have studied both (Schemske and Horvitz, 1984, 1989; Galen et al., 1987, Galen, 1989; Stanton et al., 1991; Murcia, 1990), but the only paper we know of that presents interaction terms is that of Harder and Barrett (1993) on tristylous *Pontederia*. They found that the effect of anther level and floral-tube length on pollen removal sometimes significantly depended on bee species. In other words, there seemed to be some sporadic and weak interactions. The idea of studying how animal species and floral morphology interact to determine pollination success is better appreciated in the literature of systematics than in that of population biology, but even there it has not generally been considered as a statistical problem.

What do our two studies imply about the origin of floral diversity? Essentially, they suggest a lack of underlying heterogeneity in selection for pollination success via mechanical fit. If populations of these plants were to be broken up into a series of isolated subpopulations being visited primarily by different species of bumble bees, then there is no reason to think that they would diverge to have distinct morphologies especially suited to the different pollinators. By analogy, imagine a corral of horses. We have asked a question very similar to the following: If the gate to the corral were opened, would the horses scatter in different directions? We found no evidence that populations of plants would be driven to diverge in that way. Bear in mind, however, that we deliberately made it difficult to find interactions by choosing to compare only pollinator species within *Bombus*, and to work only with natural variation among plants at one site. Had we looked across a greater range of pollinators or a greater range of floral phenotypes, we might have detected stronger interactions.

Mechanisms of Divergence

Floral diversity is immense. Some flowers are tall and slender, others are short and fat. Some have anthers and stigmas hidden deep in the recesses of floral tubes; others present these items on open platters. Size, color, odor, and texture, as well as shape and position, vary extensively from species to species, from group to group. The range of floral architecture and ornamentation is broad, and many of the primary distinguishing features of plant taxa are floral attributes (Grant, 1949).

Flowers are to plants what male genitalia are to animals—the first among organs to diverge. Below we enumerate five ways by which divergence might arise. Probably all are important (perhaps some more than others), and they are mutually exclusive only in a particular case at a particular scale. The evolutionary scale at which morphological differences are interpreted is as important to the adaptationist's program as the spatial scale at which vegetation differences are interpreted is to the ecologist's agenda. The processes of differentiation may vary from one scale to another. For instance, in Wright's shifting balance theory, genetic drift is the initiator of divergence that moves populations off adaptive peaks, and interdemic selection is the prosecutor of further divergence toward the tops of other peaks. Consider our five mechanisms and how at different scales (and in different cases) they might vary in importance.

1. *Adaptation to distinct niches*. In contrast to later niche concepts, Elton (1927) thought of niches as existing even in the absence of organisms to fill them (see Colwell 1992). Pollinators may provide a set of discrete opportunities that plants take advantage of, in which case floral differences could represent

adaptations to different pollinators. For many dramatic contrasts, a difference in niches seems to be the best explanation. For instance, while most *Dalechampia* blossoms are pollinated by female bees that collect resin, three lineages have shifted over to being pollinated by male bees that collect fragrances (Armbruster et al., 1992; Chapter 9). In each case, the shift has been associated with interpretable character evolution, as in *D. brownsbergensis* where the resin gland has become vestigial and the stigmas have taken to secreting fruity fragrances. Likewise, it is because of a shift in pollinators that the gullet-shaped flowers of jewelweeds differ from the long-spurred flowers of garden impatiens: The former are adapted to bumble bees, the latter to butterflies. It seems problematic, however, to believe that the endless numbers of minor floral differences distinguishing species that are all pollinated by the same general type of animal are universally due to specialization on different fixed pollinator types.

2. *Character displacement*. It could be that the pollinator community offers a resource base of some particular breadth and that this resource base is partitioned, via competition, among the various plants using it. Local plant populations would then diverge as a function of local plant competitors, with the arbitrary composition of communities translating into arbitrary differences between races (and eventually species). At one time, this was a very popular mechanism to invoke, especially for characters such as flowering time (Waser, 1978; Pleasants, 1980), positioning of pollen and pollinaria on bees bodies (Dressler, 1968), and usage of specific species and castes of bees (Macior, 1982). While many examples remain plausible with some supporting evidence, they also seem far from proven. One of the best studies is that of Levin (1985) on *Phlox drummondii,* a plant that has pink flowers across most of its range and red flowers where it lives sympatrically with its pink-flowered congener *P. cuspidata*. Hybrids are more or less sterile, and thus producing hybrids rather than legitimate progeny could have been selected against. Furthermore, in experimental arrays, lepidopteran pollinators move assortatively, so that the color difference effectively decreases the proportion of hybridization events. Thus, the geographic pattern, fitness benefits, and ecological mechanisms are all consistent with an interpretation of what is known as the reinforcement of isolating barriers. Reinforcement is a special case of character displacement in which the players are close relatives selected to become different to avoid the wastage of resources associated with producing dysfunctional hybrid offspring. Although there has been interest in this mode of divergence for 40 years, there is very little evidence for or against its ubiquity (reviewed by Grant, 1994). It is not clear how often divergence occurs because flowers are selected to be different per se, as opposed to being selected to fit pollinators that happen to be different or because of less adaptive processes.

3. *Adaptive wandering*. This is the term we apply to the situation in which selection does act on characters and is responsible for character evolution, but in which the direction, strength, and manner of selection varied over time scales

that were less than those through which the characters of interest diverged. The adaptive landscape is like a bean-bag chair, always changing; selection pushes characters hither and thither, but in the end (after the selective context has shifted), it would be inaccurate to say that the various forms differ because they are adapted to different pollinators or to unique plant communities. Thus, at the scale of comparing jewelweeds to glacier lilies, it would be silly to think that the differences have resulted from adaptation to New York *Bombus* vs. Colorado *Bombus,* although during the long evolutionary history separating jewelweeds and glacier lilies, it may very well be that characters have changed due to selection in particular pollinator regimes. In contrast, it seems much more reasonable to attribute regional differences in nectar spur length in *Platanthera ciliaris* to specialization on two distinct butterfly pollinators with disparate proboscis lengths (Robertson and Wyatt, 1990). Imagine, however, that these *Platanthera* populations were to remain isolated for a very long time and were to experience a series of distinct principal pollinators. They might diverge further with each step along the way being due to selection to fit a specific pollinator, but after 10,000 or 100,000 years and perhaps dozens of faunal changes, we would not attribute their divergence to the distinctness of their pollinators at that time, and their characters might not even be especially suited to those pollinators. Adaptive wandering is driven by natural selection, not by genetic drift, so although it might appear to result in nonadaptive divergence, it would not be characterized by neutral rates of evolution, nor would the path of evolution necessarily follow a random walk. Campbell (1989) has found considerable heterogeneity in selective regime as it acts on floral dimensions among sites separated by fractions of a km and from year to year. This was in *Ipomopsis aggregata,* which is pollinated in all her sites by a few species of long-proboscised bumble bees and hummingbirds. In the shallower flowers of *Lavandula latifolia,* Herrera (1988) has documented enormous variation in pollinator assemblage (at the ordinal and species level) over a 6-year period and at several spatial scales (see Chapter 3). Since he believes that the species and orders differ in their effects on fitness and breeding system (Herrera, 1987), he concludes that this "*variation* will most likely result in shifting selection regimes." If the selection regime imposed by pollinators varies at such fine scales, what should we think about the guiding hand of selection over evolutionary time?

4. *Character correlations*. Selection may act on physiology or something else and floral characters might then be dragged along by genetic correlations without being the targets of selection themselves. Such differences would then be an epiphenomenon of selection acting through organismic complexity. Darwin (1859) repeatedly urged us not to forget "correlations of growth," which he thought are "often of the most unexpected nature" (p. 134), and he noted that differences between modules within a plant can arise through correlations just as differences arise between species, genera, and families (pp. 184–185):

> With respect to the difference in the corolla of the central and exterior flowers of a head or umbel, I do not feel at all sure that C. C. [sic] Sprengel's idea that the ray-florets serve to attract insects, whose agency is highly advantageous in the fertilization of plants of these two orders, is so far-fetched. . . . But in regard to the differences both in the internal and external structure of the seeds . . . it seems impossible that they can be in any way advantageous to the plant: yet in the Umbelliferae these differences are of such apparent importance—the seeds being in some cases, according to Tausch, orthospermous in the exterior flowers and coelospermous in the central flowers—that the elder De Candolle founded his divisions of the order on analogous differences. Hence we see that modifications of structure, viewed by systematists as of high value, may be wholly due to unknown laws of correlated growth, and without being, as far as we can see, of the slightest service to the species.

It is not at all clear to what extent floral differences have arisen through selection on correlated characters. On the one hand, it seems that genetic correlations are widespread, and that response to selection on some characters will inevitably affect others (Stebbins, 1950, p. 88). On the other hand, it seems likely that between floral traits and vegetative traits the correlations are weak, polygenic, and easily broken down if and when they are disadvantageous (Berg, 1960; Conner and Via, 1993).

5. *Genetic drift*. It is possible that many characters can vary neutrally across a broad range of states, and for those characters the random processes of meiosis and fertilization in small populations will lead to divergence. In almost any group of angiosperms, a substantial proportion of the taxonomically useful characters have no obvious selective importance, characters like whether there are few or many glands on the ovary, whether the petals are ovate or obovate, the particular shade of blue in the filaments, or the number of marginal hairs on the calyx. Why should it have ever mattered? Perhaps it didn't. Still, it should be emphasized that neutrality must be fairly extreme to allow for genetic drift in the strict sense. As Lande (1976) and many others before and since have shown, even a small amount of selection is enough to override drift in all but the smallest of populations. Wright's (1943) original example of isolation by distance was corolla color (purple vs. white) in *Linanthus parryae*. At the time, there was no reason to think that corolla color was other than neutral. Recently, however, Schemske and Bierzychudek (personal communication) have found that the color morphs do sometimes differ significantly in seed production and are thus under selection, although these differences do not seem to be due to differential pollinator attraction, so flower color itself is probably not the target of selection. This example cautions that it is very hard to say that selection on a character of interest is absent. When it is, however, drift can in theory produce substantial divergence in a relatively short period of time (Lande, 1976; Lynch, 1990). In unusually specialized systems, such as orchids and euglossine bees, Kiester et al. (1984)

have shown theoretically that drift can play an important role in coevolutionary diversification, and in tristylous systems Eckert and Barrett (1992) have presented a strong case for the importance of stochastic processes in the biased loss of some style morphs over others.

To sum up then, at a gross evolutionary scale, the contrasting characters of flowers pollinated by different types of animals are almost certainly adaptive. At the other extreme, minor distinctions may have no effect on function and would thus be neutral. As evolution proceeds, the principal governor of divergence may change. One possible progression would be drift leading to minor differences, then a niche shift leading to a striking (but not particularly multifaceted) difference, then over greater time periods the adoption of a series of environments leading to an interpretation of adaptive wandering, and finally, the fine-tuning of characters such as date-of-flowering to match the local conditions that the plants find themselves in. For diversification to be adaptive, sometime during the course of evolution, selection gradients must be heterogeneous. Our results are not consistent with the view that floral differences are initiated as adaptations to particular pollinators. These data, however, are only a first attempt at grappling with a difficult question in evolutionary biology. The pollinator-species × floral-morphology interaction term is worthy of further consideration. We do not presume that Mayr or Darwin or the Grants would have disagreed or been surprised; on the other hand, they never told us that this was how to study divergence.

Acknowledgments

We thank B. Thomson, G. Weiblen, and A. Lowrance for help in the field, J. Rohlf and C. Janson for statistical advice, D. Funk and K. Omland for dialectical involvement, and D. Schemske and M. Johnston for reviewing the manuscript. Financial support was provided by the E. N. Huyck Preserve and by the National Science Foundation (grants BSR 86-14207 and 90-06380 to JT and a graduate fellowship to PW). This is contribution 890 from Ecology and Evolution at Stony Brook.

References

Armbruster, W.S. 1988. Multilevel comparative analysis of the morphology, function, and evolution of *Dalechampia* blossoms. *Ecology*, 69: 1746–1761.

Armbruster, W.S., A.L. Herzig, and T.P. Clausen. 1992. Pollination of two sympatric species of *Dalechampia* (Euphorbiaceae) in Suriname by male euglossine bees. *Am. J. Bot.*, 79: 1374–1382.

Baker, H.G. 1961. The adaptation of flowering plants to nocturnal and crepuscular pollinators. *Quart. Rev. Biol.*, 36: 64–73.

Bender, E.A., T.J. Case, and M.E. Gilpin. 1984. Perturbation experiments in community ecology: Theory and practice. *Ecology,* 65: 1–13.

Berg, R.L. 1960. The ecological significance of correlation pleiades. *Evolution,* 14: 171–180.

Brown, J.H. and A. Kodric-Brown. 1979. Convergence, competition, and mimicry in a temperate community of hummingbird pollinated flowers. *Ecology,* 60: 1022–1035.

Campbell, D.R. 1989. Measurements of selection in a hermaphroditic plant: Variation in male and female pollination success. *Evolution,* 43: 318–334.

Colwell, R.K. 1992. Niche: A bifurcation in the conceptual lineage of the term. In *Keywords in Evolutionary Biology* (E.F. Keller and E.A. Lloyd, eds.), Harvard Univ. Press, Cambridge, MA, pp. 241–258.

Conner, J. and S. Via. 1993. Patterns of phenotypic and genetic correlations among morphological and life-history traits in wild radish, *Raphanus raphanistrum. Evolution,* 47: 704–711.

Darwin, C. 1859. *On the Origin of Species*. John Murray, London.

Darwin, C. 1862, 1877. *On the Various Contrivances by Which Orchids Are Fertilised by Insects*. John Murray, London.

Devlin B., J. Clegg, and N.C. Ellstrand. 1992. The effect of flower production on male reproductive success in wild radish populations. *Evolution,* 46: 1030–1042.

Dressler, R.L. 1968. Pollination by euglossine bees. *Evolution,* 22: 202–210.

Dobzhansky, T. 1937. *Genetics and the Origin of Species*. Columbia Univ. Press, New York.

Eckert, C.G. and S.C.H. Barrett. 1992. Stochastic loss of style morphs from populations of tristylous *Lythrum salicaria* and *Decodon verticillatus* (Lythraceae). *Evolution,* 46: 1014–1029.

Elton, C.S. 1927. *Animal Ecology*. Clarendon Press, Oxford.

Feinsinger, P. 1987. Effects of plant species on each other's pollination: Is community structure influenced? *Trends Ecol. Evol.,* 2: 123–126.

Galen, C. 1989. Measuring pollinator-mediated selection on morphometric floral traits: Bumble bees and the alpine sky pilot, *Polemonium viscosum. Evolution,* 43: 882–890.

Galen, C. and M.L. Stanton. 1989. Bumble bee pollination and floral morphology: Factors influencing pollen dispersal in the alpine sky pilot, *Polemonium viscosum* (Polemoniaceae). *Am. J. Bot.,* 76: 419–426.

Galen, C., K.A. Zimmer, and M.E. Newport. 1987. Pollination in floral scent morphs of *Polemonium viscosum*: A mechanism for disruptive selection on flower size. *Evolution,* 41: 599–606.

Ghiselin, M. 1969. *The Triumph of the Darwinian Method.* Univ. California Press, Berkeley.

Gould, S.J. 1983. The hardening of the modern synthesis. In *Dimensions of Darwinism* (M. Grene, ed.), Cambridge Univ. Press, Cambridge, MA, pp. 71–93.

Gould, S.J. 1986. Evolution and the triumph of homology, or why history matters. *Am. Scientist,* 74: 60–69.

Gould, S.J. and R.C. Lewontin. 1979. The spandrels of San Marco and the Panglossian paradigm: A critique of the adaptationist programme. *Proc. R. Soc. Lond. B,* 205: 581–598.

Grant, P. 1972. Convergent and divergent character displacement. *Biol. J. Linn. Soc.,* 4: 39–68.

Grant, V. 1949. Pollination systems as isolating mechanisms in angiosperms. *Evolution,* 3: 82–97.

Grant, V. 1994. Modes and origins of mechanical and ethological isolation in angiosperms. *Proc. Natl. Acad. Sci.,* 91: 3–10.

Grant, V. and K.A. Grant. 1965. *Flower Pollination in the Phlox Family.* Columbia Univ. Press, New York.

Harder, L.D. 1990. Pollen removal by bumble bees and its implications for pollen dispersal. *Ecology,* 71: 1110–1125.

Harder, L.D. and S.C.H. Barrett. 1993. Pollen removal from tristylous *Pontederia cordata*: Effects of anther position and pollinator specialization. *Ecology,* 74: 1059–1072.

Harder, L.D. and J.D. Thomson. 1989. Evolutionary options for maximizing pollen dispersal of animal-pollinated plants. *Am. Nat.,* 133: 323–344.

Harder, L.D., J.D. Thomson, M.B. Cruzan, and R.S. Unnasch. 1985. Sexual reproduction and variation in floral morphology in an ephemeral vernal lily, *Erythronium americanum. Oecologia,* 67: 286–291.

Herrera, C.M. 1987. Components of pollinator "quality": Comparative analysis of a diverse insect assemblage. *Oikos,* 50: 79–90.

Herrera, C.M. 1988. Variation in mutualisms: The spatio-temporal mosaic of a pollinator assemblage. *Biol. J. Linn. Soc.,* 35: 95–125.

Holsinger, K.E. and J.D. Thomson. 1994. Pollen discounting in *Erythronium grandiflorum*: Mass-action estimates from pollen transfer dynamics. *Am. Nat.,* 144: 799–812.

Kiester, A.R., R. Lande, and D.W. Schemske. 1984. Models of coevolution and speciation in plants and their pollinators. *Am. Nat.,* 124: 220–243.

Lack, D. 1971. *Ecological Isolation in Birds.* Blackwell Scientific Publications, Oxford.

Lande, R. 1976. Natural selection and random genetic drift in phenotypic evolution. *Evolution,* 30: 314–334.

Lande, R. and S.J. Arnold. 1983. Measuring selection on correlated characters. *Evolution,* 37: 1210–1226.

Levin, D.A. 1985. Reproductive character displacement in *Phlox. Evolution,* 39: 1275–1281.

Lynch, M. 1990. The rate of morphological evolution in mammals from the standpoint of the neutral expectation. *Am. Nat.,* 136: 727–741.

MacArthur, R.H. 1958. Population ecology of some warblers of northeastern coniferous forests. *Ecology,* 39: 599–619.

MacArthur, R.H. 1972. *Geographical Ecology.* Harper and Row, New York.

Macior, L.W. 1982. Plant community and pollinatory dynamics in the evolution of pollination mechanisms in *Pedicularis* (Scrophulariaceae). In *Pollination and Evolution* (J.A. Armstrong, J.M. Powell, and A.J. Richards, eds.), Royal Botanic Gardens, Sydney, Australia, pp. 29–45.

Mayr, E. 1959. Isolation as an evolutionary factor. *Proc. Am. Phil. Soc.*, 103: 221–230.

Mitchell, R.J. 1993. Adaptive significance of *Ipomopsis aggregata* nectar production: Observations and experiment in the field. *Evolution*, 47: 25–35.

Müller, N.H. 1883. *The fertilization of flowers*. Translated by D. Thompson. Macmillan, London.

Murcia, C. 1990. Effects of floral morphology and temperature on pollen receipt and reward in *Ipomoea trichocarpa*. *Ecology*, 71: 1098–1109.

Nilsson, L.A. 1983. Processes of isolation and introgressive interplay between *Platanthera bifloia* (L.) Rich and *P. chlorantha* (Custer) Reichb. (Orchidaceae). *Bot. J. Linn. Soc.*, 87: 325–350.

Nilsson, L.A. 1988. The evolution of flowers with deep corolla tubes. *Nature*, 334: 147–149.

Pleasants, J.M. 1980. Competition for bumble bee pollinators in rocky mountain plant communities. *Ecology*, 61: 1446–1459.

Pleasants, J.M. 1983. Structure of plant and pollinator communities. In *Handbook of Experimental Pollination Biology* (C.E. Jones and R.J. Little, eds.), Van Nostrand Reinhold, New York, pp. 375–393.

Primack, R.B. and J.A. Silander. 1975. Measuring the relative importance of different pollinators to plants. *Nature*, 255: 143–144.

Provine, W.B. 1986. *Sewall Wright and Evolutionary Biology*. Univ. Chicago Press, Chicago.

Rabinowitz, D., J.K. Rapp, V.L. Sork, B.J. Rathcke, G.A. Reese, and J.C. Weaver. 1981. Phenological properties of wind- and insect-pollinated prairie plants. *Ecology*, 61: 49–56.

Richards, O.W. and G.C. Robson. 1926. The species problem and evolution. *Nature*, 117: 345–347, 382–384.

Robertson, J.L. and R. Wyatt. 1990. Evidence for pollination ecotypes in the yellow-fringed orchid, *Platanthera ciliaris*. *Evolution*, 44: 121–133.

Schemske, D.W. 1978. Evolution of reproductive characteristics in *Impatiens* (Balsaminaceae): The significance of cleistogamy and chasmogamy. *Ecology*, 59: 596–613.

Schemske, D.W. 1984. Population structure and local selection in *Impatiens pallida* (Balsaminaceae), a selfing annual. *Evolution*, 38: 817–832.

Schemske, D.W. and C.C. Horvitz. 1984. Variation among floral visitors in pollination ability: A precondition for mutualism specialization. *Science*, 225: 519–521.

Schemske, D.W. and C.C. Horvitz. 1989. Temporal variation in selection on a floral character. *Evolution*, 43: 461–465.

Snow, A.A. 1989. Assessing the gender role of hermaphroditic flowers. *Func. Ecol.*, 3: 249–255.

Stanton, M.L., T.L. Ashman, L.F. Galloway, and H.J. Young. 1992. Estimating male fitness of plants in natural populations. In *Ecology and Evolution of Plant Reproduction* (R. Wyatt, ed.), Chapman & Hall, New York, pp. 62–90.

Stanton, M.L., A.A. Snow, S.N. Handel, and J. Bereczky. 1989. The impact of a flower-color polymorphism on mating patterns in experimental populations of wild radish (*Raphanus raphanistrum* L.). *Evolution,* 43: 335–346.

Stanton, M.L., H.J. Young, N.C. Ellstrand, and J.M. Clegg. 1991. Consequences of floral variation for male and female reproduction in experimental populations of wild radish, *Raphanus sativus* L. *Evolution,* 45: 268–280.

Stebbins, G.L. 1950. *Variation and Evolution in Plants*. Columbia Univ. Press, New York.

Stebbins, G.L. 1970. Adaptive radiation of reproductive characteristics in angiosperms. I. Pollination mechanisms. *Ann. Rev. Ecol. Syst.,* 1: 307–326.

Stewart, S.C. and D.J. Schoen. 1987. Pattern of phenotypic viability and fecundity selection in a natural population of *Impatiens pallida*. *Evolution,* 41: 1290–1301.

Straw, R.M. 1956. Adaptive morphology of the *Penstemon* flower. *Phytomorphology,* 6: 112–119.

Strong, D.R., L.A. Szyska, and D.S. Simberloff. 1979. Tests of community-wide character displacement against null hypotheses. *Evolution,* 33: 897–913.

Strong, D.R., D. Simberloff, L.G. Abele, and A.B. Thistle. 1984. *Ecological Communities: Issues and the Evidence*. Princeton Univ. Press, Princeton, NJ.

Thomson, J.D. 1978. Effects of stand composition on insect visitation in two-species mixtures of *Hieracium*. *Am. Midl. Nat.,* 100: 431–440.

Thomson, J.D. 1980. Implications of different sorts of evidence for competition. *Am. Nat.,* 116: 719–726.

Thomson, J.D. 1982. Patterns of visitation by animal pollinators. *Oikos,* 39: 241–250.

Thomson, J.D. 1986. Pollen transport and deposition by bumble bees in *Erythronium*: Influences of floral nectar and bee grooming. *J. Ecol.,* 74: 329–341.

Thomson, J.D. and D.A. Stratton. 1985. Floral morphology and cross-pollination in *Erythronium grandiflorum* (Liliaceae). *Am. J. Bot.,* 72: 433–437.

Thomson, J.D. and B.A. Thomson. 1989. Dispersal of *Erythronium grandiflorum* pollen by bumble bees: Implications for gene flow and reproductive success. *Evolution,* 43: 657–661.

Thomson, J.D. and B.A. Thomson. 1992. Pollen presentation and viability schedules in animal-pollinated plants: Consequences for reproductive success. In *Ecology and Evolution Plant Reproduction* (R. Wyatt, ed.), Chapman & Hall, New York, pp. 1–24.

Wade, M.J. and S. Kalisz. 1990. The causes of natural selection. *Evolution,* 44: 1947–1955.

Waser, N.M. 1978. Competition for hummingbird pollination and sequential flowering in two Colorado wildflowers. *Ecology,* 59: 934–944.

Waser, N.M. 1983*a*. Competition for pollination and floral character differences among sympatric plant species: A review of evidence. In *Handbook of Experimental Pollination Biology* (C.E. Jones and R.J. Little, eds.), Van Nostrand Reinhold, New York, pp. 277–293.

Waser, N.M. 1983*b*. The adaptive nature of floral traits: Ideas and evidence. In *Pollination Biology* (L. Real, ed.), Academic Press, London, pp. 241–285.

Waser, N.M. and M.V. Price. 1983. Pollinator behavior and natural selection for flower color in *Delphinium nelsonii*. *Nature,* 302: 422–424.

Webb, C.J. and D.G. Lloyd. 1986. The avoidance of interference between the presentation of pollen and stigmas in angiosperms. II. Herkogamy. *New Zeal. J. Bot.*, 24: 163–178.

Wilson, P. in Press. Selection for pollination success and the mechanical fit of *Impatiens* flowers around bumble bee bodies. *Biol. J. Linn. Soc.*

Wilson, P. and J.D. Thomson. 1991. Heterogeneity among floral visitors leads to discordance between removal and deposition of pollen. *Ecology,* 72: 1503–1507.

Wilson, P., J.D. Thomson, M.L. Stanton, and L.P. Rigney. 1994. Beyond floral Batemania: Gender biases in selection for pollination success. *Am. Nat.,* 143: 283–296.

Wright, S. 1943. An analysis of local variability of flower color in *Linanthus parryae*. *Genetics,* 28: 139–156.

van der Pijl, L. 1961. Ecological aspects of flower evolution. II. Zoophilous flower classes. *Evolution,* 15: 44–59.

Vogel, S. 1954. *Blütenbiologishe typen als elemente der sippengliederung*. Fischer, Jena, Germany.

Young, H.J. and M.L. Stanton. 1990. Influences of floral variation on pollen removal and seed production in wild radish. *Ecology,* 71: 536–547.

Zimmerman, M. 1988. Nectar production, flowering phenology, and strategies for pollination. In *Plant Reproductive Ecology: Patterns and Strategies* (J. Lovett Doust and L. Lovett Doust, eds.), Oxford Univ. Press, New York, pp. 157–178.

5

Floral Longevity: Fitness Consequences and Resource Costs

Tia-Lynn Ashman and Daniel J. Schoen†*

Introduction

Floral longevity (the period of time from anthesis to floral senescence) plays an important role in the reproductive ecology of plants. As noted by Primack (1985), the length of time a flower is open can influence its total number of pollinator visits, which, in turn, can affect the amount and diversity of pollen a flower receives, and the amount of pollen it disseminates. Additionally, floral longevity contributes to determining the number of flowers open at any given time (floral display size), the duration of floral display, and the total number of flowers per plant. Ultimately, floral longevity influences many factors that determine the quantity and quality of progeny a plant produces. Over the period of time that a flower functions and contributes to plant fitness, it receives resources to remain alive and attractive to pollinators. Such floral maintenance expenditures may compete with future flower production or other plant functions if plant resources are limited. From an adaptive perspective, therefore, the plant's floral longevity should reflect the balance between fitness consequences and maintenance costs.

As first noted by Kerner von Marilaun (1895), flowers are displayed for only a few hours as in morning glorys (*Ipomoea* spp.) and evening primroses (*Oenothera* spp.), to many days, as in the California poppy (*Eschscholtzia californica*) and *Hepatica triloba,* or even as long as a few months, as in some orchids. The vast diversity of floral lifetimes exhibited in flowering plants (Fig. 5.1) suggests that floral longevity is a character that reflects adaptation to a variety of ecological conditions. Kerner von Marilaun suggested that this diversity is

*Department of Biological Sciences, University of Pittsburgh, Pittsburgh, PA 15260.

†Biology Department, McGill University, 1205 Avenue Docteur Penfield, Montreal PQ, H3A 1B1.

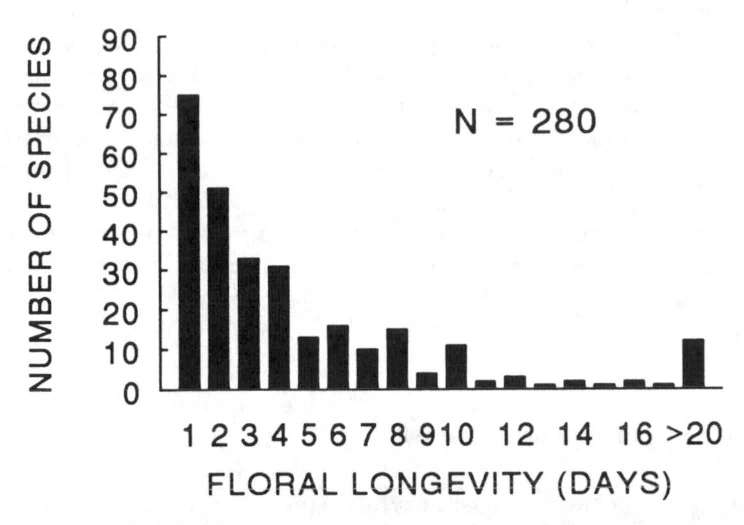

Figure 5.1. Distribution of floral longevities in 280 species. Floral longevities for taxa are represented at the species level unless data were available for more than one species, in which case the mean floral longevity for the genus is represented. (Data from Kerner von Marilaun, 1895; Bertin and Newman, 1993; Rathcke, 1988; Petanidou and Vokou, 1993; Knudsen and Olesen, 1993; R. Bertin, unpublished; Ashman and Schoen, unpublished.)

related to pollen production per flower, the number of flowers per plant, the breeding system and pollinator visitation rates. Almost a century later, Primack (1985) reviewed ecological aspects of floral longevity. He found that floral longevity is strongly influenced by habitat, but not pollinator guild. He also discovered an association between floral longevity and taxonomic class, as well as within-taxa associations with the breeding system. Primack (in conjunction with Lloyd) suggested a theoretical framework within which to interpret the diversity of floral longevity.

These earlier efforts to understand variation in floral longevity recognized many factors that may influence floral lifetimes. In this chapter, we focus on two factors; the first involves resource constraints imposed on the plant during flowering. That is, given a fixed pool of resources for flower production, the number of flowers that can be produced should depend on the costs of constructing and maintaining each flower (Primack, 1985). The second factor relates to how quickly the flower's reproductive function is fulfilled. As pollen is removed and disseminated over time, the flower's potential contribution to reproductive fitness through male function increases, and likewise, as the amount of pollen deposited on the stigma increases and ovules become fertilized, the plant gains increased fitness through female function. Taken together, these two factors determine the fitness consequences of a flower longevity strategy and allow one to evaluate floral longevities in a resource allocation framework.

In this chapter, we first develop an evolutionary stable strategy (ESS) model to formalize how fitness accrual of individual flowers, through male and female reproductive function, interacts with floral construction and maintenance costs to determine optimal floral longevities. Then using this model as a framework, we investigate published studies to evaluate whether floral longevities appear to be optimized by natural selection. We examine the validity of the model's main assumptions and our ability to rigorously test the model's predictions, explore the amount of natural variation in its primary variables, and test some predictions generated by the model. Finally, we discuss the fit between the data and model, and suggest directions to be taken in the further study of the evolution of floral longevity.

Model for Optimal Floral Longevity

In this section, a simple model to predict optimal floral longevity in plants is described. The assumptions of the model are first outlined, followed by a brief description, and a discussion of its predictions.

Model Assumptions

A basic assumption is that plants have evolved so as to control the longevity of their flowers (through allocation of resources to floral maintenance). That is, selection acting on heritable variation in floral longevity has occurred so as to maximize reproductive fitness. It is further assumed that at the time of flowering there exists a pool of limited resources that can be allocated to either the construction of new flowers or the maintenance of existing flowers. Flowers that receive resources for maintenance are kept alive and functioning for a longer period of time than those that do not. Moreover, because the resource pool is limited, allocation to floral maintenance is made at the expense of the production of new flowers. [For simplicity, the pool of resources used to provision fruits and seeds later on in the reproductive season is considered to be separate from that used for floral construction and maintenance and, therefore, does not enter into the model. Changing this assumption (e.g., so that fruits compete for resources with flowers) does not alter the qualitative behavior of the model (Schoen, unpublished)]. The daily cost of floral maintenance (e.g., nectar sugar production, floral respiration, transpiration) is assumed to be constant. One situation where modification of the latter assumption may be relevant pertains to the retention of postfunctional flowers having reduced maintenance costs later on in their lifetimes (Gori, 1983; Casper and La Pine, 1984; Cruzan et al., 1988; Delph and Lively, 1989; Weiss, 1991). This situation is discussed elsewhere (Schoen and Ashman, 1995).

We note that an alternative viewpoint as to how plants use resources for floral maintenance is possible, namely, that rather than flowers being maintained from a standing pool of resources, floral maintenance may instead be supported by the products of *daily* photosynthesis. In this case, the number of flowers produced and maintained by the plant would be constrained by the daily photosynthetic supply rate, rather than the amount of stored resources and construction costs, as assumed below. This situation may indeed occur in some plants (e.g., short-lived desert annuals), but in most species the resources available for maintenance of flowers are likely to come from both daily photosynthetic activity as well as stored resources.

Next we assume that each flower has the same potential to contribute to the overall reproductive fitness of the plant. The realization of this reproductive fitness contribution is determined by the length of time that the flower has been open (i.e., the time available for pollinators to deposit pollen on the stigma and remove pollen from the anthers). As the length of flower life increases, the contribution to the fitness of the plant made by that flower approaches a limit that occurs when the maximum potential amount of pollen has been disseminated from the anthers, and sufficient pollen has been received by the stigma to ensure maximum potential seed set per fruit. The relationship between pollen and seed fitness and floral longevity may be either gradually or rapidly increasing, and is expected to be influenced by extrinsic factors such as climate, pollinator activity, and pollinator behavior, as well as intrinsic factors, such as the maturation schedules of stigmas and anthers. Pollen and seed fitness may each accrue according to the same time schedule, or one fitness component may accrue at a more rapid rate than the other. Differences between the time course of pollen dissemination and pollen receipt are expected to occur especially in species where anther dehiscence precedes stigma maturation, or vice versa (e.g., dichogamous plants). This assumption about how pollen and seed fitness accrue over time is not meant to imply that the fitness accrual rates are not dependent on other characteristics such as flower size and amount of pollinator reward. Indeed, if these features are changed, then the pollen and seed fitness accrual rates may also change. The model presented below, however, is not concerned with the evolution of flower size and pollinator reward, but rather asks the question, given a certain flower size and maintenance cost, what is the optimal floral longevity? When phrased in this manner, it is legitimate to view fitness accrual simply as a function of flower age (time), since the different floral longevity strategies compared are assumed to be identical in their allocation to attraction and reward. A more comprehensive model might ask how all three features, attraction, maintenance, and floral longevity, are simultaneously optimized.

Model of Floral Longevity

In developing the ESS model of floral longevity, the approach used by Shaw and Mohler (1953) and Charnov (1982) was followed. This entails calculating the fitness w of a rare mutant plant with a floral longevity strategy (phenotype 1) that differs from the common floral longevity strategy (phenotype 2), as

$$w = \tfrac{1}{2}\left(\frac{m_1}{m_2} + \frac{f_1}{f_2}\right) \tag{1}$$

where m_1 and m_2 are the male fitness contributions of the mutant and common phenotypes, respectively, and f_1 and f_2 are female fitness contributions. For an individual plant, the male and female fitness contributions are obtained as the product of the respective total (male plus female) fitness contribution of *each* flower and the total number of flowers produced by the plant. The number of flowers produced per plant F is given by

$$F = \frac{R}{c(1 + tm)} \tag{2}$$

where R represents the pool of resources available for floral construction and maintenance, c is the unit cost of constructing a single flower ($c << R$), m is the daily cost of floral maintenance expressed as a proportion of the floral construction cost (e.g., when $m = 0.05$, the daily floral maintenance cost is one-twentieth the cost of constructing a new flower), and t is the number of days that a flower is kept alive (maintained) by the plant. In the present model, it is the parameter t that is assumed to evolve. Note that the total maintenance cost per flower is given by the product tmc. The expression F in Equation (2) quantifies the trade-off assumed in the model between flower number per plant and floral longevity.

Next consider the fitness contributions that each flower makes per unit time (fitness accrual rates). Starting with female fitness, it is assumed that as the time that a flower has been open increases, pollinators will have deposited increasingly more pollen on its stigma, and more ovules will have been fertilized, so that a proportion g of the ovules that were not fertilized at the beginning of the day remains unfertilized at the end of the day. After t days of flower life (if we assume that stigma receptivity begins at $t = 0$), a proportion g^t of the flower's ovules will have been unfertilized and a proportion $1 - g^t$ will have been fertilized. If each flower produces o ovules and each plant F flowers, then $f_i = F_i o\,(1 - g^{t}_{i})$. Similarly, in the case of male fitness, as the time that a flower has been open increases, pollinators will have removed increasingly more pollen from the anthers (and have added it to the pool of pollen that competes for access to ovules; Lloyd, 1984), so that a proportion p of the pollen contained in the flower at the beginning of the day remains in the flower at the end of the day. As above, it is assumed that after t days, a proportion p^t of the original pollen remains, while $1 - p^t$ has been removed by pollinators. If each flower produces P pollen grains and a fraction e of the pollen removed reaches the stigmas of other flowers, then a total of $FPe\ (1 - p^t)$ pollen grains will be dispersed to receptive stigmas after t days (if we assume all F flowers are open during that time). After pollen reaches the stigmas, it competes with pollen from N other plants for access to ovules (also from N plants), and so the male reproductive fitness of a specific donor depends on the number of available ovules in the population and proportion of that donor's pollen present on the available stigmas. For the rare mutant, therefore, the male reproductive fitness is

$$m_1 = \frac{F_1Pe(1 - p^{t_1})}{NF_2Pe(1 - p^{t_2})}\ NF_2o(1 - g^{t_2}) \qquad (3)$$

$$= F_1\frac{(1 - p^{t_1})}{(1 - p^{t_2})}\ o(1 - g^{t_2})$$

while for the common phenotype, male reproductive fitness is

$$m_2 = \frac{F_2Pe\ (1 - p^{t_2})}{NF_2Pe(1 - p^{t_2})}\ NF_2o(1 - g^{t_2}) = F_2o\ (1 - g^{t_2}) \qquad (4)$$

Subsituting Equations (2–4) into Equation (1) and simplifying yield the fitness equation for the mutant

$$w = ½\frac{(1 + t_2m)}{(1 + t_1m)}\left[\frac{(1 - p^{t_1})}{(1 - p^{t_2})} + \frac{(1 - g^{t_1})}{(1 - g^{t_2})}\right] \qquad (5)$$

From Equation (5), it can be seen that the mutant's fitness depends on the floral longevities of both phenotypes (t_1 and t_2), the daily cost of maintaining the flower relative to its construction cost (m), the proportion of remaining pollen left in the flower each day (p), as well as the proportion of unfertilized ovules that remain unfertilized each day (g). The resource pool R and absolute floral construction cost c do not enter directly into the optimal strategy—these cancel out in the fitness equation (5). Below, we refer, respectively, to the parameters $1 - p$ and $1 - g$ as the daily male and female fitness accrual rates. These will be shown to play an important role in the evolution of floral longevity.

Optimal floral longevity was found by solving $dw/dt_1 = 0$ when $t_1 = t_2 = t$ (and when $d^2w/dt_1^2 < 0$), that is, the floral longevity value that maximizes fitness. As no analytical solution could be found for these equations, numerical methods were used to randomly search the parameter space for the combinations of $1 - p$, $1 - g$, m, and t that yielded maximum fitness (absolute value of $dw/dt \leq 0.00001$).

Modifications for Protandry

Some of the plants for which literature data were available to test the model were protandrous (i.e., anthers dehisce before the start of stigma receptivity). The model described above does not consider protandry (or protogyny), but can be modified to predict optimal floral longevity in such cases. If it is assumed that protandry is selected as a means to reduce the level of self-pollination (Darwin, 1876) or interference between the dissemination and receipt of pollen (Lloyd and Yates, 1982; Lloyd and Webb, 1986; Bertin, 1993; see Chapter 6), then protandry may be considered as a constraint around which the optimal floral longevity is selected. The flower must therefore be maintained at least as long as required for the female sexual function to commence. All other things being equal, this should lead to longer flower lives in protandrous plants.

To examine how protandry influences floral longevity, consider a protandrous flower, and let stigma receptivity begin d days ($d > 0$) after flower opening. Increasing the value of d corresponds to increasing the delay in the start of the female sexual phase relative to the male phase. For simplicity, assume that the male sexual phase can last throughout the entire life of the flower, and so the pattern of sexual expression within the flower is male first, followed by bisexual. The problem is to find the optimal floral longevity, starting with d as a lower limit. Let the time period elapsing after anther dehiscence be indexed by a new variable t', so that total floral longevity is now given by $t = d + t'$ days. The fitness equation (1) is modified as follows:

$$w = \tfrac{1}{2}\frac{R/[c + (t_1' + d)\,mc]}{R/[c + (t_2' + d)\,mc]}\left[\frac{1 - p^{(t_1' + d)}}{1 - p^{(t_2' + d)}} + \frac{1 - g^{t_1'}}{1 - g^{t_2'}}\right] \tag{6}$$

To find the optimal t', the methods described above in reference to the model without protandry were used. A parallel model for protogynous species can also be formulated (Schoen and Ashman, 1995) but is not presented here.

Model Predictions

Results for different portions of the parameter space defined by different values of floral maintenance cost m can be displayed in two dimensions as "isoclines" of optimal floral longevity, that is, the combinations of daily pollen and seed fitness-accrual rates ($1 - p$ and $1 - g$) and floral maintenance cost m for which values of $t = 1, 2, \ldots, T$ days yield maximum fitness (Fig. 5.2). It can be seen from the figure that short-lived flowers ($t = 1$ day) are favored when the daily cost of maintenance is high relative to the cost of construction of new flowers (large m) and when pollen and seed fitness-accrual rates are high ($1 - p$ and $1 - g$ large). Long-lived flowers are increasingly favored as the daily floral maintenance cost and daily rate of male and female fitness accrual are reduced.

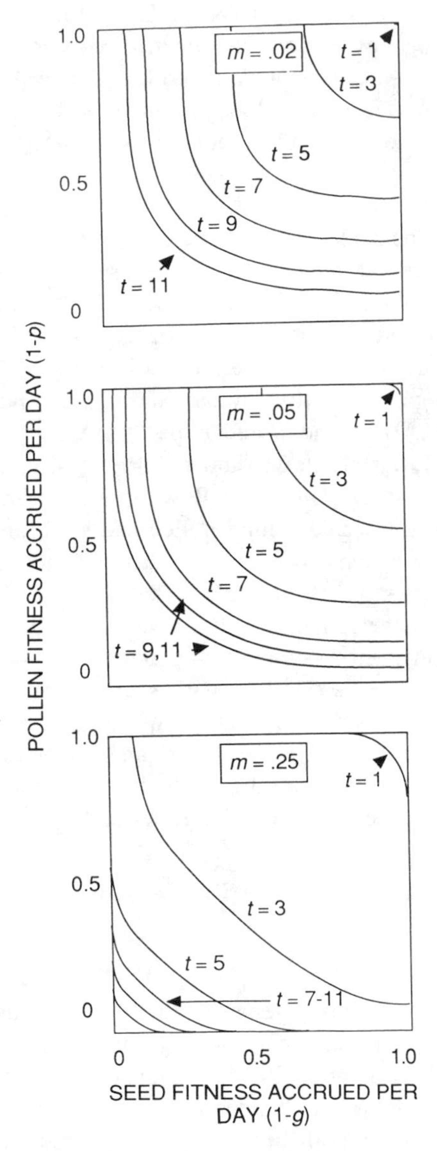

Figure 5.2. Isoclines of optimal floral longevity for homogamous flowers with respect to daily floral maintenance cost ($m = 0.02, 0.05, 0.25$), and daily seed $(1-g)$ and pollen $(1-p)$ fitness-accrual rates. Based on numerical solutions of Equation (5).

When maintenance costs are low (e.g., $m = 0.05$), small changes in the daily pollen and seed fitness-accrual rates have a relatively large effect in determining the optimal floral longevity. When maintenance costs are high, however (e.g., $m \geq 0.25$), most combinations of daily pollen and seed fitness-accrual rates select for short floral longevities. Both protandry and protogyny favor longer-lived flowers relative to the nondichogamous situation (Schoen and Ashman, 1995).

Examination of the floral longevity isoclines suggests several possible ways to test predictions of the model. First, it is clear that when the floral maintenance cost is held constant, floral longevity increases as daily pollen and seed fitness-accrual rates decline. While Fig. 5.2 is only for $m = 0.02$, 0.05, and 0.25, this statement is true over a broad range of maintenance costs and degrees of protandry (Schoen, unpublished). Thus, a qualitative test of the model's predictions could be carried out by comparing the floral longevities of plants having different levels of pollinator activity. Here one would expect a negative correlation between pollinator visitation levels per flower and floral longevity. Such a correlation is not expected to be strong, however, as flowers differ in the level of visitation by pollinators required to achieve high pollen and seed fitness. Thus, it would be more informative if pollen and seed fitness accrual could be assessed directly, for example, by measuring the rate of pollen deposition on stigmas and of pollen removal from anthers. This together with information on the cost of floral construction and maintenance would allow a more direct, quantitative test of the model. In other words, it is possible with such information to calculate the optimal floral longevity using the model, and then to compare the predicted optimal and observed floral longevities. This can be done for both dichogamous and nondichogamous species using the methods outlined above. Below both of these approaches have been used in testing the model.

Assumptions and Limitations in Testing the Model

Assumptions

The models assume that additive genetic variation for floral longevity exists or has existed at some time in the past, allowing floral longevity to evolve through natural selection. This assumption does not ignore the influence on floral longevity of postpollination physiological changes in some species (Gori, 1983). Indeed, floral senescence caused by pollination may in some cases interact with genetic variation in floral longevity and thereby introduce additional variation around the mean floral longevity of a population. We note, however, that pollination does not induce floral senescence in many species (Gori, 1983; Motten, 1986); that is, pollination or lack of pollination is not the only factor determining floral longevity. For example, plants in the genera *Ipomoea, Oenothera, Hibiscus,* and *Mirabilis* typically have short-lived flowers whether or not they have received pollen, whereas plants in other genera maintain their flowers for several days after pollen receipt (*Lupinus, Lantana, Lilium;* Gori, 1983; Weiss, 1991; Edwards and Jordan, 1992). Here we distinguish between two measures of floral longevity: "realized floral longevity" is that exhibited in the presence of pollinators, whereas "maximum floral longevity" is that exhibited in the absence of pollinators and hence pollination. Realized floral longevity corresponds closely with maximum floral longevity among all plant species for which we obtained data [maximum floral longevity (days) = 0.88 * realized floral longevity + 1.72, $P < 0.0001$, $N = 23$; Motten, 1982; Rathcke, 1988b; references in Table 5.3], suggesting that patterns and differences observed among species should remain relatively consistent when assessed by either of these measures of floral longevity.

A second assumption pertains to the finite pool of resources for flowering, and the trade-off between flower construction and maintenance. Such trade-offs can be identified by manipulating one function and observing a negatively correlated response in the other. Although no published studies have assessed the trade-off between flower construction and maintenance specifically, such a trade-off is analogous to those observed between other components of reproduction during flowering. Only a few studies have focused on trade-offs limited to the time of flowering (Solomon, 1988; de Jong and Klinkhamer, 1989). Much more evidence is available for trade-offs between flowers and fruits or vegetative organs (Table 5.1), suggesting that most plant functions can be subject to resource trade-offs. Only one study (of *Clarkia tembloriensis*) has manipulated flower longevity in order to study the trade-off between maintenance and construction (Ashman, unpublished), and it showed that increased floral lifetime led to decreased fruit weight (Table 5.1), thus suggesting that flower maintenance can be involved in resource trade-offs. In some cases, flowers (or more exactly, ovaries or fruits) may photosynthesize and contribute to their maintenance cost (Bazzaz and Carlson, 1979; Werk and Ehleringer, 1983; Galen et al., 1993). Flower maintenance, however, still represents a cost that must be met by the plant regardless of the source of photosynthates.

Limitations

The ability to rigorously test the floral longevity model presented here is limited by the precision with which one can measure the fitness and cost parameters and on the availability of this type of data. The male and female fitness-accrual parameters are the most difficult to measure directly. In lieu of exact estimates of fitness-accrual rates, one could examine variation in pollinator visitation rates. Pollinator visitation is an easily scored correlate of male and female fitness accrual, but it remains crude because pollinators differ in how much pollen they deposit and remove during a visit (Primack and Silander, 1975; Motten et al., 1981; Bertin, 1982; Schemske and Horvitz, 1984; Ashman and Stanton, 1991; Wilson and Thomson, 1991), and moreover, the number of visits required for

Table 5.1 Evidence of resource trade-offs involving flowers and other reproductive traits within a reproductive season.

	Experimental Manipulation (Treatment vs. Control)	Effect of Manipulation on*			Reference
		Flowers	Fruits or Seeds	Other†	
Trade-offs limited to the time of flowering:					
Cynoglossum officinale	Flower removal vs. none	+			de Jong and Klinkhamer (1989)
Solanum carolinense	Basal ovule removal vs. none	+s‡			Solomon (1988)
Trade-offs between flowers and other functions:					
Cucumis sativus	Flower bud removal vs. none			+s	Silvertown (1987)
Lavandula stoechas	Flower bud removal vs. none		+s		Herrera (1991)
Solanum tuberosum	Flower removal vs. none			+s	Jansky and Thompson (1990)
Lupinus luteus	Fruit removal vs. none		+s		van Stevenick (1957)
Almond	Fruit removal vs. none		+s		Baker and Brooks (1947)
Cucumis sativus	Delaying fruit production	+n			Silvertown (1987)
Raphanus raphanistrum	Fruit production series	−n			Stanton et al. (1987)
Brassica campestris	Fruit production vs. none	−n			Evans (1991)
Sidalcea oregana spicata	Fruit production vs. none	−s			Ashman (1992)
Silene vulgaris	Fruit production vs. none	−n			Colosi and Cavers (1984)
Mimulus guttatus	Fruit production vs. none	−n/s			Macnair and Cumbes (1990)
Cynoglossum officinale	Increasing no. of seed/flower	−n			Klinkhamer and de Jong (1987)
Xanthium strumarium	Increasing fruit size	−n			Farris and Lechowicz (1990)
Plantago lanceolata	Flowering vs. not		−n		Antonovics (1980)
Clarkia tembloriensis	Long- vs. short-lived flowers		−s		Ashman (unpublished)

*Notation of effects: + = increase, − = decrease, s = size, n = number.

†Vegetative or storage organs.

‡Effect seen in distal ovules.

a flower to fulfill male and female fitness may not be equivalent (Bell, 1985). A more accurate correlate of male and female fitness, and one that provides a compromise between the ease of pollinator visitation data and the rigor of genetic estimates of fitness accrual, would be to determine pollen removal and deposition over floral lifetime. Since the plant's fitness through male function is related to its ability to disperse pollen and fertilize ovules, the amount of pollen it disseminates may reflect its male fertility (see Schoen and Stewart, 1986; Devlin and Ellstrand, 1990; Broyles and Wyatt, 1990; Devlin et al., 1992; Galen, 1992). In many cases, however, this relationship is less than perfect (Stanton et al., 1992; Snow and Lewis, 1993). For example, if a substantial proportion of the pollen removed is consumed or lost, then little reaches stigmas (see Thomson and Thomson, 1989; Stanton et al., 1992). In addition, postpollination processes (e.g., pollen tube competition, incompatibility, selective abortion) can alter the paternity of resultant seeds. Pollen may also lose viability over time (see Thomson and Thomson, 1992), such that pollen removed late in the flower's life has less of a chance of siring seeds than pollen removed early. These factors combine to make pollen removal a less than perfect estimate of male fitness accrual, but for our purposes it is the most tractable one.

Estimating female fitness accrual directly can also be problematic because it is difficult to determine exactly when ovules have been fertilized. The arrival of pollen on the stigma, however, should be an indicator of ovule fertilization and, hence, of accumulating fitness through female function. Such data would be especially valuable if knowledge of the amount of pollen required for full seed set per flower was available for each species. Determining this would require a labor-intensive experiment involving a series of pollen applications (e.g., Snow, 1982), and so such data are not often available. Instead, we determined from a survey of published data that the average number of pollen grains required to fertilize an ovule is 4.6 with a range of 1.6 to 11 (Silander and Primack, 1978; Mulcahy et al., 1983; Bertin, 1982; Snow, 1982, 1986; Galen et al., 1986; Wolfe and Barrett, 1989; Young and Stanton, 1990; Spira et al., 1992; Ashman unpublished). When combined with information on the rate of pollen deposition and number of ovules produced per flower, it gives the closest approximation available for published data on female fitness accrual.

Estimating maintenance costs relative to construction costs requires that both be expressed in the same currency. There has been much debate over the appropriate currency to measure plant costs (e.g., Thompson and Stewart, 1989; Chapin, 1989). Here, we adopt carbon as the currency for construction and maintenance costs as it is common to both processes (see below). If conversion ratios were known (Bloom et al., 1985), other currencies might be useful (e.g., nutrients, H_2O). Because little data on maintenance costs exist, it was necessary to extrapolate maintenance costs for the species used, and these extrapolations probably represent underestimates of the true floral maintenance expenditure (see below).

Finally, in our tests of the model, we use data gleaned from the literature that were collected to meet objectives different from our own. Thus, the methods and conditions under which these data were collected are not always ideal for our purposes. For example, to study the evolution of floral longevity, data would be best collected on flowers in their native habitats and under typical pollination conditions. Any atypical conditions (e.g., nonnative pollinators, unusual weather conditions) during data collection could result in estimates of fitness parameters that do not represent the conditions under which the longevity phenotype evolved, and so are likely to reduce the fit between the data and model predictions.

Estimation of Model Parameters

Based on the model above for homogamous flowers, three parameters interact to determine the optimal floral longevity: the cost of maintaining flowers relative to construction costs (m), the daily rates of male and female fitness accrual ($1 - p$ and $1 - g$, respectively).

Costs

Both construction and maintenance costs must be measured to estimate the parameter m. The cost of flower construction has received some attention by researchers because it is relevant to various aspects of life history and mating system evolution in plants (Hickman and Pitelka, 1975; Harper, 1977; Smith and Evenson, 1978; Abrahamson and Caswell, 1982; Chapin, 1989; Ashman and Baker, 1992; Ashman, 1994). To express the variation in floral construction costs in a currency common to maintenance costs, we converted published data on floral biomass to estimates of carbon cost for 10 species (Table 5.2). The total cost of constructing a flower includes the amount of carbon in the flower tissue (e.g., cellulose), as well as the metabolic energy expended during flower synthesis (Chapin, 1989). Estimates of the chemical composition of flowers and carbon cost per chemical constituent were used to calculate the construction cost for each species (Lovett-Doust and Harper, 1980; Cruden and Lyon, 1985; Chapin, 1989; Ashman and Baker, 1992). Flower construction costs range from 1.06 mg of carbon per flower of wild radish (*Raphanus sativus*) to 244.01 mg for trumpet creeper (*Campsis radicans,* Table 5.2). Nectar produced prior to flower opening and functioning should be considered a construction cost as well. In the case of trumpet creeper, this adds an additional 4.38-mg carbon to flower construction cost.

The cost of floral maintenance has received much less attention than flower construction costs and so is more difficult to estimate from literature reports. Maintenance includes nectar production following anthesis, floral respiration, and transpiration. Nectar can represent a considerable investment for animal-pollinated plants (Southwick, 1984; Zimmerman and Pyke, 1988; Pyke, 1991; but see Harder and Barrett, 1992). Daily nectar investment for several nectar-

Table 5.2. Costs of floral construction and maintenance.

Species	Floral Costs			Ratio of Daily Floral Maintenance to Construction†	Reference
	Construction (mg Carbon)	Maintenance* (mg Carbon/day)		(Model Parameter *m*)	
Erythronium americanum	18.80	0.00004	n	0.000002	Motten (1986); Bell (unpublished)
Clarkia gracilis tracyi	42.52	0.199	r	0.005	Ashman (unpublished)
Trillium grandiflorum	68.43	1.065	r	0.016	Ashman (unpublished)
Campsis radicans	244.01‡	4.555	n	0.019	Bertin (1982)
Raphanus sativus	1.06	0.029	n	0.028	Stanton and Preston (1988); Stanton (unpublished)
Pontederia cordata	1.30	0.039	n	0.030	Harder and Barrett (1992)
Clintonia borealis	6.34	0.272	n	0.043	Plowright (1981); Bell (unpublished)
Ascelpias syriaca	9.85	0.609	n	0.062	Southwick et al. (1981); Bell (unpublished)
Ipomopsis aggregata	3.81	0.307	n	0.080	Pleasants (1983); Campbell (1992)
Sidalcea oregana spicata					
(H)	2.98	0.692	n	0.232	Ashman and Stanton (1991)
(F)	1.74	0.442	n	0.252	
Impatiens capensis	5.55	1.336	n	0.241	Schemske (1978)

*The maintenance cost of producing floral nectar (n) was available for a number of species, but fewer studies have measured the maintenance cost of corolla respiration (r).

†Expression of the maintenance cost relative to construction as it appears in the model's formulation.

‡Construction costs include 4.38 mg of carbon in nectar that is produced prior to flower opening.

producing species (Table 5.2) was converted from measures of sugar investment per day to carbon investment (mg) so that maintenance costs can be compared directly to construction costs (see below).

The respiration cost of maintaining a flower through its lifetime is not often measured. To assess the cost of respiratory maintenance, Ashman (unpublished) measured corolla respiration in two nectarless species (*Clarkia gracilis tracyi* and *Trillium grandiflorum*). The carbon evolved (as CO_2) from corolla tissue was measured using a portable infra-red gas analyzer (Licor #6250). Single petals, still attached to the plant, were inserted into a darkened Licor chamber and daily respiration for the whole corolla was calculated (Table 5.2). Daily respiration involved an amount of carbon investment similar to that expended on nectar in several other species (Table 5.2). This suggests that the cost of maintaining corollas during the pollination process may not be negligible.

Loss of water due to flower transpiration may also represent a maintenance cost, particularly, in dry environments. For example, flower transpiration accounts for up to 25% of reproductive water loss in *Agave* and avocado (Nobel, 1977; Whiley et al., 1988). Unfortunately, the paucity of information on this precludes its inclusion in the analysis below.

Most flowers simultaneously incur all three of these maintenance costs; however, they have not been measured concurrently for any species. The calculations of maintenance costs relative to construction costs (m) summarized in Table 5.2 therefore underestimate true floral maintenance costs. For the 10 species in Table 5.2, daily maintenance ranges from less than 1% up to 25% of construction cost.

Fitness Accrual

To estimate the male and female fitness-accrual rates, we determined both the amount of pollen removed from the flower and that deposited on the stigma over flower lifetime, as explained in more detail above. When all the pollen has been removed from the anthers and enough pollen has been received by the stigmas to fertilize all the ovules, it was assumed that the "full" male and female fitness contribution of that flower had been attained.

To illustrate contrasting temporal patterns of pollen deposition and removal, we studied flowers of two alpine species with differing floral longevities: columbine (*Aquilegia caerulea*), with 8-day flowers, and geranium (*Geranium caespitosum*), with 1- to 2-day flowers (Ashman and Schoen, unpublished). Studies were conducted in the natural habitats, flowers marked at anthesis, and subsamples harvested at 1- or 2-day intervals over floral lifetime. The amount of pollen remaining in the anthers and adhering to the styles was determined and then plotted against the number of days since anthesis (Fig. 5.3). These data were converted to the daily rate of male fitness accrual ($1 - p$) and female fitness accrual ($1 - g$) by estimating p and g from regressions of proportional fitness over time. The daily accrual rates of male and female fitness for *Aquilegia*

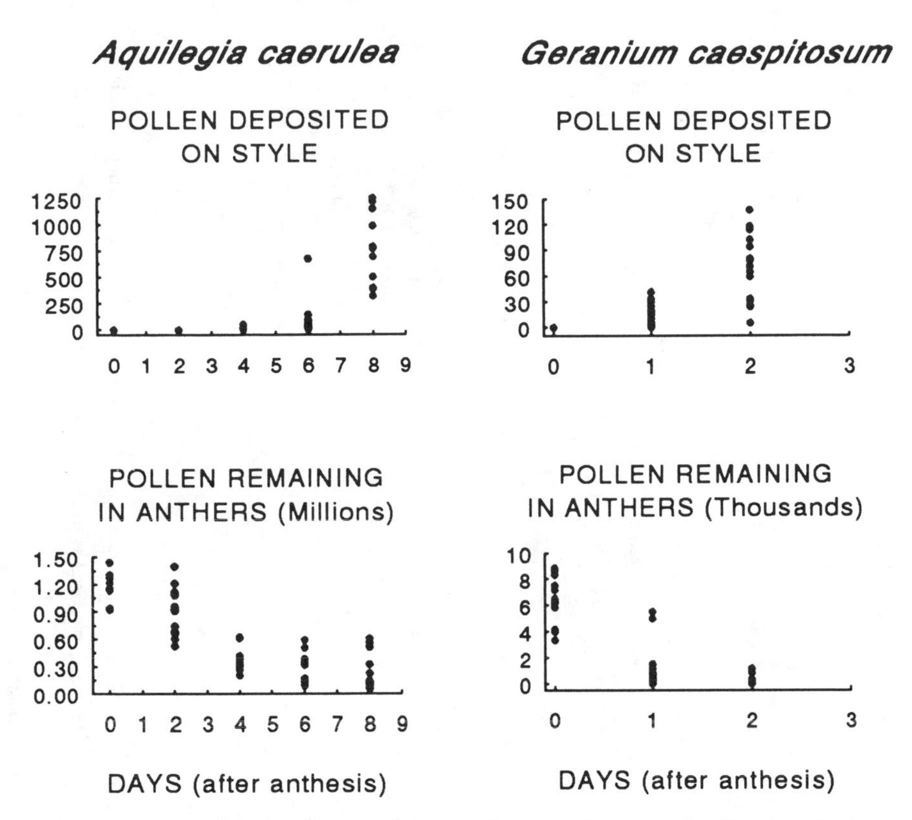

Figure 5.3. Amount of pollen remaining per flower and amount of pollen deposited per stigma of *Aquilegia caerulea* and *Geranium caespitosum* as a function of floral longevity (Ashman and Schoen, unpublished).

caerulea (0.21 and 0.07) were considerably slower than for *Geranium caespitosum* (0.86 and 0.50). For comparison, Table 5.3 presents estimates of daily fitness accrual for nine other species.

Testing Model Predictions

The optimal floral longevity model predicts that over a broad range of floral maintenance costs, floral longevity should decrease as male and female fitness-accrual rates increase (Fig. 5.2). We tested this prediction in several ways so as to use data from as many species as possible. We used data on pollinator visitation rates as a crude approximation of male and female fitness-accrual rates. To obtain these data, we surveyed the last 20 years of the *American Journal of Botany, Oecologia, Ecology,* plus selected works that reported both pollinator visitation and floral longevity. When data were available for more than one species in a

Table 5.3. Rates of daily male and female fitness accrual, and delay between onset of female and male function (d), together with observed and predicted floral longevities.

Species	Daily Fitness-Accrual Rates			Floral Longevity				Reference
				Observed*		Predicted†		
	Male $(1-p)$	Female $(1-g)$	Delay (d)	Max.	Real.	Pred.	Rev.	
Aquilegia caerulea	0.21	0.07	4[f]	na[§]	8.8	na	na	Ashman and Schoen (unpublished)
Geranium caespitosum	0.86	0.50	0	na	1.3	na	na	Ashman and Schoen (unpublished)
Erythronium americanum	0.13‡	0.13	0	10.1	8	80	22.3	Motten (1986)
Trillium grandiflorum	0.06	0.17	0	21	21	20	20	Ashman and Schoen (unpublished)
Campsis radicans	1	0.88	1[f]	4.4	3.8	2.9	2.8	Bertin (1982); Bertin (unpublished)
Raphanus sativus	0.85	0.99	0	3	1.5	1.9	1.8	Stanton et al. (1992); Ashman et al. (1993); Stanton (personal communication)
Pontederia cordata	1	1	0	1	0.33	0.8	0.8	Wolfe and Barrett (1989); Barrett (personal communication)
Ascelpias syriaca	0.5‡	0.5	0	5.0	5.0	3.9	3.7	Jennersten and Morse (1991)
Ipomopsis aggregata	0.18	0.66	1.5[f]	3.2	2.4	5.1	4.5	Waser and Fugate (1986); Campbell et al. (1991); Mitchell and Waser (1992); Mitchell (unpublished)
Sidalcea oregana spicata (H)	1	1	1[f]	4.9	2.0	1.4	1.4	Ashman and Stanton (1991); Ashman (unpublished)
Impatiens capensis	0.7	1	2[f]	3.5	3.5	2.5	2.5	Wilson and Thomson (1991); Schoen (personal communication)

[f]Delay in female function (style maturation).

*Observed floral longevity under conditions of natural pollination (realized), or that exhibited in the absence of pollinators and hence pollination (maximum).

†Predicted floral longevity using measured maintenance costs, as in Table 5.2 (pred.), or revised by estimating the unmeasured corolla respiration in addition to nectar maintenance costs (rev.).

‡Assumes equal rates (or number of visits) for pollen (or pollinaria) deposition and removal.

§Not available.

genus, we randomly selected one species to represent the genus. Pollinator visitation was standardized among species by converting all data into the form "visits per flower per hour." Floral longevity was recorded in number of days and reflects realized flower lifetime (see the "Assumptions" section). Prior to correlation analysis, both variables were natural log-transformed. The correlation between floral longevity and pollinator visitation was negative and highly significant (Fig. 5.4: $r = -0.57$, $n = 39$, $P < 0.001$), suggesting taxa that experience frequent pollinator visits have evolved shorter floral longevities relative to those that receive infrequent visitation.

A more precise test of the model is possible when we attempt to use published

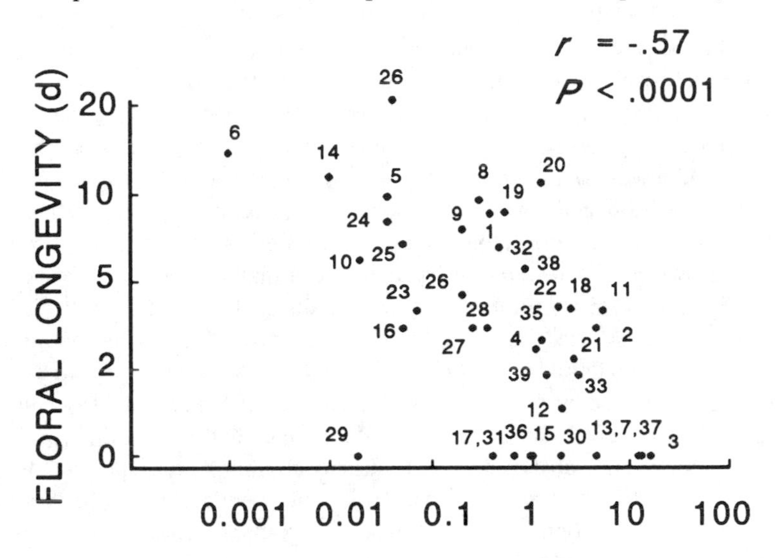

Figure 5.4. Correlation between floral longevity and pollinator visitation rate. Each observation represents one genus (see text). Note that axes have been back-transformed. References for the specific points are as follows: 1, Motten (1982); 2, Whitten (1981); 3, Stucky and Beckmann (1982); 4, de Jong and Klinkhamer (1991); 5, Nilsson (1992); 6, Rodriguez-Robles et al. (1992); 7, Spira et al. (1992); 8, Armbruster and Steiner (1992); 9, Anderson and Beare (1993); 10, Mehrhoff (1983); 11, Sugden (1986); 12, Ashman et al. (1994); 13, Dieringer (1991); 14, Gregg (1991); 15, Heithaus et al. (1974); 16, Beattie et al. (1973); 17, Schemske (1981); 18, Campbell and Motten (1985); 19–22, Motten (1986); 23, Rathcke (1988); and Real and Rathcke (1991); 24, Laverty (1992); 25, Motten (1986) and Laverty (1992); 26, Laverty (1992); and Ashman (unpublished); 27–28, Jennersten and Kwak (1991); 29, Bullock et al. (1989); 30–31, Feinsinger et al. (1988); 32, Melampy (1987); 33, Ashman and Stanton (1991); 34, Campbell et al. (1991); 35, Bertin (1982); 36, Silander and Primack (1978); 37, Schemske and Horvitz (1984); 38, Jennersten and Morse (1991); 39, McDade (1986).

data to estimate model parameters directly so as to predict optimal floral longevities (which can then be compared with observed floral longevities). We predicted floral longevity by substituting the daily accrual rates of male $(1 - p)$ and female $(1 - g)$ fitness per flower, delay times d, and maintenance costs m (Tables 5.2 and 5.3) into the first derivative of equation (5) taken with respect to floral longevity t. The optimal t was found by directly searching for t, yielding $|dw/dt| < 10^{-5}$. Predicted floral longevities were compared to observed floral longevities using a Spearman rank correlation. Observed floral longevities were correlated strongly and significantly with predicted longevities, regardless of whether maximum or realized observed longevities were used (maximum-predicted: $r_s = 0.70$, $P < 0.05$; realized-predicted: $r_s = 0.87$, $P < 0.01$). The slightly stronger relationship between predicted and realized floral longevity may indicate that the physiological response to pollination interacts with the genetically programmed (maximum) floral longevity to bring the realized longevity closer toward the optimal longevity. Hence, there may be an optimal "conditional" response to pollination that was not incorporated in the model (see below). Because we have probably underestimated maintenance costs (e.g., nectar-producing species have additional unmeasured respiratory costs), we revised the maintenance parameter to account for corolla respiration by using an empirical relationship between corolla size and respiration rate to estimate additional respiration cost (Ashman and Schoen, unpublished) and recalculated the correlations. This modification shortened predicted floral longevities, especially for plants with very small nectar costs (e.g., the predicted floral longevity for *Erythronium americanum* was changed from 80 to 22 days; see Table 5.3), and it increased the correlation between predicted and observed floral longevity slightly (revised predicted-maximum: $r_s = 0.72$ $P < 0.05$; Fig. 5.5). This is not unexpected, since the model predictions regarding optimal floral longevity are most sensitive to changes in m when m is initially near zero.

Concluding Remarks

In this chapter, we have sought to interpret natural variation in floral longevity within a theoretical framework that emphasizes two main factors: the cost of flower maintenance relative to flower construction, and the rates of male and female fitness accrual. Empirical data generally supported our basic prediction that floral longevity should decrease with increasing rates of fitness accrual over a range of relative maintenance costs.

Possible Modifications and Conceptual Extensions of the Model

Our model of optimal floral longevity makes several simplifying assumptions that may not always be warranted. Some of these assumptions could be modified without overly increasing the complexity of the model. First, we assumed that

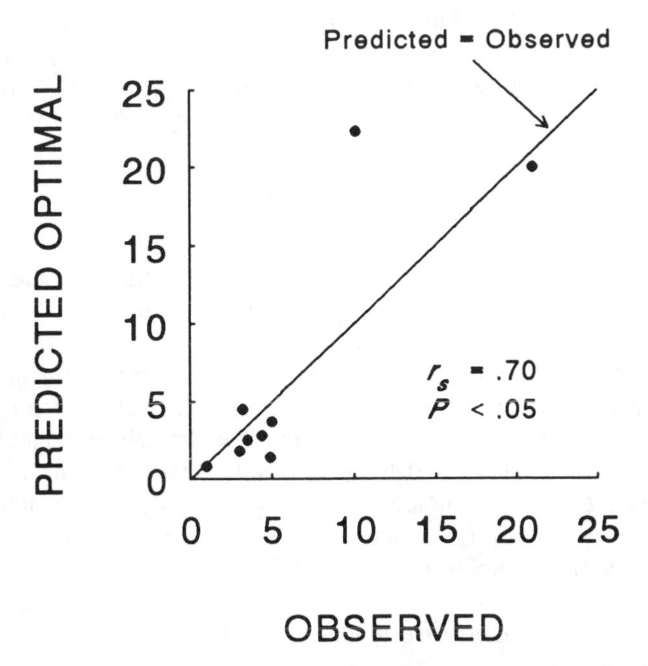

Figure 5.5. The correlation between observed maximum and predicted floral longevity. Predicted floral longevity is based on Equation (5), accrual rates for male and female fitness, and revised estimates of flower maintenance (Table 5.2).

floral maintenance costs remain constant over a flower's lifetime. To the extent that such costs decrease (e.g., if the nectar production or respiration rate declines with flower age), selection will tend to favor longer-lived flowers (Schoen and Ashman, 1995). Second, we assumed that flowers act individually in attracting insect pollinators; however, for species with large, showy inflorescences, the entire floral display may attract pollinators (see Chapters 6, 7, and 12). Hence, even when an individual flower has received sufficient pollen for full seed set and dispersed all its pollen, the plant may benefit from maintaining that flower. Alternatively, when the inflorescence rather than the flower is the unit of attraction, pollen removal and receipt may occur more rapidly, thereby reducing the optimal longevity of single flowers (Schoen and Ashman, 1995). Third, for simplicity we assumed constant rates of fitness accrual over time and, therefore, that the approach to full pollen and seed fitness contribution per flower follows a smooth, increasing function that gradually approaches an asymptote. More complex functions could be used, if required, to better represent the observed relationship of pollen and seed fitness accrual.

Our model could be further developed to account for physiological changes following pollination that initiate flower senescence. It is expected that such a conditional response should be most strongly favored in those species where the achievement of full female and male fitness correspond—that is, where floral senescence in response to pollination would not lead to the loss of opportunities for further male fitness gain by the flower (e.g., this could occur if flowers containing appreciable pollen were not maintained).

Future Directions for Empirical Work

Future studies of floral longevity should distinguish whether measurements were made in the presence or absence of pollinators (i.e., realized and maximum, respectively). Few data are available to test the model assumption that there is a trade-off between flower maintenance and construction. Manipulations could be done to either increase (via nectar removal and stimulation of nectar production) or decrease (via flower removal) flower maintenance investments, and a response in flower size or number could be monitored. Likewise, much more data are needed on male and female fitness-accrual rates, especially for plant populations that fit the criteria for use of genetic markers to determine paternity and hence male fitness accrual (see Stanton et al., 1992).

Links Between Floral Longevity and Other Floral Traits

We have treated floral longevity as a trait in relative isolation of other floral characters and selection pressures. Floral longevity, however, likely evolves in concert with other floral characters. For example, if there has been selection for the staggered release of pollen to maximize male fitness (Harder and Thomson, 1989), then there may have been concurrent selection for increased floral longevity; as such, a staggered release reduces the rate of male fitness accrual $(1 - p)$ and thereby increases optimal floral longevity. Likewise, if there has been selection for gradual maturation of the stigma and restricted pollen tube recruitment (Galen et al., 1986; Ganeshaiah and Uma Shaanker, 1988), then there could also have been selection for increased floral longevity via increased delay time d and/or reduced female fitness accrual rate $(1 - g)$. Selection for increased dichogamy [to avoid stigma-anther interference or self-pollination (Lloyd and Webb, 1986)] may have also increased the optimal floral longevity.

Selection on resource allocation patterns within individual flowers may also affect floral longevity. For example, if there has been selection (via pollinator preference) for increased flower size (Bell, 1985; Galen, 1989; Young and Stanton, 1990) or nectar production rate (Zimmerman, 1983), then the maintenance/construction ratio may be altered and thereby lead to a new optimal floral longevity. In general, floral longevity may evolve in an integrative manner with the rest of the floral phenotype.

Acknowledgments

Ashman's work was supported by a Natural Sciences and Engineering Research Council of Canada (NSERC) International Postdoctoral Fellowship and Schoen's by NSERC Operating Grant. The authors thank Dr. M. Morgan for suggestions about the model, Drs. S.C.H. Barrett and L.D. Harder for comments on the manuscript, and Drs. S.C.H. Barrett, G. Bell, R.I. Bertin, R.J. Mitchell, and M.L. Stanton for sharing their unpublished data or observations.

References

Abrahamson, W.G. and H. Caswell. 1982. On the comparative allocation of biomass, energy, and nutrients in plants. *Ecology*, 63: 982–991.

Anderson, R.C. and M.H. Beare. 1983. Breeding system and pollination ecology of *Trientialis borealis* (Primulaceae). *Am. J. Bot.*, 70: 408–415.

Antonovics, J. 1980. Concepts of Resource Allocation and Partitioning in Plants. In *Limits to Action: The Allocation of Individual Behaviour* (J.E.R. Staddon, ed.), Academic Press, New York, pp. 1–35.

Armbruster, W.S. and K.E. Steiner. 1992. Pollination ecology of four *Dalechampia* species (Euphorbiaceae) in Northern Natal, South Africa. *Am. J. Bot.*, 79: 306–313.

Ashman, T.-L. 1992. Indirect costs of seed production within and between seasons in a gynodioecious species. *Oecologia*, 92: 266–272.

Ashman, T.-L. 1994. Reproductive allocation in hermaphrodite and female plants of *Sidalcea oregana* ssp. *spicata* (Malvaceae) using four currencies. *Am. J. Bot.*, 81: 433–438.

Ashman, T.-L. and I. Baker. 1992. Variation in floral sex allocation with time of season and currency. *Ecology*, 73: 1237–1243.

Ashman, T.-L., L.F. Galloway, and M.L. Stanton. 1993. Apparent versus effective mating in an experimental population of *Raphanus sativus*. *Oecologia*, 96: 102–107.

Ashman, T.-L. and M.L. Stanton. 1991. Seasonal variation in pollination dynamics of sexually dimorphic *Sidalcea oregana* ssp. *spicata* (Malvaceae). *Ecology*, 72: 993–1003.

Baker, G.A. and R.M. Brooks. 1947. Effects of fruit thinning on almond fruits and seeds. *Botanical Gazette* 108: 550–556.

Bazzaz, F.A. and R.W. Carlson. 1979. Photosynthetic contribution of flowers and seeds to reproductive effort of an annual colonizer. *New Phytol.*, 82: 223–232.

Beattie, A.J., D.E. Breedlove, and P.R. Ehrlich. 1973. The ecology of the pollinators and predators of *Frasera speciosa*. *Ecology*, 54: 81–91.

Bell, G. 1985. On the function of flowers. *Proc. R. Soc. Lond. B*, 224: 223–265.

Bertin, R.I. 1982. Floral biology, hummingbird pollination and fruit production of trumpet creeper (*Campsis radicans*, Bignoniaceae). *Am. J. Bot.*, 69: 122–134.

Bertin, R.I. 1993. Incidence of monecy and dichogamy in relation to self-fertilization in angiosperms. *Am. J. Bot.*, 80: 557–560.

Bertin, R.I. and C.M. Newman. 1993. Dichogamy in angiosperms. *Bot. Rev.*, 59: 112–152.

Bloom, A.J., F.S. Chapin, III, and H.A. Mooney. 1985. Resource limitation in plants—an economic analogy. *Ann. Rev. Ecol. Syst.*, 16: 363–392.

Broyles, S.B. and R. Wyatt. 1990. Paternity analysis in a natural population of *Asclepias exaltata*: Multiple paternity, functional gender and the "pollen-donation hypothesis." *Evolution,* 44: 1454–1468.

Bullock, S.H., C. Martinez del Rio, and R. Ayala. 1989. Bee visitation rates to trees of *Prockia crucis* differing in flower number. *Oecologia,* 78: 389–393.

Campbell, D.R. 1992. Variation in sex allocation and floral morphology in *Ipomopsis aggregata* (Polemoniaceae). *Am. J. Bot.*, 79: 516–521.

Campbell, D.R. and A.F. Motten. 1985. The mechanism of competition for pollination between two forest herbs. *Ecology,* 66: 554–563.

Campbell, D.R., N.M. Waser, M.V. Price, E.A. Lynch, and R.J. Mitchell. 1991. Components of phenotypic selection: Pollen export and flower corolla width in *Ipomopsis aggregata*. *Evolution,* 45: 1458–1467.

Casper, B.B. and T.R. La Pine. 1984. Changes in corolla color and other floral characteristics in *Cryptantha humilis* (Boraginaceae): Cues to discourage pollinators. *Evolution,* 38: 128–141.

Chapin, F.S., III. 1989. The cost of tundra plant structures: Evaluation of concepts and currencies. *Am. Nat.,* 133: 1–19.

Charnov, E.L. 1982. *The Theory of Sex Allocation*. Princeton Univ. Press, Princeton, NJ.

Colosi, J.C. and P.B. Cavers. 1984. Pollination affects percent biomass allocated to reproduction in *Silene vulgaris* (bladder campion). *Am. Nat.,* 124: 299–306.

Cruden, R.W. and D.L. Lyon. 1985. Patterns of biomass allocation to male and female functions in plants with different mating systems. *Oecologia,* 66: 299–306.

Cruzan, M.B., P.R. Neal, and M.F. Willson. 1988. Floral display in *Phyla incisa*: Consequences for male and female reproductive success. *Evolution,* 42: 505–515.

Darwin, C. 1876. *The Effect of Cross and Self Fertilisation in the Vegetable Kingdom*. Murray, London.

de Jong, T.J. and P.G.L. Klinkhamer. 1989. Limiting factors for seed production in *Cynoglossum officinale*. *Oecologia,* 80: 167–172.

de Jong, T.J. and P.G.L. Klinkhamer. 1991. Early flowering in *Cynoglossum officinale* L.: Constraint or adaptation? *Func. Ecol.*, 5: 750–756.

Delph, L. and C.M. Lively. 1989. The evolution of floral color change: Pollinator attraction versus physiological constraints in *Fuchsia excorticata*. *Evolution,* 43: 1252–1262.

Devlin, B., J. Clegg, and N.C. Ellstrand. 1992. The effect of flower production on male reproductive success in wild radish populations. *Evolution,* 46: 1030–1042.

Devlin, B. and N.C. Ellstrand. 1990. Male and female fertility variation in wild radish, a hermaphrodite. *Am. Nat.*, 136: 87–107.

Dieringer, G. 1991. Variation in individual flowering time and reproductive success of *Agalinis strictifolia* (Scrophulariaceae). *Am. J. Bot.*, 78: 497–503.

Edwards, J. and J.R. Jordan. 1992. Reversible anther opening in *Lilium philadelphicum* (Liliaceae): A possible means of enhancing male fitness. *Am. J. Bot.*, 79: 144–148.

Evans, A.S. 1991. Whole-plant responses of *Brassica campestris* (Cruciferae) to altered sink-source relations. *Am. J. Bot.*, 78: 394–400.

Farris, M.A. and M.J. Lechowicz. 1990. Functional interactions among traits that determine reproductive success in a native annual plant. *Ecology*, 71: 548–557.

Feinsinger, P., W.H. Busby, and H.M. Tiebout, III. 1988. Effects of indiscriminate foraging by tropical hummingbirds on pollination and plant reproductive success: Experiments with two tropical treelets (Rubiaceae). *Oecologia*, 76: 471–474.

Galen, C. 1989. Measuring pollinator-mediated selection on morphometric floral traits: Bumble bees and alpine skypilot, *Polemonium viscosum*. *Evolution*, 43: 882–890.

Galen, C. 1992. Pollen dispersal dynamics in an alpine wildflower, *Polemonium viscosum*. *Evolution*, 46: 1043–1051.

Galen, C., T.E. Dawson and M.L. Stanton. 1993. Carpels as leaves: Meeting the carbon cost of reproduction in an alpine buttercup. *Oecologia*, 95: 187–193.

Galen, C., J.A. Shykoff and R.C. Plowright. 1986. Consequences of stigma receptivity schedules for sexual selection in flowering plants. *Am. Nat.*, 127: 462–476.

Ganeshaiah, K.N. and R. Uma Shaanker. 1988. Regulation of seed number and female incitation of mate competition by a pH-dependent proteinaceous inhibitor of pollen germination in *Leucaena leucocephala*. *Oecologia*, 75: 110–113.

Gori, D.F. 1983. Post-pollination phenomena and adaptive floral changes. In *Handbook of Experimental Pollination Biology* (C.E. Jones and R.J. Little, eds.), Van Nostrand Reinhold, New York, pp. 31–49.

Gregg, K.B. 1991. Reproductive strategy of *Cleistes divaricata* (Orchidaceae). *Am. J. Bot.*, 78: 350–360.

Harder, L.D. and S.C.H. Barrett. 1992. The energy cost of bee pollination for *Pontederia cordata* (Pontederiaceae). *Func. Ecol.*, 6: 226–233.

Harder, L.D. and J.D. Thomson. 1989. Evolutionary options for maximizing pollen dispersal of animal pollinated plants. *Am. Nat.*, 133: 323–344.

Harper, J.L. 1977. *Population Biology of Plants*. Academic Press, New York.

Heithaus, E.R., P.A. Opler and H.G. Baker. 1974. Bat activity and pollination of *Bauhinia pauletia*: Plant-pollinator coevolution. *Ecology*, 55: 412–419.

Herrera, J. 1991. Allocation of reproductive resources within and among inflorescences of *Lavandula stoechas* (Lamiaceae). *Am. J. Bot.*, 78: 789–794.

Hickman, J.C. and L.F. Pitelka. 1975. Dry weight indicates energy allocation in ecological strategy analysis in plants. *Oecologia*, 21: 117–121.

Jansky, S.H. and D.M. Thompson. 1990. The effect of flower removal on potato tuber yield. *Can. J. Plant Sci.*, 70: 1223–1225.

Jennersten, O. and M.M. Kwak. 1991. Competition for bumble bee visitation between *Melampyrum pratense* and *Viscaria vulgaris* with healthy and *Ustilago*-infected flowers. *Oecologia,* 86: 88–98.

Jennersten, O. and D.H. Morse. 1991. The quality of pollination by diurnal and nocturnal insects visiting common milkweed, *Asclepias syriaca. Am. Midl. Nat.,* 125: 18–28.

Kerner von Marilaun, A. 1895. *The Natural History of Plants, Their Forms, Growth, Reproduction, and Distribution.* Henry Holt, New York.

Klinkhamer, P.G.L. and T.J. de Jong. 1987. Plant size and seed production in the monocarpic perennial *Cynoglossum officinale* L. *New Phytol.,* 106: 773–783.

Knudsen, J.T. and J.M. Olesen. 1993. Buzz-pollination and patterns in sexual traits in North European Pyrolaceae. *Am. J. Bot.,* 80: 900–913.

Laverty, T.M. 1992. Plant interactions for pollinator visits: A test of the magnet species effect. *Oecologia,* 89: 502–508.

Lloyd, D.G. 1984. Gender allocations in outcrossing sexual plants. In *Perspectives on Plant Population Biology* (R. Dirzo and J. Sarukhan, eds.), Sinauer, Sunderland, MA, pp. 277–300.

Lloyd, D.G. and C.J. Webb. 1986. The avoidance of interference between the presentation of pollen and stigmas in angiosperms. I. Dichogamy. *New Zeal. J. Bot.,* 24: 135–162.

Lloyd, D.G. and J.M.A. Yates. 1982. Intersexual selection and the segregation of pollen and stigmas in hermaphroditic plants, exemplified by *Wahlenbergia albomarginata* (Campanulaceae). *Evolution,* 36: 903–915.

Lovett Doust, J. and J.L. Harper. 1980. The resource cost of gender and maternal support in an andromonoecious umbellifer *Smyrnium olusatrum. New Phytol.,* 85: 251–264.

Macnair, M.R. and Q.J. Cumbes. 1990. The pattern of sexual resource allocation in the yellow monkey flower, *Mimulus guttatus. Proc. R. Soc. Lond. B.,* 242: 101–107.

McDade, L.A. 1986. Protandry, synchronized flowering and sequential phenotypic unisexuality in neotropical *Pentagonia macrophylla* (Rubiaceae). *Oecologia* (Berlin), 68: 218–223.

Melampy, M.N. 1987. Flowering phenology, pollen flow and fruit production in the Andean shrub *Befaria resinosa. Oecologia,* 73: 293–300.

Merhroff, L.A., III. 1983. Pollination in the genus *Isotria* (Orchidaceae). *Am. J. Bot.,* 70: 1444–1453.

Mitchell, R.J. and N.M. Waser. 1992. Adaptive significance of *Ipomopsis aggregata* nectar production: Pollination success of single flowers. *Ecology,* 73: 633–638.

Motten, A.F. 1982. Autogamy and competition for pollinators in *Hepatica americana* (Ranunculaceae). *Am. J. Bot.,* 69: 1296–1305.

Motten, A.F. 1986. Pollination ecology of the spring wildflower community of a temperate deciduous forest. *Ecol. Monogr.,* 56: 21–42.

Motten, A.F., D.R. Campbell, D.E. Alexander and H.L. Miller. 1981. Pollination effectiveness of specialist and generalist visitors to a North Carolina population of *Claytonia virginica. Ecology,* 62: 1278–1287.

Mulcahy, D.L., P.S. Curtis and A.A. Snow. 1983. Pollen competition in a natural population. In *Handbook of Experimental Pollination Biology* (L.E. Jones and R.J. Little, eds.), Van Nostrand Reinhold, New York, pp. 330–337.

Nilsson, L.A. 1992. Long pollinia on eyes: Hawk-moth pollination of *Cynorkis uniflora* Lindley (Orchidaceae) in Madagascar. *Bot. J. Linn. Soc.*, 109: 145–160.

Nobel, P.S. 1977. Water relations of flowering in *Agave deserti*. *Bot. Gaz.*, 138: 1–6.

Petanidou, T. and D. Vokou. 1993. Pollination ecology of Labiatae in a Phryganic (East Mediterranean) ecosystem. *Am. J. Bot.*, 80: 892–899.

Pleasants, J.M. 1983. Nectar production patterns in *Ipomopsis aggregata* (Polemoniaeceae). *Am. J. Bot.*, 70: 1469–1475.

Plowright, R.C. 1981. Nectar production in the boreal forest lilly, *Clintonia borealis*. *Can. J. Bot.*, 59: 156–160.

Primack, R.B. 1985. Longevity of individual flowers. *Ann. Rev. Ecol. Syst.*, 16: 15–37.

Primack, R.B. and J.A. Silander. 1975. Measuring the relative importance of different pollinators to plants. *Nature*, 255: 143–144.

Pyke, G.H. 1991. What does it cost a plant to produce floral nectar? *Nature*, 350: 58–59.

Rathcke, B. 1988*a*. Interactions for pollination among coflowering shrubs. *Ecology*, 69: 446–457.

Rathcke, B. 1988*b*. Flowering phenologies in a shrub community: Competition and constraints. *J. Ecol.*, 76: 975–994.

Real, L.A. and B.J. Rathcke. 1991. Individual variation in nectar production and its effect on fitness in *Kalmia latifolia*. *Ecology*, 71: 149–155.

Rodriguez-Robles, J.A., E.J. Melendez, and J.D. Ackerman. 1992. Effects of display size, flowering phenology, and nectar availability on effective visitation frequency in *Comparettia falcata* (Orchidaceae). *Am. J. Bot.*, 79: 1009–1017.

Schemske, D.W. 1978. Evolution of reproductive characteristics in *Impatiens* (Balsaminaceae): The significance of cliestogamy and chasmogamy. *Ecology*, 59: 596–613.

Schemske, D.W. 1981. Floral convergence and pollinator sharing in two bee-pollinated tropical herbs. *Ecology*, 62: 946–954.

Schemske, D.W. and C.C. Horvitz. 1984. Variation among floral visitors in pollination ability: A precondition for mutualism specialization. *Science*, 225: 519–521.

Schoen, D.J. and T.-L. Ashman. 1995. The evolution of floral longevity: A resource allocation to maintenance versus construction of repeated structures in modular organisms. *Evolution* 49: 131–139.

Schoen, D.J. and S.C. Stewart. 1986. Variation in male reproductive investment and male reproductive success in white spruce. *Evolution*, 40: 1109–1120.

Shaw, R.F. and J.D. Mohler. 1953. The selective advantage of the sex ratio. *Am. Nat.*, 87: 337–342.

Silander, J.A. and R.B. Primack. 1978. Pollination intensity and seed set in the evening primrose (*Oenothera fruiticosa*). *Am. Midl. Nat.*, 100: 213–216.

Silvertown, J. 1987. The evolution of hermaphroditism: An experimental test of the resource model. *Oecologia,* 72: 157–159.

Smith, C.A. and W.E. Evenson. 1978. Energy distribution in reproductive structures of *Amaryllis*. *Am. J. Bot.,* 65: 714–716.

Snow, A.A. 1982. Pollination intensity and potential seed set in *Passiflora vitifolia*. *Oecologia,* 55: 231–237.

Snow, A.A. 1986. Pollination dynamics in *Epilobium canum* (Onagraceae): Consequences for gametophytic selection. *Am. J. Bot.,* 73: 139–151.

Snow, A.A. and P.O. Lewis. 1993. Reproductive traits and male fertility in plants: Empirical approaches. *Ann. Rev. Ecol. Syst.,* 24: 331–351.

Solomon, B.P. 1988. Patterns of pre- and postfertilization resource allocation within an inflorescence: Evidence for interovary competition. *Am. J. Bot.,* 75: 1074–1079.

Southwick, E.E. 1984. Photosynthate allocation to floral nectar: A neglected energy investment. *Ecology,* 65: 1775–1779.

Southwick, E.E., G.M. Loper, and S.E. Sadwick. 1981. Nectar production, composition, energetics and pollinator attractiveness in spring flowers of western New York. *Am. J. Bot.,* 68: 994–1002.

Spira, T.P., A.A. Snow, D.F. Whigham, and J. Leak. 1992. Flower visitation, pollen deposition and pollen-tube competition in *Hibiscus moscheutos* (Malvaceae). *Am. J. Bot.,* 79: 428–433.

Stanton, M.L., T.-L. Ashman, L.F. Galloway, and H.J. Young. 1992. Estimating male fitness in natural populations. In *Ecology and Evolution of Plant Reproduction: New Approaches* (R. Wyatt, ed.), Chapman & Hall, New York, pp. 62–90.

Stanton, M.L., J.K. Bereczky, and H.D. Hasbrouck. 1987. Pollination thoroughness and maternal yield regulation in wild radish, *Raphanus raphanistrum* (Brassicaceae). *Oecologia,* 74: 68–76.

Stanton, M.L. and R.E. Preston. 1988. Ecological consequences and phenotypic correlates of petal size variation in wild radish, *Raphanus sativus* (Brassicaceae). *Am. J. Bot.,* 75: 528–539.

Stucky, J.M. and R.L. Beckmann. 1982. Pollination biology, self-incompatibility, and sterility in *Ipomoea pandurata* (L.) G.F.W. Meyer (Convolvulaceae). *Am. J. Bot.,* 69: 1022–1031.

Sugden, E.A. 1986. Anthecology and pollinator efficacy of *Styrax officinale* subsp. *redivivum* (Styracaceae). *Am. J. Bot.,* 73: 919–930.

Thompson, K. and A.J.A. Stewart. 1981. The measurement and meaning of reproductive effort in plants. *Am. Nat.,* 117: 205–211.

Thomson, J.D. and B.A. Thomson. 1989. Dispersal of *Erythronium grandiflorum* pollen by Bumble bees: Implications for gene flow and reproductive success. *Evolution,* 43: 657–661.

Thomson, J.D. and B.A. Thomson. 1992. Pollen presentation and viability schedules in animal-pollinated plants: Consequences for reproductive success. In *Ecology and*

Evolution of Plant Reproduction: New Approaches (R. Wyatt, ed.), Chapman & Hall, New York, pp. 1–24.

van Steveninck, R.F.M. 1957. Factors affecting the abscission of reproductive organs in yellow lupins (*Lupinus luteus* L.) *J. Exp. Bot.,* 8: 373–381.

Waser, N.M. and M.L. Fugate. 1986. Pollen precedence and stigma closure, a mechanism of competition for pollination between *Delphinium nelsonii* and *Ipomopsis aggregata. Oecologia,* 70: 573–577.

Weiss, M.R. 1991. Floral colour changes as cues for pollinators. *Nature,* 354: 221–223.

Werk, K.S. and J.R. Ehleringer. 1983. Photosynthesis by flowers of *Encelia farinosa* and *Encelia californica* (Asteraceae). *Oecologia,* 57: 311–315.

Whiley, A.W., K.R. Chapman, and J.B. Saranah. 1988. Water loss by floral structures of avocado (*Persea americana* cv. *fuerte*) during flowering. *Aust. J. Agric. Res.,* 39: 457–467.

Whitten, W.M. 1981. Pollination ecology of *Monarda didyma, M. clinopodia,* and hybrids (Lamiaceae) in the Southern Appalachian Mountains. *Am. J. Bot.,* 68: 435–442.

Wilson, P. and J.D. Thomson. 1991. Heterogeneity among floral visitors leads to discordance between removal and deposition of pollen. *Ecology,* 72: 1503–1507.

Wolfe, L.M. and S.C.H. Barrett. 1989. Patterns of pollen removal and deposition in tristylous *Pontederia cordata* L. *Biol. J. Linn. Soc.,* 36: 317–329.

Young, H.J. and M.L. Stanton. 1990. Influence of floral variation on pollen removal and seed production in wild radish. *Ecology,* 71: 536–547.

Zimmerman, M. 1983. Plant reproduction and optimal foraging: Experimental nectar manipulations in *Delphinium nelsonii. Oikos,* 41: 57–63.

Zimmerman, M. and G.H. Pyke. 1988. Pollination ecology of Christmas bells (*Blandfordia nobilis*): Patterns of standing crop of nectar. *Aust. J. Ecol.,* 13: 301–309.

6

Pollen Dispersal and Mating Patterns in Animal-Pollinated Plants

Lawrence D. Harder and Spencer C.H. Barrett†*

Immobility complicates mating by angiosperms because the transfer of male gametes between individuals requires pollen vectors. Although abiotic and biotic vectors can transport pollen considerable distances (Bateman, 1947*a*; Squillace, 1967; Kohn and Casper, 1992; Godt and Hamrick, 1993), the resulting pattern of pollen dispersal does not intrinsically maximize the number and quality of matings. Consequently, floral evolution generally involves two classes of adaptations that promote mating success. The morphological traits that characterize floral design and display modify the actions of pollen vectors so as to enhance fertility (see below). In contrast, physiological traits mitigate unsatisfactory pollen dispersal by rejecting unsuitable male gametophytes (Jones, 1928; de Nettancourt, 1977; Marshall and Ellstrand, 1986; Seavey and Bawa, 1986; Barrett, 1988; Snow and Spira, 1991; Walsh and Charlesworth, 1992) or zygotes (Stephenson, 1981; Casper, 1988; Becerra and Lloyd, 1992; Montalvo, 1992). As a result of postpollination processes, the realized mating pattern does not simply mirror the pattern of pollination (e.g., Campbell, 1991; also see Waser and Price, 1993). However, these processes can only filter the incipient mating pattern established during pollination, so that pollination fundamentally determines the maximum frequency and diversity of mating opportunities. Consequently, the role of pollination in governing the scope for mating inextricably links the evolution of pollination and mating systems.

Most plant species employ animals as pollen vectors because the behavioral flexibility of animals disposes them to manipulation by plant characteristics. Animal pollination produces a characteristic mating pattern that generally reflects both the tendency of pollinators to move among neighboring flowers and plants

*Department of Biological Sciences, University of Calgary, Calgary, Alberta, Canada T2N 1N4.
†Department of Botany, University of Toronto, Toronto, Ontario, Canada M5S 3B2.

(reviewed by Levin and Kerster, 1974; Handel, 1983; Richards, 1986) and the residence of some of a flower's pollen on a pollinator during succeeding visits to other flowers (pollen carryover; reviewed by Robertson, 1992). These features typically result in local cross-pollination (Schaal, 1980; Handel, 1982; Thomson and Thomson, 1989; Devlin and Ellstrand, 1990; Meagher, 1991; Campbell, 1991) and self-pollination, due partly to pollen transport among flowers on the same plant (geitonogamous pollination; reviewed by de Jong et al., 1993; see also Chapter 7). For self-compatible species, such a pollination pattern often results in a mixed-mating system (reviewed by Schemske and Lande, 1985; Barrett and Eckert, 1990), including some biparental inbreeding (Ritland and Ganders, 1985; Waller and Knight, 1989). In addition, stigmas of animal-pollinated plants often receive many pollen grains from each of several different donors, so that maternal plants produce more full sibs (correlated mating) than would be expected from random mating (Schoen and Clegg, 1984; Schoen, 1985; Ritland, 1989). Hence, animal pollination delivers a complex mixture of pollen from many sources, which in turn facilitates the implementation of diverse mating systems.

Plants govern their mating opportunities through the effects of floral design and display on pollinator attraction and pollen dispersal. Enhanced attraction obviously favors both male and female function when pollination is insufficient to fertilize as many ovules as plants can develop into seeds (see also Chapter 12). Even if pollen receipt does not limit seed production, increased pollinator attraction can benefit male function if increased pollen removal decreases the proportion of removed grains that fertilize ovules and floral mechanisms restrict removal by individual pollinators (Harder and Thomson, 1989; Harder and Wilson, 1994). Attraction of many pollinators and restricted pollen removal also increase potential mate diversity because, with many pollinators following dissimilar foraging paths, an individual plant imports pollen from and exports pollen to a larger sample of the population. Floral characteristics influence pollinator attraction by providing resources of value to animals (e.g., nectar, pollen, floral oils, etc.; reviewed by Simpson and Neff, 1983) and by signaling the location of these resources with visual and/or olfactory displays (e.g., showy perianths, nectar guides, floral odors, many-flowered inflorescences; see Waser and Price, 1983*a*; Galen, 1985; Galen and Newport, 1987; Thomson, 1988; Stanton et al., 1989).

Once a pollinator has been attracted, floral traits affect the success of pollen dispersal by determining the amount of pollen exchanged between each flower's sexual organs and the pollinator. Flowers influence pollen removal by each pollinator through the schedule of pollen presentation and dispensing (Harder and Thomson, 1989; Armstrong, 1992; Harder and Barclay, 1994; LeBuhn and Anderson, 1994; Harder and Wilson, 1994) and by controlling reward availability, which affects the duration of pollinator visits (e.g., Harder, 1990*a*, Young and Stanton, 1990). The site of contact between the pollinator and the flower's pollen-presenting structures also affects pollen removal (Harder, 1990*a*; Murcia, 1990; Wilson and Thomson, 1991; Harder and Barrett, 1993) and additionally determines the susceptibility of transported pollen to pollinator grooming or burial by pollen from subsequently visited plants. Recipient flowers influence pollen deposition through the duration and site of contact between the stigma and pollinator, which depend on the flower's control of pollinator position and on stigma size and position (Galen and Plowright, 1985; Murcia, 1990; Johnston, 1991). Thus, although flowers interact with pollinators only briefly, the diverse influences of flower characters enable flowers to manipulate patterns of pollen dispersal.

Even though the pattern of pollen dispersal determined by the interaction between pollinator and floral characteristics circumscribes the opportunities for mating between plants, the relation between pollination and mating is seldom the focus of ecological and evolutionary studies of plant reproduction (for exceptions, see Levin and Kerster, 1974; Waser and Price, 1983*b*; Brown and Clegg, 1984; Richards, 1986; Abbott and Irwin, 1988; Campbell, 1991; Stanton et al., 1991; Holsinger, 1992; Lloyd and Schoen, 1992; Waser, 1993; Kohn and Barrett, 1994). On one hand, pollination biologists emphasize floral characters and their influence on pollinator visitation, pollen export and seed production by individual plants, but they typically pay little attention to who has mated with whom. On the other hand, mating-system biologists generally focus on mating patterns at the population level, especially the relative frequency of selfing vs. outcrossing, with less regard for the processes producing those patterns. Consequently, modern studies of the ecology and evolution of plant reproduction often tend to disarticulate reproductive processes from their outcomes.

The relative isolation of studies of pollination and mating systems reflects the separate development of ecological and genetic perspectives that has characterized the history of evolutionary biology in general. Pollination biology developed from the natural-history tradition established by Sprengel, Darwin, F. and H. Müller, Delpino, Knuth, and others during the nineteenth century (reviewed by Baker, 1983) and maintains a primarily empirical approach to analyzing ecological aspects of plant-pollinator interactions and their consequences for reproduction under natural conditions. In contrast, mating-system biology developed during the middle of the twentieth century as a component of population genetics, and uses laboratory-intensive techniques to examine formal theoretical problems. Hence, pollination and mating-system biology arose from different, but complementary, perspectives on plant reproduction. Nevertheless, the isolation of pollination and mating-system biology hinders comprehensive understanding of the function and evolution of floral characters that influence fertility.

Our objective in this chapter is to illustrate how analysis of the relation between pollination and mating can clarify the functional significance of floral design and display. As our primary example, we consider the effects of the number of flowers that a plant exposes simultaneously (daily inflorescence size) on pollinator attraction, pollen dispersal, and the resulting mating pattern, especially the relative frequency of selfing and outcrossing. We begin with a brief overview of the influences of daily inflorescence size on pollination and mating. We then temporarily step back from the inflorescence perspective and examine pollen dispersal among individual flowers to illustrate theoretically and empirically how characteristics of flowers and their pollinators determine the dispersion of donor pollen among recipient flowers. Given this background in pollen dispersal, we then return to the question of how the arrangement of flowers into inflorescences influences mating patterns. In particular, we predict the influence of inflorescence size on various mating parameters, based on a simplified model of pollen carryover, and we then test many of these predictions with an empirical study of the role of daily inflorescence size on mating in experimental arrays of *Eichhornia paniculata* (Pontederiaceae). To illustrate some evolutionary consequences of the interaction between pollination and mating, we also present a cost-benefit analysis of daily inflorescence size. Based on our analysis of the dependence of mating on pollination, we propose novel interpretations of several aspects of floral design and display, including sterile flowers, dicliny, dichogamy, and heterostyly.

General Influences of Daily Inflorescence Size on Pollination and Mating

Most plants display several flowers at one time, so that their mating patterns depend on the collective contributions of individual flowers to pollination and seed production. Such aggregated flowering complicates the influences on plant fertility because flowers do not function in isolation from other flowers within an inflorescence. These complications arise primarily from the relation of pollinator attraction and geitonogamous pollen transfer to daily inflorescence size.

Large inflorescences often attract more pollinators than small inflorescences (Schaffer and Schaffer, 1979; Augspurger, 1980; Paton and Ford, 1983; Schmitt, 1983; Bell, 1985; Geber, 1985; Andersson, 1988; Cruzan et al., 1988; Thomson, 1988; Klinkhamer et al., 1989; Klinkhamer and de Jong, 1990; Eckhart, 1991; see also Chapter 7) because the proximity of many flowers reduces pollinator flight costs. However, individual flowers receive more visits only when each pollinator visits a fixed or increasing proportion of the available flowers. Many pollinators do not increase the number of flowers visited per inflorescence in proportion to increases in daily inflorescence size, so that the number of pollinator visits per flower often declines with increasing inflorescence size (e.g., Schmitt, 1983; Geber, 1985; Andersson, 1988; Klinkhamer et al., 1989; Robertson, 1992). Consequently, if daily inflorescence size does not affect pollen removal and deposition during each flower visit, then increased flower display could reduce mating frequency.

Regardless of whether the proportion of visited flowers changes, increases in the number of flowers visited per pollinator with increasing daily inflorescence size generally increase geitonogamous self-pollination (reviewed by de Jong et al., 1993; also see Barrett et al., 1994; Hodges, 1995). The mating consequences of geitonogamy depend on whether such self-pollination directly reduces the amount of pollen reaching other plants (pollen discounting; Holsinger et al., 1984), the relative susceptibility of self pollen to self-incompatibility (reviewed by de Nettancourt, 1977; Barrett, 1988), and the severity of inbreeding depression suffered by selfed offspring (reviewed by Charlesworth and Charlesworth, 1987).

Pollen-Dispersal Between Individual Flowers

The roles of floral design and display in mating are most readily appreciated by first identifying floral influences on pollen dispersal. Unfortunately, although pollen carryover determines the incipient mating pattern, most studies of this process have focused on the extent of carryover, with less emphasis on its fundamental influences (although see Lertzman and Gass, 1983; Waser and Price, 1984; Galen and Plowright, 1985; Thomson, 1986; Feinsinger and Busby, 1987). Despite the absence of explicit functional studies of pollen dispersal, published descriptions of the likely underlying mechanisms can be formalized mathematically to examine how specific interactions between flowers and their pollinators might affect the pattern of pollen dispersal. We therefore briefly describe two models of dispersal. The first model proposes that the pollen on a pollinator constitutes a single, homogeneous pool (single-compartment model; see Fig. 6.1a). This model captures general features of pollen dispersal and is the most commonly used theoretical description of pollen carryover (e.g., Bateman, 1947*b*; Plowright and Hartling, 1981; Crawford, 1984; Geber, 1985; de Jong et al., 1992; Robertson, 1992; Barrett et al., 1994). However, observed pollen carryover is often more extensive than predicted by the single-compartment model (Thomson, 1986, Morris et al., 1994; Harder and Wilson, unpublished), implying that pollen dispersal is not merely a simple decay process. Therefore, the second model that we consider describes one mechanism that complicates pollen dispersal, namely, frequent grooming by pollinators (two-compartment model; Fig. 6.1b). To evaluate these alternative perspectives of pollen dispersal, we also compare the patterns of dispersal predicted by the one- and two-compartment models with observed dispersal of *Pontederia cordata* L. (Pontederiaceae) pollen by bumble bees. This example also illustrates that floral morphology strongly affects patterns of pollen dispersal.

Single-Compartment Model

The simplest representation of pollen dispersal considers pollen on the pollinator's body as a single, completely mixed pool (Fig. 6.1a). While visiting a flower,

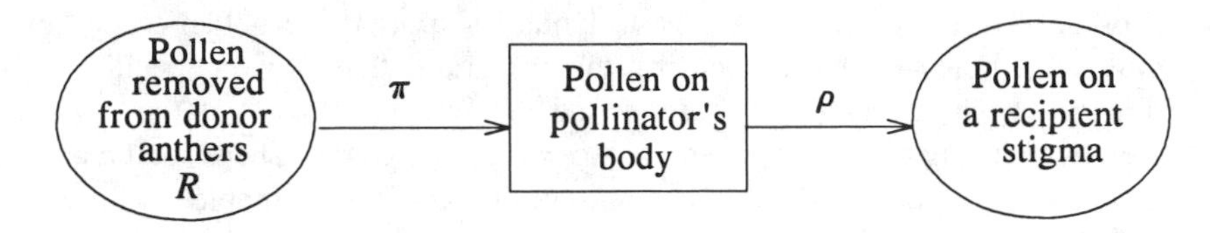

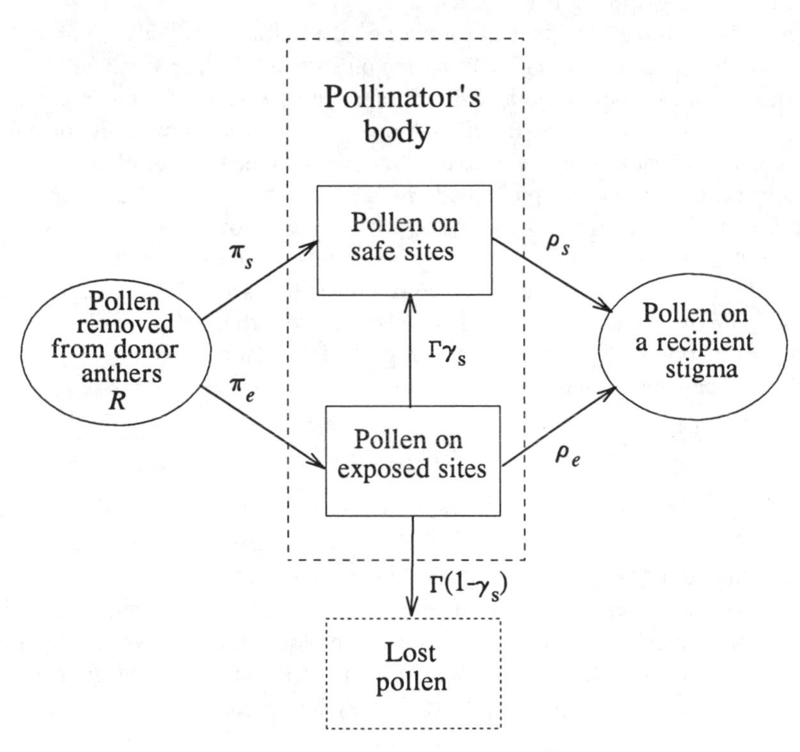

Figure 6.1. Two models of pollen dispersal: (a) Each pollinator carries a single, mixed pool of pollen; or (b) each pollinator carries pollen in a pool exposed to grooming and a pool safe from grooming. Greek symbols indicate transfer probabilities between states during each flight between flowers.

the pollinator removes R pollen grains, of which a fraction π remains on the pollinator as it moves to the next flower. Simultaneously, the flower's stigma removes a fixed proportion ρ (deposition fraction) of the pollen already on the pollinator's body, so that the fraction $1 - \rho$ is carried over for deposition on a subsequently visited recipient. These dynamics apply to all pollen carried by a pollinator; however, let us focus on the πR grains that leave a specific donor flower. The first recipient flower visited receives $d_1 = \rho\pi R$ of these grains, thereby reducing the donor pollen pool to $\pi R(1 - \rho)$ grains. The second recipient flower receives $d_2 = \rho\pi R(1 - \rho)$ donor grains and further depletes the donor pollen pool to $\pi R(1 - \rho)^2$ grains, so that the third recipient receives $d_3 = \rho\pi R(1 - \rho)^2$ donor grains, etc. In general, recipient k receives

$$d_k = \rho\pi R(1 - \rho)^{k-1} \tag{1}$$

pollen grains from the donor of interest.

The pollen from a specific donor flower on the pollinator's body is not replenished, so that repeated deposition of pollen on stigmas causes a continual decline in both the donor pollen pool and the number of donor grains deposited on successive recipient flowers. Based on Equation (1), deposition of donor pollen on the kth recipient flower (d_k) increases linearly with the number of donor grains initially picked up by the pollinator (πR). In contrast, d_k increases and then decreases with increases in ρ, with the kth recipient flower receiving the most donor pollen when $\rho = 1/k$(i.e., $\partial d_k/\partial\rho = 0$ at this value of ρ). Because of these relations, few recipient flowers receive donor pollen when either the pollinator initially removes few donor grains, or a large proportion of the pollen on the pollinator's body is deposited on each stigma. The total number of donor grains reaching recipient stigmas is

$$\begin{aligned} D &= \sum_k \rho\pi R(1 - \rho)^{k-1} \\ &= \pi R \end{aligned} \tag{2}$$

so that this model proposes that all pollen carried away from the donor flower is successfully dispersed (see the appendix to this chapter for the general solution to the sum of a geometric series). Pollen can be lost from this system, but only during removal from the donor flower when the fraction $1 - \pi$ of the R grains displaced from the anthers is not carried away by the pollinator.

Two-Compartment Model: Pollinator Grooming

Unlike the process depicted by the single-compartment model, pollen dispersal by most animals probably involves pollen loss during transport. This loss can occur either because pollen is dislodged from the pollinator's body during grooming (e.g., bees and flies; see Holloway, 1976; Macior, 1967; Michener et al., 1978; Kimsey, 1984; Thomson, 1986; Harder, 1990*b*) or proboscis coiling (e.g., butterflies and moths; see Levin and Berube, 1972), or because the pollen on the pollinator's body is not mixed and so becomes buried under layers of pollen from more recently visited flowers (e.g., hummingbirds; see Price and Waser, 1982; Lertzman and Gass, 1983; Feinsinger and Busby, 1987). Although the mechanisms involved differ, these processes should generally produce multiple pools of pollen on the pollinator's body. To illustrate how multiple pollen pools affect pollen dispersal, we now develop a model of pollen transport by grooming pollinators following Harder and Wilson (unpublished). For a corresponding model of pollen layering, see Lertzman and Gass (1983) and Harder and Wilson (unpublished).

Grooming generally mixes pollen, as assumed by the single-compartment model; however, if some "exposed" sites are more susceptible to this behavior than other "safe" sites (e.g., see Kimsey, 1984), then the pollen on the pollinator's body will act as two (or more) linked pools (see Fig. 6.1b for a diagram of the modeled pollen dynamics). We assume that both the safe and exposed sites contact a flower's sexual organs during a pollinator visit. In particular, the pollinator removes a total of R pollen grains from each flower, of which the fraction π_s is added to the safe site and π_e is added to the exposed site (a third fraction, $1 - \pi_s - \pi_e$, falls from the pollinator and is lost). Simultaneously, the flower's stigma receives fractions ρ_s and ρ_e of the pollen from other flowers currently on the safe and exposed sites, respectively. As the pollinator flies between flowers, it grooms and displaces a fraction Γ of the pollen from the exposed site. This groomed pollen suffers one of two fates: a fraction γ_s is moved onto the safe site, whereas the remaining fraction $1 - \gamma_s$ is lost from pollination because it either falls from the pollinator or is groomed into pollen-carrying structures.

Now consider pollen from a specific donor flower. Immediately after visiting $k - 1$ recipient flowers, the pollinator carries s_{k-1} and e_{k-1} donor grains on its safe and exposed sites, respectively. As the pollinator flies to the next flower, grooming moves $\Gamma\gamma_s e_{k-1}$ donor grains from the exposed site to the safe site, leaving $(1 - \Gamma)e_{k-1}$ donor grains on the exposed site. Consequently, the kth recipient flower receives

$$d_k = \rho_s(s_{k-1} + \Gamma\gamma_s e_{k-1}) + \rho_e(1 - \Gamma)e_{k-1}$$

pollen grains from the donor flower. Incorporation of definitions for e_{k-1} and s_{k-1} (see Harder and Wilson, unpublished, for derivations) yields

$$d_k = R(\omega\chi^{k-1} + \theta\xi^{k-1}) \tag{3}$$

where

$$\omega \equiv \rho_s\left[\pi_s - \frac{\pi_e\Gamma\gamma_s(1-\rho_s)}{\rho_s - \rho_e - \Gamma(1-\rho_e)}\right]$$

$$\chi \equiv 1 - \rho_s$$

$$\theta \equiv \pi_e\left\{\rho_e(1-\Gamma)+\rho_s\Gamma\gamma_s\left[1 + \frac{1-\rho_s}{\rho_s - \rho_e - \Gamma(1-\rho_e)}\right]\right\}$$

$$\xi \equiv (1-\Gamma)(1-\rho_e)$$

Clearly, pollen dispersal on grooming pollinators is much more complicated than the simple decay process described by the single-compartment model [Equation (1)].

Like the single-compartment model, Equation (3) predicts that the number of pollen grains deposted from a specific donor flower declines with visits to successive recipient flowers. However, grooming quickly depletes donor pollen on the exposed site and therefore causes a faster decline in deposition of donor pollen on the first few recipients. In contrast, distant recipients receive relatively more donor pollen than if pollen were carried in a single compartment, because grooming shifts some exposed pollen to the safe site, so that more pollen passes through the safe site than was originally deposited there by the donor anthers.

Based on the grooming model for pollen dispersal [Equation (3)], the total number of donor grains reaching recipient stigmas is

$$\begin{aligned} D &= R\sum_k(\omega\chi^{k-1} + \theta\xi^{k-1}) \\ &= R\left(\frac{\omega}{1-\chi} + \frac{\theta}{1-\xi}\right) \\ &= R\left[\pi_s + \pi_e\frac{\rho_e(1-\Gamma) + \gamma_s\Gamma}{\rho_e(1-\Gamma) + \Gamma}\right] \end{aligned} \quad (4)$$

This model therefore proposes that, because of grooming, the fraction

$$\frac{\pi_e\Gamma(1-\gamma_s)}{(\pi_s + \pi_e)[\rho_e(1-\Gamma) + \Gamma]}$$

of the $(\pi_s + \pi_e)R$ pollen grains that the pollinator carries away from each flower never reaches stigmas.

The characteristics of total pollen dispersal for both the single-[Equation (2)] and two-compartment models [Equation (4)] were explicitly derived from the donor (male) perspective; however, they also represent the recipient (female) perspective of the total number of pollen grains deposited by each pollinator. This symmetry of male and female perspectives on pollen dispersal can be demonstrated for the single-compartment model by first identifying the equilibrium number of pollen grains carried by each pollinator. The number of grains carried by a pollinator just before arriving at flower x, L_x, is the sum of the number of pollen grains on the pollinator before visiting flower $x - 1$ that were not deposited on that flower, $(1 - \rho) L_{x-1}$, and the number of grains the pollinator removed and carried away from the preceding flower, πR, or

$$L_x = (1 - \rho) L_{x-1} + \pi R$$

At equilibrium, $L_x = L_{x-1} = L^*$, so that $L^* = \pi R/\rho$. When a pollinator visits a flower, it deposits the fraction ρ of the pollen it carries on the stigma. Consequently, each flower receives $\rho L^* = \pi R$ pollen grains, which equals the number of grains dispersed from each flower [see Equation (2)]. A similar equivalence between male and female perspectives can be demonstrated for the two-compartment model (Harder and Wilson, unpublished). This symmetry between dispersal and deposition should be a general feature (on average) of plants whose flowers function simultaneously as pollen donors and recipients.

The preceding models formalize pollen removal, transport, and deposition as the interaction of parameters that specify the probability of pollen moving from one location to another. Although these parameters describe processes, they also represent specific floral and pollinator characteristics. Both the one- and two-compartment models include parameters that summarize the interaction between the pollinator, the pollen presented in a flower (R and π and its variants), and the flower's stigma (ρ and its variants). These parameters generally depend on the size and placement of the respective sexual organ(s), the size and orientation of the pollinator, and the duration of the pollinator's visit, which in turn depends on reward availability. In addition, the grooming model includes pollinator-specific parameters that summarize the details of grooming behavior (Γ and γ_s).

An Example of the Influences of Floral Characters on Pollen Carryover

The preceding models of animal pollination propose that the pattern of pollen dispersal depends on stamen characteristics of the donor flower and pistil characteristics of recipient flowers. We now illustrate these influences by describing some results of a study of pollen dispersal for *Pontederia cordata,* a heterostylous species (Harder and Barrett, unpublished). Many heterostylous species are particularly well suited to studies of the influences of flower structure on pollen dispersal for two reasons. First, heterostyly involves two (distyly) or three morphs (tristyly) that differ in the placement of anthers and stigmas, so that this unusual reproductive system explicitly involves pronounced intraspecific variation in floral structure. Second, in many species the two or three different anther levels produce pollen of distinctly different sizes (Darwin, 1877; Ganders, 1979; Dulberger, 1992), so that pollen from one morph can be readily identified on the stigmas of another. This feature enables studies of pollen dispersal between sexual organs at the same level (legitimate pollination) without resorting to manipulative techniques (such as emasculation) that can alter pollen dispersal (see Price and Waser, 1982; Thomson, 1986; Morris et al., 1994; Harder and Wilson, unpublished).

The observations we present involve flowers of the mid-styled morph of *P. cordata,* which produce both short-level anthers (included within the perianth tube) and exserted, long-level anthers. Given this arrangement of anthers, the legitimate recipients for this morph are long-styled plants (for long-level pollen) and short-styled plants (for short-level pollen). Pollen produced by the two anther levels of mid-styled flowers differs significantly in size (long-level pollen, mean±SE diameter = 35.6±0.24 μm, n = 19; short-level pollen, 18.7±0.07 μm, n = 19) and can be readily distinguished from the mid-level pollen of other morphs (long-styled morph, 28.5±0.15 μm, n = 22; short-styled morph, 28.0±0.13 μm, n = 17; Harder and Barrett, 1993). No morph produces pollen at the same level as its stigma, so that pollen from a donor plant of one morph can be unequivocally identified on the stigma of a legitimate recipient if the pollinator does not visit other plants of the donor's morph before it visits the recipients. Measurement of pollen carryover therefore involved picking donor and recipient inflorescences during early morning before the single-day flowers opened; enclosing these inflorescences to preclude pollinator visits until the flowers opened and anthers dehisced; allowing a clean bumble bee (*Bombus vagans* Smith) to visit a donor inflorescence with 10 flowers followed by up to six recipient inflorescences of a specific legitimate morph; and counting the appropriate pollen on the stigmas of visited recipient flowers. Because bumble bees begin visiting *P. cordata* flowers before anthesis, we could collect clean, but experienced, bees as we picked inflorescences.

Several aspects of pollen carryover are immediately obvious from the results for mid-styled *P. cordata* flowers (Fig. 6.2). First, if a single-compartment model appropriately described pollen carryover for this species, then log(pollen deposition + 1) should decline linearly in Fig. 6.2. Instead, deposition declines relatively quickly for initial recipients and then more slowly for later recipients (also see Thomson, 1986; Morris et al., 1994). This pattern is more consistent with a two-compartment model of pollen dispersal [Equation (3)], which is not surprising given that the bees involved groomed frequently while flying between flowers (based on videotaped records of experimental trials; Harder and Barrett, unpublished). Indeed, nonlinear regression indicates that the general form of Equation (3) fits the data for both anther levels better than the single-compartment model (based on the absolute sizes of the error mean squares). The regression estimate of $\chi \equiv 1 - \rho_s$ for Equation (3) implies that approximately 97% of the pollen on safe sites remains there between flower visits (long-level anthers, 96.8%; short-level anthers, 97.4%). In contrast, the estimate of $\xi \equiv (1 - \Gamma)(1 - \rho_e)$ suggests that approximately 81% of the pollen on exposed sites is not displaced between visits (long-level anthers, 81.6%; short-level anthers, 80.5%).

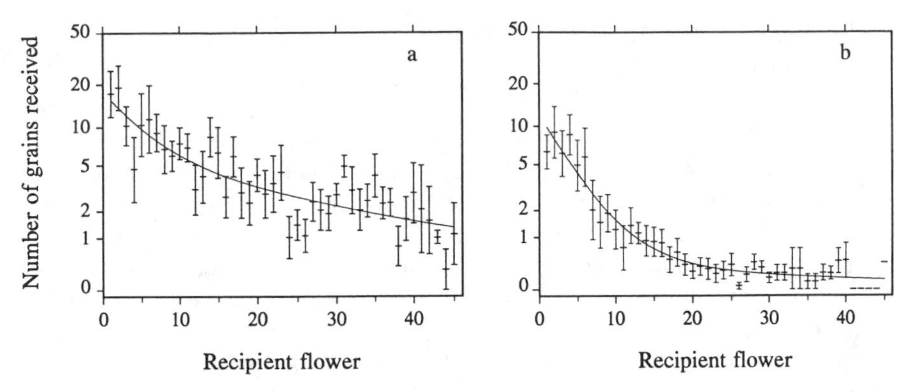

Figure 6.2. Patterns of legitimate pollen deposition (mean±SE) for pollen from (a) long- and (b) short-level anthers of mid-styled *Pontederia cordata* flowers. Recipient flowers are identified by the order in which they were visited after the bee left the donor inflorescence. The curves in each panel are the least-squares, nonlinear regression fits of Equation (3). Based on eight trials per anther level.

This apparent imbalance in residence probabilities for the two pollen pools illustrates the need for considering two-compartment models when describing pollen dispersal.

Second, the position of sexual organs in *P. cordata* flowers clearly affects legitimate pollen dispersal. Comparison of Figs. 6.2a and 6.2b indicates that more pollen reaches a given recipient flower from long-level anthers than short-level anthers. For example, deposition declines so rapidly for short-level pollen that flowers beyond the 13th recipient receive fewer than one pollen grain, on average (Fig. 6.2b), whereas more than 45 recipient flowers receive an average of at least one grain from long-level anthers (Fig. 6.2a). This more extensive dispersal of pollen from long-level anthers occurs despite a 4.4-fold advantage in pollen production for short-level anthers in mid-styled flowers. Such a pronounced difference in successful pollen dispersal suggests that the position of sexual organs primarily affects the susceptibility of pollen to grooming, especially the fraction of exposed pollen that is groomed into safe sites (γ_s).

Finally, Fig. 6.2 illustrates considerable variability around the generally declining trend in average pollen receipt. Such variation is a common feature of pollen dispersal (see Lertzman and Gass, 1983; Geber, 1985; Thomson, 1986; Feinsinger and Busby, 1987; Galen and Rotenberry, 1988; Waser, 1988; Wolfe and Barrett, 1989; Robertson, 1992; Stanton et al., 1992) and implies that animal pollination is highly stochastic. Although many aspects of the interaction between pollinators and flowers could produce such variation, stochastic versions of the grooming model [Equation (3)] indicate that most of the recipient-to-recipient variation probably arises during deposition of pollen on stigmas (i.e., variation in ρ_e and ρ_s; Harder and Wilson, unpublished).

Pollen Dispersal by Multiflowered Inflorescences and Implications for Mating

As outlined above (see "General Influences of Daily Inflorescence Size on Pollination and Mating), the simultaneous display of more than one flower bears diverse implications for pollen dispersal from the plant's (rather than the flower's) perspective and for mating patterns. In contrast to its role in pollinator attraction, the effects of daily inflorescence size on pollen dispersal and mating have been little studied. To illustrate these effects, we now examine the consequences of pollen carryover for selfing and outcrossing from both theoretical and empirical perspectives. Based on these results, we then construct a cost-benefit model that identifies how the somewhat antagonistic roles of floral display in pollinator attraction and pollen dispersal can be balanced to maximize a plant's fertility.

Expected Frequency of Selfing and Outcrossing

To appreciate qualitatively the mating implications of displaying more than one flower, consider an example (Fig. 6.3) in which the pollinator visits V=4 flowers per plant, carries away $D = \pi R = 200$ grains per flower, and deposits 10% of the pollen it carries on each flower ($\rho = 0.1$) according to the single-compartment model [Equation (1)]. The four bold lines in Fig. 6.3 depict pollen dispersal from the flowers visited on plant 0, whereas the remaining diagonal lines portray pollen received by plant 0 (and 3) and from other plants (four curves per donor). Consideration of pollen receipt by plant 0 (i.e., between the dotted vertical lines) indicates several aspects of mating. First, movement of the pollinator between flowers on the donor plant results in geitonogamous deposition of pollen from the first three flowers visited. Second, each plant receives pollen from seven outcross donors. Third, these donors make disparate pollen contributions to plant 0, with most recently visited plants donating more pollen than more distant plants. This unequal pollen donation is further clarified by considering pollen export from plant 0 to plant 3 (i.e., between the dashed-dotted vertical lines). Fourth, each donor plant contributes many pollen grains to each of several recipients, producing a correlation in the identity of male gametes received by individual stigmas and by different flowers on a plant. Finally, a donor's perspective of pollen dispersal (e.g., following the bold curves) is equivalent to the recipient's perspective (e.g., considering the four flowers on a given plant), in the number of mates and relative intensity of mating with those mates.

To formalize these influences of daily inflorescence size on pollen dispersal and mating parameters, we now set the single-compartment model for individual flowers [Equation (1)] into the whole-plant context. Although the model represents the processes involved in pollen dispersal less completely than multicompartment models, it serves our purpose of illustrating the general pattern of pollen dispersal among plants. The models that follow incorporate several assumptions.

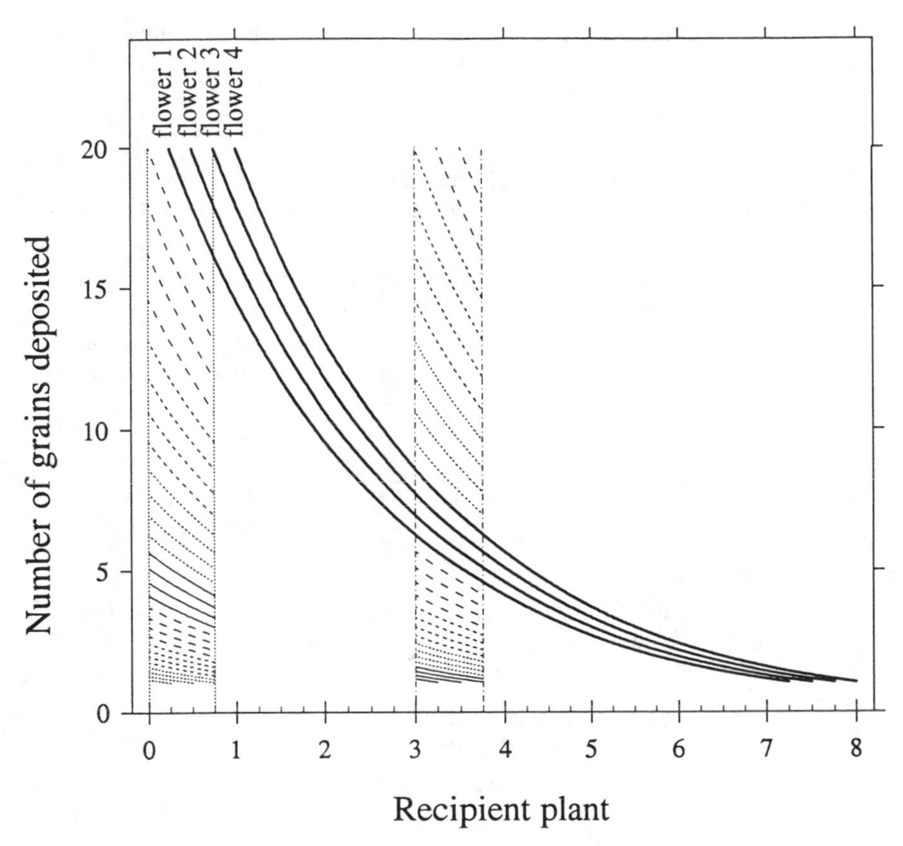

Figure 6.3. Pollen export and import predicted by a single-compartment model of pollen dispersal [Equations (5) and (6)] in which the pollinator carries away 200 pollen grains from each of the four flowers visited per plant and deposits 10% of the pollen it carries on each flower. Large ticks on the abscissa indicate the first flower visited on each plant; small ticks indicate the remaining flowers visited. The bold lines illustrate pollen dispersal from the four flowers visited on plant 0. Vertical blocks depict pollen receipt from all donors by plants 0 and 3. Within these blocks, adjacent lines of the same type represent pollen from flowers on a particular donor plant.

As for Equation (1), we assume that pollen on the pollinator occupies a single, completely mixed pool, each flower contributes $D = \pi R$ grains to this pool, and the stigma of each flower removes a fixed proportion of the pool (ρ). We additionally assume that the pollinator visits V flowers on each plant (without revisits) and pollen transfer between flowers on the same plant does not differ from that between flowers on different plants. In fact, geitonogamous transfer may be more likely than transfer between plants because bees groom more often while flying between plants (Harder, 1990*b*). For heuristic purposes, it is easier to develop this model from the perspective of the recipient plant, rather than the donor plant.

Self-Pollination

Two modes of self-pollination result from the actions of pollinators: movement of pollen between the anthers and stigma of the same flower (facilitated intrafloral selfing) and pollen movement between flowers of the same plant (geitonogamous selfing; Lloyd and Schoen, 1992). In the most general case, each of the V flowers visited on a plant receives I grains through facilitated intrafloral selfing and $V-1$ flowers also receive geitonogamously transferred self-pollen (e.g., see plant 0 in Fig. 6.3). To quantify geitonogamous selfing, consider the jth flower visited on the recipient of interest. Based on Equation (1), flower j receives

$$G_j = \sum_{i=1}^{j-1} \rho D(1-\rho)^{i-1} = D[1-(1-\rho)^{j-1}]$$

geitonagamous pollen grains (for $j=1$, $G_j=0$). For all flowers visited on the plant of interest, the cumulative number of self-grains received is

$$S = \sum_{j=1}^{V} (I + G_j) = V(I+D) - \frac{D[1-(1-\rho)^V]}{\rho} \quad (5)$$

The last term in Equation (5) represents the cumulative number of pollen grains removed from flowers on this plant that are dispersed to other plants. According to Equation (5), self-pollination for the entire plant increases in an accelerating manner with the number of flowers visited by each pollinator.

Cross-Pollination

Now consider the number of outcrossed pollen grains received by the recipient plant from the kth donor plant visited before the recipient. Begin by focusing on the jth flower visited on the recipient and the ith-last flower visited on the donor (e.g., consider plant 0 in Fig. 6.3 as the donor and plant 3 as the recipient). If the pollinator visits V flowers on each plant, then $(k-1)V+i+j-1$ visits separate these two flowers. Based on Equation (1), flower i donates a total of

$$\sum_{j=1}^{V} \rho D(1-\rho)^{(k-1)V+i+j-2} = D[1-(1-\rho)^V](1-\rho)^{(k-1)V+i-1}$$

pollen grains to the recipient plant. Summing over all flowers on the donor plant indicate that

$$O_k = \frac{D[1 - (1 - \rho)^V]^2(1 - \rho)^{(k-1)V}}{\rho} \tag{6}$$

grains from donor k reach the recipient plant. This expression for outcrossed pollen dispersal is the product of three terms: the number of donor grains that reach all other plants via a single pollinator, $\sum_k O_k = D[1 - (1 - \rho)^V]/\rho$; the fraction of donor pollen remaining on the pollinator just before it visits the recipient, $(1 - \rho)^{(k-1)V}$; and the fraction of pollen on the pollinator's body that is deposited on each inflorescence, $1 - (1 - \rho)^V$. Equation (6) is merely the original single-compartment model expressed in terms of recipient plants, rather than flowers [note that Equation (6) equals Equation (1) when $V = 1$)]. Hence, the incorporation of more than one flower visit per plant does not alter our earlier conclusions for that model concerning the dependence of pollen dispersal on the number of grains carried from the donor by the pollinator ($D = \pi R$) or the deposition fraction ρ. As the complement to self-pollination, outcrossing by the entire plant increases in a decelerating manner with the number of flowers visited per plant.

Mating Parameters

Mating integrates both pollination and postpollination processes. Typically, studies of mating do not distinguish between these processes, but rather assess their cumulative effects, which are summarized by parameters such as s (fraction of selfed seeds), t (fraction of outcrossed seeds), r_s and r_p (correlated selfing and outcrossing, respectively). In this chapter, we are primarily interested in the pollination component of mating (for reviews of postpollination influences, see Lyons et al., 1989; Marshall and Folsom, 1991). Therefore, to avoid confusion with the standard symbols, we define the pollination components of these mating parameters by unique symbols (identified below).

Number of Mates

The number of plants that receive at least one pollen grain from a particular donor plant on a single pollinator, m, is found by setting Equation (6) equal to 1 and solving for $k{=}m$, which yields

$$m = 1 + \frac{\log \rho - \log D - 2 \log[1 - (1 - \rho)^V]}{V \log(1 - \rho)} \tag{7}$$

Figure 6.4 illustrates that the number of mates decreases with increases in the deposition fraction, ρ, and the number of flowers visited per plant. In addition,

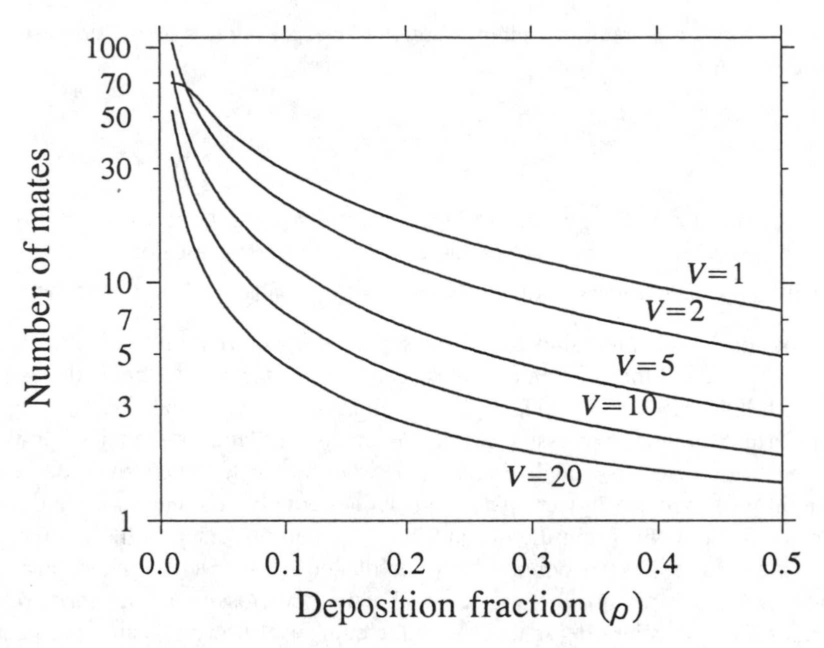

Figure 6.4. Relation of the number of plants contributing pollen to each stigma as a result of pollen dispersal by a single pollinator to the number of flowers visited per plant (V) and the fraction of pollen on the pollinator's body deposited on each flower (ρ). Based on Equation (7), with D=200 grains.

the number of mates increases at a decelerating rate with increases in the number of grains exported, D.

Pollination Component of Selfing

Three aspects of self-pollination warrant attention. First, for the jth flower visited on a plant, the fraction of pollen grains that arrive via self-pollination is

$$\psi_j = \frac{I + G_j}{I + D}$$

$$= 1 - \frac{D(1 - \rho)^{j-1}}{I + D} \qquad (8)$$

As Barrett et al. (1994) demonstrated, this fraction increases asymptotically with each successive flower visited. Second, the selfing fraction for the entire plant is

$$\Psi = \frac{S}{V(I + D)}$$
$$= 1 - \frac{D[1 - (1 - \rho)^V]}{\rho V(I + D)} \qquad (9)$$

which is also the average fraction of self-grains received by the V flowers visited (Barrett et al., 1994, referred to this average as $\overline{\psi}$). According to Equation (9), aspects of pollen dispersal that promote extensive carryover between plants [small ρ, small V [see Fig. 6.5b] or large D) reduce the fraction of self-pollen received by stigmas. In contrast, Ψ is positively related to the intensity of intrafloral selfing I. Third, any self-deposited pollen that would otherwise have been dispersed to other plants represents lost outcrossed pollen (pollen discounting; Holsinger et al., 1984). Equation (9) illustrates that pollen discounting can involve two components: one geitonogamous, the other intrafloral. All geitonogamously deposited pollen is discounted, because it is included in the pool of pollen on the pollinator's body that is destined to be deposited on other flowers (Lloyd, 1992). In contrast, intraflorally deposited pollen is discounted only if it can be considered lost to outcrossing.

Correlated Mating

Ritland (1989) defined correlated mating as the proportion of all possible pairs of seeds that are full sibs. To quantify correlated mating from the perspective of the pollen received by a plant, one simply considers the fraction of all possible pairs of received grains for which both grains are derived from the same donor. If the plant receives N grains, then there are

$$\binom{N}{2} = \frac{N(N-1)}{2}$$

possible pairs of grains. Similarly, if plant k donated n_k of these grains, then $n_k(n_k - 1)/2$ of the pairs of grains will have originated solely from this donor. As a result, the proportion of con-paternal pollen pairs donated by plant k is

$$\frac{n_k(n_k - 1)/2}{N(N-1)/2} = \frac{n_k(n_k - 1)}{N(N-1)}$$

If plant k is also the recipient plant, then this equation estimates correlated self-pollination. Correlated cross-pollination from all m donors that contributed pollen is

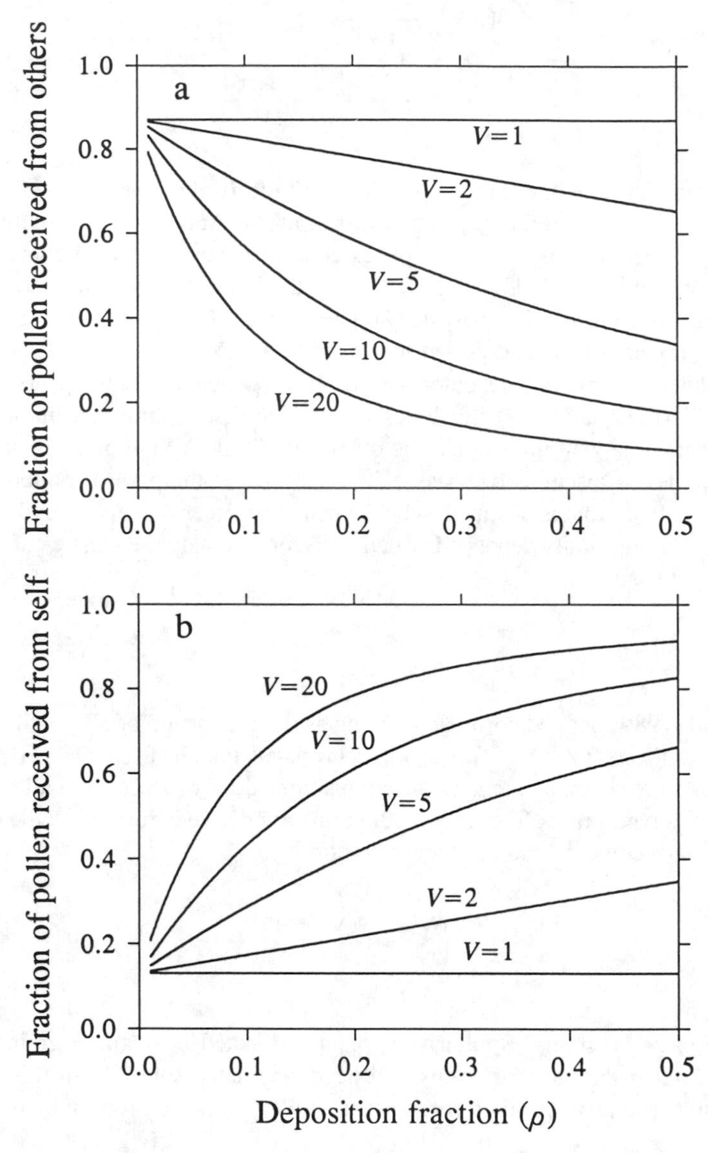

Figure 6.5. Relation of (a) the fraction of outcrossed pollen received and (b) the fraction of self-pollen, ψ, received to the number of flowers visited per plant (V) and the fraction of pollen on the pollinator's body deposited on each flower (ρ). Based on Equation (9), with $D=200$ grains and $I=30$ grains.

$$\frac{\sum_{k=1}^{m} n_k(n_k - 1)}{N(N - 1)}$$

We can relate correlated self- and cross-pollination to pollen dispersal by recognizing that $N = V(I + D)$, $n_k = S$ [see Equation (5)] for self-pollination, and $n_k = O_k$ [see Equation (6)] for cross-pollination. Hence, correlated self-pollination is

$$\Phi_s = \frac{S(S - 1)}{V(I + D)[V(I + D) - 1]} \tag{10}$$

Because Equation (10) approaches Ψ^2 as the number of self grains increases, correlated selfing is influenced by pollen dispersal in much the same manner as the selfing fraction, Ψ [see Equation (9)], being negatively related to parameters that extend carryover and positively related to the intensity of intrafloral selfing (see Fig. 6.5b).

Correlated cross-pollination is

$$\Phi_o = \frac{\sum_{k=1}^{m} O_k(O_k - 1)}{V(I + D)[V(I + D) - 1]} \tag{11}$$

which differs from correlated selfing in that it does not depend on dispersal parameters in a simple monotonic fashion (Fig. 6.6). When pollinators visit few flowers per plant, decreases in carryover (large ρ or small D) reduce the number of mates (see Fig. 6.4) and increase the proportion of pollen contributed to those mates that are involved. On the other hand, when pollinators visit more than a few flowers per plant, correlated outcrossing is maximal at intermediate values of the deposition fraction ρ. This peak in correlated outcrossing occurs because low and high values of ρ lead to relatively equitable contributions of pollen per mate, although for very different reasons. Low values of ρ produce relatively flat carryover curves, resulting in many mates each receiving similar contributions of pollen per mate. At high values of ρ, increased pollen discounting resulting from geitonogamous self-pollination removes most donor pollen from the pollinator before it can move to an unrelated plant (see Fig. 6.5b), so that outcrossing entails the relatively flat tail of the carryover curve, which involves little pollen per donor.

Ritland (1989) derived a simple relation between correlated outcrossing and pollen carryover [$\Phi_O = \rho/(2 - \rho)$ in terms of the parameter definitions of this chapter], which he proposed could be applied to estimate ρ once correlated outcrossing had been estimated from electrophoretic data. This relation is a

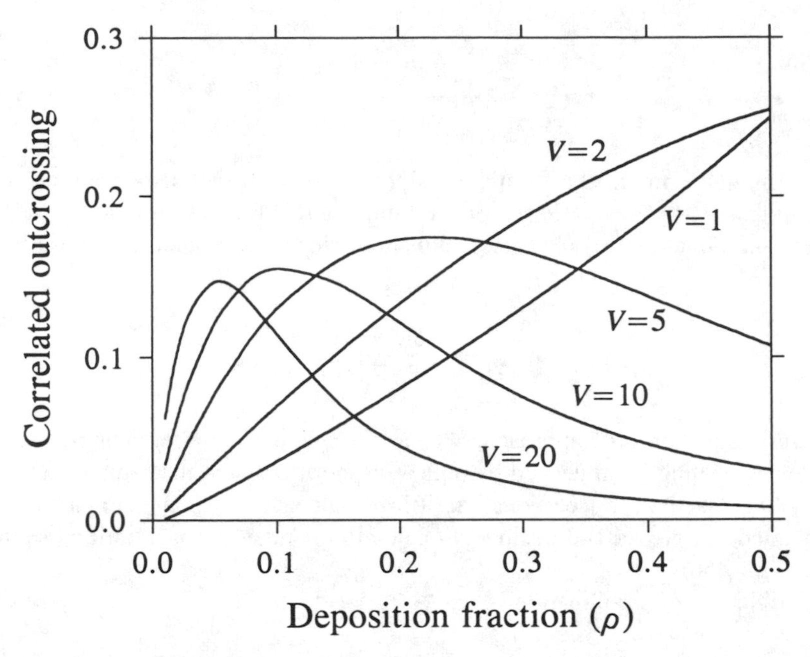

Figure 6.6. Relation of correlated outcrossing to the number of flowers visited per plant (V) and the fraction of pollen on the pollinator's body deposited on each flower (ρ). Based on Equations (5), (6), (10), and (11), with D=200 grains and I=30 grains.

special case of Equation (11), with $V = 1$, $I = 0$ and large D and m. However, when pollinators visit more than one flower per plant and/or their visits result in intrafloral selfing, the relation of correlated outcrossing and pollen carryover is more complex than Ritland suggested (see Fig. 6.6). Unfortunately, such complexity implies that estimating ρ requires more information than a measure of correlated outcrossing, even if a single-compartment model [Equation (6)] adequately describes pollen dispersal.

Applicability of Model

Even if individual pollinators dispersed pollen according to the single-compartment model, the patterns of mate diversity for flowers, inflorescences, and plants will typically differ from those described in the preceding sections. Repeated pollinator visits complicate mating patterns for individual flowers whenever pollinators follow different flight paths and consequently arrive with pollen from different donors and carry pollen to different recipients. Furthermore, a plant's flowers experience distinct visit chronologies if each pollinator does not visit the same subset of flowers. Because successive pollinator visits to a flower remove different amounts of pollen (Galen and Stanton, 1989; Harder, 1990*a*; Young and Stanton, 1990; Klinkhamer et al., 1991; Harder and Wilson, 1994; LeBuhn and Anderson, 1994), flowers with dissimilar histories will contribute different amounts of pollen to the pool on the pollinator's body, so that the relative amount of pollen received from different flowers (and plants) will deviate from the scenario outlined in Fig. 6.3. The extent of such deviation will depend on the schedule and pattern of anthesis within and between inflorescences, the schedules of pollen removal and receipt, inflorescence architecture, and pollinator behavior. Repeated visits will generally disrupt the symmetry of male and female perspectives on pollen dispersal associated with dispersal by individual pollinators. The symmetry of mating (as opposed to pollination) will be further altered by postpollination processes that favor particular pollen donors in determining seed paternity. In spite of these concerns, the following empirical results illustrate that the predicted roles of pollen dispersal in mating are quite robust (also see Barrett et al., 1994).

Observations of the Effects of Daily Inflorescence Size on Pollination and Mating

To examine the diverse influences of daily inflorescence size on pollination and mating, we conducted an experiment involving artificial arrays of *Eichhornia paniculata* (Spreng.) Solms-Laubach (Pontederiaceae). This fully self-compatible, tristylous species produces up to 20 single-day flowers per inflorescence per day. Although the inflorescence is structurally a panicle, it functions as a raceme as only one flower opens per panicle branch on a given day. All plants used in this experiment were long-styled, so that the results are not complicated by heterostyly. The arrays included 35 or 36 plants (depending on the treatments involved) arranged in a 6 × 6 grid with approximately 30-cm spacings. Experimental plants produced flowers on a single inflorescence and excess flowers were removed from inflorescences to produce 3-, 6-, 9-, or 12-flowered plants. During an individual trial, an array included either all plants of one inflorescence size (pure arrays), or plants of two sizes (pairwise arrays) with the number of plants per treatment adjusted so that both treatments involved the same total number of flowers within the array (i.e., equivalent pollen and ovule production per treatment). The experiment included all four pure arrays and all six possible pairwise arrays, with four replicates per combination.* For several weeks prior to the experiment, we placed *E. paniculata* plants at the array locations so that the resident bees were familiar with this species.

During each experimental trial, we recorded pollinator activity and sampled fruit. An array was set up in the morning before anthesis. Once flowers opened, we observed pollinator behavior for 15 min during each of the first 3 h of

*Replication involved arrays in two locations that were sampled on each of 2 days. Details of the effects of location will be presented elsewhere.

flowering. These observations included counts of the entries and exits to the array by bees [primarily *Bombus vagans* and *B. fervidus* (Fabricius)] and records of the sequence of inflorescence visits (including position in the array and number of flowers visited) by focal bees. Once flowers wilted in late afternoon, one flower from each of the top, middle, and bottom thirds of the inflorescence was marked with paint. Capsules produced by marked flowers were collected once the seeds had ripened (11–12 days) and stored in separate envelopes until the seeds could be counted and assayed electrophoretically.

The plants used in this experiment contained known electrophoretic markers so that we could both count the seeds sired by plants of each inflorescence size and estimate the fraction of selfed seeds and correlated mating. Plants in all arrays were polymorphic for *PGI-2*. In pairwise arrays, plants with one inflorescence size were homozygous for the fast *AAT-3* allele and plants in the competing treatment were homozygous for the slow *AAT-3* allele. Pure arrays included equal numbers of plants with one of these homozygous *AAT-3* genotypes.

Pollinator Preferences

Assessment of pollinators' preferences for particular inflorescence sizes in our array experiment is complicated because the protocol for constructing pairwise arrays resulted in plants with larger inflorescences being less abundant than those with smaller inflorescences. To accommodate this inequality in availability, we calculated the following preference index:

$$\text{Preference} = \ln\left(\frac{\text{odds that the bee visited a large inflorescence}}{\text{odds that any inflorescence was large}}\right)$$

for each focal bee that we observed visit more than 10 inflorescences.† If a bee visits large inflorescences in proportion to their relative abundance, this index equals 0, whereas preference for large inflorescences results in a positive index and preference for small inflorescences results in a negative index.

Based on this index, bees' preferences varied with the ratio of the two inflorescence sizes in an array (Fig. 6.7). For example, the mean preference for 6-flowered plants in arrays with 3- and 6-flowered plants is virtually identical to the mean preference for 12-flowered plants in arrays with 6- and 12-flowered plants (Fig. 6.7, ratio of inflorescence sizes = 2). Trend analysis (Kirk, 1982) of these results indicates that increases in the ratio of inflorescence sizes increased bees' preference for large inflorescences (linear contrast, $F_{1,21} = 31.00$, $P < 0.001$), but reduced the incremental advantage of larger inflorescences (quadratic component of linear+quadratic contrast, $F_{1,21} = 9.04$, $P < 0.01$). These

---

†The odds of an event is the ratio of the probability of the event occurring to the probability of the event not happening.

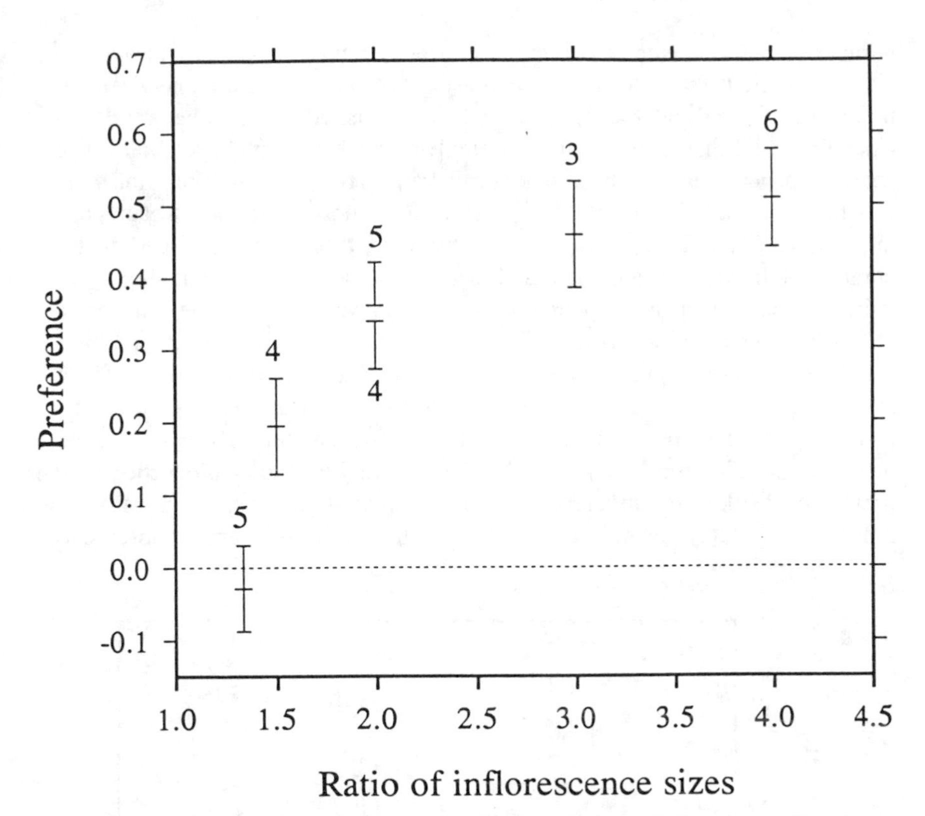

Figure 6.7. Mean (±SE) preference by bumble bees for larger inflorescences in pairwise arrays of *Eichhornia paniculata* as affected by the relative sizes of available inflorescences. See text for a definition of preference. The dotted horizontal line indicates indifference. Numbers adjacent to error bars indicate the numbers of bees sampled. For two treatment combinations, one inflorescence size included twice as many flowers as the competing size. Therefore to avoid confusion, the results for three-flowered vs. six-flowered plants do not include a lower error bar, whereas the results for six-flowered vs. 12-flowered plants lack an upper error bar. (After Harder and Barrett 1995).

results imply that flower number per se does not affect a bee's preferences for particular inflorescences; instead, the attractive advantages of producing more flowers arise in competitive situations involving a variation in daily inflorescence size. Indeed, the frequency of bee entries to pure arrays did not vary significantly with inflorescence size (Barrett et al., 1994).

Number of Flowers Visited per Inflorescence

Our assessment of whether the number of flowers visited per inflorescence increases in proportion to the number of open flowers considered the median number of flowers visited by each bee. Earlier analysis of the pure arrays alone (Barrett et al., 1994) indicated an increasing decelerating relation between daily inflorescence size and the number of flowers visited per pollinator (also see Chapter 7). Further analysis of the complete experiment indicates that a bee's behavior depends on both the number of flowers on the inflorescence being visited and the sizes of other available inflorescences (flower number × competitor interaction, $F_{9,137}=2.73$, $P<0.01$). In particular, these results indicate that the number of flowers visited increased with inflorescence size, but the greatest increase occurred when competing inflorescences were considerably smaller than the inflorescence being visited (Fig. 6.8). Such context-dependent pollinator behavior greatly complicates the selective influences on pollination, especially as they relate to daily inflorescence size. Regardless of the pollination environment, the increase in the number of flowers visited did not keep pace with increases in inflorescence size, so that bees visited a smaller proportion of the flowers on 12-flowered inflorescences (1/3–1/2) than they visited on 3-flowered inflorescences (2/3). Overall, the increased attractiveness of large inflorescences

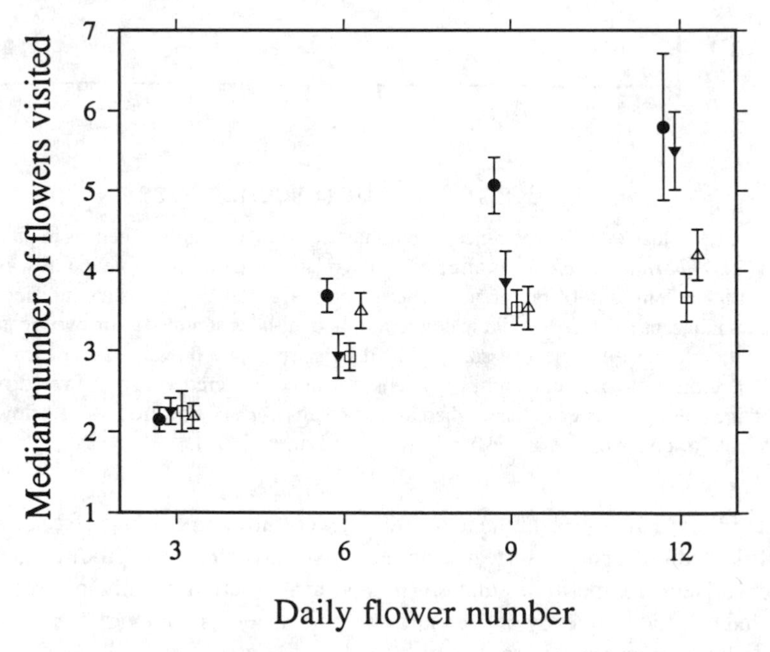

Figure 6.8 Mean (±SE) median number of flowers visited per inflorescence by *Bombus fervidus* workers in experimental arrays of *Eichhornia paniculata* in relation to inflorescence size and the size of competing inflorescences (solid circles, 3-flowered competitors; solid downward-pointing triangles, 6-flowered competitors; open squares, 9-flowered competitors; open upward-pointing triangles, 12-flowered competitors). Based on 3–12 bees per treatment combination. (After Harder and Barrett 1995).

counterbalanced the declining proportion of flowers visited per inflorescence per pollinator, so that the cumulative number of visits per flower did not differ significantly among inflorescence sizes ($F_{5,21} = 2.40$, $P > 0.05$, based on an analysis similar to that presented in the "Pollinator preferences" Section).

Seed Production

Overall, seed production in this experiment was pollen-limited, as the average flower produced 56.5 (SE = 1.38, $n = 184$) seeds, whereas a sample of flowers in the pure arrays that received supplemental pollination produced an average of 90 seeds (see Barrett et al., 1994). This pollen limitation probably resulted because the arrays involved only long-styled flowers of a tristylous species, so that anthers and stigmas contacted very different positions on pollinators' bodies. In spite of the preferences by bees for larger inflorescences in competitive situations, flower number per inflorescence did not significantly affect seed set ($P > 0.75$). Consequently, differences in relative fertility (see below) did not arise from a variation in seed production. Furthermore, because seed production was not resource-limited, the mating events recorded by seed genotypes should accurately reflect the pollination component of mating, especially given that for *E. paniculata,* self and intramorph pollen have equivalent pollen-tube growth and siring ability (Cruzan and Barrett, 1993) and this species seldom aborts fertilized seeds (Morgan and Barrett, 1939; Toppings, 1989).

Frequency of Self-Fertilization for Entire Inflorescences

Equation (9) proposes that the cumulative incidence of selfing for all flowers on an inflorescence increases with daily inflorescence size. To test this prediction, we determined the electrophoretic genotypes of three seeds from three flowers for every inflorescence in all arrays. Based on these data, we estimated the total fraction of selfed seeds, s, and the associated standard error (based on 100 bootstrap samples) for each inflorescence size in an array with Ritland's (1990) multilocus selfing rate program. Influences on s were assessed with ANOVA, in which individual estimates of s were weighted by the inverse of their squared standard error.

As predicted, the fraction of selfed seeds increased with daily inflorescence size (Fig. 6.9a; $F_{3,16} = 3.30$, $P < 0.05$). Trend analysis of this effect indicated that almost all the variation in s associated with inflorescence size could be explained by a linear relation ($F_{1,16} = 9.47$, $P < 0.01$). In contrast, Equation (9) describes a decelerating relation; however, consideration of only four inflorescence classes allows little power for assessing this more complex hypothesis. The effects of inflorescence size were not influenced by the size of competing inflorescences in an array ($P > 0.75$ for both the main competition effect and its interaction with inflorescence size of the focal treatment).

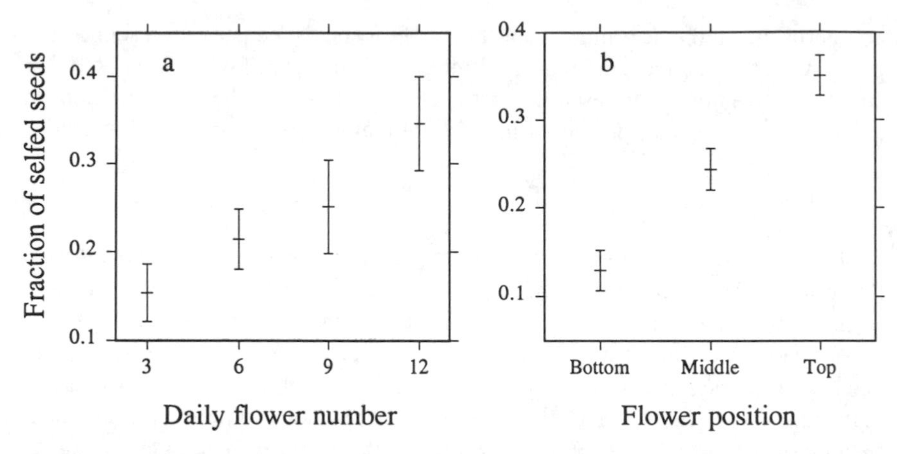

Figure 6.9. Mean (±SE) proportion of selfed seeds per inflorescence-size treatment in relation to (a) daily inflorescence size and (b) flower position within the inflorescence for *Eichhornia paniculata* in experimental arrays. Each mean is based on 16 replicates. (After Harder and Barrett 1995).

Frequency of Self-Fertilization and Flower Position

The preceding analysis amalgamates the fraction of selfed seeds for all flowers within an inflorescence; however, Equation (8) proposes that the first flowers visited by a pollinator on an inflorescence should experience less selfing than those visited later in the same bout. The vertical structure of *E. paniculata* inflorescences and the tendency of the bees involved to visit low flowers first and then move upward (Barrett et al., 1994; also see Manning, 1956; Percival and Morgan, 1965; Waddington and Heinrich, 1979; Corbet et al., 1981; Best and Bierzychudeck, 1982) implies that the fraction of selfed seeds should increase from bottom to top flowers. To test this prediction, the three flowers involved in the estimates of selfing described in the preceding section were collected from the bottom, middle, and top of the inflorescence, respectively. We could therefore estimate the fraction of selfed seeds with respect to flower position in the ANOVA.

Position within an inflorescence strongly affected a flower's fraction of selfed seeds (Fig. 6.9b; $F_{2,32} = 23.75$, $P < 0.001$). This relation did not vary among inflorescence sizes (position×size interaction, $F_{6,32} = 0.99$, $P > 0.25$), perhaps because the many bees visiting an inflorescence neither start on the same flower nor visit flowers in the same sequence. In addition, position effects did not vary with the size of competing inflorescences in an array ($P > 0.25$ for all interactions involving position and the size of competing inflorescences).

Correlated Outcrossing

Probably the least intuitive result of our theoretical analysis of the effects of pollen dispersal on mating parameters involved correlated outcrossing. In particular, given low to moderate pollen carryover, the fraction of seeds that are full sibs via outcrossing increases, peaks, and then decreases with increasing inflorescence size (Fig. 6.6). We tested this prediction with estimates of correlated outcrossing (Ritland, 1990) for the pure arrays. A weighted ANOVA of these data indicates significantly higher correlated outcrossing for 6-flowered plants than for 3-, 9-, or 12-flowered plants (Fig. 6.10); $F_{3,20} = 5.11$, $P < 0.01$). The maximal correlated outcrossing at an intermediate inflorescence size is consistent with Equation (11); however, the similarity of correlated outcrossing for 9- and 12-flower plants is not expected. The remarkable consistency in estimates of correlated outcrossing for 3-, 9-, or 12-flowered plants suggests the occurrence of a lower limit.

Outcrossed Siring Success

Daily inflorescence size could affect the number of seeds that a plant sires by two complementary mechanisms. First, pollen discounting resulting from geitonogamous pollination reduces the amount of pollen available for outcrossing by large inflorescences [see Equation (9)]. Second, the increased pollen removal resulting from a bee visiting more flowers per inflorescence could stimulate more thorough grooming by the bee, thereby increasing the amount of pollen lost during transport (see Harder and Thomson, 1989; Harder, 1990*b*). As a consequence of

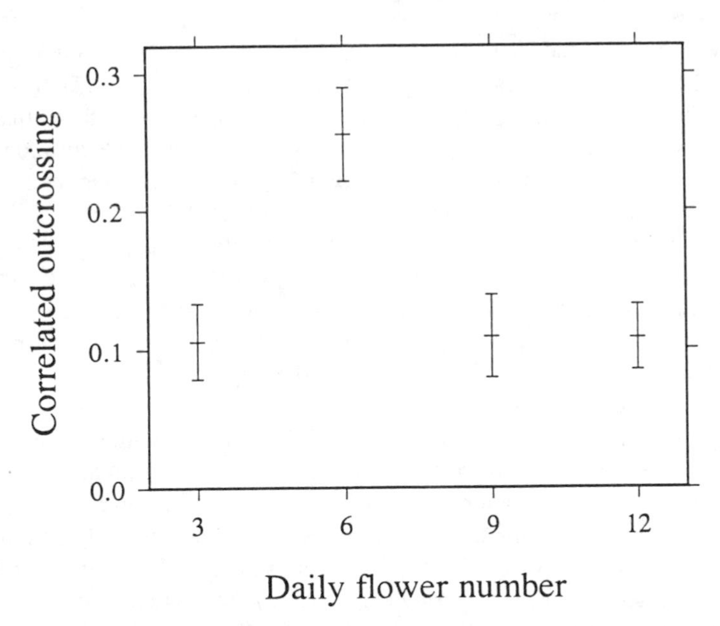

Figure 6.10. Mean (±SE) proportion of pairs of seeds that are full sibs via outcrossing in relation to daily inflorescence size for *Eichhornia paniculata* in pure arrays. Each mean is based on 16 replicates.

these mechanisms, large inflorescences should sire a smaller proportion of the outcrossed seeds produced by an array.

Estimation of the proportion of all outcrossed seeds sired by inflorescences of a particular size in pairwise arrays required identification of outcrossed seeds and their paternity. Based on our previous estimates of the fraction of selfed seeds per inflorescence size s (see the "Frequency of Self-Fertilization for Entire Inflorescences" section above and Harder and Barrett, 1995), we estimated the number of outcrossed seeds on inflorescences of size i, $(1-s_i)T_i$, where T_i is the total number of seeds produced by those inflorescences. This *maternal* perspective on outcrossing differs from the *paternal* perspective in that it includes seeds produced by treatment i inflorescences that were sired by treatment j plants, B_i, and excludes seeds sired by treatment i plants on treatment j plants, B_j. These seeds sired by pollen transfer *between* the inflorescence-size treatments (B_i and B_j) could be identified unequivocally because they were heterozygous for *AAT-3*, whereas all parental plants were homozygous at this locus. Given estimates of the number of outcrossed seeds on maternal plants and the numbers of seeds sired by between-treatment pollination, the fraction of outcrossed seeds sired by treatment i plants (outcrossed siring success) is

$$o_i = \frac{(1 - s_i)T_i - B_i + B_j}{(1 - s_i)T_i + (1 - s_j)T_j}$$

All else being equal, the competing inflorescence-size treatments in pairwise arrays should sire the same numbers of seeds, so that $o_i = 0.5$. In contrast, small inflorescences sired a significantly greater proportion of seeds than large inflorescences (paired t-test, $t_{23} = 2.74$, $P < 0.02$; mean ± SE difference in $o_i = 0.077 \pm 0.028$). Furthermore, outcrossed siring success declined with the frequency of self-fertilization (Fig. 6.11; $\hat{o}_i = 0.573 – 0.254 s_i$; $F_{1,22} = 18.43$, $P < 0.001$, $R^2 = 0.239$). These results indicate that the increased geitonogamous selfing that accompanies large inflorescences (Fig. 6.9a) causes significant pollen discounting and thereby reduces siring success through outcrossing (also see Harder and Barrett, 1995). Whether changes in pollen loss associated with pollinator grooming additionally affected outcrossed siring success is less clear, as we did not measure grooming losses.

In this experiment, we used unequal numbers of plants with competing inflorescence sizes so that both inflorescence treatments in a pairwise array were represented by the same total number of flowers. As a result, each size presented equivalent numbers of gametes, but they were deployed differently on individual plants. Such a situation would occur if all plants produced the same numbers of flowers during their entire flowering period (i.e., equal total inflorescence size), but they differed in their daily inflorescence sizes and consequently flowered for different periods. Because few-flowered plants sired more seeds than many-flowered plants, our experimental results imply that individual plants would

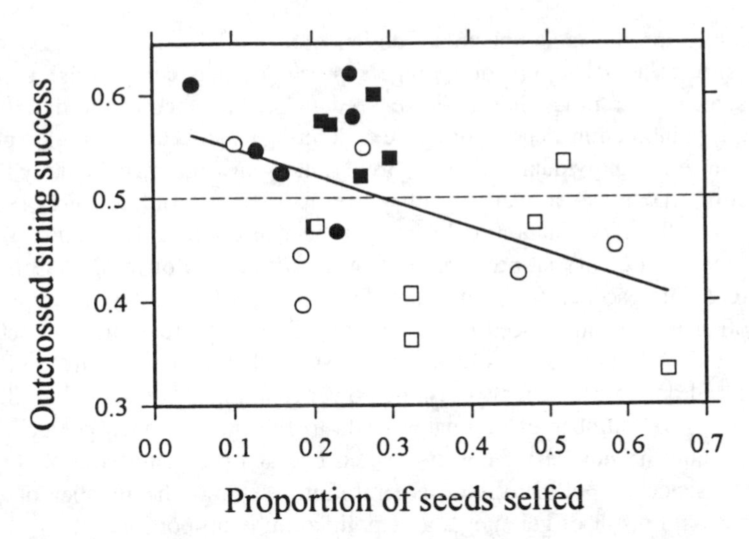

Figure 6.11. Effect of the frequency of self-fertilization, s_i, on relative outcrossed siring success, o_i, in pairwise arrays. The solid line illustrates the regression between o_i and s_i, whereas the dashed line indicates equivalent outcrossed siring success for competing inflorescence-size treatments. This analysis excluded pure arrays because outcrossing is the exact complement of selfing when only one treatment is involved. We also included data for only one inflorescence treatment per array (randomly selected) because the outcrossed siring success of one treatment is the complement of that for the competing treatment in an array. Symbols indicate inflorescence sizes: solid circles, 3 flowers; solid squares, 6 flowers; open circles, 9 flowers; open squares, 12 flowers. (After Harder and Barrett 1995).

maximize their male fertility by presenting few flowers each day (i.e., small daily inflorescence size) and flowering for a protracted period. This conclusion applies specifically to plants with the (unknown) carryover characteristics at play during this experiment: other characteristics can promote more liberal floral display (see the following section).

A Cost-Benefit Analysis of Daily Flower Production

During the course of this chapter, we have outlined the variety of influences of daily inflorescence size on pollination and mating. The benefits of simultaneously displaying many flowers primarily relate to enhanced attraction of pollinators to the entire plant. Increased inflorescence size could additionally reduce self-pollination for species that self as flowers wilt (delayed selfing; Lloyd and Schoen, 1992) if increased attractiveness resulted in more visits per flower, leaving less pollen in exhausted flowers (Schoen and Dubuc, 1990). However, large floral displays often result in fewer visits per flower (Schmitt, 1983; Geber, 1985; Andersson, 1988; Klinkhamer et al., 1989; Robertson, 1992) and promote pollen discounting through geitonogamous pollen transfer, which can greatly reduce siring success if self-pollination triggers processes that compromise a pollen grain's chance of producing a viable seed (e.g., self-incompatibility, higher abortion, or inbreeding depression). The conflicting benefits and costs of exposing many flowers to individual pollinators imply that some intermediate daily flower production maximizes a plant's fertility (see Klinkhamer and de Jong, 1993).

To assess how pollination and mating influence optimal daily inflorescence size, consider a plant that produces F flowers during its flowering season (i.e., $F \equiv$ total inflorescence size), of which f are displayed each day. If we assume that pollen receipt limits seed production by all flowers, then the symmetry of female and male perspectives on mating assures that every seed has only one mother and one father. We also assume power functions between daily inflorescence size, the number of pollinator visits to the inflorescence per day (i.e., $n = af^b$) and the number of flowers visited by each pollinator (i.e., $V = f^c$).* In the absence of pollinator-induced intrafloral selfing, the number of seeds produced as a result of geitonogamous pollination is proportional to

$$(1 - \delta)\left[VD - \frac{D[1 - (1 - \rho)^V]}{\rho}\right]$$

where δ is the reduction in the success of self pollen relative to outcrossed pollen [see Equation (5)]. Correspondingly, genetic contributions through outcrossing are proportional to

$$\frac{D[1 - (1 - \rho)^V]}{\rho}$$

Half of these contributions comprise the plant's own outcrossed seeds and half involve seeds sired on other plants by the focal plant. In sum, a plant makes genetic contributions to

$$\frac{D}{\rho}\left\{(1 - \delta)\rho V + \delta\,[1 - (1 - \rho)^V]\right\}$$

seeds as a result of a single pollinator visit.

To quantify the fertility resulting from all visits on a given day, we must specify some pattern of pollen removal during successive pollinator visits (i.e., define D for each of the n pollinators) and sum the resulting fertilities. For

*de Jong et al.'s (1992) model of daily inflorescence size differs from the model presented here in assuming that each pollinator visits all exposed flowers. Our model describes this situation if $c=1$.

simplicity, we assume that each pollinator removes a fixed proportion, P, of the pollen remaining in a flower, so that the total number of viable seeds per day, T, is proportional to

$$T \propto \frac{h(fPA_0)^g[1-(1-P)^{gn}]}{\rho[1-(1-P)^g]}\left\{(1-\delta)\rho V + \delta[1-(1-\rho)^V]\right\}$$

where A_0 is the initial pollen production per flower, and h and g describe the relation between pollen removal and pollen dispersal (D) per pollinator [see Harder and Thomson, 1989; Harder and Wilson (1994) assess the assumption of constant proportional removal].† Finally, a plant that produces f flowers per day until it has produced F flowers blooms for F/f days, so that the total number of viable matings per inflorescence is $M=FT/f$, or

$$M \propto \frac{Ff^{g-1}h(PA_0)^g\,[1-(1-P)^{gn}]}{\rho[1-(1-P)^g]}\left\{(1-\delta)\rho V + \delta\,[1-(1-\rho)^V]\right\} \quad (12)$$

The optimal daily inflorescence size‡ is found by setting the partial derivative of Equation (12) with respect to f equal to 0 and solving for f^*, yielding

$$\begin{aligned} 0 = {} & (g-1)[1-(1-P)^{gn}]\,\{(1-\delta)\rho V + \delta[1-(1-\rho)^V]\} \\ & - bgn(1-P)^{gn}\ln(1-P)\,\{(1-\delta)\rho V + \delta[1-(1-\rho)^V]\} \\ & + cV[1-(1-P)^{gn}][(1-\delta)\rho - \delta(1-\rho)^V\ln(1-\rho)] \end{aligned} \quad (13)$$

Although Equation (13) lacks a direct solution (recall that n and V are functions of f), this expression exposes several conclusions. First, F, A_0, and h have canceled, so that according to this model, the optimal daily inflorescence size does not depend on total inflorescence size (as long as $F > f$), pollen production per flower, or the proportionality constant of the relation between pollen removal and subsequent dispersal. Rather, f^* depends on the details of pollinator attraction and the number of flowers visited per pollinator (i.e., a, b, and c), pollen removal per pollinator, P, pollen carryover, ρ, and the relative fertility of selfed pollen, δ. Second, the optimal daily inflorescence size does not generally maximize the number of pollinator visits per flower ($nV/f = af^{b+c-1}$). In particular, although the total number of matings is maximized at some intermediate flower production,

---

†de Jong et al.'s (1992) model of daily inflorescence size differs from the model presented here in not incorporating a decelerating relation between pollen removal and dispersal (see Harder and Thomson, 1989). Our model describes this situation if $g=1$.

‡This model of optimal daily inflorescence size explicitly maximizes absolute fitness. The corresponding ESS model that maximizes fitness of a rare phenotype with daily inflorescence size of f_1 flowers relative to the fitness of the predominant phenotype with f_0 flowers (i.e., $w=M_1/M_0$) identifies the same optimal flower production.

visits per flower is maximal for either a single flower, when $b + c < 1$, or infinite flowers, when $b + c > 1$. Third, when selfing is not detrimental, so that geitonogamous pollen discounting affects only the seed genotypes (i.e., $\delta = 0$), Equation (13) reduces to

$$0 = (c + g - 1)[1 - (1 - P)^{gn}] - bgn(1 - P)^{gn} \ln(1 - P)$$

which does not include ρ. Hence, in the absence of a cost of selfing, the details of pollen carryover do not affect optimal daily inflorescence size.

Numerical solution of Equation (13) for specific parameter values reveal further details of the influences on optimal daily inflorescence size. When self-pollen has a low probability of producing viable seeds (i.e., $\delta \rightarrow 1$), the consequences of pollen discounting are severe for many-flowered inflorescences. As a result, the optimal daily inflorescence size is smaller for given levels of pollen removal P, and carryover, ρ, than when selfing is not detrimental (i.e., $\delta \rightarrow 0$; compare Figs. 6.12a and b). Large inflorescences can be optimal even with a large cost of selfing given two conditions (see Fig. 6.12). First, if relatively little of the pollen carried by a pollinator is deposited on each flower (i.e., extensive carryover; small ρ), then comparatively little pollen is deposited geitonogamously, resulting in limited pollen discounting. Second, restricted pollen removal during

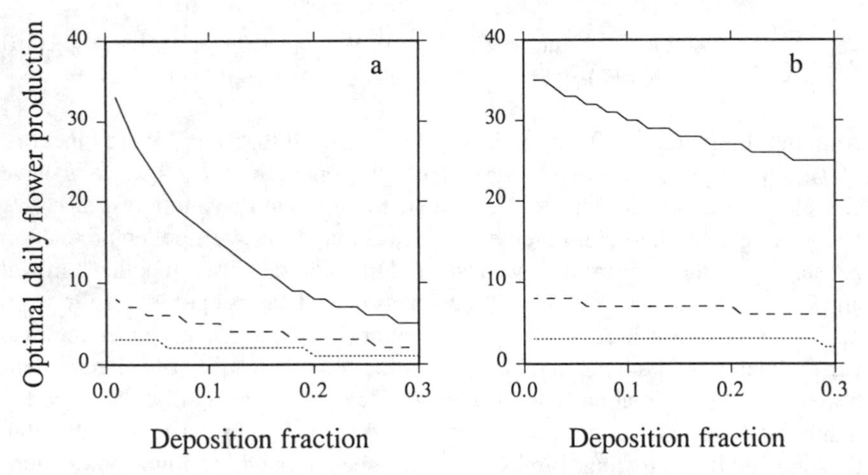

Figure 6.12. Relation of optimal daily inflorescence size to the proportion of remaining pollen removed per pollinator (P), the fraction of pollen on the pollinator's body deposited on each flower (deposition fraction ρ), and the reduction in the value of a selfed seed relative to an outcrossed seed (δ). The different lines in each panel relate the optimal flower display when $P=0.1$ (solid lines), 0.2 (dashed lines), or 0.3 (dotted lines). Panel (a) depicts complete inbreeding depression (i.e., $\delta=1$), whereas for panel (b) self-pollen is 80% as likely to fertilize seeds as outcrossed pollen (i.e., $\delta=0.2$). Based on numerical solutions to Equation (13), with $a=5$, $b=0.5$, $c=0.4$, and $g=0.3$.

individual pollinator visits (i.e., small P) ameliorates the diminishing returns (i.e., $g < 1$) associated with animal pollination and results in greater cumulative pollen dispersal as long as many pollinators are involved (Harder and Thomson, 1989; Harder and Wilson 1994). Hence, large flower displays are feasible with restricted pollen removal in spite of pollen discounting because the increased pollen dispersal resulting from attracting more pollinators exceeds discounting losses. This conclusion further implies that the geitonogamous self-pollination associated with the optimal inflorescence size (see Fig. 6.13b) represents a cost of maximizing pollen dispersal and receipt, rather than a selected feature in its own right (also see Lloyd, 1992).

Schoen and Dubuc (1990) previously derived a model of inflorescence size and number that complements our model in several ways. Schoen and Dubuc concentrated on total inflorescence size and incorporated the resource costs of producing more flowers. Their model implicitly assumed a direct correlation between total F and daily inflorescence size f to incorporate pollination influences.* In contrast, our model assumes that total flower production has been resolved (i.e., F is fixed) and identifies the pollination and mating influences on optimal daily flower production. In doing so, we assumed that resource constraints are less relevant to optimal daily inflorescence size. Any such constraint will further increase the costs of simultaneously displaying many flowers and reduce the optimal daily inflorescence size.

Evolutionary Implications

Pollination affects fertility through both the diversity of outcrossed matings and the relative frequency of selfing. The functional linkage between pollination and mating implies that selection that increases average fertility will often result in correlated evolution of floral characters and mating systems. As a result, some combinations of floral characters and mating systems should be more prevalent than others, such as the relation between intrafloral self-pollination and the proximity of anthers and stigmas within a flower (Breese, 1959; Rick et al., 1978; Ennos, 1981; Schoen, 1982; Thomson and Stratton, 1985; Barrett and Shore, 1987; Barrett and Husband, 1990; Murcia, 1990; Robertson and Lloyd, 1991; also see Chapter 14). We believe that many more associations between floral characters and mating systems remain to be discovered.

The consequences of pollen discounting promise to be a particularly rich area for future analyses of the interaction between pollination and mating. As defined by Holsinger et al. (1984), pollen discounting describes the extent to which self-

*The applicability of this assumption apparently varies between species as Harder and Cruzan (1990) found intraspecific correlations ranging from 0.150–0.676 for nine legume species (significant correlations for seven species) and from 0.229–0.900 for eight ericad species (significant correlations for seven species).

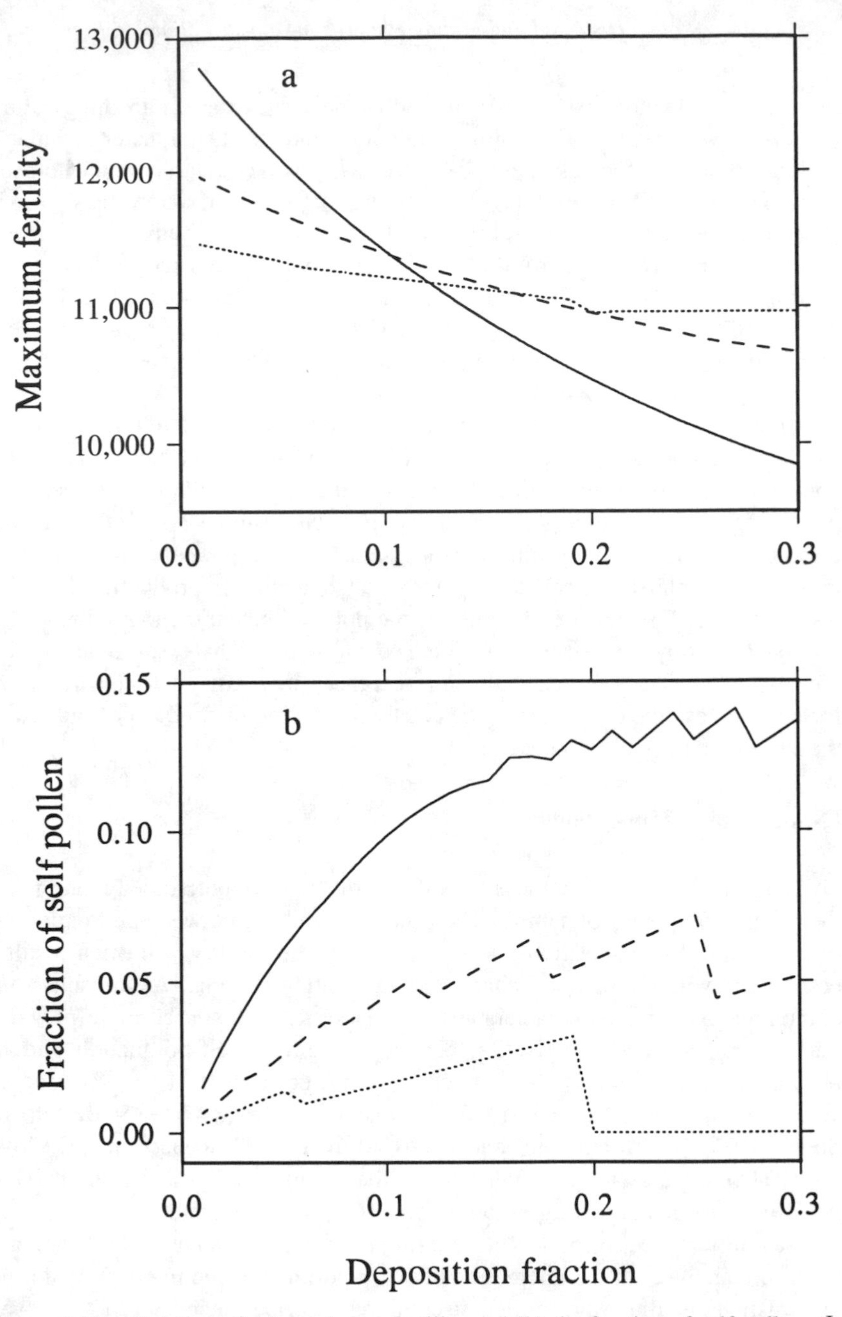

Figure 6.13. Relation of (a) maximum fertility and (b) the fraction of self-pollen, Ψ, to the proportion of remaining pollen removed per pollinator (P) and the fraction of pollen on the pollinator's body deposited on each flower (deposition fraction ρ). The different lines relate the optimal flower display when P=0.1 (solid lines), 0.2 (dashed lines), or 0.3 (dotted lines). Based on Equations (12) and (9), respectively, with f^* determined by numerical solution of Equation (12), for A_0=50,000 grains, a=5, b=0.5, c=0.4, δ=1, g=0.3, h=1, and F=150 flowers.

pollination reduces the number of pollen grains involved in cross-pollination. Such reduction affects the evolution of self-fertilization because discounting lessens the advantage of facultatively selfing individuals (Nagylaki, 1976; Wells, 1979; Holsinger et al., 1984; Holsinger, 1988), which, in the absence of discounting, contribute an average of three gene copies to the next generation (two through each selfed seed and one through paternal contribution to outcrossed seed) for every two copies contributed by outcrossing individuals (Fisher, 1941). Our model of daily inflorescence size identified additional implications of pollen discounting when discounting is coupled with postpollination processes that compromise the fertilization success of self-pollen. Indeed, it is helpful to recognize that the consequences of pollen discounting for male fertility range from increased average homozygosity of offspring when selfed and outcrossed offspring are equally successful ("cross-discounting") to the absolute failure of self-pollen with complete incompatibility or inbreeding depression ("fatal discounting"; also see de Jong et al., 1992, 1993). Because the complete absence of inbreeding depression is probably uncommon (see Jarne and Charlesworth, 1993), theoretical analyses of the evolution of selfing should realistically incorporate the interaction between pollen discounting and postpollination losses (Uyenoyama et al., 1993). Furthermore, whenever postpollination processes discriminate against self-pollen or selfed zygotes, pollen discounting reduces male fertility and should strongly influence floral evolution.

Whether pollen discounting represents a significant loss of outcrossed-mating opportunities, and hence is a significant process in plant ecology, genetics, and evolution, has been questioned (Lloyd, 1992; Holsinger and Thomson, 1994). This doubt arises from the observation that "(p)ollen grains are much more numerous than ovules, and . . . (c)onsequently, an increase in self-fertilization can often be achieved with a minimal effect on the outcrossing pollen pool" (Lloyd, 1992, p. 372; also see Holsinger and Thomson, 1994). Although it may be true that pollen grains are plentiful and individually inexpensive, the important consideration is not production, but the number of mating opportunities. Given that $<1\%$ of a plant's pollen typically reaches stigmas (e.g., Levin and Berube, 1972; Galen and Stanton, 1989; Harder and Thomson, 1989; Wolfe and Barrett, 1989; Young and Stanton, 1990), the relevant question is not "how much pollen is involved in selfing?", but rather "how much of the pollen that would otherwise reach other plants is involved in selfing?" Because geitonogamous pollen transfer is functionally equivalent to interplant transfer (Fægri and van der Pijl, 1979; Lloyd and Schoen, 1992), especially when pollinators fly between flowers in the inflorescence, geitonogamous selfing will generally draw pollen directly from the restricted pollen pool carried by the pollinator that is destined for deposition on other flowers. Hence, although intrafloral self-pollination may result in little pollen discounting, particularly for herkogamous species, geitonogamous self-pollination probably commonly limits opportunities for outcrossing (Lloyd, 1992). It is particularly striking that the two studies reporting little evidence of discounting involved single-flowered plants (Rausher et al., 1993; Holsinger and Thomson 1994), whereas demonstrations of extensive pollen discounting involved plants that simultaneously display many flowers (Ritland, 1991; Holsinger, 1992; Kohn and Barrett, 1994; this study).

Mating Patterns and the Evolution of Floral Design and Display

Our analysis of the mating consequences of pollination emphasizes that the open flowers in an inflorescence do not act independently during a plant's interaction with individual pollinators. Given the fundamental role of geitonogamous pollination in determining the incidence of self-pollination and pollen discounting (Figs. 6.5, 6.9, and 6.11), the flower cannot be considered the operational unit of either male or female function for animal-pollinated plants; rather, this role belongs to the entire floral display. This conclusion influences ecological, genetic, and evolutionary interpretations of the design of individual flowers, because the mating consequences of characters that affect the pattern of pollen dispersal depend on how many flowers a pollinator visits on the same plant. Hence, complete assessment of the functional significance of flower shape and size, the size and position of sexual organs, reward characteristics, and the schedules of male and female function requires consideration of their effects on the entire plant's mating success.

In our model of daily inflorescence size, we treated pollen removal and carryover as evolutionarily fixed; however, the details of both processes depend on various floral characteristics (e.g., anther and stigma placement, corolla shape; pollen-dispensing mechanisms, stigma size and structure) that are undoubtedly subject to selection (e.g., Campbell, 1989; Galen, 1989; Johnston, 1991). Because of the functional correlations between such characters, the evolution of daily inflorescence size likely occurs in concert with the modification of flower form. In general, large inflorescences should accompany restricted pollen removal and extensive carryover, because fertility is globally maximized under these conditions (see Fig. 6.13a). Alternatively, if either restricted removal or extensive carryover is not evolutionarily feasible, then fertility is maximized by exposing few flowers at any time (also see de Jong et al., 1992; Robertson, 1992). In spite of the expectation of functional correlations between floral and inflorescence characters, few studies have assessed their occurrence through comparative interspecific analysis (although see Harder and Cruzan, 1990). Such studies are necessary to appreciate fully the evolutionary impact of interactions between floral design and display. Unfortunately, our current limited understanding of the role of specific floral characters in pollen carryover precludes the formulation of detailed hypotheses.

Although geitonogamous pollen discounting can constrain the evolutionary opportunities for plants to enhance pollinator attraction through larger displays, several floral adaptations relieve this constraint. The simplest means of increasing display size without paying discounting costs involves the incorporation of flowers that neither donate nor receive pollen in the display. Some species accomplish this by producing sterile flowers at the periphery of the inflorescence (e.g., many Asteraceae with radiate capitula, some *Viburnum* spp., *Hydrangea* spp.), whereas other species maintain flowers for protracted periods after their primary sexual roles are complete. Such flowers contribute to long-distance signaling of the location of floral rewards but, because they typically differ in morphology or color from rewarding flowers, they indicate which flowers are unrewarding to attracted pollinators (Gori, 1983; Cruzan et al., 1988; Weiss, 1991).

Dichogamy, the temporal separation of male and female function within individual flowers (reviewed by Lloyd and Webb, 1986), also enables large floral displays with reduced discounting costs, particularly when inflorescence structure and pollinator behavior interact so that individual pollinators visit female-phase flowers before male-phase flowers. Dichogamy has traditionally been interpreted as a mechanism discouraging self-fertilization (see Chapter 14), even though many dichogamous species also exhibit self-incompatibility (reviewed by Lloyd and Webb, 1986), and dichogamy is equally prevalent among self-compatible and self-incompatible species (Bertin and Newman, 1993). As a complement to this "antiselfing" explanation, Lloyd and Webb (1986) proposed that dichogamy promotes outcrossing by reducing interference between male and female organs in individual flowers. Although dichogamy necessarily eliminates intrafloral interference, we believe that it probably plays a more important role in reducing the interference *between* flowers on the same inflorescence that results in geitonogamous pollen discounting. By reducing within-plant discounting, dichogamy allows plants to disperse more pollen to other plants on individual pollinators, while allowing the simultaneous presentation of many flowers (some female and the remainder male) to attract more pollinators. Because both self-incompatible species and self-compatible species subject to inbreeding depression suffer from the lost mating opportunities associated with geitonogamous pollination, the incidence of a "cross-promotion" mechanism such as dichogamy should not vary with the occurrence of physiological antiselfing mechanisms, as Bertin and Newman's (1993) survey indicates.

Heterostyly probably represents the epitome of floral mechanisms that promote outcrossing by restricting pollen discounting. For heterostylous species, anthers and stigmas occupy dissimilar positions within flowers on individual plants, but other plants in the population produce the reciprocal arrangement(s) of sexual organs (see Barrett, 1992*a*). In a monomorphic species, the extreme herkogamy exhibited by an individual plant of a heterostylous species would preclude efficient pollen transfer. For example, our array experiments with the long-styled morph of *Eichhornia paniculata* resulted in pollen limitation of seed production even though the plants were frequently visited, with often three or four bees simultaneously visiting an array of 36 plants (also see Kohn and Barrett, 1992). Although such herkogamy clearly limits pollen transfer within morphs, our demonstration of the extensive pollen carryover between morphs of *Pontederia cordata* (Fig. 6.2) illustrates that the reciprocal arrangement of sexual organs on different plants promotes crossing between morphs (also see Lloyd and Webb, 1992). Such cross-pollination undoubtedly results because stigmas of one morph receive comparatively little of the pollen produced by noncorresponding anther levels, thereby enabling more pollen to reside on the pollinator until it visits a legitimate recipient. Like dichogamy, heterostyly enables large flower displays; however, because heterostyly limits pollen dispersal between inflorescencess of the same morph [pollen wastage; see Bateman, 1952; Baker, 1954 (cited in Lloyd and Webb, 1992); Yeo, 1975; Glover and Barrett, 1986], it will reduce pollen discounting for clonal species more effectively than would dichogamy, which functions only within inflorescences.

The fertility advantages of inflorescence-level mechanisms that enhance pollinator attraction while limiting pollen discounting are clearly illustrated by reinterpretation of Podolsky's (1992, 1993) analysis of andromonoecy in *Besleria triflora*. Podolsky (1992) demonstrated that daily inflorescence size positively affected the frequency of hummingbird visits per inflorescence and per flower. Because seed production by *B. triflora* is pollen-limited, such increased attraction directly benefits female fertility. Indeed, Podolsky (1992) concluded ". . . that enhanced male function is not necessary for the evolution of andromonoecy" (p. 2259) because although staminate flowers increase attractiveness, they ". . . were less effective than perfect flowers both in the length of time pollen was presented and in the quantity of pollen dispersed per visit" (p. 2258). However, this emphasis on individual flowers ignores the fact that andromonoecy is an inflorescence characteristic, not a floral characteristic. Consequently, the relevant question is "Does the presence of staminate flowers benefit the *plant's* male fertility more than the presence of additional perfect flowers?" Such an advantage for male fertility seems likely for *B. triflora* because donor flowers dispersed more pollen to recipient flowers when a hummingbird visited intervening staminate flowers than when the intervening flowers were perfect and therefore acted as recipients (Podolsky, 1992). Hence, although the addition of staminate or perfect flowers equally increases an inflorescence's attractiveness (Podolsky, 1992), the addition of staminate flowers would promote male fertility more effectively because the addition of perfect flowers increases geitonogamous pollination and *B. triflora* suffers strong inbreeding depression (see Podolsky, 1992). In summary, andromonoecy in *B. triflora* probably increases both male and female fertility, so that this breeding system exemplifies an additional mechanism enabling increased pollinator attraction without aggravating pollen discounting.

Pollinator Service and Mating-System Evolution

Unsatisfactory pollinator service is frequently invoked as a mechanism promoting diverse evolutionary outcomes, ranging from the evolution of self-fertilization

(Henslow, 1879; Müller, 1883; Baker, 1955; recent studies reviewed by Barrett, 1988; also see Chapter 14) to shifts between outcrossing breeding systems (e.g., Ganders, 1978; Delph, 1990; Weller and Sakai, 1990; Barrett, 1992*b*). The seeming paradox of a single mechanism producing very different outcomes (Barrett et al., 1992) actually arises because pollinator service can be unsatisfactory for two reasons. *Insufficient* pollinator service arises whenever plants receive too few visits, resulting in pollen limitation of seed production, regardless of quality. In contrast, *inferior* pollinator service results in sufficient pollination to overcome pollen limitation, but the pattern of pollen dispersal limits opportunities for gametophytic competition and mate choice to produce the highest-quality offspring. These two sources of unsatisfactory pollinator service favor very different evolutionary responses by both floral characters and mating systems, so that assessing their importance requires different experimental approaches.

Insufficient pollination results because pollinators are either rare or have been attracted by other plant species, and it promotes characters that result in self-pollination and the assurance of seed production. In general, the amount of reproductive assurance required to favor the spread of a selfing variant depends on both the mode of self-pollination (i.e., how and when self-pollination occurs) and frequency of pollinator visits (Lloyd, 1992). The few studies that have examined both the fraction of selfed seeds and pollinator activity in different populations reported correlations that support the reproductive assurance hypothesis (Rick et al., 1978; Wyatt, 1986; Piper et al., 1986; Husband and Barrett, 1992). However, no studies have examined how the fraction of selfed seeds varies with the number of visits received by individual flowers and/or plants, let alone considered how this relation depends on the mode of selfing. Such experiments are central to testing the reproductive assurance hypothesis.

Inferior pollination and its effects on offspring quality arise because pollinators either disperse pollen poorly among conspecific plants (intraspecific inefficiency) or they do not consistently visit a single species, thereby transferring pollen between species (interspecific pollination). Intraspecific inefficiency results because inadequate pollen carryover or excessive pollen discounting (either intrafloral or geitonogamous) increases self-pollination and limits the diversity of potential mates. In contrast, interspecific pollination results in receipt of foreign pollen that can interfere with fertilization by conspecific pollen (Thomson et al., 1981; Armbruster and Herzig, 1984; Stucky, 1985; Waser and Fugate; 1986, Harder et al., 1993) and the loss on foreign stigmas of pollen that would otherwise have reached conspecific stigmas (Campbell, 1985; Campbell and Motten, 1985; Feinsinger and Busby, 1987; Feinsinger et al., 1988; Feinsinger and Tiebout, 1991). Like intraspecific inefficiency, interspecific pollination can increase the fraction of selfed seed; however, selfing increases indirectly with interspecific pollination because outcross pollen is intercepted by heterospecific stigmas, thereby diminishing its abundance on conspecific stigmas relative to self-pollen. To our knowledge, no studies to date have examined whether the fraction of selfed seeds varies either among types of pollinators (although see Chapter 14) or with the relative abundance of other plant species that share the same pollinators. Analysis of evolutionary changes in breeding systems would benefit from such information.

Jarne and Charlesworth (1993) drew attention to a phenomenon that provides our final illustration of the interaction of evolutionary influences on floral and mating-system characteristics. They presented observations of " . . . large and showy flowers found in the arctic . . . (and) a high incidence of self-incompatibility at high altitudes in Patagonia, where insect visitation is low" (p. 454) as examples contradicting the reproductive assurance hypothesis. However, reproductive assurance is only a reasonable evolutionary option if the inbreeding depression that accompanies increased selfing in an originally outcrossing population diminishes over generations as recessive deleterious alleles are purged from the population (see Barrett and Charlesworth, 1991). Such evolutionary reduction of inbreeding depression may be more difficult in rigorous environments, such as those at high latitudes and altitudes, where traits that would be neutral or mildly disadvantageous in more benign conditions become detrimental [see Dudash (1990) for an example of environmental effects on inbreeding depression]. Hence, we propose that an interaction between environmental conditions and the intensity of inbreeding depression constrains the evolution of selfing in stressful environments and favors floral traits that promote outcrossing. This example provides further indication that the outcome of mating (who has mated with whom) cannot be logically separated from the ecological and physiological processes that determine how two specific gametes come to form a zygote (also see Richards, 1986; Lloyd and Schoen, 1992; Lloyd, 1992; Waser, 1993). Given that the ecology and evolution of plant fertility integrate the processes and outcome of mating, we expect that the isolation of pollination and mating-system biology that has characterized studies of plant reproduction during much of the twentieth century will not be sustained.

Appendix: Sum of a Geometric Series

Repeatedly in this chapter, we require the sum of a geometric series, such as

$$\begin{aligned} S &= \sum_{i=1}^{n} ax^{i-1} \\ &= a + ax + ax^2 + ax^3 + \cdots + ax^{n-1} \end{aligned} \tag{A.1}$$

The solution to this sum can be found by multiplying Equation (A.1) by x

$$xS = ax + ax^2 + ax^3 + \cdots + ax^{n-1} + ax^n \tag{A.2}$$

subtracting Equation (A.2) from Equation (A.1)

$$S - Sx = a - ax^n$$

and solving for S, yielding

$$S = \frac{a(1 - x^n)}{1 - x} \tag{A.3}$$

If $-1 < x < 1$ and Equation (A.1) involves an indefinite series (i.e., $n = \infty$), then the x^n term in Equation (A.3) equals zero.

Acknowledgments

We gratefully acknowledge W.W. Cole, L.J. Mydynski, S.A. Rasheed, C.E. Thomson, and M.J. Vonhof for electrophoretic assays and pollen counting; M.B. Vander Meulen, J.R. Kohn, S.W. Graham, and W.W. Cole for assistance in the field; K. Ritland and W.G. Wilson for clarifying theoretical issues; M.T. Morgan, K. Ritland, D.J. Schoen, J.D. Thomson, and N.M. Waser for comments on the manuscript; and the Natural Sciences and Engineering Research Council of Canada for financial support.

References

Abbott, R.J. and J.A. Irwin. 1988. Pollinator movements and the polymorphism for outcrossing rate at the ray floret locus in Groundsel, *Senecio vulgaris* L. *Heredity,* 60: 295–298.

Andersson, S. 1988. Size dependent pollination efficiency in *Anchusa officinalis* (Boraginaceae): Causes and consequences. *Oecologia* 76: 125–130.

Armbruster, W.S. and A.L. Herzig. 1984. Partitioning and sharing of pollinators by four sympatric species of *Dalechampia* (Euphorbiaceae) in Panama. *Ann. Missouri Bot. Gard.,* 71: 1–16.

Armstrong, J.E. 1992. Lever action anthers and the forcible shedding of pollen in *Torenia* (Scrophulariaceae). *Am. J. Bot.,* 79: 34–40.

Augspurger, C.K. 1980. Mass-flowering of a tropical shrub (*Hybanthus prunifolius*): Influence on pollinator attraction and movement. *Evolution,* 34: 475–488.

Baker, H.G. 1955. Self-incompatibility and establishment after "long-distance" dispersal. *Evolution,* 9: 347–349.

Baker, H.G. 1983. An outline of the history of anthecology, or pollination biology. In *Pollination Biology* (L. Real, ed.), Academic Press, New York, pp. 7–28.

Barrett, S.C.H. 1988. The evolution, maintenance, and loss of self-incompatibility systems. In *Plant Reproductive Ecology: Patterns and Strategies* (J. Lovett Doust and L. Lovett Doust, eds.), Oxford Univ. Press, pp. 98–124.

Barrett, S.C.H. (ed.) 1992*a*. *Evolution and Function of Heterostyly*. Springer-Verlag, Berlin.

Barrett, S.C.H. 1992*b*. Gender variation and the evolution of dioecy in *Wurmbea dioica* (Liliaceae). *J. Evol. Biol.*, 5: 423–444.

Barrett, S.C.H. and D. Charlesworth. 1991. Effects of a change in the level of inbreeding on the genetic load. *Nature,* 352: 522–524.

Barrett, S.C.H. and C.G. Eckert. 1990. Variation and evolution of mating systems in seed plants. In *Biological Approaches and Evolutionary Trends in Plants* (S. Kawano, ed.), Academic, New York, pp. 229–254.

Barrett, S.C.H. and B.C. Husband. 1990. Variation in outcrossing rates in *Eichhornia paniculata:* The role of demographic and reproductive factors. *Plant Spec. Biol.,* 5: 41–55.

Barrett, S.C.H. and J.S. Shore. 1987. Variation and evolution of breeding systems in the *Turnera ulmifolia* L. complex (Turneraceae). *Evolution,* 41: 340–354.

Barrett, S.C.H., J.R. Kohn, and M.B. Cruzan. 1992. Experimental studies of mating-system evolution: The marriage of marker genes and floral biology. In *Ecology and Evolution of Plant Reproduction* (R. Wyatt, ed.), Chapman & Hall, New York, pp. 192–230.

Barrett, S.C.H., L.D. Harder, and W.W. Cole. 1994. Effects of flower number and position on self-fertilization in experimental populations of *Eichhornia paniculata* (Pontederiaceae). *Func. Ecol.,* 8: 526–535.

Bateman, A.J. 1947*a*. Contamination in seed crops. I. Insect pollination. *J. Gen.,* 48: 257–275.

Bateman, A.J. 1947*b*. Contamination in seed crops. III. Relation with isolation distance. *Heredity,* 1: 303–336.

Bateman, A.J. 1952. Self-incompatibility systems in angiosperms. 1. Theory. *Heredity,* 6: 285–310.

Becerra, J.X. and D.G. Lloyd. 1992. Competition-dependent abscission of self-pollinated flowers of *Phormium tenax* (Agavaceae): A second action of self-incompatibility at the whole flower level? *Evolution,* 46: 458–469.

Bell, G. 1985. On the function of flowers. *Proc. Roy. Soc. Lond. B,* 224: 223–265.

Bertin, R.I. and C.M. Newman. 1993. Dichogamy in angiosperms. *Bot. Rev.*, 59: 112–152.

Best, L.S. and P. Bierzychudeck. 1982. Pollinator foraging on foxglove (*Digitalis purpurea*): A test of a new model. *Evolution,* 36: 70–79.

Breese, E. L. 1959. Selection for differing degrees of out-breeding in *Nicotiana rustica*. *Ann. Bot.*, 23: 331–344.

Brown, B.A. and M.T. Clegg. 1984. Influence of flower color polymorphism on genetic transmission in a natural population of the common morning glory, *Ipomoea purpurea*. *Evolution,* 38: 796–803.

Campbell, D.R. 1985. Pollen and gene dispersal: The influences of competition for pollination. *Evolution,* 39: 419–431.

Campbell, D.R. 1989. Measurements of selection in a hermaphroditic plant: Variation in male and female pollination success. *Evolution,* 43: 318–334.

Campbell, D.R. 1991. Comparing pollen dispersal and gene flow in a natural population. *Evolution,* 45: 1965–1968.

Campbell, D.R. and A.F. Motten. 1985. The mechanism of competition for pollination between two forest herbs. *Ecology,* 66: 554–563.

Casper, B.B. 1988. Evidence for selective embryo abortion in *Cryptantha flava*. *Am. Nat.*, 132: 318–326.

Charlesworth, D. and B. Charlesworth. 1987. Inbreeding depression and its evolutionary consequences. *Ann. Rev. Ecol. Syst.*, 18: 237–268.

Corbet, S.A., I. Cuthill, M. Fallows, T. Harrison, and G. Hartley. 1981. Why do nectar-foraging bees and wasps work upwards on inflorescences? *Oecologia*, 51: 79–83.

Crawford, T.J. 1984. What is a population? In *Evolutionary Ecology* (B. Shorrocks, ed.), Blackwell, Oxford, pp. 135–173.

Cruzan, M.B. and S.C.H. Barrett. 1993. Contribution of cryptic incompatibility to the mating system of *Eichhornia paniculata* (Pontederiaceae). *Evolution*, 47: 925–938.

Cruzan, M.B., P.R. Neal, and M.F. Willson. 1988. Floral display in *Phyla incisa:* Consequences for male and female reproduction success. *Evolution,* 42: 505–515.

Darwin, C.R. 1877. *The Different Forms of Flowers on Plants of the Same Species*. Murray, London.

Delph, L.F. 1990. The evolution of gender dimorphism in New Zealand *Hebe* (Scrophulariaceae) species. *Evol. Trends Plants,* 4: 85–97.

Devlin, B. and N.C. Ellstrand. 1990. Male and female fertility variation in wild radish, a hermaphrodite. *Am. Nat.*, 136: 87–107.

Dudash, M.R. 1990. Relative fitness of selfed and outcrossed progeny in a self-compatible, protandrous species, *Sabatia angularis* L. (Gentianaceae): A comparison in three environments. *Evolution,* 44: 1129–1139.

Dulberger, R. 1992. Floral polymorphisms and their functional significance in the heterostylous syndrome. In *Evolution and Function of Heterostyly* (S.C.H. Barrett, ed.), Springer-Verlag, Berlin, pp. 41–84.

Eckhart, V.M. 1991. The effects of floral display on pollinator visitation vary among populations of *Phacelia linearis* (Hydrophyllaceae). *Evol. Ecol.*, 5: 370–384.

Ennos, R.A. 1981. Quantitative studies of the mating system in two sympatric species of *Ipomoea* (Convolvulaceae). *Genetica,* 57: 93–98.

Fægri, K. and L. van der Pijl. 1979. *The Principles of Pollination Ecology,* 3rd ed. Pergamon, Oxford.

Feinsinger, P. and W.H. Bushby. 1987. Pollen carryover: Experimental comparisons between morphs of *Palicourea lasiorrachis* (Rubiaceae), a distylous, bird-pollinated tropical treelet. *Oecologia,* 73: 231–235.

Feinsinger, P. and H.M. Tiebout, III. 1991. Competition among plants sharing hummingbird pollinators: Laboratory experiments on a mechanism. *Ecology,* 72: 1946–1952.

Feinsinger, P., W.H. Busby, and H.M. Tiebout, III. 1988. Effects of indiscriminate foraging by tropical hummingbirds on pollination and plant reproductive success: Experiments with two tropical treelets. *Oecologia,* 76: 471–474.

Fisher, R.A. 1941. Average excess and average effect of a gene substitution. *Ann. Eugenics,* 11: 53–63.

Galen, C. 1985. Regulation of seed-set in *Polemonium viscosum:* Floral scents, pollination, and resources. *Ecology,* 66: 792–797.

Galen, C. 1989. Measuring pollinator-mediated selection on morphometric floral traits: Bumblebees and the alpine sky pilot, *Polemonium viscosum*. *Evolution,* 43: 882–890.

Galen, C. and M.E.A. Newport. 1987. Bumble bee behavior and selection on flower size in the sky pilot, *Polemonium viscosum*. *Oecologia,* 74: 20–23.

Galen, C. and R.C. Plowright. 1985. The effects of nectar level and flower development on pollen carry-over in inflorescences of fireweed (*Epilobium angustifolium*) (Onagraceae). *Can. J. Bot.*, 63: 488–491.

Galen, C. and J.T. Rotenberry. 1988. Variance in pollen carryover in animal-pollinated plants: Implications for mate choice. *J. Theor. Biol.*, 135: 419–429.

Galen, C. and M.L. Stanton. 1989. Bumble bee pollination and floral morphology: Factors influencing pollen dispersal in the alpine sky pilot, *Polemonium viscosum* (Polemoniaceae). *Am. J. Bot.*, 76: 419–426.

Ganders, F.R. 1978. The genetics and evolution of gynodioecy in *Nemophila menziesii* (Hydrophyllaceae). *Can. J. Bot.*, 56: 1400–1408.

Ganders, F.R. 1979. The biology of heterostyly. *New Zeal. J. Bot.*, 17: 607–635.

Geber, M.A. 1985. The relationship of plant size to self-pollination in *Mertensia ciliata*. *Ecology*, 66: 762–772.

Glover, D.E. and S.C.H. Barrett. 1986. Stigmatic pollen loads in populations of *Pontederia cordata* from the southern U.S. *Am. J. Bot.*, 73: 1607–1612.

Godt, M.J.W. and J.L. Hamrick. 1993. Patterns and levels of pollen-mediated gene flow in *Lathyrus latifolius*. *Evolution*, 47: 98–110.

Gori, D.F. 1983. Post-pollination phenomena and adaptive floral changes. In *Handbook of Experimental Pollination Biology* (C. E. Jones and R. J. Little, eds.), Van Nostrand Reinhold, New York, pp. 31–49.

Handel, S.N. 1982. Dynamics of gene flow in an experimental population of *Cucumis melo* (Cucurbitaceae). *Am. J. Bot.*, 69: 1538–1546.

Handel, S.N. 1983. Pollination ecology, plant population structure, and gene flow. In *Pollination Biology* (L. Real, ed.), Academic, New York, pp. 163–211.

Harder, L.D. 1990*a*. Pollen removal by bumble bees and its implications for pollen dispersal. *Ecology,* 71: 1110–1125.

Harder, L.D. 1990*b*. Behavioral responses by bumble bees to variation in pollen availability. *Oecologia,* 85: 41–47.

Harder, L.D. and R.M.R. Barclay. 1994. The functional significance of poricidal anthers and buzz pollination: Controlled pollen removal from *Dodecatheon*. *Func. Ecol.*, 8: 509–517.

Harder, L.D. and S.C.H. Barrett. 1993. Pollen removal from tristylous *Pontederia cordata:* Effects of anther position and pollinator specialization. *Ecology,* 74: 1059–1072.

Harder, L.D. and S.C.H. Barrett. 1995. Mating cost of large floral displays in hermaphrodite plants. *Nature,* 373: 512–515.

Harder, L.D. and M.B. Cruzan. 1990. An evaluation of the physiological and evolutionary influences of inflorescence size and flower depth on nectar production. *Func. Ecol.*, 4: 559–572.

Harder, L.D. and J.D. Thomson. 1989. Evolutionary options for maximizing pollen dispersal of animal-pollinated plants. *Am. Nat.*, 133: 323–344.

Harder, L.D. and W.G. Wilson. 1994. Floral evolution and male reproductive success: Optimal dispensing schedules for pollen dispersal by animal-pollinated plants. *Evol. Ecol.,* 8: 542–559.

Harder, L.D., M.B. Cruzan, and J.D. Thomson. 1993. Unilateral incompatibility and the effects of interspecific pollination for *Erythronium americanum* and *Erythronium albidum* (Liliaceae). *Can. J. Bot.*, 71: 353–358.

Henslow, G. 1879. On the self-fertilization of plants. *Trans. Linn. Soc., 2nd Series, Bot.*, 1: 317–398.

Hodges, S.A. 1995. The influence of nectar production on hawkmoth behavior, self pollination, and seed production in *Mirabilis multiflora* (Nyctaginaceae). *Am. J. Bot.* 82: 197–204.

Holloway, B.A. 1976. Pollen-feeding in hover-flies (Diptera: Syrphidae). *New Zeal. J. Zool.*, 3: 339–350.

Holsinger, K.E. 1988. Inbreeding depression doesn't matter: The genetic basis of mating system evolution. *Evolution,* 42: 1235–1244.

Holsinger, K.E. 1992. Ecological models of plant mating systems and the evolutionary stability of mixed mating systems. In *Ecology and Evolution of Plant Reproduction* (R. Wyatt, ed), Chapman & Hall, New York, pp. 169–191.

Holsinger, K.E. and J.D. Thomson. 1994. Pollen discounting in *Erythronium grandiflorum:* Mass-action estimates from pollen transfer dynamics. *Am. Nat.* 144: 799–812.

Holsinger, K.E., M.W. Feldman, and F.B. Christiansen. 1984. The evolution of self-fertilization in plants: A population genetic model. *Am. Nat.*, 124: 446–453.

Husband, B.C. and S.C.H. Barrett. 1992. Pollinator visitation in populations of tristylous *Eichhornia paniculata* in northeastern Brazil. *Oecologia,* 89: 365–371.

Jarne, P. and D. Charlesworth. 1993. The evolution of the selfing rate in functionally hermaphrodite plants and animals. *Ann. Rev. Ecol. Syst.*, 24: 441–466.

Johnston, M.O. 1991. Natural selection on floral traits in two species of *Lobelia* with different pollinators. *Evolution,* 45: 1468–1479.

Jones, D.F. 1928. *Selective Fertilization.* Univ. Chicago Press, Chicago.

de Jong, T.J., P.G.L. Klinkhamer, and M.J. van Staalduinen. 1992. The consequences of pollination biology for selection of mass or extended blooming. *Func. Ecol.*, 6: 606–615.

de Jong, T.J., N.M. Waser, and P.G.L. Klinkhamer. 1993. Geitonogamy: The neglected side of selfing. *Trends Ecol. Evol.*, 8: 321–325.

Kimsey, L.S. 1984. The behavioural and structural aspects of grooming and related activities in euglossine bees (Hymenoptera: Apidae). *J. Zool.*, 204: 541–550.

Kirk, R.E. 1982. *Experimental Design*. Brooks/Cole, Monterey, CA.

Klinkhamer, P.G.L. and T.J. de Jong. 1990. Effects of plant size, plant density and sex differential nectar reward on pollinator visitation in the protandrous *Echium vulgare* (Boraginaceae). *Oikos*, 57: 399–405.

Klinkhamer, P.G.L. and T.J. de Jong. 1993. Attractiveness to pollinators: A plant's dilemma. *Oikos,* 66: 180–184.

Klinkhamer, P.G.L., T.J. de Jong, and G.-J. de Bruyn. 1989. Plant size and pollinator visitation in *Cynoglossum officinale*. *Oikos,* 54: 201–204.

Klinkhamer, P.G.L., T.J. de Jong, and R.A. Wesselingh. 1991. Implications of differences between hermaphrodite and female flowers for attractiveness to pollinators and seed production. *Neth. J. Zool.*, 41: 130–143.

Kohn, J.R. and S.C.H. Barrett. 1992. Experimental studies on the functional significance of heterostyly. *Evolution,* 46: 43–55.

Kohn, J.R. and S.C.H. Barrett. 1994. Pollen discounting and the spread of a selfing variant in tristylous *Eichhornia paniculata:* Evidence from experimental populations. *Evolution* 48: 1576–1594.

Kohn, J.R. and B.B. Casper. 1992. Pollen-mediated gene flow in *Cucurbita foetidissima* (Cucurbitaceae). *Am. J. Bot.*, 79: 57–62.

LeBuhn, G. and G.J. Anderson. 1994. Anther tripping and pollen dispensing in *Berberis thunbergii*. *Am. Mid. Nat.*, 131: 257–265.

Lertzman, K.P. and C.L. Gass. 1983. Alternate models of pollen transfer. In *Handbook of Experimental Pollination Biology* (C.E. Jones and R.J. Little, eds.), Van Nostrand Reinhold, New York, pp. 474–489.

Levin, D.A. and D.E. Berube. 1972. *Phlox* and *Colias:* The efficiency of a pollination system. *Evolution,* 26: 242–250.

Levin, D.A. and H.W. Kerster. 1974. Gene flow in seed plants. *Evol. Biol.* 7, 139–220.

Lloyd, D.G. 1984. Gender allocations in outcrossing cosexual plants. In *Perspectives on Plant Population Ecology* (R. Dirzo and J. Sarukhán, eds.), Sinauer, Sunderland, MA, pp. 277–300.

Lloyd, D.G. 1992. Self- and cross-fertilization in plants. II. The selection of self-fertilization. *Internatl. J. Plant Sci.*, 153: 370–380.

Lloyd, D.G. and D.J. Schoen. 1992. Self- and cross-fertilization in plants. I. Functional dimensions. *Internatl. J. Plant Sci.*, 153: 358–369.

Lloyd, D.G. and C.J. Webb. 1986. The avoidance of interference between the presentation of pollen and stigmas in angiosperms. I. Dichogamy. *New Zeal. J. Bot.*, 24: 135–162.

Lloyd, D.G. and C.J. Webb. 1992. The selection of heterostyly. In *Evolution and Function of Heterostyly* (S.C.H. Barrett, ed.), Springer-Verlag, Berlin, pp. 179–207.

Lyons, E.E., N.M. Waser, M.V. Price, J. Antonovics, and A.F. Motten. 1989. Sources of variation in plant reproductive success and implications for concepts of sexual selection. *Am. Nat.*, 134: 409–433.

Macior, L.W. 1967. Pollen-foraging behavior of *Bombus* in relation to pollination of nototribic flowers. *Am. J. Bot.*, 54: 359–364.

Manning, A.A. 1956. Some aspects of the foraging behaviour of bumble-bees. *Behaviour*, 9: 164–201.

Marshall, D.L. and N.C. Ellstrand. 1986. Sexual selection in *Raphanus sativus:* Experimental data on nonrandom fertilization, maternal choice, and consequences of multiple paternity. *Am. Nat.*, 127: 446–461.

Marshall, D.L. and M.W. Folsom. 1991. Mate choice in plants: An anatomical to population perspective. *Ann. Rev. Ecol. Syst.*, 22: 37–63.

Meagher, T.R. 1991. Analysis of paternity within a natural population of *Chamaelirium luteum*. II. Patterns of male reproductive success. *Am. Nat.*, 137: 738–752.

Michener, C.D., M.L. Winston, and R. Jander. 1978. Pollen manipulation and related activities and structures in bees of the family Apidae. *Univ. Kansas Sci. Bull.*, 51: 575–601.

Montalvo, A.M. 1992. Relative success of self and outcross pollen comparing mixed- and single-donor pollinations in *Aquilegia caerulea*. *Evolution*, 46: 1181–1198.

Morgan, M.T. and S.C.H. Barrett. 1989. Reproductive correlates of mating system evolution in *Eichhornia paniculata* (Spreng.) Solms (Pontederiaceae). *J. Evol. Biol.* 2: 182–203.

Morris, W.F., M.V. Price, N.M. Waser, J.D. Thomson, B.A. Thomson, and D.A. Stratton. 1994. Systematic increase in pollen carryover and its consequences for geitonogamy in plant populations. *Oikos*, 71: 431–440.

Müller, H. 1883. *The Fertilisation of Flowers*. Macmillan, London.

Murcia, C. 1990. Effect of floral morphology and temperature on pollen receipt and removal in *Ipomoea trichocarpa*. *Ecology*. 71: 1098–1109.

Nagylaki, T. 1976. A model for the evolution of self-fertilization and vegetative reproduction. *J. Theor. Biol.*, 58: 55–58.

de Nettancourt, D. 1977. *Incompatibility in Angiosperms*. Springer-Verlag, Berlin.

Paton, D.C. and H.A. Ford. 1983. The influences of plant characteristics and honeyeater size on levels of pollination in Australian plants. In *Handbook of Experimental Pollination Biology* (C.E. Jones and R.J. Little, eds.), Van Nostrand Reinhold, New York, pp. 235–248.

Percival, M.S. and P. Morgan. 1965. Observations on the floral biology of *Digitalis* species. *New Phytol.*, 64: 1–22.

Piper, J.G., B. Charlesworth, and D. Charlesworth. 1986. Breeding system evolution in *Primula vulgaris* and the role of reproductive assurance. *Heredity*, 56: 207–217.

Plowright, R.C. and L.K. Hartling. 1981. Red clover pollination by bumble bees: A study of the dynamics of a plant-pollinator relationship. *J. Appl. Ecol.*, 18: 639–647.

Podolsky, R.D. 1992. Strange floral attractors: Pollinator attraction and the evolution of plant sexual systems. *Science,* 258: 791–793.

Podolsky, R.D. 1993. Evolution of a flower dimorphism: How effective is pollen dispersal by "male" flowers? *Ecology,* 74: 2255–2260.

Price, M.V. and N.M. Waser. 1982. Experimental studies of pollen carryover: Hummingbirds and *Ipomopsis aggregata*. *Oecologia,* 54: 353–358.

Rausher, M.D., D. Augustine, and A. VanderKooi. 1993. Absence of pollen discounting in a genotype of *Ipomoea purpurea* exhibiting increased selfing. *Evolution,* 47: 1688–1695.

Richards, A.J. 1986. *Plant Breeding Systems*. Allen and Unwin, London.

Rick, C.M., M. Holle, and R.W. Thorp. 1978. Rates of cross-pollination in *Lycopersicon pimpinellifolium:* Impact of genetic variation in floral characters. *Plant Syst. Evol.*, 129: 31–44.

Ritland, K. 1989. Correlated matings in the partial selfer *Mimulus guttatus*. *Evolution,* 43: 848–859.

Ritland, K. 1990. A series of FORTRAN computer programs for estimating plant mating systems. *J. Heredity,* 81: 235–237.

Ritland, K. 1991. A genetic approach to measuring pollen discounting in natural plant populations. *Am. Nat.*, 138: 1049–1057.

Ritland, K. and F.R. Ganders. 1985. Variation in the mating system of *Bidens menziesii* (Asteraceae) in relation to population substructure. *Heredity,* 47: 35–52.

Robertson, A.W. 1992. The relationship between floral display size, pollen carryover and geitonogamy in *Myosotis colensoi* (Kirk) Macbride (Boraginaceae). *Biol. J. Linn. Soc.*, 46: 333–349.

Robertson, A.W. and D.G. Lloyd. 1991. Herkogamy, dichogamy and self-pollination in six species of *Myosotis* (Boraginaceae). *Evol. Trends Plants,* 5: 53–63.

Schaal, B.A. 1980. Measurement of gene flow in *Lupinus texensis*. *Nature*, 283: 450–451.

Schaffer, W.M. and M.W. Schaffer. 1979. The adaptive significance of variations in reproductive habit in the Agavaceae II. Pollinator foraging behavior and selection for increased reproductive expenditure. *Ecology,* 60: 1051–1069.

Schemske, D.W. and R. Lande. 1985. The evolution of self-fertilization and inbreeding depression in plants. II. Empirical observations. *Evolution,* 39: 41–52.

Schmitt, J. 1983. Flowering plant density and pollinator visitation in *Senecio*. *Oecologia,* 60: 97–102.

Schoen, D.J. 1982. The breeding system of *Gilia achilleifolia:* Variation in floral characteristics and outcrossing rate. *Evolution,* 36: 352–360.

Schoen, D.J. 1985. Correlation between classes of mating events in two experimental plant populations. *Heredity,* 55: 381–385.

Schoen, D.J. and M.T. Clegg. 1984. Estimation of mating system parameters when outcrossing events are correlated. *Proc. Natl. Acad. Sci. USA,* 11: 5258–5262.

Schoen, D.J. and M. Dubuc. 1990. The evolution of inflorescence size and number: A gamete-packaging strategy in plants. *Am. Nat.*, 135: 841–857.

Seavey, S.R. and K.S. Bawa. 1986. Late-acting self-incompatibility in angiosperms. *Bot. Rev.*, 52: 195–219.

Simpson, B.B. and J.L. Neff. 1983. Evolution and diversity of floral rewards. In *Handbook of Experimental Pollination Biology* (C.E. Jones and R.J. Little, eds.), Van Nostrand Reinhold, New York, pp. 142–159.

Snow, A.A. and T.P. Spira. 1991. Pollen vigour and the potential for sexual selection in plants. *Nature*, 352: 796–797.

Squillace, A.E. 1967. Effectiveness of 400-foot isolation around a slash pine seed orchard. *J. Forestry,* 65: 823–824.

Stanton, M.L., T.-L. Ashman, L.F. Galloway, and H.J. Young. 1992. Estimating male fitness of plants in natural populations. In *Ecology and Evolution of Plant Reproduction* (R. Wyatt, ed.), Chapman & Hall, New York, pp. 62–90.

Stanton, M.L., A.A. Snow, S.N. Handel, and J. Bereczky. 1989. The impact of a flower-color polymorphism on mating patterns in experimental populations of wild radish (*Raphanus raphanistrum* L.) *Evolution*, 43: 335–346.

Stanton, M.L., H.J. Young, N.C. Ellstrand, and J.M. Clegg. 1991. Consequences of floral variation for male and female reproduction in experimental populations of wild radish, *Raphanus sativus* L. *Evolution,* 45: 268–280.

Stephenson, A.G. 1981. Flower and fruit abortion: Proximate causes and ultimate functions. *Ann. Rev. Ecol. Syst.*, 12: 253–279.

Stucky, J.M. 1985. Pollination systems of sympatric *Ipomoea hederacea* and *I. purpurea* and the significance of interspecific pollen flow. *Am. J. Bot.*, 72: 32–43.

Thomson, J.D. 1986. Pollen transport and deposition by bumble bees in *Erythronium:* Influences of floral nectar and bee grooming. *J. Ecol.*, 74: 329–341.

Thomson, J.D. 1988. Effects of variation in inflorescence size and floral rewards on the visitation rates of traplining pollinators of *Aralia hispida. Evol. Ecol.*, 2: 65–76.

Thomson, J.D. and D.A. Stratton. 1985. Floral morphology and cross-pollination in *Erythronium grandiflorum* (Liliaceae). *Am. J. Bot.*, 72: 433–437.

Thomson, J.D. and B.A. Thomson. 1989. Dispersal of *Erythronium grandiflorum* pollen by bumblebees: Implications for gene flow and reproductive success. *Evolution,* 43: 657–661.

Thomson, J.D., B.J. Andrews, and R.C. Plowright. 1981. The effect of a foreign pollen on ovule development in *Diervilla lonicera* (Caprifoliaceae). *New Phytol.*, 90: 777–783.

Toppings, P. 1989. The significance of inbreeding depression to the evolution of self-fertilization in *Eichhornia paniculata* (Spreng.) Solms (Pontederiaceae). M.S. thesis, Univ. Toronto, Canada.

Uyenoyama, M.K., K.E. Holsinger, and D.M. Waller. 1993. Ecological and genetic factors directing the evolution of self-fertilization. *Oxford Surv. Evol. Biol.*, 9: 327–381.

Waddington, K.D. and B. Heinrich. 1979. The foraging movements of bumblebees on vertical inflorescences: An experimental analysis. *J. Compar. Physiol.*, 134: 113–117.

Waller, D.M. and S.E. Knight. 1989. Genetic consequences of outcrossing in the cleistogamous annual, *Impatiens capensis*. II. Outcrossing rates and genotypic correlations. *Evolution*, 43: 860–869.

Walsh, N.E. and D. Charlesworth. 1992. Evolutionary interpretations of differences in pollen tube growth rates. *Quar. Rev. Biol.*, 67: 19–37.

Waser, N.M. 1988. Comparative pollen and dye transfer by pollinators of *Delphinium nelsonii*. *Func. Ecol.*, 2: 41–48.

Waser, N.M. 1993. Population structure, optimal outbreeding, and assortative mating in angiosperms. In *The Natural History of Inbreeding and Outbreeding: Theoretical and Empirical Perspectives* (N.W. Thornhill, ed.), Univ. Chicago Press, Chicago, pp. 173–199.

Waser, N.M. and M.L. Fugate. 1986. Pollen precedence and stigma closure: A mechanism of competition for pollination between *Delphinium nelsonii* and *Ipomopsis aggregata*. *Oecologia*, 70: 573–577.

Waser, N.M. and M.V. Price. 1983*a*. Pollinator behaviour and natural selection for flower colour in *Delphinium nelsonii*. *Nature*, 302: 422–424.

Waser, N.M. and M.V. Price. 1983*b*. Optimal and actual outcrossing in plants, and the nature of plant-pollinator interaction. In *Handbook of Experimental Pollinator Biology* (C.E. Jones and R.J. Little, eds.), Van Nostrand Reinhold, New York, pp. 341–372.

Waser, N.M. and M.V. Price. 1984. Experimental studies of pollen carryover. Effects of floral variability in *Ipomopsis aggregata*. *Oecologia*, 62: 262–268.

Waser, N.M. and M.V. Price. 1993. Crossing distance effects on prezygotic performance in plants: An argument for female choice. *Oikos*, 68: 303–308.

Wells, H. 1979. Self-fertilization: Advantageous or deleterious? *Evolution*, 33: 252–255.

Weiss, M.R. 1991. Floral colour changes as cues for pollinators. *Nature*, 354: 227–229.

Weller, S.G. and A.K. Sakai. 1990. The evolution of dicliny in *Schiedea* (Caryophyllaceae), an endemic Hawaiian genus. *Plant Spec. Biol.*, 5: 83–96.

Wolfe, L.M. and S.C.H. Barrett. 1989. Patterns of pollen removal and deposition in tristylous *Pontederia cordata* L. (Pontederiaceae). *Biol. J. Linn. Soc.*, 36: 317–329.

Wilson, P. and J.D. Thomson. 1991. Heterogeneity among floral visitors leads to discordance between removal and deposition of pollen. *Ecology*, 72: 1503–1507.

Wyatt, R. 1986. Ecology and evolution of self-pollination in *Arenaria uniflora* (Caryophyllaceae). *J. Ecol.*, 74: 403–418.

Yeo, P.F. 1975. Some aspects of heterostyly. *New Phytol.*, 75:147–153.

Young, H.J. and M.L. Stanton. 1990. Influences of floral variation on pollen removal and seed production in wild radish. *Ecology*, 71: 536–547.

7

The Ecology of Geitonogamous Pollination

Allison A. Snow, Timothy P. Spira,[†] Rachel Simpson,[‡] and Robert A. Klips**

Introduction

Working as a natural historian in the 1700s, C.K. Sprengel wrote a pioneering book demonstrating that many hermaphroditic species require pollinator visits in order to produce seed (see Chapters 1 and 2). He did not provide a scientific explanation as to why cross-pollination is important, but in the next century Darwin, H. and F. Müller, and others proposed that various outcrossing mechanisms have evolved to avoid selfing and the consequences of inbreeding (Darwin, 1876; see Baker, 1983). Darwin also recognized that the potential for selfing is greatest in species with massive floral displays because having many flowers promotes the transfer of self-pollen to other flowers on the same genetic individual (*geitonogamy*). Following Darwin's lead, many authors have suggested that the avoidance of selfing has been a major factor in the evolution of traits such as dioecy, self-incompatibility, monoecy, temporal separation of male and female organs (*dichogamy*), spatial separation of anthers and stigmas within flowers (*herkogamy*), and having few open flowers per day (see reviews by Arroyo, 1976; Lloyd, 1979; Bawa and Beach, 1981; Willson, 1983; Wyatt, 1983; Richards, 1986; Charlesworth and Charlesworth, 1987; Thomson and Brunet, 1990; de Jong, et al., 1992*a*; Harder and Barrett 1995; Hodges, 1995; also see Chapters 6, 8, and 14).

More recently, evolutionary ecologists have suggested that self-pollination is

*Department of Plant Biology, Ohio State University, 1735 Neil Ave., Columbus, OH 43210-1293.

[†]Department of Biological Sciences, Clemson University, Clemson, SC 29634-1903.

[‡]Department of Biology, University of Michigan, Ann Arbor, MI 48109.

Send correspondence to A. A. Snow (Tel.: 614-292-3445; Fax: 292-6345; E-mail: snow.1@osu.edu).

also detrimental in species that are completely *self-incompatible* (Lloyd and Webb, 1986; Webb and Lloyd, 1986; Bertin, 1993; Bertin and Newman, 1993; Klinkhamer and de Jong, 1993; also see Chapter 13). In these species, self-pollination can potentially reduce seed set by (1) diluting the amount of compatible pollen arriving on stigmas, (2) taking up limited space on the stigmatic surface, or (3) preventing outcross pollen tubes from fertilizing ovules. Self-pollination may also limit *male* reproductive success by wasting pollen that could otherwise sire seeds on conspecific neighbors (Holsinger et al., 1984; Lloyd and Webb, 1986; Charlesworth and Charlesworth, 1987; Schoen and Lloyd, 1992; de Jong et al., 1993; Klinkhamer and de Jong, 1993; see Harder and Barrett 1995, Chapter 6). Therefore, whether a species is self-compatible (SC) or self-incompatible (SI), self-pollination may reduce an individual's fitness. In fact, some features that have been considered adaptive mechanisms to reduce inbreeding (e.g., dichogamy) are equally common in SC and SI species, suggesting that avoidance of inbreeding is not the only reason for plants to avoid self-pollination (Bertin, 1993; Bertin and Newman, 1993).

Within-flower self-pollination can be lessened by spatial or temporal separation of anthers and stigmas, but neither of these traits prevents geitonogamous self-pollination. Protandry and protogyny are seldom synchronized within a plant or even within an inflorescence, so self-pollen may be transferred from male to female phase flowers on the same plant. Exceptions include some monoecious and andromonoecious species that are functionally unisexual at any given time and change sex within or between years (e.g., Stout, 1928; Cruden and Hermann-Parker, 1977; Bierzychudek, 1982; Thomson and Barrett, 1981; Gleeson, 1982; Cruden, 1988; Schlessman et al., 1990). For example, umbellifers achieve synchronous dichogamy by delaying stigma receptivity within an umbel until the staminate phase of all early-blooming flowers has passed, and by having synchronous development within umbel orders (i.e., primary, secondary, tertiary; Cruden and Hermann-Parker, 1977). This flexibility of sex expression is seldom possible in plants with hermaphroditic flowers (but see McDade, 1986; Cruden, 1988). Thus, the only species that can completely avoid geitonogamy are those with synchronized dichogamy, a single open flower per day, or dioecy.

Several authors have suggested that geitonogamy leads to high rates of self-fertilization in SC species, and reduces reproductive success in those that are SI (e.g., Lloyd and Webb, 1986; Hessing, 1988; de Jong et al., 1993; Klinkhamer and de Jong, 1993). In contrast, others have argued that the costs associated with geitonogamy are relatively minor because pollinators rarely visit more than a few flowers per plant, and pollen carryover allows some outcross pollen to reach most of these flowers' stigmas (e.g., Geber, 1985; Robertson, 1992). The extent of geitonogamous pollination undoubtedly varies a great deal within and among species, depending on factors such as daily flower number (e.g., Stephenson, 1982; Schoen and Dubuc, 1990), plant densities (e.g., Beattie, 1976; Heinrich, 1979; Klinkhamer and de Jong, 1990; Murawski and Hamrick, 1991), and pollinator behavior (e.g., Frankie et al., 1976; Waddington and Heinrich, 1979; Schmitt, 1980; Best and Bierzychudek, 1982). Few investigators have attempted to quantify geitonogamous pollination, but interest in this area appears to be growing, as indicated by a recent spate of papers on the subject (Klinkhamer et al. 1989, 1993; Dudash, 1991; de Jong et al., 1992*b*, 1993; Lloyd, 1992; Lloyd and Schoen, 1992; Robertson, 1992; Schoen and Lloyd, 1992; Leclerc-Potvin and Ritland, 1993; Morse, 1994; Barrett, et al., 1994).

The purpose of this chapter is to review ecological factors that affect the frequency of geitonogamy in plants with >1 flower per day and assess the impact of this process on individual fitness. To keep the topic manageable, we focus on geitonogamy in animal-pollinated species with hermaphroditic flowers. We begin with a review of how pollinators respond to variation in flower number, followed by a discussion of how much self-pollination occurs due to consecutive within-plant visits. We then show how geitonogamous pollination affects selfing rates and seed set, emphasizing experimental studies of both SC and SI species. Surprisingly, few field studies document the effects of daily flower number on selfing rates or overall reproductive success, and throughout this chapter we evaluate methods by which this relationship can be explored empirically.

Pollinator Behavior and Daily Flower Number

Pollinators often forage in restricted areas by moving from one flower or inflorescence to its immediate neighbor (e.g., Pyke 1978*a, b;* Schmitt, 1980), and this behavior is expected to lead to frequent geitonogamy. In nature, however, pollinators' foraging movements at a given plant are often curtailed after a few visits. Lloyd and Schoen (1992) noted that geitonogamy would be far more common were it not for "the fortunate and still largely unexplained habit shared by virtually all flower visitors of visiting only a fraction of the available flowers on a plant before moving on to the next plant." Possible reasons for short visitation sequences include satiation, depleted floral rewards, competition with other pollinators, the need to find other types of food, predator avoidance, and behaviors such as searching for mates. Two major questions, then, are how many flowers on a plant are visited consecutively, and how do these within-plant visits affect the numbers of self and outcross pollen grains deposited on a plant's stigmas?

Many studies have shown that pollinators visit fewer than about 10–20 flowers on a plant even when far more flowers are present (see Robertson, 1992). In a Costa Rican population of *Combretum farinosum,* hummingbirds probed an average of only 12 flowers per plant when more than 1000 flowers were available (Schemske, 1980). Augspurger (1980) found that the mass-flowering shrub *Hybanthus prunifolius* produced approximately 200 flowers per day at peak flowering, yet social bees visited an average of about 10 flowers per plant before leaving. In populations of *Geranium caespitosum,* Hessing (1988) observed that most plants had <20 flowers, and bees visited about half of the available flowers on each plant. Visitation sequences were even shorter at large individuals of *Sabatia angularis:* small bees, wasps, and hover flies visited an average of two flowers out of a mean of 35 available flowers on each plant (Dudash, 1991). Waser (1982) compared visit lengths of five types of pollinators foraging on three subalpine wildflower species in Colorado (Table 7.1). Hawkmoths and hummingbirds visited more flowers per plant (about two to seven flowers) than did bumble bees, halictid bees, or butterflies, which visited fewer than three flowers per plant. Visitation sequences were considerably longer in a fly-pollinated perennial herb with clusters of prostrate ramets (Fig. 7.1; Robertson, 1992). In a New Zealand population of *Myosotis colensoi,* Robertson found that tachinid flies visited an average of 19 flowers on the largest individuals (about 100 flowers per plant).

Clonal plants such as *M. colensoi* may be especially prone to geitonogamy if pollinators make consecutive visits to ramets of the same genetic individual (Handel, 1985; Morse, 1994). Frequent geitonogamy is also expected when one or a few plants comprise most of a pollinator's foraging area. For example,

*Table 7.1. Visit lengths (numbers of flowers per plant) of different pollinators of three protandrous, self-incompatible, subalpine wildflower species.**

Plant and Pollinator Species	Mean Number of Flowers Visited per Plant	N
Delphinium nelsonii (2–8 flowers/day)**		
Hummingbird (*Selasphorous platycerus*)	2.2	537
Bumblebee queen (*Bombus appositus*)	1.6	319
Halictid bee (*Halictus* spp.)	1.4	9
Delphinium barbeyi (10–20 flowers/day)†		
Hummingbird (*Selasphorous rufus*)	3.4	132
Bumblebee queen (*Bombus appositus*)	2.7	130
Bumblebee worker (*Bombus appositus*)	2.7	23
Ipomopsis aggregata (10–20 flowers/day)‡		
Hummingbird (*Selasphorous* spp.)	6.5	275
Bumblebee queen§ (*Bombus appositus*)	2.8	151
Hawkmoth§ (*Hyles lineata*)	4.7	239
Butterfly§ (*Papilio rutulus*)	2.5	45

*From Table 1 in Waser (1982); *N* is number of observations.

**Schulke, personal communication to Snow; 1993 data, mean = 4.6, *N* = 76.

†Waser, personal communication to Snow; 1976–1981 data.

‡Flower number from deJong et al. (1992); mean at peak flowering was 12.3 open flowers.

§Not a major pollinator.

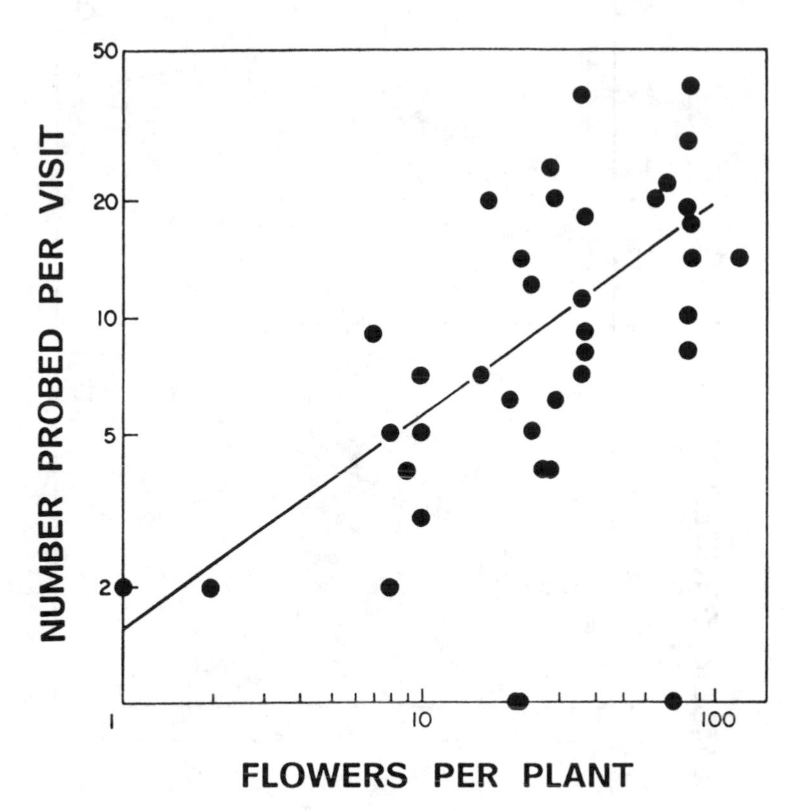

Figure 7.1. The relationship between daily flower number and the number of flowers probed per visit by tachinid flies on *Myosotis colensoi,* a clonal plant (reprinted with permission from Robertson, 1992). The linear regression line does not include insects that left after a single visit.

Schemske (1980) noted that large individuals of *Combretum farinosum* were defended by territorial hummingbirds, and the proportion of flowers that set fruit declined as flower number increased (also see Carpenter, 1976).

Several studies have demonstrated that pollinators visit more flowers in succession as flower number increases, but the *proportion* of flowers visited declines dramatically on larger plants (Figs. 7.1 and 7.2; Pyke 1978*a, b;* Geber, 1985; Anderson and Symon, 1988; Hessing, 1988; Schmid-Hempel and Speiser, 1988; Klinkhamer and de Jong, 1990). The shape of this response curve has rarely been determined, but for plants with very large floral displays a decelerating curve might be expected (e.g., Fig. 7.2a). Thus, if geitonogamy is a detrimental side-effect of large floral displays, the costs associated with self-pollination may decrease relative to the benefits gained by producing many flowers (but see de Jong et al., 1993, and Chapter 6).

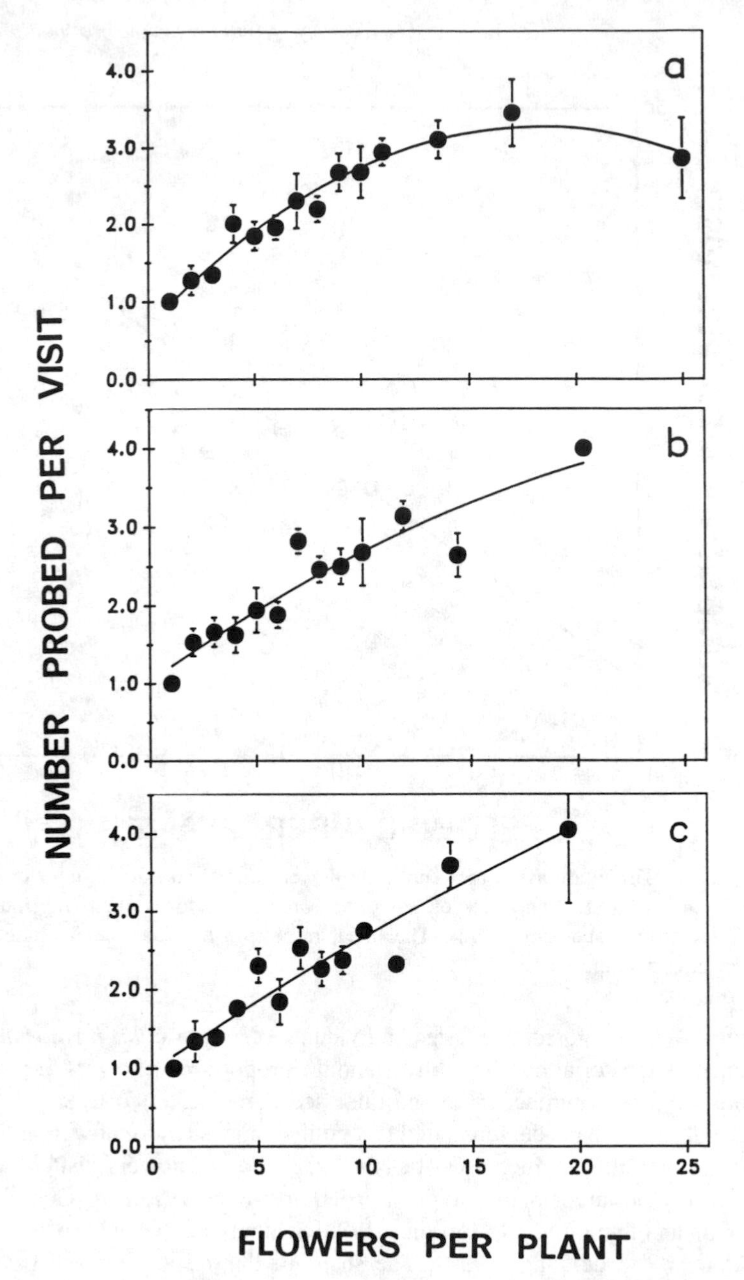

Figure 7.2. The relationship between daily flower number and the number of flowers probed by bumble bees on *Aconitum columbianum* (reprinted with permission from Pleasants and Zimmerman, 1990). Data are means ± 1 SD from a total of 393 inflorescences (each plant had one inflorescence) observed on 3 days (labeled a, b, and c). Polynomial regression lines are shown.

Pollen Carryover

The frequency of geitonogamous pollination also depends on how much self- vs. outcross pollen is deposited on a plant's stigmas during consecutive within-plant visits. The dynamics of pollen accumulation and turnover on a pollinator's body are still poorly understood due to the difficulty of observing this process directly (Lertzmann and Gass, 1983; Robertson, 1992; Chapter 6). When pollen carryover is limited, most of the pollen picked up at one flower is deposited on the next, so visits to a sequence of flowers on the same plant result in a great deal of self-pollination. In contrast, extensive pollen carryover allows some outcross pollen grains to be deposited throughout a pollinator's foraging bout at a given plant, especially if the pollinator visits only a few flowers before departing.

Direct observations of pollen carryover (or self-pollination in general) are rare because it is usually impossible to distinguish between self- and outcross pollen in stigmatic pollen loads. An exception is heterostylous species in which pollen size differs among flower morphs. Mixtures of pollen grains from different morphs can be observed on stigmas, and carryover can be studied by observing the decay curve for outcross pollen loads during visitation sequences to previously unvisited flowers (e.g., Feinsinger and Busby, 1987; Wolfe and Barrett, 1989; Chapter 6). Other species that allow the identification of self- vs. outcross pollen are those in which outcross pollen always germinates and self-pollen does not produce more than a short tube (as occurs in species such as *Passiflora vitifolia,* which has sporophytic self-incompatibility; Snow, personal observation). In this case, Alexander's stain can be used to discriminate between germinated outcross pollen and self-incompatible pollen grains containing cytoplasm (Alexander, 1969). However, in many species a substantial number of outcross pollen grains fail to germinate, making them indistinguishable from self-pollen (Snow, personal observation). Furthermore, in species with gametophytic self-incompatibility, self-pollen grains produce pollen tubes that resemble those from outcross pollen, so it is not possible to quantify the extent of self-pollination by direct observations of stigmas or styles.

Some investigators have estimated carryover of outcross pollen by counting the number of pollen grains deposited on stigmas in a sequence of visits to emasculated, previously unvisited recipient flowers (see refs. in Robertson, 1992, and Fig. 7.3). A major problem with this technique, however, is that there is no opportunity for self-pollen to be picked up and deposited on subsequent stigmas in place of outcross pollen, or for self-pollen to interfere with the performance of outcross pollen. In addition, pollen-collecting bees might visit emasculated flowers less often or for shorter time periods than normal flowers.

Researchers have also quantified pollen dispersal by using fluorescent dye powders as pollen analogues (e.g., Waser and Price, 1982, 1984; Thomson et al., 1986; Hessing, 1988; Waser, 1988; Morris et al., 1994). Pollen analogues are useful for comparing qualitative differences in geitonogamy due to variables

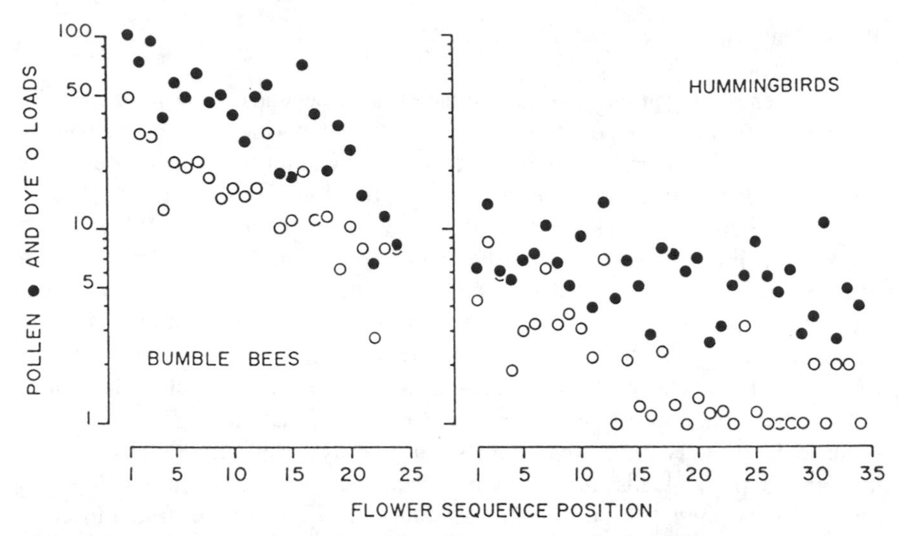

Figure 7.3. Pollen and dye loads on stigmas of emasculated flowers of *Delphinium nelsonii,* shown in order of their position in visitation sequences (reprinted with permission from Waser, 1988). Each value is a mean of at least three observations.

such as daily flower number (e.g., Dudash, 1991; de Jong et al., 1992*b*), although dispersal patterns of powders and pollen grains can be sufficiently different to preclude using dye particles for accurate estimates of pollen dispersal (Thomson et al., 1986; Fig. 7.3). For example, bumble bees deposited about twice as much pollen compared to dye particles in visits to sequences of *Delphinium* flowers (Fig. 7.3). In species with pollinia, this problem can be avoided by labeling the entire pollen dispersal unit (e.g., Pleasants, 1991; Nilsson et al., 1993). Peakall (1989) used this technique to show that 22% of the pollinia on stigmas of the orchid *Prasophyllum fimbria* were the result of self-pollination. In populations of *Erythronium grandiflorum,* a pollen color polymorphism was used to measure pollen carryover (Thomson, 1986). Unfortunately, the color difference disappeared when pollen germinated on the stigma, introducing a good deal of error into estimates of pollen dispersal (Thomson, 1986). Despite limitations of various techniques for quantifying carryover and self-pollination, the methods described above have been useful for demonstrating that geitonogamy is common and generally increases with daily flower number (e.g., Hessing, 1988; Dudash, 1991; de Jong et al., 1992*b*). Furthermore, estimates of pollen carryover can be used to model pollen dispersal and investigate the effects of variation in traits such as daily flower number on self-pollination (see reviews by Robertson, 1992; de Jong et al., 1993; Chapter 6). Because most models rely on untested assumptions and rather rough estimates of pollen carryover, however, empirical studies of geitonogamous pollination in natural populations are essential.

Geitonogamy in Self-Compatible Species

In this section we focus on the effects of geitonogamy on selfing rates, realizing that selfing can also be influenced by postpollination processes such as differential pollen-tube growth, flower abortion, or seed abortion following self- vs. outcross pollination (e.g., Bateman, 1956; Bowman, 1987; Bertin and Sullivan, 1988; Casper et al., 1988; Aizen et al., 1990; Becerra and Lloyd, 1992; Montalvo, 1992; Cruzan and Barrett, 1993). If self-pollen tubes grow more slowly than outcross or if the resulting embryos are more likely to abort, then using genetic markers to determine selfing rates based on seed or seedling genotypes can lead to underestimates of the true extent of self-pollination. Nonetheless, in many species self- and outcross pollen tubes grow at similar rates and/or are equally effective at siring viable seeds (e.g., Waser and Price, 1991, Snow and Spira, 1993; Johnston, 1993). When this is the case and seed set per fruit is *pollen-limited,* the relative proportion of self-pollen on the stigma should be closely correlated with the proportion of selfed seeds in a fruit. If seed set per fruit is not pollen-limited (e.g., Snow, 1986), differences in the timing of self- and outcross pollen deposition will also affect the selfing rate, because earlier arriving pollen is favored (Epperson and Clegg, 1987; Barrett et al. 1992; Lloyd and Schoen, 1992). For example, early within-plant visits to flowers with freshly dehisced anthers will result in a higher proportion of selfed seeds than later visits that occur when much of a plant's pollen has been removed.

One way to assess the extent of selfing due to geitonogamy is to determine the relationship between daily flower number and the proportion of self- vs. outcross progeny produced from those flowers. This is often difficult to carry out in natural populations because many genetic markers are needed to distinguish between self- and outcross progeny; hence, most research on plant mating systems to date deals with population-level rather than individual-level measures of outcrossing rates (e.g., Brown and Allard, 1970; Clegg, 1980; Ritland and Jain, 1981; Sun and Ganders, 1988; Waller and Knight, 1989; but see Ritland and El-Kassaby, 1985; Morgan and Barrett, 1990; Murawski et al., 1990; Motten and Antonovics, 1992; Cruzan et al., 1994).

To address this problem, Schoen and Lloyd (1992) developed statistical methods for estimating the mating systems of distinct classes of plants, such as those with one vs. many flowers per day, and demonstrated how these estimates can be used to quantify autodeposition and geitonogamous pollination, respectively. In a natural population of *Impatiens pallida,* they found that fruits from singly displayed flowers had a selfing rate of 6%, whereas the selfing rate of flowers on unmanipulated plants with three or more flowers each day was much higher, 44%. Thus, most of the selfing in this protandrous, self-compatible annual was due to geitonogamy. This may be the case in many self-compatible species with spatial or temporal separation of anthers and stigmas, but more studies of this kind are needed.

Other investigators have studied geitonogamous selfing in artificial arrays of experimental plants. This strategy allows the investigator to maintain control over genetic markers and variables such as spatial arrangement, population size, plant size, and daily flower number. A possible disadvantage is that more selfing is expected in small, artificial populations than larger ones, where pollinators are likely to forage over a greater area. In a study of a protandrous herbaceous perennial, *Malva moschata,* Crawford (1984) established an array consisting of a central plant surrounded by six others that were homozygous at a distinct allozyme locus. He then labelled flowers on the central plant and recorded the number of other flowers in anthesis on the same day. Electrophoresis of the central plant's progeny showed that the selfing rate was strongly correlated with daily flower number (Fig. 7.4). The details of Crawford's study have not been published, but this work indicates that when daily flower number increased from 10–20, the selfing rate increased from about 10–40%. Honeybees and bumble bees were the major pollinators, and they apparently visited more flowers in succession on days when more flowers were available on the central plant. Similar results were obtained when much larger plants were used (Crawford, 1984), suggesting that perhaps pollinators did not leave the plants as quickly as in other studies cited above.

We conducted a similar experiment to determine the effect of daily flower number on visit length and selfing rate in wild rose mallow, *Hibiscus moscheutos* (Snow et al., "unpublished data"). The pollination biology of this large, marsh-dwelling perennial is described in detail in Spira (1989), Spira et al. (1992), and Snow and Spira (1993). Briefly, bumble bees and a specialist anthophorid bee, *Ptilothrix bombiformis,* make frequent visits to the showy single-day flowers, resulting in strong competition among pollen tubes for ovules. *Hibiscus moscheutos* is self-compatible, but automatic selfing is prevented by spatial separation of anthers and stigmas. We detected no consistent differences in the growth rates of self- and outcross pollen tubes or in the number of seeds sired by each pollen type (Snow and Spira, 1993; R. Klips et al., unpublished data). Therefore, selfing rates should reflect the relative amounts of self- and outcross pollen grains arriving on stigmas.

Population-wide estimates of the mating system have been hindered because the only polymorphic allozyme locus we have found to date is glucophosphoisomerase (GPI), which has three common alleles at a population in Edgewater, Maryland. Using GPI, we estimated that the selfing rate at this population was 36%, based on electrophoresis of 434 seeds from 38 maternal families (calculated as in Ritland, 1990).

Individual plants produce multiple shoots from a fibrous rootstock (no lateral spreading), and at peak flowering up to 20 or more flowers are open at once on the largest plants. At a population in Ohio, the average number of open flowers per plant was 6.1 ± 4.3 (mean, SD, N = 70) during the height of flowering. In a Maryland population sampled during the latter part of the blooming period,

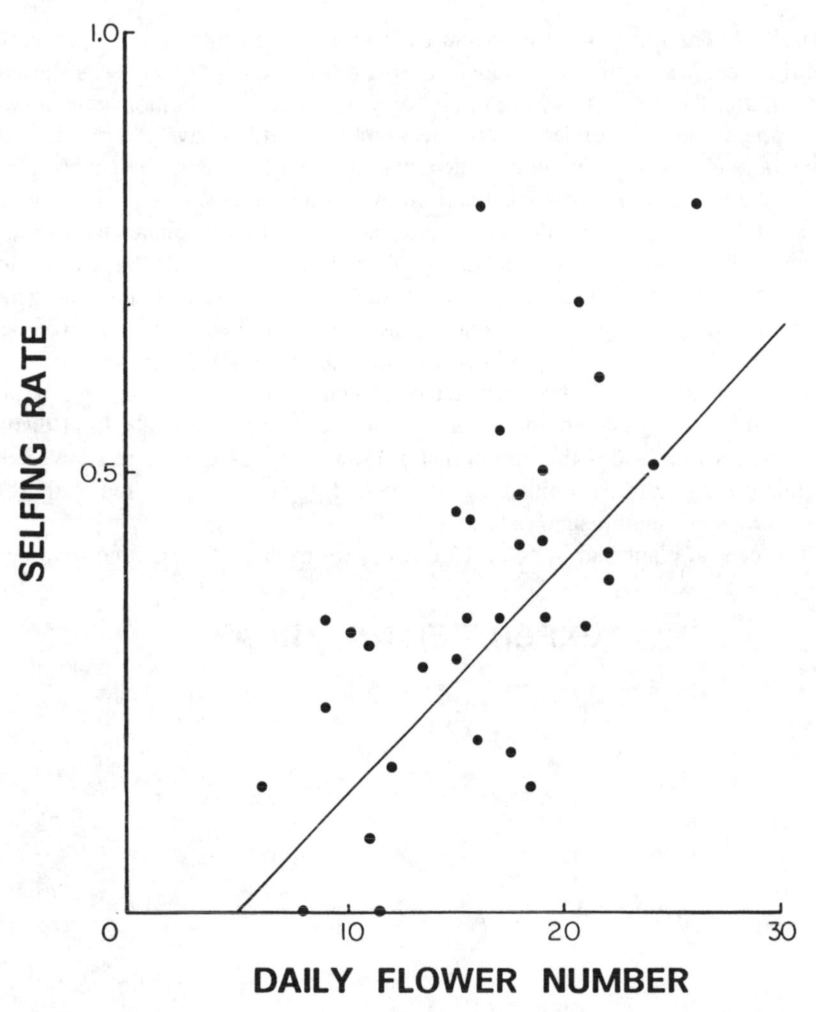

Figure 7.4. The relationship between daily flower number and selfing rate of a single plant in an experimental array of *Malva moschata*, a protandrous, bee-pollinated weedy herb (reprinted with permission from Crawford, 1984). A weighted regression line is shown.

only 3.4 ± 1.8 (N = 47) flowers were open per plant. Given this range in daily flower numbers, we conducted an experiment to compare the selfing rates of plants with 3, 6, or 12 flowers open simultaneously.

Methods

In 1992 we established an artificial array of potted plants in an open field located about 3 km from the nearest natural or artificial population in Edgewater,

Maryland. The plants used in this array were collected from a local population (Mill Swamp) and grown outdoors in 20-L tubs. The experiment was started 2 weeks after the first plants began to bloom. A central plant homozygous for one GPI allele was surrounded by six near neighbors (3 m away) and six outer neighbors (6 m away), none of which shared GPI alleles with the central plant, as shown in Fig. 7.5. Flower number on the central plant was adjusted each day to a total of 3, 6, or 12 flowers, while each neighboring plant always had 3 flowers. To maintain the designated number of flowers, we placed two or more plants side by side at a given position in the array so as to resemble one genet. (For simplicity, the plants at each position are described as if they were one plant, but they actually represented two or more genets.) When necessary, we also taped freshly cut flowers in water-filled vials onto the shoots to control daily flower number. Every attempt was made to achieve a "normal" floral display (all flowers were accessible to pollinators), and a pilot experiment showed that pollinators did not discriminate against picked flowers as compared to those *in situ* (Simpson, unpublished data).

The central plant had 3, 6, or 12 flowers for each of 4 days, and treatments

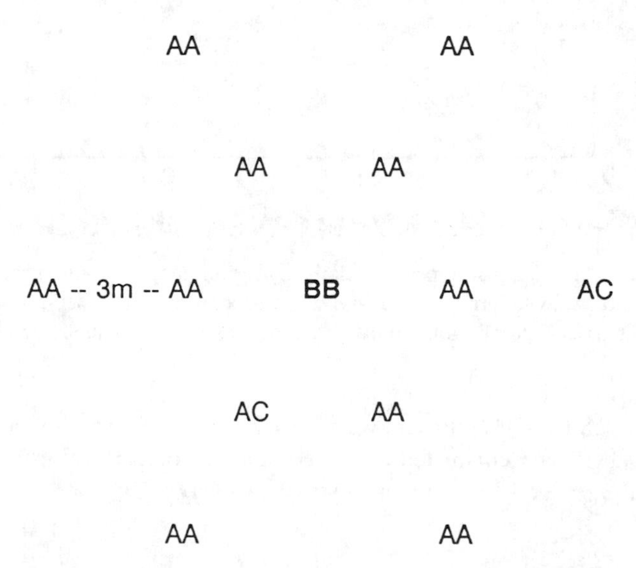

Figure 7.5. Diagram of the experimental array of potted individuals of *Hibiscus moscheutos*. Distance between plants is shown, and letters indicate GPI genotypes; see text for details.

were alternated among days (pollen survives for only 1 day, so contamination between days was not a problem). All flowers on the central plant were labeled appropriately for later fruit collection. On 3 other days when all central flowers were removed, we sampled fruits from near neighbors in order to check for unintended gene flow from other populations. Allozyme analysis revealed that 4% of 359 seeds from 28 of these fruits were sired by "alien" pollen with the same allele as the central plant. Actual levels of unintended gene flow may have been somewhat higher due to immigration of pollen with the same genetic markers as those of other plants in the array.

Both anthophorid bees and bumble bees visited the flowers often, resulting in a mean fruit set of 91% on the central plant (averaged over the 12 experimental days). Pollinator abundance varied somewhat from day to day, although the weather was uniformly hot, humid, and without rain throughout the experiment. To quantify the potential for geitonogamy, we recorded the number of consecutive flowers a bee visited per foraging bout at the central plant. A total of 18–25 foraging bouts were observed on 2–3 days for each flower number treatment. Selfing rates were determined by sampling an equal number of seeds from each available fruit (90–115 seeds per day; 390–432 seeds per flower number treatment). A seed with the same homozygous GPI genotype as the central plant was assumed to be the result of selfing, and the mean selfing rate for each treatment was based on $N = 4$ days.

Results and Discussion

Pollinator responses provided a good indicator of how flower number affected selfing rates of the central plant (Figs. 7.6a and 7.6b). When only three flowers were displayed, pollinators visited an average of 1.5 flowers per visit and the selfing rate averaged 25%. When six flowers were present, visit length approximately doubled as did the selfing rate, presumably due to geitonogamous pollination. However, increasing the floral display to 12 flowers had little additional effect on either visit length or selfing rate. We did not detect a significant effect of flower number on the proportion of selfed seed using ANOVA, but the power of this analysis was limited because the number of replicates was small ($N = 4$ days). Our results indicate that geitonogamy can indeed lead to higher selfing rates, in this case up to 50% selfing. Furthermore, the functional relationship between flower number and selfing rate appeared to be nonlinear, as expected based on pollinators' responses to flower number.

Barrett et al. (1994) conducted a similar experiment with *Eichhornia paniculata,* a bee-pollinated, tristylous annual native to Brazil. This species produces up to 20 single-day, nondichogamous flowers at once and is self-compatible. Long-styled plants with 3, 6, 9, or 12 flowers were exposed to bumble bee pollinators in Ontario, Canada (see also Chapter 6). In the long-styled morph, self- and outcross pollen do not differ in pollen-tube growth rates or the proportion

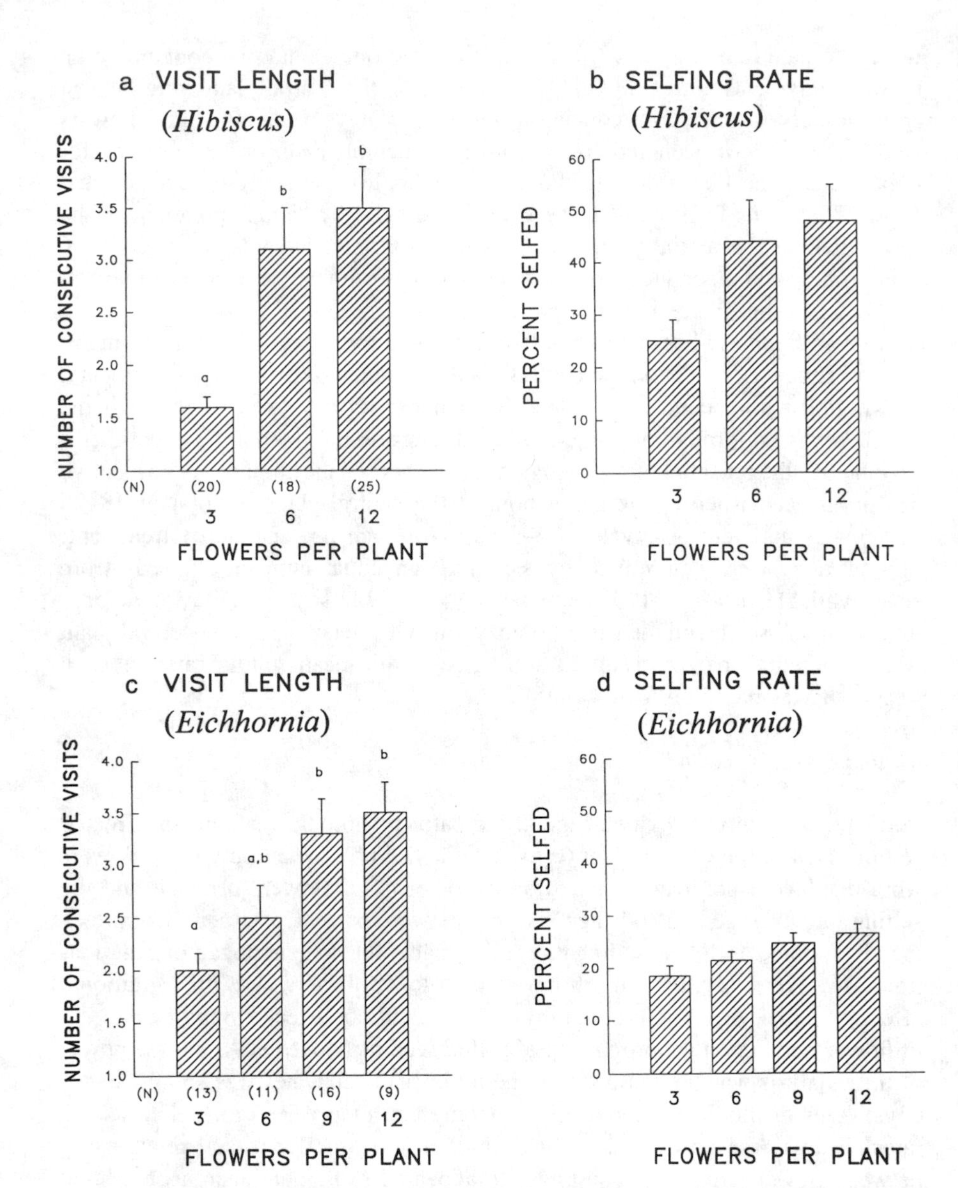

Figure 7.6. The relationship between daily flower number and visit length or selfing rate in *Hibiscus moscheutos* (a and b from Snow et al., in preparation) and *Eichhornia paniculata* (c and d from Barrett et al., 1994; reprinted with permission). Means ± 1 SE; means with the same superscripts are not significantly different ($P > 0.05$; Tukey tests). $N = 4$ for percent selfing (b and d); see text for further details.

of seeds sired (Cruzan and Barrett, 1993), so setting rates should reflect the extent of self-pollination.

Four groups of 36 potted plants with uniform flower numbers were placed outdoors, and the mating system for each group was calculated using data from two allozyme loci. As in the *Hibiscus* study, pollinators visited more flowers in succession on larger plants, but the difference between the largest size classes, in this case 9 vs 12 flowers, was not significant (Fig. 7.6c). Selfing rates increased with flower number, as expected, to a maximum of 26% (Fig. 7.6d). Somewhat higher selfing rates (35%) were reported for 12-flowered *Eichhornia* plants in other arrays (pooled data from three types of arrays in which half of the plants had 3, 6, or 12 flowers per day, respectively; see Chapter 6).

Taken together, these studies of *Impatiens, Malva, Hibiscus,* and *Eichhornia* demonstrate that geitonogamy can have a major effect on selfing rates, even in plants with as few as six flowers per day. The relationship between daily flower number and the proportion of selfed progeny may be linear or asymptotic, depending on pollinator behavior and the extent of pollen carryover. The artificial array experiments with *Hibiscus* and *Eichhornia* are noteworthy because they included observations of pollinator movements as well as data on selfing rates. Pollinators responded similarly to daily flower number in the two studies, visiting a maximum of three to four flowers on the largest plants, but higher selfing rates were seen in *Hibiscus* than *Eichhornia*.

As mentioned earlier, geitonogamous pollination may be more frequent in small experimental arrays than larger natural populations [however, note that the single-locus selfing rate at the Mill Swamp population (36%) was within the range shown in Fig. 7.6b]. Other factors that can affect the potential for geitonogamous selfing include average nectar levels (e.g., Heinrich, 1979; Galen and Plowright, 1985; Kadmon and Shmida, 1992), distance to flowering neighbors (e.g., Ellstrand et al., 1978; Crawford, 1984; Murawski and Hamrick, 1991; van Treuren et al., 1993), and a plant's size relative to its neighbors (e.g., Crawford, 1984; Dudash, 1991; also see Chapter 6). For example, Harder and Barrett (Chapter 6) reported longer visitation sequences when large plants were surrounded by smaller plants as opposed to those of comparable size.

The fitness consequences of geitonogamous selfing depend on the extent of inbreeding depression, which is largely determined by the genetic load of a given population (Charlesworth and Charlesworth, 1987). Obtaining ecologically meaningful estimates of inbreeding depression is difficult (Schemske, 1983; Schoen, 1983; Dudash, 1990), but it is clear that the relative performance of selfed progeny varies among individuals (e.g., Dudash, 1990; Snow and Spira, 1993), populations (Holtsford and Ellstrand, 1990; see Chapter 14), and, most markedly, among species (see refs. in Ågren and Schemske, 1993). When inbreeding depression is especially strong, selfed embryos may abort prior to maturation (e.g., Denti and Schoen, 1988; Levin, 1991). In other species inbreeding depression appears to be negligible (e.g., Flores, 1990; Barrett and

Charlesworth, 1991). Frequent selfing is thought to result in diminished inbreeding depression, as confirmed by some empirical studies (Holtsford and Ellstrand, 1990; Barrett and Charlesworth, 1991; Barrett and Kohn, 1991; but see Ågren and Schemske, 1993). In conclusion, the cost of self-pollination may range from none at all to a severe loss of fitness, depending on the extent of inbreeding depression.

Geitonogamy in Self-Incompatible Species

Self-pollination within and among flowers of SI species has two potential costs: reduced pollen dispersal, which could limit male reproductive success, and reduced female fecundity due to dilution of outcross pollen or interference by self-pollen grains. Few empirical studies have attempted to quantify these costs or determine the extent to which geitonogamy vs. autodeposition are responsible for them. Studies of pollen export have been hindered by the well-known difficulties of carrying out paternity analysis under natural conditions (reviewed by Snow and Lewis, 1993). Likewise, determining whether self-pollen dilutes the amount of outcross pollen on stigmas is problematic because it is usually impossible to distinguish between self- and outcross pollen grains. More is known about interference, also referred to as stigma clogging, which is described below.

Hand-Pollination Experiments

Evidence for interference has been found in some species but not others (see references in Shore and Barrett, 1984; Bertin and Sullivan, 1988; Galen et al., 1989; Scribailo and Barrett, 1994; see also Chapter 13), but most of these hand-pollination studies involve unrealistic pollination treatments. For example, when large mixtures of self- and outcross pollen are applied to stigmas, with pure outcross pollen as a control, seed set may be similar in the two treatments because ample outcross pollen was used (e.g., Shore and Barrett, 1984). Smaller pollen loads, such as those found in some pollen-limited populations (e.g., Snow, 1982, 1986), might reveal a negative effect of self-pollen on seed set.

Galen et al. (1989) conducted a detailed hand-pollination study of the extent to which self-pollen interferes with outcross pollen in a SI alpine perennial, *Polemonium viscosum*. Previous hand-pollination with self-pollen had no effect on the adhesion of outcross pollen that was applied later, but outcross pollen was less likely to germinate if self-pollen was already present. This may have been due to the physical barrier created by a layer of incompatible pollen that prevented outcross pollen from contacting the stigma directly. Self-pollen grains rarely germinate in this species, so other effects in the style or ovary were not anticipated. Because natural populations of *P. viscosum* are often pollen-limited, lower germination of outcross pollen is likely to lead to reduced seed production (Galen, 1985). A key question in need of further work is whether large doses of self-pollen arrive prior to outcross pollen in the field. Results from hand-pollination experiments such as these are difficult to interpret without knowing whether the sizes and timing of artificial pollen loads are representative of natural ones.

The consequences of SI pollination are greatest when ovules are usurped by SI pollen tubes, as occurs in *Ipomopsis aggregata* (Waser and Price, 1991; also see Seavey and Bawa, 1986; Kahn and Morse, 1991; Broyles and Wyatt, 1993; Chapter 13). In *I. aggregata* applications of self-pollen to one stigma lobe and outcross pollen to two others led to a 30–40% decrease in seed set relative to that from emasculated flowers receiving only outcross pollen (Waser and Price, 1991). Autodeposition of self-pollen was considerable: Averages of 50–100 self-pollen grains were deposited on stigmas of greenhouse-grown plants, and this occurred even when male-phase flowers were probed once a day to mimic hummingbird visits. The cost of autodeposition was determined by comparing seed set from unaltered vs. emasculated flowers that were hand-pollinated. Removal of undehisced anthers led to significant increases in seed set (from 8 to 11.5 seeds per fruit when hand-pollinated early in the day, and from 4.5 to 8.5 when pollinated 9 hrs later). Autodeposition may be somewhat less of a problem in the field, where pollen may be lost prior to stigma receptivity, but it probably does occur. In the field, the combination of autodeposition and geitonogamous pollination significantly reduced seed set (de Jong et al., 1992*b*), as described further below.

Another potential effect of SI pollen is that in some species the presence of self-pollen results in stigma closure and/or early flower senescence (i.e., senescence prior to the arrival of outcross pollen; see Stephenson, 1982; Waser and Fugate, 1986; Waser and Price, 1991; Aizen, 1993; Scribailo and Barrett, 1994). In flowers of *Ipomopsis aggregata,* for example, the stigma lobes closed when pollinated with either self- or outcross pollen (Waser and Price, 1991). Furthermore, unpollinated flowers of *I. aggregata* remained in the female phase for 3.6 days, whereas corollas abscised after only 1 day when pollinated with either self- or outcross pollen (Snow, unpublished data; 20–31 flowers per treatment, pollinated in an unheated greenhouse on the first day of anthesis). Thus, when autodeposition or geitonogamy occur early in a flower's lifespan, the flower may senesce before it has a chance to receive outcross pollen or export its own pollen to other individuals.

Field Studies

Self-pollination has the potential to interfere with female reproductive success, but for the reasons stated above this has rarely been documented. Morse (1994) suggested that most pollinations in large clones of a self-incompatible milkweed were the result of self-pollination, and found that hand-pollinated flowers had a greater chance of setting fruit when they were bagged as opposed to unbagged.

This suggests that geitonogamy can reduce seed production in milkweeds (also see Broyles and Wyatt, 1993). Another method for measuring the cost of geitonogamy is to determine whether seed set increases when all flowers on a plant are emasculated. For hummingbird-pollinated species, in particular, removing anthers is unlikely to affect pollinators' foraging behavior. A positive effect of emasculation on seed set, as seen in hummingbird-pollinated *Passiflora vitifolia* (Snow, 1982) and *Ipomopsis aggregata* (de Jong et al., 1992*b*), indicates that self-pollination does entail fecundity costs. The extent to which geitonogamy contributes to these costs can be determined by varying daily flower number.

To our knowledge, the only study that directly addresses this question was conducted by de Jong et al. (1992*b*). They first used fluorescent dyes to show that geitonogamous pollination increased with flower number in *Ipomopsis aggregata*. To determine whether geitonogamy reduced seed set more in large plants as compared to small ones, they selected pairs of plants of various sizes and emasculated all flowers on one plant in each pair. On average, whole-plant emasculations raised seed set per flower by 25%, but the beneficial effects of emasculation were not greater for larger plants. Also, in a different group of unmanipulated plants, fruit set and seeds per fruit were not higher for smaller plants, as had been predicted. However, plant size may be an indicator of overall vigor and ability to mature seeds, so results from the unmanipulated plants are inconclusive. In the experiment with paired plants, small sample sizes and variable daily flower numbers (which were not recorded) could account for the lack of significance. Repeating this whole-plant emasculation experiment would be worthwhile, particularly because a cost of geitonogamy seems likely when self-pollen causes premature stigma closure and usurps ovules.

Conclusions

Although geitonogamy has received relatively little attention from plant reproductive biologists, studies involving pollinator observations, pollen-tracking techniques, emasculation experiments, and genetic markers suggest that geitonogamous pollination is common enough to affect mating success and, in the case of self-compatible species, mating system and progeny vigor as well. For many animal-pollinated species, geitonogamy may be an unavoidable cost of requiring a large floral display and abundant rewards to attract pollinators. In addition, producing the same number of flowers over a longer time interval may not be an option if the plant's growing season is relatively short. Of particular interest, then, is the relationship between daily flower number and the frequency of self-pollination.

In self-compatible species, individual selfing rates provide a useful minimum estimate of the extent of self-pollination. For example, in *Malva moschata* a 2-fold increase in flower number, from 10–20 flowers per plant, led to a 3-fold increase in the proportion of selfed progeny (from about 10–40%; Fig. 7.4). Similar but much smaller effects of daily flower number were seen in *Eichhornia paniculata* and *Hibiscus moscheutos* (Fig. 7.6; see also Chapter 6). In the latter, both visit lengths and selfing rates were similar for 6- and 12-flowered plants, indicating that there may be a threshold daily flower number above which there is little further increase in selfing rates. Clearly, more studies of this kind are needed in order to fully understand the effect of daily flower number on individual selfing rates.

The consequences of geitonogamous pollination are more elusive in self-incompatible species, especially with regard to whether pollen that is wasted on a plant's own stigmas leads to a loss of male reproductive success (e.g., de Jong et al., 1993). Dilution of the amount of outcross pollen in stigmatic pollen loads, coupled with possible negative effects of self-pollen on the success of outcross pollen, could inhibit female function. Interference from self-incompatible pollen is particularly likely in species such as milkweeds in which self-pollinia produce hundreds of rapidly growing pollen tubes that appear to usurp ovules (Kahn and Morse, 1991; Broyles and Wyatt, 1993). Morse (1994) inferred that geitonogamy was responsible for low fruit set in two milkweed clones, and in *Ipomopsis aggregata,* removing all of a plant's anthers to prevent both autodeposition and geitonogamy resulted in significantly higher seed set (de Jong et al., 1992*b*). To our knowledge, however, no study has conclusively shown that geitonogamy alone reduces female reproductive success in a self-incompatible species.

In summary, more research is needed on the extent of geitonogamous pollination and its consequences. As noted by de Jong et al. (1993), new techniques for tracking local pollen and gene flow should improve efforts to model the dynamics of pollen transfer within and among plants. Pollen analogues such as fluorescent dyes have proven useful for determining how flower number affects geitonogamous pollination (e.g., de Jong et al., 1992*b*), but we concur with Robertson (1992), who cautioned against relying too heavily on indirect estimates of pollen carryover and self-pollination, and emphasized the value of using genetic markers to measure selfing rates directly (at least in SC species with comparable seed set from self- and outcross pollen). Further studies that focus on the relationships between daily flower number and (1) pollinator behavior, (2) reproductive success, and (3) selfing rates (in SC species) will allow quantitative rather than qualitative assessments of possible costs of geitonogamous pollination. These kinds of quantitative empirical studies are needed to rigorously evaluate the role that geitonogamy may play in the evolution of floral traits that inhibit self-pollination.

Acknowledgments

We thank Kathy Cochrane, Jennifer Tressler, Dennis Whigham, and the staff at the Smithsonian Environmental Research Center for help with the *Hibiscus* experiment; Bradd Schulke and Nickolas Waser for providing unpublished data; and Spencer Barrett, Michele Dudash, Daniel Schoen, Nickolas Waser, and two anonymous reviewers for comments on the manuscript. This research was supported by NSF grants BSR 8906959 to T. P. Spira, and BSR 8906667 and DEB 9306393 to A. A. Snow.

References

Ågren, J. and D.W. Schemske. 1993. Outcrossing rate and inbreeding depression in two annual monoecious herbs, *Begonia hirsuta* and *B. semiovata*. *Evolution,* 47: 125–135.

Aizen, M.A. 1993. Self-pollination shortens flower lifespan in *Portulaca umbraticola* H.B.K. (Portulacaceae). *Int. J. Plant Sci.,* 154: 412–415.

Aizen, M.A., K.B. Searcy, and D.L. Mulcahy. 1990. Among- and within-flower comparisons of pollen-tube growth following self- and cross-pollinations. *Am. J. Bot.,* 77: 671–676.

Alexander, M.P. 1969. Differential staining of aborted and nonaborted pollen. *Stain Technol.,* 44: 117–122.

Anderson, G.J. and D. Symon. 1988. Insect foragers on *Solanum* flowers in Australia. *Ann. Mo. Bot. Gard.,* 75: 842–852.

Arroyo, M.T.K. 1976. Geitonogamy in animal pollinated tropical angiosperms: A stimulus for the evolution of self-incompatibility. *Taxon,* 25: 543–548.

Augspurger, C.K. 1980. Mass-flowering in a tropical shrub (*Hybanthus prunifolius*): Influence on pollinator attraction and movement. *Evolution,* 34: 475–488.

Baker, H.G. 1983. An outline of the history of anthecology, or pollination biology. In *Pollination Biology* (L. Real, ed.), Academic Press, New York, pp. 7–30.

Barrett, S.C.H. and D. Charlesworth. 1991. Effects of a change in the level of inbreeding on the genetic load. *Nature,* 352: 522–524.

Barrett, S.C.H. and J.R. Kohn. 1992. Genetic and evolutionary consequences of small population size in plants: Implications for conservation. In *Genetics and Conservation of Rare Plants* (D.A. Falk and K.E. Holsinger, eds.), Oxford Univ. Press, New York, pp. 3–30.

Barrett, S.C.H., J.R. Kohn, and M.B. Cruzan. 1992. Experimental studies of mating system evolution: The marriage of marker genes and floral biology. In *Ecology and Evolution of Plant Reproduction* (R. Wyatt, ed.), Chapman & Hall, New York, pp. 192–230.

Barrett, S.C.H., L.D. Harder, and W.W. Cole. 1994. Effects of flower number and position on self-fertilization in experimental populations of *Eichhornia paniculata* (Pontederiaceae). *Func. Ecol.,* 8: 526–535.

Bateman, A.J. 1956. Cryptic self-incompatibility in the wallflower: *Cheiranthus chieri* L. *Heredity,* 10: 257–261.

Bawa, K.S. and J.H. Beach. 1981. Evolution of sexual systems in flowering plants. *Ann. Missouri Bot. Gard.,* 68: 254–274.

Beattie, A.J. 1976. Plant dispersion, pollination and gene flow in *Viola*. *Oecologia,* 25: 291–300.

Becerra, J. and D.G. Lloyd. 1992. Competition-dependent abscission of self-pollinated flowers of *Phormium tenax* (Agavaceae): A second action of self-incompatibility at the whole flower level. *Evolution,* 46: 458–469.

Bertin, R.I. 1993. Incidence of monoecy and dichogamy in relation to self-fertilization in angiosperms. *Am. J. Bot.,* 80: 557–560.

Bertin, R.I. and C.M. Newman. 1993. Dichogamy in angiosperms. *Bot. Rev.,* 59: 112–152.

Bertin, R.I. and M. Sullivan. 1988. Pollen interference and cryptic self-incompatibility in *Campsis radicans*. *Am. J. Bot.,* 75: 1140–1147.

Best, L.S. and P. Bierzychudek. 1982. Pollinator foraging on foxglove (*Digitalis purpurea*): A test of a new model. *Evolution,* 36: 70–79.

Bierzychudek, P. 1982. The demography of jack-in-the-pulpit, a forest perennial that changes sex. *Ecol. Monogr.,* 52: 335–351.

Bowman, R.N. 1987. Cryptic self-incompatibility and the breeding system of *Clarkia unguiculata* Lindl. (Onagraceae). *Am. J. Bot.,* 74: 471–486.

Brown, A.D.H. and R.W. Allard. 1970. Estimation of the mating system in open-pollinated maize populations using isozyme polymorphisms. *Genetics,* 66: 133–145.

Broyles, S.D. and R. Wyatt. 1993. The consequences of self-pollination in *Asclepias exaltata,* a self-incompatible milkweed. *Am. J. Bot.,* 80: 41–44.

Carpenter, F.L. 1976. Plant-pollinator interactions in Hawaii: Pollination energetics of *Metrosideros collina* (Myrtaceae). *Ecology,* 57: 1125–1144.

Casper, B.B., L.S. Sayigh, and S.S. Lee. 1988. Demonstration of cryptic incompatibility in distylous *Amsinckia douglasiana*. *Evolution,* 42: 248–253.

Charlesworth, D. and B. Charlesworth. 1987. Inbreeding depression and its evolutionary consequences. *Ann. Rev. Ecol. Syst.,* 18: 237–268.

Clegg, M.T. 1980. Measuring plant mating systems. *Bioscience,* 30: 814–818.

Crawford, T.J. 1984. What is a population? In *Evolutionary Ecology* (B. Shorrocks, ed.), 23rd Symposium British Ecological Society, Blackwell Scientific, London, pp. 135–173.

Cruden, R.W. 1988. Temporal dioecism: Systematic breadth, associated traits, and temporal patterns. *Bot. Gaz.,* 149: 1–15.

Cruden, R.W. and S.M. Hermann-Parker. 1977. Temporal dioecism: An alternative to dioecism? *Evolution,* 31: 863–866.

Cruzan, M.B. and S.C.H. Barrett. 1993. Contribution of cryptic incompatibility to the mating system of *Eichhornia paniculata* (Pontederiaceae). *Evolution,* 47: 925–934.

Cruzan, M.B., J.L. Hamrick, M.L. Arnold, and B.D. Bennett. 1994. Mating system variation in hybridizing irises: Effects of phenology and floral densities on family outcrossing rates. *Heredity,* 72: 95–105.

Darwin, C. 1876. *The Effects of Cross and Self Fertilisation in the Vegetable Kingdom*. Murray, London.

Denti, D. and D.J. Schoen. 1988. Self-fertilization rates in white spruce: Effect of pollen and seed production. *J. Hered.*, 79: 284–288.

Dudash, M.R. 1990. Relative fitness of self and outcrossed progeny in a self-compatible, protandrous species, *Sabatia angularis* L. (Gentianaceae): A comparison in three environments. *Evolution,* 44: 1129–1139.

Dudash, M.R. 1991. Plant size effects on female and male function in hermaphroditic *Sabatia angularis* (Gentianaceae). *Ecology,* 72: 1003–1012.

Ellstrand, N., A.M. Torres, and D.A. Levin. 1978. Density and the rate of apparent outcrossing in *Helianthus annuus*. *Syst. Bot.*, 3: 403–407.

Epperson, B.K. and M.T. Clegg. 1987. First-pollination primacy and pollen selection in morning glory, *Ipomoea purpurea*. *Heredity,* 58: 5–14.

Feinsinger, P. and W.H. Busby. 1987. Pollen carryover: Experimental comparisons between morphs of *Palicourea lasiorrachis* (Rubiaceae), a distylous, bird-pollinated, tropical treelet. *Oecologia,* 73: 231–235.

Flores, S.I. 1990. Reproductive Biology of *Ludwigia peploides peploides* and *L. peploides glabrescens*. Ph.D. diss., Univ. Chicago.

Frankie, G.W., P.A. Opler, and K.S. Bawa. 1976. Foraging behavior of solitary bees: Implications for outcrossing of a neotropical forest tree species. *J. Ecol.*, 64: 1049–1057.

Galen, C. 1985. Regulation of seed set in *Polemonium viscosum*: Floral scents, pollination, and resources. *Ecology,* 66: 792–797.

Galen, C. and R.C. Plowright. 1985. The effects of nectar level and flower development on pollen carryover in inflorescences of fireweed (*Epilobium angustifolium*) (Onagraceae). *Can. J. Bot.*, 63: 488–491.

Galen, C., T. Gregory, and L.F. Galloway. 1989. Costs of self-pollination in a self-incompatible plant, *Polemonium viscosum*. *Am. J. Bot.*, 76: 1675–1680.

Geber, M.A. 1985. The relationship of plant size to self-pollination in *Mertensia ciliata*. *Ecology,* 66: 762–772.

Gleeson, S.K. 1982. Heterodichogamy in walnuts: Inheritance and stable ratios. *Evolution,* 36: 892–902.

Handel, S.N. 1985. The intrusion of clonal growth patterns on plant breeding systems. *Am. Nat.*, 125: 367–384.

Harder, L.D., and S.C.H. Barrett. 1995. Mating costs of large floral displays in hermaphrodite plants. *Nature,* 373: 512–514.

Heinrich, B. 1979. Resource heterogeneity and patterns of movement in foraging bumble bees. *Oecologia,* 40: 235–245.

Hessing, M.B. 1988. Geitonogamous pollination and its consequences in *Geranium caespitosum*. *Am. J. Bot.*, 75: 1324–1333.

Hodges, S.A. 1995. The influence of nectar production on hawkmoth behavior, self pollination, and seed production in *Mirabilis multiflora* (Nyetaginaceae). *Amer. J. Botany,* 82: 197–204.

Holsinger, K.E., M.W. Feldman, and F.B. Christiansen. 1984. The evolution of self-fertilization in plants: A population genetic model. *Am. Nat.*, 124: 446–453.

Holtsford, T.P. and N.C. Ellstrand. 1990. Inbreeding effects in *Clarkia tembloriensis* (Onagraceae) populations with different natural outcrossing rates. *Evolution*, 44: 2031–2046.

Johnston, M.O. 1993. Tests of two hypotheses concerning pollen competition in a self-compatible, long-styled species (*Lobelia cardinalis*: Lobeliaceae). *Am. J. Bot.*, 80: 1400–1406.

de Jong, T.J., P.G.L. Klinkhamer, and M.J. van Staalduinen. 1992*a*. The consequences of pollination biology for selection of mass or extended blooming. *Func. Ecol.*, 6: 606–615.

de Jong, T.J., N.M. Waser, and P.G.L. Klinkhamer. 1993. Geitonogamy: The neglected side of selfing. *Trends Ecol. & Evol.*, 8: 321–325.

de Jong, T.J., N.M. Waser, M.V. Price, and R.M. Ring. 1992*b*. Plant size, geitonogamy and seed set in *Ipomopsis aggregata*. *Oecologia*, 89: 310–315.

Kadmon, R. and A. Shmida. 1992. Departure rules used by bees foraging for nectar: A field test. *Evol. Ecol.*, 6: 142–151.

Kahn, A.P. and D.H. Morse. 1991. Pollen germination and fertilization in self and cross-pollinated common milkweed, *Asclepias syriaca*. *Am. Midl. Nat.*, 126: 61–67.

Klinkhamer, P.G.L. and T.J. de Jong. 1993. Attractiveness to pollinators: A plant's dilemma. *Oikos*, 66: 180–184.

Klinkhamer, P.G.L., T.J. de Jong, and G. DeBruyn. 1989. Plant size and pollinator visitation in *Cynoglossum officinale*. *Oikos*, 54: 201–204.

Leclerc-Potvin, C. and K. Ritland. 1993. Modes of self-fertilization in *Mimulus guttatus* (Scrophulariaceae): A field experiment. *Am. J. Bot.*, 81: 199–205.

Lertzmann, K.P. and C.L. Gass. 1983. Alternative models of pollen transfer. In *Handbook of Experimental Pollination Biology* (C.E. Jones and R.J. Little, eds.), Scientific and Academic Editions, New York, pp. 474–489.

Levin, D.A. 1991. The effect of inbreeding on seed survivorship in *Phlox*. *Evolution*, 45: 1047–1049.

Lloyd, D.G. 1979. Some reproductive factors affecting the selection of self-fertilization in plants. *Am. Nat.*, 113: 67–79.

Lloyd, D.G. 1992. Self- and cross-fertilization in plants. II. The selection of self-fertilization. *Int. J. Plant Sci.*, 153: 370–380.

Lloyd, D.G. and D.J. Schoen. 1992. Self- and cross-fertilization in plants. I. Functional dimensions. *Int. J. Plant Sci.*, 153: 358–369.

Lloyd, D.G. and C.J. Webb. 1986. The avoidance of interference between the presentation of pollen and stigmas in angiosperms. I. Dichogamy. *New Zeal. J. Bot.*, 24: 134–162.

McDade, L.A. 1986. Protandry, synchronized flowering and sequential phenotypic unisexuality in neotropical *Pentagonia macrophylla* (Rubiaceae). *Oecologia*, 68: 218–223.

Montalvo, A.M. 1992. Relative success of self and outcross pollen comparing mixed- and single-donor pollinations in *Aquilegia caerulea*. *Evolution*, 46: 1181–1198.

Thomson, J.D. and S.C.H. Barrett. 1981. Temporal variation of gender in *Aralia hispida* Vent. (Araliaceae). *Evolution,* 35: 1094–1107.

Thomson, J.D. and J. Brunet. 1990. Hypotheses for the evolution of dioecy in seed plants. *Trends Ecol. Evol.,* 5: 11–16.

Thomson, J.D., M.V. Price, N.M. Waser, and D.A. Stratton. 1986. Comparative studies of pollen and fluorescent dye transport by bumble bees visiting *Erythronium grandiflorum. Oecologia,* 69: 561–566.

van Treuren, R. Bulsma, N.J. Ouborg, and W. van Delden. 1993. The effects of plant density and population size on outcrossing rates in locally endangered *Salvia pratensis. Evolution,* 47: 1094–1104.

Waddington, K.D. and B. Heinrich. 1979. The foraging movements of bumble bees on vertical "inflorescences": An experimental analysis. *J. Comp. Physiol.,* 134: 113–117.

Waller, D.M. and S.E. Knight. 1989. Genetic consequences of outcrossing in the cleistogamous annual, *Impatiens capensis.* II. Outcrossing rates and genotypic correlations. *Evolution,* 43: 860–869.

Waser, N.M. 1982. A comparison of distances flown by different visitors to flowers of the same species. *Oecologia,* 55: 251–257.

Waser, N.M. 1988. Comparative pollen and dye transfer by pollinators of *Delphinium nelsonii. Func. Ecol.,* 2: 41–48.

Waser, N.M. and M.L. Fugate. 1986. Pollen precedence and stigma closure: A mechanism of competition for pollination between *Delphinium nelsonii* and *Ipomopsis aggregata* (Polemoniaceae). *Oecologia,* 70: 573–577.

Waser, N.M. and M.V. Price. 1982. A comparison of pollen and fluorescent dye carryover by natural pollinators of *Ipomopsis aggregata* (Polemoniaceae). *Ecology,* 63: 1168–1172.

Waser, N.M. and M.V. Price. 1984. Experimental studies of pollen carryover: Effects of floral variability in *Ipomopsis aggregata. Oecologia,* 62: 262–268.

Waser, N.M. and M.V. Price. 1991. Reproductive costs of self-pollination in *Ipomopsis aggregata* (Polemoniaceae): Are ovules usurped? *Am. J. Bot.,* 78: 1036–1043.

Webb, C.J. and D.G. Lloyd. 1986. The avoidance of interference between the presentation of pollen and stigmas in angiosperms II. Herkogamy. *New Zeal. J. Bot.,* 24: 163–178.

Willson, M.F. 1983. *Plant Reproductive Ecology.* John Wiley & Sons, New York.

Wolfe, L. and S.C.H. Barrett. 1989. Patterns of pollen removal and deposition in tristylous *Pontederia cordata* L. (Pontederiaceae). *Biol. J. Linn. Soc.,* 36: 317–329.

Wyatt, R. 1983. Plant-pollinator interactions and the evolution of breeding systems. In *Pollination Biology* (L. Real, ed.), Academic Press, New York, pp. 51–96.

8

Flower Size Dimorphism in Plants with Unisexual Flowers

*Lynda F. Delph**

Introduction

Angiosperm species exhibit variation in every aspect of their flowers, from obvious differences in color, size, shape, and number to less conspicuous differences such as nectar production and the schedule of events that take place within each flower. Sprengel (1793) was perhaps the first to realize that floral characters are shaped by their function. Recent studies have shown that, for some species, characters that enhance male function may be different from those that enhance female function. For example, characters such as color (Stanton et al., 1986), color change (Cruzan et al., 1988), and shape (Campbell, 1989) have varying consequences for the success of the two gamete types. Campbell (1989) found that selection intensities through male function favored wide corollas, whereas narrow corollas were favored via female function. Studies such as this one indicate that selection through male function may actually be opposite in direction to selection through female function. Hence, investigations into how selection is operating on floral characters are made difficult in hermaphroditic plants because of the problems associated with quantifying male fitness (see Stanton et al., 1992). One approach to understanding how traits associated with male or female function are shaped by selection, therefore, is to investigate secondary floral structures (e.g., petals) in species with unisexual flowers (and see Chapter 11). In such species, the sex functions are separated between flowers, and if selection favors different traits via each sex function, then sexual dimorphism in floral traits may evolve. I show here that sexual dimorphism in corolla size is a common phenomenon and investigate some of the possible underlying causes.

The traditional view regarding sexual dimorphism in flower size has been that

*Department of Biology, Indiana University, Bloomington, IN 47405.

corollas on male flowers are always or nearly always larger than those on female flowers. This agrees with the suggestion that the evolution of floral traits such as corolla size and shape, and flower number has primarily been driven by selection for effective pollen dispersal, rather than pollen receipt, because success as a male is thought to be limited by access to mates, whereas success as a female is limited primarily by resources (this difference has been termed Bateman's principle; Bateman, 1948; Willson, 1979; Sutherland and Delph, 1984; Bell, 1985). However, this reasoning came long after the view of males being larger had emerged. To show how the thinking on this topic has itself evolved, I start with a historical perspective of flower size in plants. Following that, I focus on three aspects of flower size dimorphism in plants with unisexual flowers. By flower size I am referring to petal size in most cases, but also include other perianth parts (e.g., sepals and tepals) whenever they are the primary attractive structure or petals are entirely absent. First, the pattern of dimorphism is discussed in conjunction with a review of a large data set ($N > 900$) of dioecious and monoecious species. These data present a somewhat different picture than previously suggested, showing that the nature of sexual dimorphism, rather than being highly conserved, is one that appears to vary substantially, even at the generic level. Second, data on flower size dimorphism in gynodioecious and gynomonoecious species are presented and interpreted in light of the predictions of the causal mechanisms. Finally, several recent studies evaluating the effects of floral dimorphism on pollinator visitation are discussed for gynodioecious species.

Historical Perspective

The view that male (staminate) flowers generally have larger corollas than female (pistillate) flowers has a historical basis that began in the 1880s with the work of Müller and Darwin. However, the mechanisms thought to be responsible for the difference in corolla size between the sexes changed through time from a non-adaptive to an adaptive view.

Darwin (1877) was struck by the fact that whenever he observed sexual dimorphism in corolla size, it was always the female flowers that had relatively small corollas, especially in gynodioecious species. In contrast, he noted that corollas of male flowers of andromonoecious species did not differ in size from those of hermaphrodite flowers. From these observations, he concluded that "the decreased size of the female corollas . . . is due to a tendency to abortion spreading from the stamens to the petals." He even went so far as to doubt whether the dimorphism could be considered an adaptation caused by natural selection. Instead, Darwin viewed the reduction of the corolla of female flowers as a developmental consequence of the lack of pollen.

One of Darwin's contemporaries, H. Müller, considered flower size, along with odor and color, as having a significant influence on pollinator visitation.

He concluded (as reviewed by Knuth, 1906) that corollas on male flowers are larger than those on females in order to make the males more conspicuous to pollinators, thereby attracting "the earlier visits of insects, so that the female flowers may be pollinated." This appears to be the first functional hypothesis regarding why a difference might exist between the sexes, namely, that the relative size of the corolla may be related to its attractive nature. However, as I discuss below, it was not until 1985 (Bell, 1985) that the attractive function of the corolla was incorporated into a hypothesis regarding sexual dimorphism in corolla size.

One of the first data sets on the issue was compiled by Baker (1948, drawing primarily from reports by Darwin and Knuth), and it showed that in 76 gynodioecious and gynomonoecious species, female flowers have relatively small corollas as compared to hermaphrodite flowers. He noted that this phenomenon was not easily explained, and focused on the exceptions rather than the rule in trying to understand the underlying causes. For example, he noted that wind-pollinated species did not have this dimorphism in corolla size, and that contrary to what Darwin said, data from three andromonoecious species showed that corollas on male flowers were indeed occasionally smaller than their hermaphrodite counterparts. From these and other observations, he concluded that a decrease in corolla size could not be attributed solely to the abortion of pollen. He went on to show that the "general rule" that female flowers have smaller corollas extended to six recently derived dioecious species, and somewhat prophetically stated that this might not be the case in well-established cases of dioecy because of other, unspecified, factors. Nevertheless, he concluded by agreeing with Darwin that natural selection was unlikely to have been responsible for sexual dimorphism in corolla size.

Baker suggested that experimentation would be needed to understand why female flowers have smaller corollas, and Plack responded by bringing to light a series of experiments using the gynodioecious species, *Glechoma hederacea*. Plack (1957) found that the removal of anthers from buds of the hermaphrodite flowers resulted in a decrease in the size of the corolla that was roughly proportional to the number of anthers removed: Removal of just one of the four anthers slightly reduced corolla size relative to unmanipulated flowers, whereas removal of all four anthers resulted in corollas that were approximately the size of those of the naturally occurring female flowers. In a subsequent paper, she reported success in restoring the corollas of emasculated flowers to their normal size by applying gibberellic acid (Plack, 1958). Hence, her experiments indicated a link between hormone production by pollen and corolla size, and provided a mechanism for Darwin's idea concerning the effect of the lack of pollen. These experiments may have led some readers to conclude that developmental control of this type precluded an adaptive explanation for the smaller size of female flowers. However, as pointed out by many authors (e.g., Richards, 1986, and Stanton and Galloway, 1990), the effect of hormones on phenotypic features may have an adaptive basis.

Prior to 1975, most, if not all, of the emphasis and conclusions had been based on temperate species. Then, Bawa and Opler (1975) reported flower size data for tropical species that showed the opposite trend from that reported for temperate species: Out of 20 dioecious tropical trees, 14 had larger corollas on female than male flowers. They speculated that this sexual dimorphism might be caused because the petals on the females flowers would have to enclose the relatively large ovaries and nectaries. Here then, was the suggestion of a novel functional cause underlying corolla size dimorphism, although they did not present the idea as such.

The first functional hypothesis, that of attraction, was presented and tested by Bell (1985). He stated that "the flower is primarily a male organ, in the sense that the bulk of allocation to secondary floral structures is designed to procure the export of pollen rather than the fertilization of ovules." The underlying assumption of this hypothesis is that male function normally requires more visits by pollinators, and hence allocation to attraction may be more important for male function than female function. In other words, male flowers are larger because there is a strong positive correlation between the number of visits received and male fitness, but not female fitness. Consequently, flowers on male plants should be selected to have larger corollas than those on females, and/or be more numerous, for adaptive reasons. Bell (1985) suggested that there has been directional selection in the past for increasing the attractiveness of male flowers based on fitness differences among individuals.

By drawing together Knuth's and Baker's reports on corolla size, Bell (1985) presented the largest data set up to that time on the relative size of corollas in dimorphic species: Of the 79 species reported, 74 had smaller corollas on female than male or hermaphroditic flowers, leading him to conclude that corollas of male flowers are generally larger than those of female flowers. Exceptions included the dioecious species *Silene latifolia* (also called *S. alba*), which is well known for having larger female than male flowers (see Meagher, 1992). Furthermore, data on flower number indicated that male plants produce more flowers than female plants, hence further increasing their potential attractiveness.

The conclusion that pollen-bearing flowers have larger corollas than female flowers has not changed over the past century, in spite of the tropical data presented by Bawa and Opler (1975). In contrast, views concerning the cause of sexual dimorphism in corolla size did change, from nonfunctional aspects of the presence or absence of pollen (and consequently, the hormones that the pollen produced), to two functional aspects of the corollas themselves, namely, their attractive function and enclosing function.

Reassessment of Flower Size Dimorphism in Dioecious and Monoecious Species

What is the nature of the dimorphism in species with strictly unisexual flowers, that is, are male flowers the larger of the two sexes, or are females? Three studies have addressed this question in the past and reached equivocal conclusions. Of these, two contained information on 7 and 16 temperate species, respectively (Baker, 1948, and Bell, 1985), and the other information on 20 tropical species (Bawa and Opler, 1975). Given that approximately 10% of angiosperms exhibit gender dimorphism (Lloyd and Bawa, 1984), a larger data set is needed to provide a more comprehensive assessment. Hence, data were compiled for 552 dioecious and 367 monoecious species from 102 families, from a variety of sources, including unpublished data, measurements on fresh and preserved material, herbarium sheets, and published data in papers and flora (Delph et al. unpublished). The results indicated that the relative size of male and female flowers is highly labile, both phylogenetically and geographically.

Methods

The relative size of parts of male and female flowers was determined for each species, and one of three categories was assigned: male > female, male = female, or female > male. For those cases in which petals were present, petal size (length and/or width) was compared, whereas sepals (or other structures such as tepals) were compared when petals were lacking. Hence, comparisons of "flower size" refer to the size of a particular perianth part, usually petals. In addition to flower size, the type of pollination of each species was categorized as either animal or wind, and the locality was categorized as being either temperate or tropical. When fresh, preserved, or herbarium material was used, a range of 2 to 25 flowers, each from a separate plant, were measured for each sex and a mean calculated. The sexes were categorized as being different in size if the means differed by more than one standard error. Information on flower size from floras varied from explicit measurements to statements regarding relative size. Often, a range of sizes for each sex was given. For example, *Maquira costaricana* perianths were said to range from 2–2.5 mm in female flowers and to 1 mm in males (Croat, 1978), and hence, this species was categorized as having larger female than male flowers. In other cases, measurements were given for one sex and a statement of the relative size of the other sex was provided without measurements (e.g., sepals and petals of female flowers of *Abuta panamensis* are similar to those of males flowers, which have inner sepals 2–2.8 mm long and lack petals; Croat, 1978). Possible misinterpretations of information given in floras may have caused some species to be categorized as dimorphic, when in fact the two flower types are not statistically different in size. However, this should not have resulted in any cases being falsely categorized as being female > male when they are really male > female, or vice versa.

It is certain that not all 919 points are statistically independent of each other for phylogenetic reasons (Felsenstein, 1985). In other words, all 10 species within a hypothetical genus might show the pattern of male flowers being larger than female flowers if the ancestor showed this pattern and flower size dimorphism were a conserved trait. Hence, a method was needed to correct for the phylogeny, so that only independent points would be included in the contingency tables. I applied a modification of the method suggested by Harvey and Mace (1982), which assumed ancestry within genera based on the category of size dimorphism. This is best explained by example. If, in a genus for which I had information on 20 species, 10 were categorized as having male > female and 10 as female > male, then only one evolutionary change (i.e., change in character state) was assumed. The 10 species having male > female were assumed to comprise one group of close relatives, and the other 10 species another. Hence, such a genus contributed two points to the contingency tables. In addition, if all the genera within a family exhibited the same character state, then the entire family contributed only one point to the analysis. This method should underestimate the number of character state changes, is therefore a fairly conservative estimate of the number of independent points, and is the method used in the analyses presented here. In addition to this method, two other methods for calculating independence were applied to the data. For one, each species was considered independent, resulting in 919 cases. The other was highly conservative and considered all genera within families with the same character state to be the most closely related. This reduced the number of independent points to 296. Results from these latter two methods are not shown; however, it is worth noting that, for the statistical comparisons discussed in the next section, all three methods gave the same *qualitative* results.

Patterns of Flower Size Dimorphism in Dioecious and Monoecious Species

Application of the first method described above for estimating the number of independent points reduced the number of cases to 436. Character state changes occurred in 48% of the genera for which data on more than one species had been collected, in 75% of the families with more than one genus, in all of the orders with more than one family, and in all superorders. Hence, a substantial number of changes occur at the lower taxonomic levels, indicating that the polarity of size dimorphism is highly labile. For example, in *Sagittaria australis,* male flowers are larger than females, whereas in *S. montevidensis,* females are larger than males.

The overall pattern showed that a size dimorphism occurred in 85% of the 436 cases (Delph et al., unpublished). When we considered the relationship between the type of pollination and pattern of dimorphism, the results indicated that there was no difference between animal- vs. wind-pollinated species for those cases involving a dimorphism ($\chi^2 = 0.63$, $P = 0.43$, Fig. 8.1). For species of either pollination type, male flowers were larger than female flowers slightly more often than females were larger than males, but this difference was only marginally significant in both cases (animal-pollinated: $\chi^2 = 3.68$, $P = 0.06$; wind-pollinated: $\chi^2 = 2.86$, $P = 0.09$). This was in contrast to past analyses, each of which suggested that one or the other of the two sexes was almost always

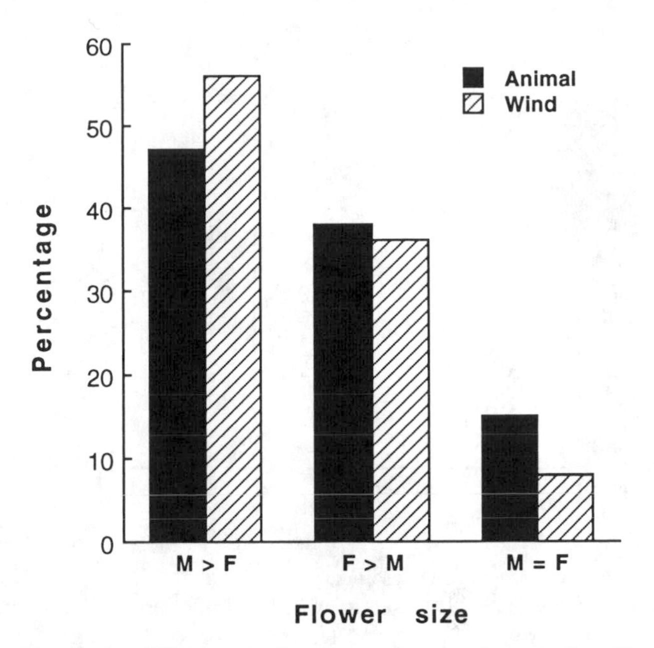

Figure 8.1. Flower size differences in dioecious and monoecious species with contrasting modes of pollination. Percentage of cases in which male flowers are larger than (M>F), smaller than (F>M), or equal to (M=F) female flowers of the same species. Flower size refers to petal size in most cases (see text for further explanation). A total of 372 "independent" animal-pollinated and 64 wind-pollinated cases were included. For both pollination modes, there was no significant difference in the likelihood of M>F or F>M (Delph et al., unpublished).

larger than the other. A factor contributing to this disagreement became clear when locality was taken into account (Fig. 8.2). The tropical vs. temperate distributions were strikingly different ($\chi^2 = 10.5$, $P = 0.001$), with a size dimorphism existing in 84 and 89% of the cases, respectively. In the temperate zone, species were significantly more likely to have larger male flowers than female flowers ($\chi^2 = 17.1$, $P = 0.001$), whereas in the tropics, there was almost equal likelihood of either sex having larger flowers than the other ($\chi^2 = 0.23$, $P = 0.63$). This difference between the temperate and tropic distributions was caused by a significant difference in their animal-pollinated ($\chi^2 = 15.5$, $P < 0.001$), but not their wind-pollinated species ($\chi^2 = 0.05$, $P = 0.82$): A higher percentage of female > male cases occurred for animal-pollinated tropical cases (42%) than animal-pollinated temperate cases (22%).

Causes of Flower Size Dimorphism in Dioecious and Monoecious Species

These results can be used to address the question of which factors control sexual dimorphism in flower size, that is, developmental or functional causes. The fact

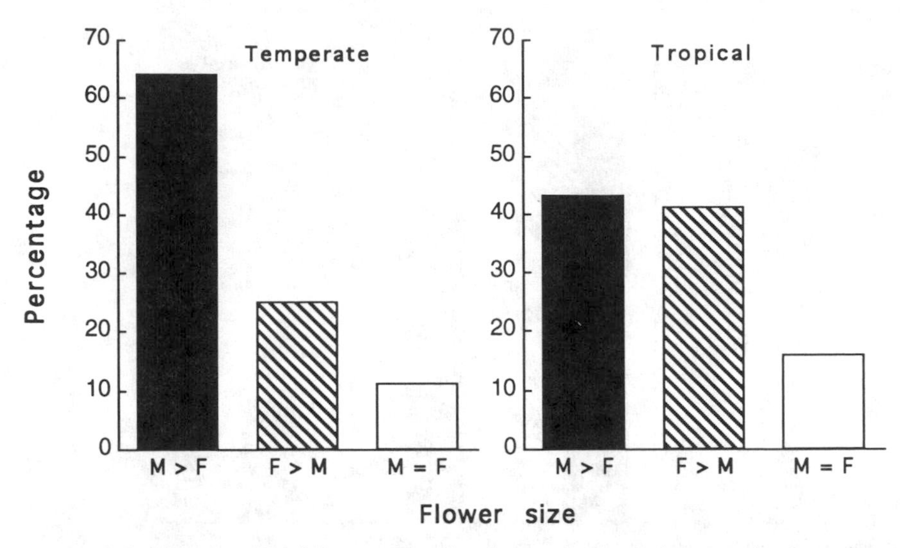

Figure 8.2. Flower size differences in dioecious and monoecious species from contrasting locations. Percentage of cases in which male flowers are larger than (M>F), smaller than (F>M), or equal to (M=F) female flowers of the same species, in either the temperate (N=100) or tropical (N=336) zone. The distributions of the two locations differ significantly (see text; Delph et al., unpublished).

that, of those cases involving dimorphism, 46% of them have female flowers larger than males negates the view that there is a strong developmental homology between pollen and perianth size. Clearly, the lack of pollen does not necessitate that female flowers will be smaller than males. This is not to say that hormones do not have an effect on perianth size; rather, it suggests that the response of the plant to the hormones (or lack thereof) can be modified.

These results cannot address the question of which of the two functional causes of dimorphism, the attraction and enclosing functions, is relatively more important in the evolution of flower size dimorphism. To address this, information relevant to each of the functions needs to be collected. The attractive function could be assessed given information on the relative size of attractive parts, and the enclosing function could be assessed given the relative size of the parts that were enclosed by the petals and/or sepals. Support for the attraction hypothesis would include patterns such as species with attractive perianth parts having relatively large male flowers more often than species with nonattractive perianth parts. Similarly, male flowers should be the relatively larger of the two in animal-pollinated, but not wind-pollinated species. This would agree with Bell's contention (Bell, 1985) that males have been selected to allocate more to attraction than females, because fulfilling male function requires that they be more attractive to pollinators. If, on the other hand, the relative size of the enclosed parts seems to be a good predictor of the relative size of the petals or sepals on the two sexes, then this would support the hypothesis that the enclosing function of the perianth has influenced its size.

A difference between temperate and tropical species in the pattern of dimorphism has been uncovered, and several possibilities exist for this difference. If, for example, females of dioecious tropical species have relatively large reproductive parts as compared to their males (as suggested by Bawa and Opler, 1975), then this might select directly for larger perianths on females than males, because of the enclosing function of the perianth. However, it is also possible that larger reproductive parts make each female flower more costly to produce than each male flower. The result of this greater cost might be to reduce the overall number of flowers produced per plant, as females of dioecious species have been shown to vary their investment in reproduction primarily by varying the number of flowers they produce rather than the number they mature into fruit (Sutherland and Delph, 1984; Delph and Lloyd, 1991). Reduction in flower number would, in turn, cause a disparity in the number of flowers present on males and females, making males more attractive. One way to counterbalance this difference in attractiveness would be to produce larger flowers on female plants. This explanation may be plausible for two temperate species, *Silene latifolia* and *Ilex opaca,* in which flowers on females are larger than those on males. In both species, male plants produce over 15 times more flowers than females (Delph, unpublished), and in *S. latifolia,* it has been shown that there is a genetically based size/number trade-off (Meagher, 1992). However, knowing that floral sex ratios are skewed toward maleness in many species (see Lloyd and Webb, 1977) is not enough to conclude that the size/number trade-off is responsible for the tropical/temperate difference. For this we need to know whether the sex ratios are, on average, more skewed among tropical as compared to temperate species, and such data are not available.

Other differences between tropical and temperate dioecious species, unrelated to the size of the ovaries, may also contribute to the disparity of flower size dimorphism. One such difference would be whether or not rewards differed markedly between male and female flowers in such a way as to make females relatively more unrewarding in the tropics, and hence selected for greater attractiveness through an increase in flower size (see Ågren and Schemske, 1991; also see Chapter 11). One-half of the data needed to test this hypothesis is available. Data from 48 species indicated that, in the tropics, one-third of the females produce no rewards and are pollinated by deceit (Renner and Feil, 1993). If this turns out to be more often the case in the tropics than the temperate zone, then this hypothesis would be supported, but once again, the data do not yet exist.

Additional, though somewhat anecdotal, information on the tropical, monoecious genus *Begonia* can be considered with regard to the connection between unrewarding flowers and relative flower size. In this genus, female flowers are unrewarding. Some species have larger female than male flowers, whereas others have larger male than female flowers (Ågren and Schemske, 1991; see Chapter

11). Schemske (personal communication) has noted that those species with relatively large female flowers are those in which plant size is relatively large and individual plants are sparsely distributed. The number of flowers open at any one time on these plants is relatively high. Insect visitors tend to visit more flowers before moving on to another plant than when visiting other, smaller species of *Begonia* that grow clumped together, thereby potentially increasing the degree of geitonogamy (within plant self-fertilization). It is possible that female flowers in large species are larger than males to increase the chances that a female will be among the first flowers visited by a pollinator, and hence outcrossed rather than geitonogamously selfed (Schemske, personal communication).

I discuss the *Begonia* case not because it has been well worked out (in fact, the observations described have not been quantified), but because it points to the need for data relating to pollinator behavior. Behavioral patterns influenced by the rewards of the two sexes, as well as other traits (such as average plant size and the overlap between the production of the two sexes of flowers in monoecious species, just to mention two), could greatly affect the relative flower size of the two sexes. Although data sets of large numbers of species from multiple families can provide information on the overall pattern of sexual dimorphism, the selective forces underlying any particular instance may be somewhat idiosyncratic, and an understanding of these forces will require observation and experimentation at the population level. One place to start might be a genus such as *Begonia,* in which the direction of the dimorphism varies between species, allowing for comparative studies.

Corolla Size in Gynodioecious and Gynomonoecious Species

In gynodioecious species, two types of plants exist: one that produces hermaphroditic flowers (hence referred to as polleniferous plants) and one that produces female flowers (female plants). In gynomonoecious species, these two flower types are both produced on every plant (i.e., there is only one plant type). As mentioned above, until recently, the largest data set on flower size dimorphism was that of Baker's (1948) on gynodioecious and gynomonoecious species. In this section, I present data showing that the trend toward hermaphrodite flowers being larger than female flowers still holds. I also briefly discuss the predictions of various hypotheses regarding which morph should be larger, and review several studies that have investigated how size dimorphism affects pollinator visitation.

Patterns of Flower Size Dimorphism and Flower Number in Gynodioecious and Gynomonoecious Species

A list of gynodioecious or gynomonoecious species is presented in Table 8.1, with information on the direction of the flower size dimorphism. Of 131 species

Table 8.1. Relative flower size in gynodioecious and gynomonoecious species. Comparisons are based on literature surveys, measurements taken from fresh or pickled material, and unpublished data.

Species in which the female flowers are smaller than the hermaphrodite flowers:		
Anacardiaceae	Compositae	Hydrophyllaceae
Rhus cotinus	*Cirsium acaule*	*Nemophila menziesii*
integrifolia	*Petasites frigida*	Iridaceae
ovata	*officinalis*	*Iris douglasiana*
Boraginaceae	Crassulaceae	Labiatae
Echium vulgare	*Sedum rhodiola*	*Calamintha alpina*
Eritrichum aretioides	Cruciferae	*clinopodium*
Caryophyllaceae	*Cardamine amara*	*grandiflora*
Agrostemma githago	Cucurbitaceae	*nepeta*
Alsine verna	*Cucurbita foetidissima*	*Dracocephalum*
Arenaria ciliata	Dipsacaceae	*moldavicum*
Cerastium arvense	*Knautia arvensis*	*Horminium pyrenaicum*
Dianthus carthusianorum	*Scabiosa atropurpurea*	*Hyssopus officinalis*
superbus	*columbaria*	*Lycopus europaeus*
Lychnis coronaria	*succisa*	*Melissa clinopodium*
flos-cuculi	Escalloniaceae	*officinalis*
Minuartia obtusiloba	*Carpodetus serratus*	*Mentha aquatica*
Sagina nodosa	Gentianaceae	*arvensis*
Silene acaulis	*Gentiana bellidifolia*	*gentilis*
inflata	Geraniaceae	*hirsuta*
maritima	*Erodium cicutarium*	*sylvestris*
noctiflora	*Geranium cinereum*	*vulgaris*
nutans	*columbinum*	*Nepeta cataria*
vulgaris	*maculatum*	*glechoma*
Spergularia media	*molle*	*Origanum vulgare*
rubra	*palustre*	*Prunella grandiflora*
Stellaria glauca	*phaeum*	*vulgaris*
gruminea	*pratense*	*Salvia pratensis*
longipes	*pyrenaicum*	*sylvestris*
nemorum	*sanguineum*	*verticillata*
Viscaria alpina	*sylvaticum*	*Satureja hortensis*
vulgaris		*Stachys germanica*
Celastraceae		*Thymus serpyllum*
Euonymus europaeus		

continued

from 30 families, an overwhelming majority (98%) exhibit larger hermaphrodite than female flowers. Two genera in the Caryophyllaceae, *Cerastium* and *Dianthus,* contain one or more species in which female flowers are relatively small and, as reported by Knuth (1906), one species in which female flowers do not differ in size from their hermaphroditic counterparts. In addition, one out of nine species of *Hebe* (Scrophulariaceae) shows no flower size dimorphism. In no cases are female flowers larger than hermaphrodite flowers.

Quantitative measurements of flower size for both morphs of 14 species show

Table 8.1. (continued)

Species in which the female flowers are smaller than the hermaphrodite flowers:

Lauraceae	Ranunculaceae	Thymelaeaceae
Ocotea tenera	*Ranunculus arvensis*	*Pimelea traversii*
Liliaceae	Rosaceae	*prostrata*
Chionographis japonica	*Fragaria vesca*	Umbelliferae
var. *kurohimensis*	Rubiaceae	*Gingidia baxteri*
Lobeliaceae	*Sherardia arvensis*	*decipiens*
Lobelia siphilitica	Saxifragaceae	*enysii*
Malvaceae	*Ribes alpinum*	*flabellata*
Sidalcea oregana	*Saxifraga granulata*	*montana*
Oleaceae	Scrophulariaceae	*trifolialata*
Syringa persica	*Digitalis ambigua*	*Lignocarpa carnosula*
Onagraceae	*lutea*	*Scandia geniculata*
Fuchsia colensoi	*purpurea*	*rosaefolia*
excorticata	*Hebe epacridea*	Valerianaceae
thymifolia	*fruticeti*	*Valeriana officinalis*
lycioides	*haastii*	*montana*
microphylla	*petriei*	*saxatilis*
perscandens	*ramosissima*	*tripteris*
Polygonaceae	*strictissima*	
Polygonum viviparum	*subalpina*	
Rumex alpinus	*traversii*	
crispus	*Chionohebe thomsonii*	

Species in which the female flowers are equal in size to the hermaphrodite flowers:

Caryophyllaceae	Scrophulariaceae
Cerastium semidecandrum	*Hebe stricta* var.
Dianthus caesius	*atkinsonii*

that, on average, the petals (or attractive structures) of hermaphrodite flowers are 1.3 times the size of those of female flowers (see Table 8.2). Hermaphrodite flowers of these species range from being equal in size to females in *Hebe stricta*, to being twice as large as females in *Fuchsia excorticata*.

Causes of Flower Size Dimorphism in Gynodioecious and Gynomonoecious Species

The three hypotheses that have been identified as possibly explaining flower size dimorphism all give the same prediction with regard to the pattern of dimorphism in gynodioecious species: Hermaphrodite flowers should be larger than female flowers. For example, hermaphrodite flowers contain viable pollen, whereas females do not; the perianth of hermaphrodite flowers must enclose both anthers and ovaries, whereas those on females only enclose ovaries; and, the fitness of polleniferous plants is affected by their ability to both donate and receive pollen, whereas the fitness of females is only affected by the receipt of pollen. Hence,

Table 8.2. Quantitative measurements of flower size for the sexual morphs of gynodioecious species.

Species	Part Measured	⚥:♀ Ratio	Flower Size (mm)		Significantly Different?	Ref.
			Hermaphrodite	Female		
Carpodetus serratus	Diameter of corolla	1.2	7.8	6.6	Yes	Delph and Lloyd (unpublished)
Dianthus superbus	Petal length	1.2	52.1	42.7	Yes	Sugawara (1993)
Fucshia excorticata	Tube + sepal length	2.0	51.0	25.0	Yes	Delph and Lively (unpublished)
Geranium maculatum	Petal length	1.4	19.2	13.5	Yes	Ågren and Willson (1991)
Glechoma hederacea	Petal length	1.5	16.5	11.3	—	Plack (1957)
Hebe stricta	Diameter of corolla	1.0	4.5	4.5	No	Delph and Lively (1992)
subalpina	Diameter of corolla	1.4	5.6	4.1	Yes	Delph (1990)
Iris douglasiana	Petal length	1.3	4.7	3.5	Yes	Uno (1982)
Lobelia siphilitica	Corolla length	1.0	1.21	1.17	Yes	Yetter (1988)
Ocotea tenera	Petal length	1.1	2.4	2.2	$P<0.07$	Gibson (unpublished)
Phacelia linearis	Diameter of corolla	1.3	13.9	11.0	Yes	Eckhart (1991)
Pimelea traversii	Sepal length	1.3	8.0	6.4	Yes	Stanton (unpublished)
Sidalcea oregana	Petal length	1.5	8.8	5.7	Yes	Ashman and Stanton (1991)
Silene acaulis	Petal length	1.2	11.8	9.5	Yes	Delph and Carroll (unpublished)

petals on hermaphrodite flowers may be larger than those on females because of the hormones produced by the pollen, the need to enclose a larger volume of reproductive parts, or the need to be more attractive than the females. It is not possible to attribute the larger size of hermaphrodite flowers to any single cause.

However, experiments, as well as information on other aspects of the two morphs, can be applied to the question of what the underlying causes of the dimorphism are. For example, Plack's emasculation experiments (1957), which caused a reduction in corolla size and suggested a lack of pollen as being responsible for the small size of females, were performed on hermaphrodite flowers of a gynodioecious species. In addition, natural variation in the number of functional anthers suggests a similar cause and effect. For example, within some populations of gynodioecious *Geranium maculatum,* there is a low frequency of plants that have only partially reduced male fertility, that is, only a portion of their anthers are functional (Ågren and Willson, 1991). Within these "intermediates," a positive correlation was found between petal size and the number of functional anthers.

Although the proximate cause of flower size dimorphism may be a hormonal signal produced by pollen, the ultimate cause may be selection for greater attraction acting to increase the male fitness of polleniferous plants. A comparison of flower number and flower costs suggests the importance of attraction in determining the pattern of dimorphism. Given that female flowers are generally smaller than hermaphrodite flowers, each flower is likely to contain less biomass, thereby allowing female plants to produce more flowers per plant if the two morphs are allocating equally to flower production. Equal biomass allocation of this sort has been shown for checker mallow, *Sidalcea oregana* (Ashman, 1994). However, Table, 8.3 shows that contrary to this reasoning, in general, polleniferous plants make either more flowers or equal numbers of flowers per plant as compared to female plants (16 out of 21 cases or 76% significantly greater than expected, $P < 0.001$). This argues that polleniferous plants are investing more biomass in flowers than females [e.g., *Cucurbita foetidissima* (Kohn, 1989) and *Hebe subalpina* (Delph, 1990)] and supports the hypothesis that greater allocation to flowering enhances male fitness, either through increasing attraction and/or the amount of pollen produced at the expense of seed production.

Consequences of Flower Size Dimorphism on Pollinator Visitation in Gynodioecious Species

In all seven studies that have investigated pollinator visitation to the morphs of gynodioecious species, pollinators showed a preference for hermaphrodite flowers (Assouad et al., 1978; Atsatt and Rundel, 1982; Uno, 1982; Bell, 1985; Ashman and Stanton, 1991; Eckhart, 1991; Delph and Lively, 1992). A strong relationship between corolla size and the probability of receiving a visit was shown by Bell (1985). He observed visitation to *Fragaria virginiana* flowers and subsequently

Table 8.3. Number of flowers on the sexual morphs of gynodioecious species.

Species	Information on Flower Number	Ref.
More flowers on polleniferous plants than on females:		
Cortaderia richardii	More spikelets per panicle, but not more flowers per spikelet	Conner (1965)
Cirsium arvense	More inflorescences per stem	Lloyd and Myall (1976)
Geranium sylvaticum	More flowers per plant	Vaarama and Jääskeläinen (1967)
Gingidia flabellata montana	More flowers per plant	Webb (1981); Lloyd and Webb (1977)
Lignocarpa carnosula	More flowers per inflorescence	Lloyd and Webb (1977)
Ocotea tenera	More flowers per plant	Gibson (unpublished)
Scandia geniculata rosaefolia	More flowers per inflorescence	Lloyd and Webb (1977)
Stellaria longipes	More flowers per plant	Philipp (1980)
Equal numbers of flowers:		
Dianthus superbus	Equal number of flowers per stem	Sugawara (1993)
Euonymus europaeus	Equal number of flowers	Webb (1979)
Geranium maculatum	Equal number of flowers	Ågren and Willson (1991)
Hebe stricta	Equal numbers of flowers per inflorescence and inflorescences per plant	Delph and Lively (1992)
subalpina	Equal number of flowers per stem	Delph (1990)
Thymus vulgaris	Equal number of inflorescences	Assouad et al. (1978); Dommée and Jacquard (1985)
More flowers on females than on polleniferous plants:		
Glechoma hederacea	More flowers	Willis (1892)
Phacelia linearis	More flowers per plant	Eckhart (1991)
Plantago lanceolata	Same number of inflorescences in 1-year-old plants, but for multiyear plants females have more	Krohne et al. (1980)
Satureja hortensis	More flowers per plant	Correns (1928)
Sidalcea oregana	More flowers per plant	Ashman (1994)

determined the wet mass of the petals. Figure 8.3 shows that the probability of visitation was highly correlated with petal mass, resulting in fewer visits to the smaller female flowers.

However, differences exist other than flower size differences that may affect the attractiveness of the flowers: In four separate studies, hermaphrodite flowers were shown to produce more nectar than female flowers (Atsatt and Rundel, 1982; Uno, 1982; Ashman and Stanton, 1991; Delph and Lively, 1992; but see Eckhart, 1992). Hence, although the greater visitation might, in part, be caused by the relatively greater attractiveness of the larger hermaphrodite flowers, it is also associated with a difference in nectar production between the morphs. For example, Ashman and Stanton (1991) found a correlation between petal length and nectar production for *Sidalcea oregana,* and concluded that pollinator visitation rates were influenced more by differences in petal length than those in flower number as a consequence of this correlation. Nevertheless, there was no evidence of females being pollen-limited, in part because female flowers remained open and receptive longer than hermaphrodite flowers.

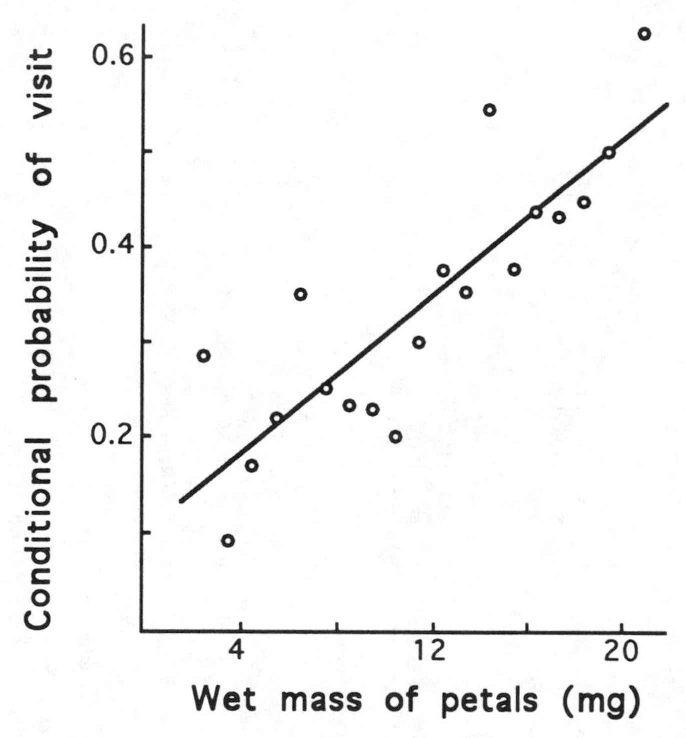

Figure 8.3. Flower size, as measured by the wet mass of the petals, and rate of visitation in the gynodioecious species *Fragaria virginiana* (after Bell, 1985). This relationship shows why females flowers, whose petals are only half as big as those of hermaphrodite flowers, receive fewer visits than hermaphrodite flowers.

In *Fuchsia lycioides,* which is pollinated by nectar-seeking hummingbirds, the difference in nectar production by the two morphs was so extreme (hermaphrodites produced six times more nectar on average than females) that investigators questioned why females were visited at all (see Atsatt and Rundel, 1982). They suggested that two aspects of these flowers maintained visits to females: Nectar production is highly variable within plants and the two sexes show a difference in flower position and size, which make it easier for floral visitors to see nectar in the female flowers. Hermaphrodite flowers are pendulous and the anthers block the tube from visual assessment of nectar at the base of the tube, whereas female flowers are not pendulous and lack the large anthers. Hence, foraging efficiency may be equalized: Although there is less nectar in female than hermaphrodite flowers, it is easier to find.

Rewards other than nectar, such as pollen, also differ between the morphs and may contribute to the preference for hermaphrodite flowers. In *Hebe stricta,* the hermaphrodite flowers were preferred overall, in part because they produce more nectar than the females, but also because the majority of visits were made by pollen-collecting female bees (Delph and Lively, 1992). These bees preferred to visit hermaphrodite flowers in the morning when anthers were dehiscing and presenting their pollen, whereas they shifted to preferring female flowers in the afternoon when the standing nectar crop of the two morphs had been equalized, in accordance with an ideal free distribution. Afternoons visits were plentiful enough to ensure fruit set: Hand-pollinations did not increase fruit set over that obtained with natural levels of pollination, suggesting that it was not limited by pollen availability.

In a population-level study of *Phacelia linearis,* Eckhart (1991) proposed that variation in the pollinator fauna among populations might alter the selective pressures on flower size. More visits were made to hermaphrodite flowers in all three populations studied, showing the importance of pollen as a reward, because the two morphs did not differ in nectar production. However, the effect of flower size was confounded with that of flower number across populations. Contrary to many species (see Table 8.3), female plants produced more flowers than polleniferous plants. Moreover, flower visitation rates were highly dependent on flower number in two out of three populations. Differences among populations occurred because differences in the reaction to size vs. number varied among insect species. Beetles and large-bodied bees in particular were more influenced by the relative size of flowers than small-bodied insects. Eckhart (1991) therefore suggested that selection on flower size would differ among populations if the relative abundance of the different pollinators varied. In contrast, the production of large numbers of flowers was found to be advantageous in all populations.

This study points to a potential conflict: Should a plant make larger flowers or more flowers? Different conclusions were reached for the two morphs, because although seed set was shown experimentally to be resource-limited, pollen donation was assumed to be visitation-limited. Eckhart took into account the relative costs (in biomass) and benefits (in visitation) of flower size vs. flower number, and concluded that "producing fewer, larger flowers may often be the better strategy" for polleniferous plants, because of the high cost associated with additional flower production. For females, he took into account the trade-off between investing in seeds vs. attraction, and concluded that, at least for some populations, smaller flowered plants would be able to produce more seeds than larger flowered plants. Hence, he concluded that the relative flower sizes appear to be adaptive.

In combination, these empirical studies support at least two hypotheses. First, given that fruit set did not appear to be pollinator-limited in the studies that investigated this question, they lend support to Bateman's principle; that is, female fitness may be saturated with relatively fewer visits than male fitness. Hence, selection would be less likely to operate to increase the attractiveness of the females. Second, they suggest that flower size is often used by pollinators as a cue to how rewarding a flower may be. This cue is likely to be a reliable one in gynodioecious species, as hermaphrodite flowers are the only ones to provide pollen, and they usually contain more nectar than female flowers of the same species.

Conclusions

Contrary to popular belief, male flowers are not always larger than female flowers. In fact, among dioecious and monoecious species, female flowers are larger than male flowers nearly half the time that a dimorphism exists (Delph et al., unpublished). The past belief in larger male flowers was probably caused by a bias toward considering temperate species, as these are significantly more likely to have relatively large male flowers, whereas in the tropics neither sex is more likely than the other to be the larger of the two. The fact that female flowers can and often do have larger petals than male flowers suggests that even though pollen has been shown to produce hormones that affect the growth and expansion of the petals, either (1) this is not the case in most species, or more likely, (2) selection is able to uncouple the response from the signal in such a way as to modify the developmental tie between the presence of pollen and growth of the petals. As pointed out by Stanton and Galloway (1990), any additive genetic variation for the response to or production of hormonal signals would allow selection to alter the developmental relationship between pollen and petal growth.

Four lines of inquiry for the future should greatly aid our understanding of what controls flower size dimorphism in species with unisexual flowers. First, more experimental emasculations and hormone additions of the type that Plack (1957, 1958) performed on *Glechoma* should be attempted and include a manipulation of female as well as male or hermaphrodite flowers. With regard to the adaptive nature of the dimorphism, comparative studies of genera in which the direction of the size dimorphism varies between species would undoubtedly be useful, especially if they involved observations of pollinator behavior toward natural and manipulated flowers combined with quantification of this behavior on components of plant fitness. In addition, the affect of the two functional aspects of the perianth, attraction and protection, on the evolution of flower size dimorphism should be evaluated with data on the relative size of attractive structures and the volume of the structures enclosed by the perianth parts from a large variety of species. Finally, the adaptive significance of larger male flowers of many animal-pollinated species has been assumed on the basis that there is a strong positive correlation (though not necessarily linear relationship) between the number of visits received and male fitness. Direct measures of male fitness curves utilizing either the natural variation in flower size or experimentally induced variation would be helpful to test this assumption.

Acknowledgments

Many thanks to David Lloyd for his inspiration and help in collecting flowers, and to Laura Galloway and Maureen Stanton for sharing their ideas and unpublished data with me. Thanks are also due to Phil Gibson and Doug Schemske for sharing unpublished data and to Curt Lively, Jon Ågren, and an anonymous reviewer for comments on an earlier version of the manuscript. Research funds were provided by NSF grant BSR-9010556.

References

Ågren, J. and D.W. Schemske. 1991. Pollination by deceit in a neotropical monoecious herb, *Begonia involucrata*. *Biotropica*, 23: 235–241.

Ågren, J. and M.F. Willson. 1991. Gender variation and sexual differences in reproductive characters and seed production in gynodioecious *Geranium maculatum*. *Am. J. Bot.*, 78: 470–480.

Ashman, T.-L. 1994. Reproductive allocation in hermaphrodite and female plants of *Sidalcea oregana* spp. *spicata* using four currencies. *Am. J. Bot.*, 81: 433–438.

Ashman, T.-L. and M.L. Stanton. 1991. Seasonal variation in pollination dynamics of sexually dimorphic *Sidalcea oregana* ssp. *spicata* (Malvaceae). *Ecology*, 72: 993–1003.

Assouad, M.W., B. Dommee, R. Lumaret, and G. Valdeyron. 1978. Reproductive capacities in the sexual forms of the gynodioecious species *Thymus vulgaris* L. *Bot. J. Linn. Soc.*, 77: 29–39.

Atsatt, P.R. and P.W. Rundel. 1982. Pollinator maintenance vs. fruit production: Partitioned reproductive effort in subdioecious *Fuchsia lycioides*. *Ann. Missouri Bot. Gard.*, 69: 199–208.

Baker, H.G. 1948. Corolla size in gynodioecious and gynomonoecious species of flowering plants. *Proc. Leeds Phil. Soc.*, 5: 136–139.

Bateman, A.J. 1948. Intrasexual selection in *Drosophila*. *Heredity,* 2: 349–369.

Bawa, K.S. and P.A. Opler. 1975. Dioecism in tropical trees. *Evolution,* 29: 167–179.

Bell, G. 1985. On the function of flowers. *Proc. Roy. Soc. Lond. B.,* 224: 223–265.

Campbell, D. 1989. Measurements of selection in a hermaphroditic plant: Variation in male and female pollination success. *Evolution,* 43: 318–334.

Connor, H.E. 1965. Breeding systems in New Zealand grasses V. Naturalized species of *Cortaderia*. *New Zeal. J. Bot.,* 3: 17–23.

Correns, C. 1928. Bestimmung, Vererbung und Verteilung des Geschlechtes bei den hoheren Pflanzen. *Handbuch der Vererbungswissenschaft,* 2: 1–138.

Croat, T.B. 1978. *Flora of Barro Colorado Island*. Stanford Univ. Press, Stanford, CA.

Cruzan, M.B., P.R. Neal, and M.F. Willson. 1988. Floral display in *Phyla incisa*: Consequences for male and female reproductive success. *Evolution,* 42: 505–515.

Darwin, C. 1877. *The Different Forms of Flowers on Plants of the Same Species*. Murray, London.

Delph, L.F. 1990. Sex-differential resource allocation patterns in the subdioecious shrub *Hebe subalpina*. *Ecology,* 71: 1342–1351.

Delph, L.F. and C.M. Lively. 1992. Pollinator visitation, floral display, and nectar production of the sexual morphs of a gynodioecious shrub. *Oikos,* 63: 161–170.

Delph, L.F. and D.G. Lloyd. 1991. Environmental and genetic control of gender in the dimorphic shrub *Hebe subalpina*. *Evolution,* 45: 1957–1964.

Dommee, B. and P. Jacquard. 1985. Gynodioecy in thyme, *Thymus vulgaris* L.: Evidence from successional populations. In *Genetic Differentiation and Dispersal in Plants* (P. Jacquard et al., eds.), NATO ASI Series, Vol. G5, Springer-Verlag, Berlin, pp. 141–164.

Eckhart, V.M. 1991. The effects of floral display on pollinator visitation vary among populations of *Phacelia linearis* (Hydrophyllaceae). *Evol. Ecol.,* 5: 370–384.

Eckhart, V.M. 1992. The genetics of gender and the effects of gender on floral characters in gynodioecious *Phacelia linearis* (Hydrophyllaceae). *Am. J. Bot.,* 79: 792–800.

Felsenstein, J. 1985. Phylogenies and the comparative method. *Am. Nat.,* 125: 1–15.

Harvey, P.H. and G. Mace. 1982. Comparisons between taxa and adaptive trends: Problems of methodology. In *Current Problems in Sociobiology* (King's College Sociobiology Group ed.), Cambridge Univ. Press, Cambridge, pp. 343–361.

Knuth, P. 1906. *Handbook of Flower Pollination,* Vol. 1. Clarendon Press, Oxford.

Kohn, J.R. 1989. Sex ratio, seed production, biomass allocation, and the cost of male function in *Cucurbita foetidissima* HBK (Cucurbitaceae). *Evolution,* 43: 1424–1434.

Krohne, D.T., I. Baker, and H.G. Baker. 1980. The maintenance of the gynodioecious breeding system in *Plantago lanceolata* L. *Am. Midl. Nat.,* 103: 269–279.

Lloyd, D.G. and K.S. Bawa. 1984. Modification of the gender of seed plants in varying conditions. *Evol. Biol.,* 17: 255–338.

Lloyd, D.G. and A.J. Myall. 1976. Sexual dimorphism in *Cirsium arvense* (L.) Scop. *Ann. Bot.,* 40: 115–123.

Lloyd, D.G. and C.J. Webb. 1977. Secondary sex characters in plants. *Bot. Rev.*, 43: 177–216.

Meagher, T.R. 1992. The quantitative genetics of sexual dimorphism in *Silene latifolia* (Caryophyllaceae). I. Genetic variation. *Evolution*, 46: 445–457.

Philipp, M. 1980. Reproductive biology of *Stellaria longipes* Goldie as revealed by a cultivation experiment. *New Phytol.*, 85: 557–569.

Plack, A. 1957. Sexual dimorphism in Labiatae. *Nature*, 180: 1218–1219.

Plack, A. 1958. Effect of gibberellic acid on corolla size. *Nature*, 182: 610.

Renner, S.S. and J.P. Feil. 1993. Pollinators of tropical dioecious angiosperms. *Am. J. Bot.*, 80: 1100–1107.

Richards, A.J. 1986. *Plant Breeding Systems*. George Allen & Unwin, London.

Sprengel, C.K. 1793. *Das entdeckte Geheimnis der Natur im Bau und in der Befruchtung der Blumen*. Vieweg, Berlin.

Stanton, M.L. and L.F. Galloway. 1990. Natural selection and allocation to reproduction in flowering plants. *Lect. Mat. Life Sci.*, 22: 1–50.

Stanton, M.L., A.A. Snow, and S.N. Handel. 1986. Floral evolution: Attractiveness to pollinators increases male fitness. *Science*, 232: 1625–1627.

Stanton, M.L., T.-L. Ashman, L.F. Galloway, and H.J. Young. 1992. Estimating male fitness of plants in natural populations. In *Ecology and Evolution of Plant Reproduction* (R. Wyatt ed.), Chapman & Hall, New York, pp. 62–90.

Sugawara, T. 1993. Gynodioecy in *Dianthus superbus* L. XV International Botanical Congress Abstracts, p. 280. Yokohama, Japan.

Sutherland, S. and L.F. Delph. 1984. On the importance of male fitness in plants: Patterns of fruit-set. *Ecology*, 65: 1093–1104.

Uno, G.E. 1982. Comparative reproductive biology of hermaphroditic and male-sterile *Iris douglasiana* Herb. (Iridaceae). *Am. J. Bot.*, 69: 818–823.

Vaarama, A. and O. Jääskeläinen. 1967. Studies on gynodioecism in the Finnish populations of *Geranium silvaticum* L. *Ann. Acad. Scient. Fenn. A*, 108: 1–39.

Webb, C.J. 1979. Breeding system and seed set in *Euonymus europaeus* (Celastraceae). *Plant Syst. Evol.*, 132: 299–303.

Webb, C.J. 1981. Gynodioecy in *Gingidia flabellata* (Umbelliferae). *New Zeal. J. Bot.*, 19: 111–113.

Willis, J.C. 1892. On gynodioecism in the Labiatae. *Proc. Phil. Soc. Camb.*, 7: 349–352.

Willson, M.F. 1979. Sexual selection in plants. *Am. Nat.*, 113: 777–790.

Yetter, R.B. 1988. The Expression of Male-Sterility in *Lobelia siphilitica* L. (Campanulaceae): A Life History Approach. Ph.D. thesis, Miami Univ., Oxford, OH.

PART 3

Model Systems

9

Evolution of Floral Morphology and Function: An Integrative Approach to Adaptation, Constraint, and Compromise in *Dalechampia* (Euphorbiaceae)

*W. Scott Armbruster**

Introduction

In his treatise on the structure and fertilization of flowers, Sprengel (1793) described in great detail how the structures of flowers are designed for efficient function in attracting pollinating insects and placing pollen on, and receiving it from, them. Sprengel also noted that there may be conflicting requirements on floral design: Flowers also need to protect pollen and nectar from rain. Flower design must therefore represent a compromise between these competing functions. Thus, Sprengel anticipated an area of modern interest: the role of constraints and adaptive compromise in the evolution of morphology (see review by Arnold, 1992). The focus of Sprengel's book, however, was on the optimality of floral design for efficient pollination.

Darwin (1859, 1877) reviewed Sprengel's ideas and extended his observations to many other species, especially in the Orchidaceae. He considered the topic of flower design from the more modern perspective of adaptation through natural selection, rather than design by the Creator as had Sprengel. In his treatment of floral evolution, Darwin (1877) developed the concept, now referred to as preadaptation (or "exaptation"; Gould and Vrba, 1982), that some features may have originated for selective reasons unrelated to the present function of those features. Darwin argued that what we now call preadaptations may have been important in the evolution of floral structures such as nectaries, commented on the existence of useless, vestigial floral structures, and noted that not all floral structures necessarily have functions. However, his general thesis was that flowers are highly adapted organ systems and an apparent lack of adaptive function for

*Department of Biology and Wildlife, and Institute of Arctic Biology, University of Alaska Fairbanks, Fairbanks, AK 99775-018O.

any particular structure usually reflects our lack of knowledge rather than absence of function. Thus, the focus of Darwin's writings was also on the optimal adaptation of floral form and function.

Most subsequent research in the area of floral form and function has similarly focused on discovering the adaptive function of floral structures and elucidating the dynamics of interactions with pollinators (see reviews in Stebbins, 1971, 1974). The variety of flowers and their pollination systems has inspired a large literature on floral form and pollination ecology. In the past decade or two, however, there has been renewed interest in the role of constraint and selective "compromise" in the evolution of organisms (e.g., Gould and Lewontin, 1979; Arnold, 1992), and this perspective has also emerge in the plant-reproduction literature (e.g., Primack, 1987; see also Chapters 3, 4, and 10).

Although recognizing the important role of adaptive evolution, this new approach also explicitly examines how phenomena other than adaptation may affect evolution. These include genetic correlations, developmental constraints, and conflicts between selection generated by pollination ecology vs. other factors (e.g., seed dispersal or sexual selection). For example, in a recent review Primack (1987) suggested that the developmental and functional relationships between flowers, fruits, and seeds strongly constrain the course of evolution of any one of the organs. Selection for seed size, for example, may strongly affect flower size and may, in turn, affect the pollination ecology of populations. Selection on floral morphology generated by sexual selection may often differ from selection imposed by pollinators (e.g., Mulcahy, 1983; Willson and Burley, 1983; Willson, 1991). It is thus clear that future research should focus on interactions among the multiple selective pressures and constraints that affect floral morphology.

There are several notable examples of recent research on floral form and function that have taken an integrated approach to the roles of natural selection and other factors in floral evolution. These studies include consideration of multiple conflicting or fluctuating selective forces generated by diverse pollinators and seed predators (Schemske and Horvitz, 1984, 1988, 1989; Galen, 1985, 1989; Galen and Newport, 1988), the conflicting selective pressures reflected in male vs. female reproductive success (Stanton and Preston, 1988; Galen and Stanton, 1989; Campbell, 1989; Campbell et al., 1991), and measured selection on heritable floral traits (Waser and Price, 1981; Stanton et al., 1986; Campbell, 1989, 1991; Galen, 1989; see also, Chapters 10 and 11).

Several new tools have been developed since Darwin's time that allow us to address with greater precision Sprengel's and Darwin's original questions about the evolution of form and function of flowers. These include statistical phenotypic, quantitative genetic, developmental, and selection studies of floral form (e.g., Berg, 1960; Guerrant, 1982; Lande and Arnold, 1983; Stanton et al., 1986; Campbell, 1989, 1991; Galen, 1989; Harvey and Pagel, 1991; Diggle, 1992; Conner and Via, 1993). Another relatively new tool is the use of phylogenetic information inferred from cladistic studies to reconstruct the evolutionary history of both floral form and reproductive function (see Donoghue, 1989; Cox, 1990; Thomson and Brunet, 1990; Weller et al., 1990; Armbruster, 1992; Chase and Hills, 1992; McDade, 1992; Rieseberg et al., 1992; Wagner et al., 1994) or correct for phylogenetic biases in statistical tests of adaptive (or other) hypotheses (Ridley, 1983; Felsenstein, 1985; Brooks and McLennan, 1991; Harvey and Pagel, 1991; Armbruster, 1992). I anticipate that these relatively new approaches will contribute significantly in the near future to our understanding of the evolution of floral form and function.

In the present contribution, I use the the neotropical species of the euphorb vine *Dalechampia* as a model system for the study of floral evolution. I have chosen this system because it comprises a tractable number of closely related species exhibiting considerable variation in pollination ecology. In this essay I present an integrative approach, evaluating, within a phylogentic context, both adaptive and nonadaptive influences on the evolution of floral form and function. My goal is to use phylogeny and comparative analysis to examine the relative importance of several evolutionary factors. I focus on the roles of pollinator-mediated selection, genetic constraints, genetic context (i.e., the complex of all genes affecting traits under selection), and adaptive compromise to disparate selective pressures in generating the floral variation expressed in *Dalechampia*. To illustrate this integrative approach to floral evolution, I will discuss five topics: (1) the functional significance of intraspecific and interspecific variation in blossom form, (2) how natural selection may have generated covariance between genetically independent traits, (3) how genetic correlation may have constrained the course of floral evolution, (4) how selection operating in different genetic contexts may have led to a nonadaptive diversification in blossom color, and (5) how the evolution of style form and function may reflect adaptive compromise to two divergent selective pressures.

Neotropical *Dalechampia*

Dalechampia L. (Euphorbiaceae) comprises approximately 120 species and occurs throughout most of the lowland tropics, except Australia and Oceania (Pax and Hoffman, 1919). About 90 species are found in the neotropics. *Dalechampia* flowers first caught the attention of pollination ecologists over a hundred years ago. The strange, functionally bisexual pseudanthia (blossom inflorescences, which I will call blossoms; Fig. 9.1), inspired speculation about the plants' pollination ecology (Müller, 1879; Cammerloher, 1931; Webster and Webster, 1972), but up until recently only a few casual observations had been made (Gueinzius, 1858; Müller, 1879; Cammerloher, 1931). Detailed studies of the pollination ecology of *Dalechampia* have been conducted in the last two decades (e.g., Armbruster and Webster, 1979; Armbruster, 1984, 1988; Armbruster and Herzig, 1984; Sazima et al., 1985). The blossoms of most *Dalechampia* species

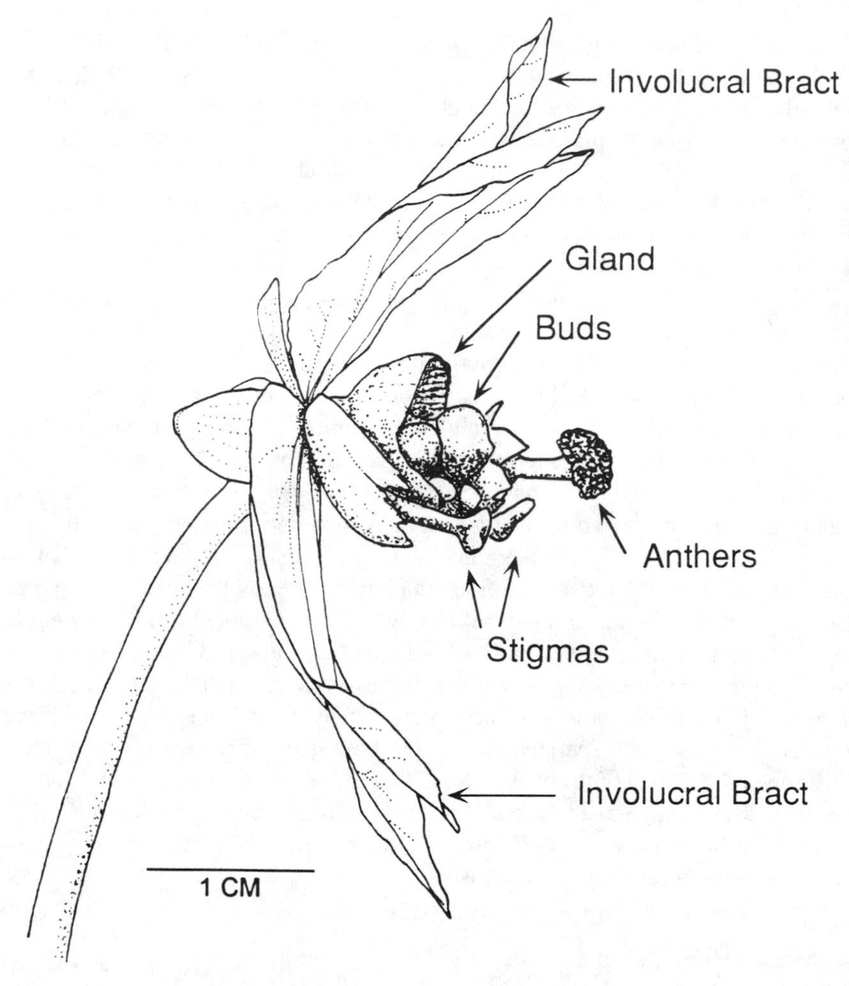

Figure 9.1. Schematic drawing of typical *Dalechampia* blossom inflorescence (pseudanthium).

secrete a triterpene resin that is collected by bees that use resin in nest construction. The only other plant group known to offer such a reward is *Clusia* (Guttiferae; see review in Armbruster, 1984).

About a dozen neotropical species of *Dalechampia* are pollinated by fragrance-collecting male euglossine bees (Armbruster and Webster, 1979; Armbruster et al., 1989, 1992). These bees visit *Dalechampia* flowers, as well as orchid flowers and other sources, to collect fragrances that are apparently used as precursors to sex pheromones (Whitten et al., 1989). A few neotropical *Dalechampia* species are apparently pollinated by pollen-collecting bees; no reward other than pollen is offered to visiting bees (Armbruster, 1993). In Madagascar, pollen-eating cetoniine beetles and pollen-collecting bees are the exclusive pollinators of *Dalechampia* (Armbruster et al., 1993).

Although much of the variation in pollination ecology can be explained by differences in floral chemistry, variation in the shape and size of blossoms is also an important determinant of which insects are pollinators (Armbruster, 1988). Species with large blossoms are pollinated primarily by larger euglossine bees such as *Eulaema* (Apidae, Euglossini; Armbruster and Herzig, 1984). Species with smaller blossoms are pollinated by the smaller *Hypanthidium* (Megachilidae, Anthidiini) and, sometimes, the even smaller *Trigona* (Apidae, Meliponini; Armbruster and Herzig, 1984; Armbruster, 1988, 1990). Thus, *Dalechampia* species in the neotropics appear to specialize on resin-collecting bees of different sizes.

Phylogenetic studies of neotropical *Dalechampia* have been combined with data on the pollination ecology of extant species (Armbruster, 1992, 1993). This analysis reveals some of the general historical patterns in the evolution of pollination systems in the genus. The earliest *Dalechampia* were probably pollinated by pollen-collecting bees or fragrance-collecting male euglossine bees. The resin-reward system apparently evolved by modification of preexisting resin-secretion system that defended the flowers from herbivores (Armbruster et al., unpublished). Male euglossine pollination has evolved two to three times independently from resin-reward ancestors. Among the resin-reward species, there have been repeated parallelisms and reversals between pollination by large female euglossine bees and pollination by small female megachilids. In general, reward systems and pollination ecology have been evolutionarily labile and prone to homoplasy, compared with other aspects of the plants' morphology (Armbruster, 1993). Species that are close phylogenetically are not constrained to having identical pollination systems.

Blossom Form and Function

Sprengel (1793) and Darwin (1859, 1877) were the first to relate variation in blossom form to variation in function. The method they used was generally qualitative: Differences in floral function between two or several species were attributed to differences in their floral form, or vice versa (but see Ridley, 1992). More recent researchers have realized that the possibility of chance associations between form and function exists when the sample sizes are small, and these authors have taken a statistical approach to establishing such relationships (e.g., Armbruster, 1984, 1988, 1990; Schemske and Horvitz, 1984; Thomson and Stratton, 1985; Epperson and Clegg, 1987).

Recently, several authors have pointed out that species are not usually independent samples, as is assumed in most comparative statistical analyses, but are instead clustered by their phylogenetic relationships (Ridley, 1983; Felsenstein,

1985; see reviews in Harvey and Pagel, 1991; Armbruster, 1992; Miles and Dunham, 1993). Members of a single clade can be expected to have the same morphological trait because they share recent ancestors, rather than because they share some ecological feature. Hence, any comparative analysis that ignores phylogeny risks inflating the degrees of freedom and making type-I errors (rejecting the null hypothesis when it is true; i.e., finding a relationship when there is none; Felsenstein, 1985).

Comparative analyses of intraspecific and interspecific variation in *Dalechampia* blossom size and shape have suggested the possible adaptive significance of several blossom features. Three apparent relationships between blossom form and pollination function were found to be particularly strong in earlier studies. A strong positive correlation between the area of the resin gland (GA) and the size of the principal pollinators was observed (Armbruster, 1984, 1988). This suggested that the variation in GA reflects adaptation to pollinators of different sizes. This makes sense because we should expect plant populations adapted to pollination by larger bees to offer more reward (and have a larger resin gland) to keep the insects interested and faithful; similarly, we should expect populations adapted to pollination by smaller bees to offer less reward (Heinrich and Raven, 1972; Armbruster, 1984).

Previous analyses suggested that the distances between the gland and stigmas (GSD) and the gland and anthers (GAD) have evolved to "match" the length of the principal pollinators. The focus of the bees' attention is the gland that secretes resin or fragrance. If the stigmas and/or anthers are sufficiently close, the bees will stand on them while collecting the reward; if too far away from the gland, the bees will not contact them at all. The distance between the gland and stigmas or anthers must be considerably less than the length of the principle pollinator for pollen to be deposited regularly on stigmas (GSD) or picked up regularly from anthers (GAD) (Armbruster, 1988, 1990). Results of regression analyses supported these expectations (Armbruster, 1988).

Previous analyses, however, were conducted on population and species means without taking into account phylogentic structuring of the species. Because phylogenetic information is now available for neotropical *Dalechampia* (Armbruster, 1994), I have reanalyzed the form-function data using two phylogenetically corrected method.

Methods

I used a modified form of Ridley's (1983) independent comparisons method to test for association, while completely eliminating the influence of phylogenetic bias on the associations between pollinator size and GA, GSD, and GAD. This method is a good conservative check for the existence of a positive or negative relationship between two traits, but it does not allow the estimation of correlation and regression coefficients. I used parsimony to map the pollinators GAD, GSD, and GAD onto a reconstructed phylogenetic tree reported previously (Armbruster, 1993). I then inspected the tree for unambiguous changes in pollinators: from smaller bees to larger, or larger bees to smaller. At each point of change, I scored the change in GA, GSD, and GAD as consistent (positive) or inconsistent (negative) with the direction of change in pollinators. I then tabulated the results and tested, using a binomial sign test (Sokal and Rohlf, 1981), whether the number of consistent changes for each blossom character was greater than expected by chance.

In order to get estimates of strengths of correlations and direction of slopes, I used a modification of nested analysis of covariance developed in an earlier study (Armbruster, 1988; also described by Bell, 1989); this technique allows independent estimates of relationships at different levels in a taxonomic or phylogenetic hierarchy (Armbruster, 1988, 1991; Bell, 1989). In the reanalysis, I included 11 clades as a level of nesting to statistically remove most of the effect generated by phylogenetic clustering of species (Felsenstein, 1985) and allow separate analysis of the relationship between variables within and among clades.

Nested analysis of covariance (nested ANCOVA) is analogous to the more familiar nested analysis of variance (nested ANOVA): It partitions sums of products, however, rather than sums of squares as in nested ANOVA (Table 9.1). The technique developed by Armbruster (1988, 1991) and Bell (1989) uses both nested ANOVA and nested ANCOVA to obtain estimates of the variance and covariance components for each level in the taxonomic or phylogenetic hierarchy. The variance and covariance components are then used to calculate the component correlation and/or regression coefficients for each level. The component coefficients at one level in the hierarchy are statistically independent of the component coefficients at other levels, which is not true of conventional (mean square) correlation and regression coefficients. Component coefficients estimate the unique, independent relationship between variables operating at that level in the heirarchy. This technique is therefore suited to the statistical control of phylogeny while estimating the relationship between form and function. In addition to allowing for statistical control of most phylogenetic effects, it allows independent estimation of the among-clade and within-clade relationships.

I used SAS PROCNESTED (SAS Institute, 1985) to estimate the variance

Table 9.1. Nested analysis of covariance for population means nested within species and species nested within clades of Dalechampia. *≡ E indicates expected value; k_1, k_2, k_3 are estimated as in the analysis of variance.*

Source	DF	Mean Cross-Products (MCP)	*E*(MCP)
Clade (*C*)	*C*-1	MCP_c	$cov_{error} + k_2\ covs + k_3\ cov_c$
Species (*S*)	*S*-1	MCP_s	$cov_{error} + k_1\ covs$
Population (error)	$n-S$	MCP_{error}	cov_{error}

and covariance components from three levels: populations, species, and terminal clades. Component correlation coefficients were calculated for each level as

$$\frac{\mathrm{cov}_{xy}'}{(\mathrm{var}_x' \cdot \mathrm{var}_y')^{1/2}} \tag{1}$$

where cov′ and var′ are covariance and variance components, respectively, and subscripts x and y refer to the two traits (see Armbruster, 1988, 1991, for details).

Results and Discussion

The independent comparisons test showed a tight positive relationship between pollinators size and the values of GA, GSD, and GAD. Most changes in GA, GSD, and GAD were in the same direction as those in pollinator size (i.e., positively related; Table 9.2). The probability that chance generated the preponderance of consistent changes was very small (GA: $P = 0.002$; GSD: $P = 0.012$; GAD: $P = 0.021$). Thus, one can safely rule out the significance of the correlation between pollination ecology and blossom morphology being a phylogenetic artifact.

The nested analysis of covariance, with terminal clade included as a level, also indicated that the positive relationship between pollinator size and blossom size characters is real and not a phylogenetic artifact. Although conventional (mean square) correlations between the length of the principal pollinator and gland area were quite large at all levels, the stongest component correlations between these variables occurred at the population level (Table 9.3). This is primarily the result of most covariance occurring at the population level and negligible additional variance in pollinator size being added at the species and clade levels (i.e., as much variation in pollinator size among conspecific populations as among species and clades). Covariance components at the level of clades were negligibly small (negative components are assumed to be zero). Thus,

Table 9.2. Independent comparisons test for relationship between direction of change in pollinator size and inflorescence characters associated with speciation in Dalechampia. ≡ *Numbers in columns 2–7 are independent occurrences observed when the pollinators and characters were mapped onto the reconstructed phylogenetic tree.*

	Blossom Character					
	GA		GSD		GAD	
Pollinator	Positive	Negative	Positive	Negative	Positive	Negative
Small to large	6	0	6	1	6	1
Large to small	4	0	4	0	3	0
Total	10	0	10	1	9	1
P-value	0.002		0.012		0.021	

Table 9.3. Nested analysis of covariance between the length of the principal pollinator and the gland area (GA), gland-stigma distance (GSD), and gland-anther distance (GAD) across 42 populations of 23 species of Dalechampia *belonging to 11 phylogenetic groups (clades). The phylogenetic groupings were derived from cladistic analysis of morphological characters not associated with pollination ecology (Armbruster, 1993).*

Covariance Source	Degrees of Freedom	Covariance Component			Component Correlation			Mean Square (Conventional) Correlation		
		GA	GSD	GAD	GA	GSD	GAD	GA	GSD	GAD
Clade	7	0.04NS	−0.04NS	−0.07NS	0.00	0.00	0.00	0.75**	0.45NS	0.62*
Species	4	−0.12NS	0.08NS	0.16*	0.00	0.00	0.00	0.52*	0.78***	0.83***
Population (=error)	4	0.36	0.06	0.04	0.95	0.52	0.32	0.95***	0.52***	0.32*
Total	15	0.29	0.10	0.13	0.74	0.52	0.62	0.74***	0.53*	0.62**

*$P<0.05$.

**$P<0.01$.

***$P<0.001$.

NS = not significant.

divergence of clades has not generated any significant additional variance in pollinators or covariance between pollinators and blossom morphology, and the correlation between morphology and ecology cannot be a phylogenetic artifact.

Together, these observations are consistent with the interpretation that large-glanded populations (and species) have adapted to pollination by large resin-collecting bees, whereas those with small glands have adapted to pollination by smaller bees. The observation that both GSD and GAD are correlated with pollinator size after controlling for phylogeny is consistent with the interpretation that populations (and species) with large GSD and GAD have adapted to pollination by large bees, whereas those with small GSD and GAD have adapted to pollination by smaller bees. The specialization in pollination ecology reflected in all three characters appears to be the consequence of selection against sharing pollinators with sympatric congeners and the resulting loss of pollen to alien stigmas and loss of pollinator service (Armbruster, 1985, 1990).

Covariation of Floral Traits

There has been a long-standing interest in causes of covariation of phenotypic traits within and among populations and species (e.g., Stebbins, 1950, 1974; Berg, 1960; Dobzhansky, 1959, 1970). In recent years there has been renewed interest in the evolutionary significance of trait covariation (e.g., Lande, 1979; Lande and Arnold, 1983; Cheverud, 1982, 1984; Riska, 1986, 1989; Mitchell and Shaw, 1992; Conner and Via, 1993). At least three factors may commonly generate covariation of traits among populations: (1) the direct effects of natural selection (selective correlation; Darwin, 1877; Tedin, 1925; Stebbins, 1950; Armbruster, 1991), (2) pleiotropy and/or linkage disequilibrium (genetic correlation; Lande and Arnold, 1983; Falconer, 1989), (3) genetic drift (drift covariance; Armbruster, 1991; Armbruster and Schwaegerle, 1995). However, the relative importance of these three causes of trait covariation has never been established. Relatively few authors have discussed the ecological and evolutionary significance of the covariation of floral characters (but see Darwin, 1877; Berg, 1960; Armbruster, 1991; Conner and Via, 1993; Campbell et al., 1994). In this section I bring together the results of several previous studies of *Dalchampia* blossom morphology to illustrate how one can address the causes of floral trait covariation among populations.

Selective Correlation of Functionally Related Traits

Darwin (1877, p. 284) was the first to point out that natural selection is likely to cause the covariation of interacting parts of a flower. Tedin (1925) and Stebbins (1950) also discussed the role of natural selection in generating among-population correlations between genetically independent traits and called this phenomenon "selective correlation" (see Armbruster, 1991; Armbruster and Schwaegerle, 1995).

The three blossom characters discussed above (GA, GSD, and GAD) are positively intercorrelated when conspecific populations and species are compared. Because these characters appear to play major roles in pollination ecology and hence be of selective importance, it seems worthwhile to understand what causes the observed integration of the floral characters.

A study that traced the development of the blossoms of *Dalechampia scandens* showed that GSD and GAD are developmentally independent of GA (Armbruster, 1991, unpublished). Hierarchical analyses of covariance within genets, among genets, and among populations in *D. scandens* showed that virtually all covariance among these traits resides at the level of among populations and above. Because genetic correlation should usually be expressed within populations (as well as among), these observations suggest that GA, GSD, and GAD are probably genetically independent (Armbruster, 1991, unpublished) and the covariance between GA, GSD, and GAD is probably not caused by pleiotropy or gene linkage.

The likelihood that genetic drift has generated among-population covariance between traits can be addressed by considering the statistical significance of the association. Genetic drift is a random sampling effect much like the statistical sampling effects usually tested for in statistics (expressed by the *P*-value). Therefore, if populations are independent (and traits are not genetically correlated), the *P*-value of the correlation coefficient is a measure of the likelihood that drift (and other sampling effects) has generated among-population covariance.

However, populations are usually structured by phylogenetic history, and ignoring this structure in the data can lead to type-I error (concluding there is a significant relationship when there is none). In the absence of molecular data, a reasonable estimate of phylogeny for many populations is their geographic distribution: Nearby populations are likely the most closely related. In the case of *D. scandens,* imagine that drift generated covariance between GA, GSD, and GAD among three general regions in northern South America, where the study was conducted. Subsequent population fission increased population number, of which eight were sampled, with each population being similar to others in its region. If this were how significant among-population correlations arose (assuming $N = 8$, when the true $N = 3$), we would expect to see covariance expressed among regions rather than within regions. However, the strengths of the component correlations within regions and among regions were similar, indicating that genetic drift probably did not generate the among-population covariance (Armbruster, 1991).

This leaves natural selection as the most likely generator of among-population covariance among the three floral characters (GA, GSD, and GAD). Is it possible that these traits interact functionally in some way so that the value of one trait influences the selection gradient acting on other characters? Indeed, this is consistent with our understanding of how these traits operate in pollination. In species with resin glands, gland area sets an upper limit on the size of bees that will visit the inflorescences (Armbruster, 1984, 1988, 1990). For a population with a particular gland area, bees larger than the upper size limit will not visit the flowers because the rewards are inadequate to justify their energy and time expenditure in foraging (see Heinrich and Raven, 1972). Smaller bees, however, will find sufficient reward. *Dalechampia* populations with smaller glands are therefore attractive only to smaller bees.

The distance between the gland and stigmas affects the probability that the bee will contact the stigmas and potentially deposit pollen. If the bee is smaller than this distance, it will only rarely contact the stigmas; if much larger, it will regularly contact them. Thus, GSD effectively sets the minimum size of the pollinator. Any bee smaller than this minimum will not contact stigmas; bees larger will. Similarly, GAD sets the minimum size of bees that will contact the anthers and transport pollen (Armbruster, 1985, 1988). It is obvious that, in order for pollination to occur, the minimum bee size for effective pollination, as determined by GSD and GAD, must be smaller than the size of the largest bees attracted to the flowers, as determined by GA. Thus, if drift or selection causes the gland area to decrease, optimal GSD and GAD would also decrease. Hence, selection should generate covariance between the three traits (Armbruster, 1990, 1991).

Genetic Correlation as a Constraint on Phenotypic Evolution

Genetic correlation between traits occurs when (1) both traits are affected by a single locus (pleiotropy), (2) when the loci controlling the traits are located on the same chromosome or, (3) when the loci controlling the traits are statistically associated during chromosome assortment (the last two conditions are called linkage disequilibium; Falconer, 1989). Two divergent views have emerged in the literature about the role of genetic correlation in evolution. Some authors have argued that genetic correlations constitute serious constraints on the course of evolution; certain combinations of trait values may be genetically impossible to acheive, regardless of their potential selective values (Lande, 1979; Lande and Arnold, 1983; Cheverud, 1984; Via and Lande, 1985; Clarke, 1987; Campbell et al., 1994). Alternatively, selection may break down or generate new genetic correlations, and hence genetic correlations may not impose constraints on long-term evolution (Clarke, 1987; Zeng, 1988). Of course, the importance of genetic correlation as an evolutionary constraint will be affected by the cause of the genetic correlation (pleiotropy vs. linkage disequilibrium) and the strength of the selective pressures. In this section I review results from previous studies on covariation of floral characters to illustrate the potential importance of genetic correlation as a constraint on the evolution of *Dalechampia* blossoms.

The length of the involucral bracts (BL) covaries with GA, GSD, and GAD. I used nested ANCOVA to analyze the covariance between BL and GA in

Dalechampia scandens to determine whether selection, drift, or genetic correlation has generated the covariance (Armbruster, 1991). The first difference between the relationship between BL and GA and that among the other characters is that BL and GA are developmentally related; both increase with inflorescence age. As a result, there is significant covariance between these traits among blossoms of different age within genets. There was also significant component covariance among genets within populations (i.e., after the developmental state has been controlled for statistically). This observation, in combination with the fact that the broad-sense among-population heritability of these characters is near unity (Armbruster, 1985; Armbruster and Schwaegerle, 1995), suggests that the two traits are genetically correlated. The among-genet and among-population relationships have roughly the same slope, and both lie on the developmental trajectory. Covariance both within and among populations thus appears to reflect pleiotropy. Hence, genetic correlation appears to have constrained the course of population divergence with respect to these two characters in *D. scandens* and probably other species (Armbruster, 1991).

Genetic correlations may originate by natural selection favoring genetic correlation between traits that function together in some way (Dobzhansky, 1959, 1970; Cheverud, 1982, 1984; Wagner, 1984; Conner and Via, 1993). Thus, finding that two traits are genetically correlated does not eliminate the possibility that natural selection was ultimately responsible for the relationship. However, unless there is a compelling case for the functional relationship between two genetically correlated characters, I would favor as more parsimonious the explanation that such a relationship is a nonadaptive genetic constraint. Because there is no compelling functional relationship between BL and GA, I interpret their covariance as the result of a simple pleiotropic relationship. Because bract length is easy to manipulate, it should be possible to test this provisional conclusion by conducting field experiments on pollinator response to, and fitness consequences of varying bract size (see Campbell et al., 1994).

Evolution of Floral Colors: Nonadaptive Divergence?

One of the major cornerstones in the development of the concept of "pollination syndromes" in plants has been associations observed between pollinator type and flower color. For example, bird-pollinated flowers are commonly red, moth-pollinated flowers are commonly white, and bat-pollinated flowers are commonly green or dull white (Faegri and van der Pijl, 1979; Wyatt, 1983). These associations have led to the expectation that a group of related plants pollinated by different animal species will often exhibit diverse flower colors (Grant and Grant, 1965). Similarly, variation in floral color exhibited by a group of related plants should often reflect adaptation to diverse pollinators. As early as Darwin's day, however, it was recognized that the same selective pressure operating in different (genetic) contexts can lead to a diversity of adaptive responses (Darwin, 1859, 1877; Stebbins, 1950, 1974). Thus, variation in floral color may be the result of adaptation to different pollinators (adaptive divergence) or of differences in other aspects of the species' biology that are genetically correlated with floral color (i.e., nonadaptive divergence with respect to pollination). To my knowledge, the relative frequency of adaptive vs. nonadaptive divergence as a cause of variation in floral colors (or any trait) has never been investigated.

Species of *Dalechampia* exhibit considerable diversity in blossom color, as well as pollination ecology (Table 9.4). Webster and Webster (1972) suggested that this diversity in blossom color reflects adaptation to diverse pollinators, which is consistent with the "traditional" view of flower color evolution. However, a detailed analysis of the relationship between pollination ecology and blossom color is needed to choose betwen the alternative hypotheses of adaptive and nonadaptive divergence in blossom color.

Methods

To test for an association between pollination system and blossom colors, I classified 35 *Dalechampia* taxa (species and pollination ecotypes) into three basic pollination modes [pollination by fragrance-collecting male euglossine bees (Apidae), female *Hypanthidium* (Megachilidae), or female euglossine bees], five classes of bract color (green to pale green, white, pink to magenta, yellow, brown), and four classes of resin color (clear to whitish, yellow to orange, blue to green, maroon). I then tabulated frequencies and tested for associations using the *G*-statistic (Sokal and Rohlf, 1981). I used two procedures for tabulating. First, I counted all species or species groups with each combination of pollination mode and color. Second, to control for phylogenetic inflation of degrees of freedom (Felsenstien, 1985), I counted only independent origins of each association in the phylogeny.

Results and Discussion

The results of the contingency table analyses are presented in Table 9.5. They show that there is at most a weak association between bract color and pollination mode ($P = 0.10$ when phylogeny has been controlled for; Table 9.5). This association, if it exists, reflects a weak tendency for species pollinated by female euglossine bees to have white or pink bracts and those species pollinated by smaller bees such as *Hypanthidium* to have green bracts. There are several exceptions in both directions. Other variation in bract color appears to occur at random with respect to pollination mode. There is no significant relationship between pollination mode and resin color (Table 9.5), or between pollination mode and the color of the stigmas or sepals (Table 9.4). It is noteworthy that, in contrast to the "traditional" view, all the variation in color is associated with

Table 9.4. Haphazard sample of neotropical Dalechampia *species, their pollinators, and the color of their involucral bracts, resin, staminate sepals, and stigmas.*

Species (or Species Group)	Bract Color	Resin/Gland Color	Sepal Color	Stigma Color	Pollinators/Syndrome
D. fragrans	Green	*	Green	Green	Male *Euglossa*/male fragrance-collecting bees[5]
D. pentaphylla	White	Clear/white	Green	Green	Female *Eulaema*/large resin-collecting bees[3]
D. leutzelbergii	White	Yellow	Green	Green	Female *Euglossa*/medium resin-collecting bees[3]
D. megacarpa	Green	Bluish	Green	Green	Female *Euglossa*/medium resin-collecting bees[3]
D. hutchisoniana	Brown	Clear/white	Green	Green	Female *Hypanthidium*/small resin-collecting bees[3]
D. papillistigma	Green	Yellow	Green	Green	Female *Hypanthidium*/small resin-collecting bees[3]
D. websteri	White	Yellow	Green	Green	Female *Euglossa*/medium resin-collecting bees[2]
D. spathulata	Pink	Yellow	White	Green	Male *Eulaema*/male fragrance-collecting bees[6]
D. dioscoreifolia	Pink	Maroon	Pink	Green	Female *Eulaema*/large resin-collecting bees[2]
D. sp. nov. "bella"	Magenta	Greenish	Magenta	Magenta	Female *Eulaema*/large resin-collecting bees[2]
D. aristilochiifolia	Magenta	Yellow	Magenta	Magenta	Female *Eulaema*/large resin-collecting bees[2]
D. schottii	White	Blue	Green	Green	Female *Hypanthidium*/small resin-collecting bees[1,2]
D. canescens	White	Yellow	Green	Green	Female *Euglossa*/medium resin-collecting bees[2]
D. osana	White	Whitish	Green	Green	Female *Euglossa*/medium resin-collecting bees[2]
D. shankii	White	*	White	Green	*Trigona*/pollen-collecting bees[3]
D. ficifolia	White	Whitish	Green	Green	Female *Euglossa*/medium resin-collecting bees[3]
D. schippii	Pale pink	Orange	Green	Green	Female *Euglossa*/medium resin-collecting bees[1]
D. caperonioides	Pink	Orange	Green	Green	Female *Hypanthidium*/small resin-collecting bees[2]
D. humilis	Yellow	Yellow	Green	Yellow	Female *Hypanthidium*/small resin-collecting bees[2]
D. triphylla	Green	Clear	Green	Green	Female *Hypanthidium*/small resin-collecting bees[2]
D. cissifolia	Green	Clear	Green	Green	Female *Hypanthidium*/small resin-collecting bees[4]
D. arenalensis	Green	Whitish	Green	Green	Female *Euglossa*/medium resin-collecting bees[2]
D. tenuaramea	Green	Clear	Green	Green	Female *Hypanthidium*/small resin-collecting bees[4]
D. tiliifolia	White	Orange	White	Green	Female *Eualema*/large resin-collecting bees[4]
D. ilheotica A	White	Yellow	Green	Green	Female *Euglossa*/medium resin-collecting bees[2]

continued

Table 9.4. (continued)

Species (or Species Group)	Bract Color	Resin/Gland Color	Sepal Color	Stigma Color	Pollinators/Syndrome
D. ilheotica B	White	Yellow	Green	Green	Female and male *Euglossa*/medium resin-collecting bees and male fragrance-collecting bees[3]
D. scandens A	White/pale green	Whitish	Green	Green	Female *Hypanthidium*/small resin-collecting bees[1]
D. scandens B	White/pale green	Whitish	Green	Green	Female *Euglossa*/medium resin-collecting bees[1]
D. scandens C	White/pale green	Whitish	Green	Green	Female *Eulaema*/large resin-collecting bees[1]
D. armbrusteri	Pale green	Whitish	Green	Green	Female *Euglossa*/medium resin-collecting bees[2]
D. brownsbergensis	Pale green	*	Green	Green	Male *Euglossa*/male fragrance-collecting bees[5]
D. stipulacea	Pale green	Yellow	Green	Green	Female *Eulaema*/large resin-collecting bees[2]
D. stipulacea var. *minor*	White and green	Whitish	Green	Green	Female *Euglossa*/medium resin-collecting bees[7]
D. magnistipulata	Pale green	Yellow	Green	Green	Female *Eufriesea*/large resin-collecting bees[6]
D. pernambucensis	Pale green	Orange	Green	Green	Female *Hypanthidium*/small resin-collecting bees[3]

[1]Armbruster (1985).

[2]Armbruster (1988).

[3]Armbruster (1993).

[4]Armbruster and Herzig (1984).

[5]Armbruster et al. (1992).

[6]Armbruster and Webster (1979).

[7]Sazima et al. (1985).

*Resin gland absent.

Table 9.5. Association between color of involucral bracts, floral resin, and pollination mode in neotropical Dalechampia. *Numbers in columns 2–10 are frequencies of species or species groups; numbers in parentheses are the times the association originated in the inferred phylogeny. There is weak evidence for an association between resin color and pollination mode* ($G = 6.46$, $P = 0.012$*; corrected for phylogeny:* $G = 2.82$, $P = 0.10$*). There is no evidence for an association between resin color and pollination mode* ($G = 0.580$; $P = 0.50$*; corrected for phylogeny:* $G = 0.02$, $P > 0.50$*).*

	Bract Color					Resin Color			
Pollination	Green/Pale Green	White	Pink/Magenta	Yellow	Brown	Clear/Whitish	Yellow/Orange	Blue/Green	Maroon
Male fragrance-collecting bees	2(1)	1(1)	1(1)	0(0)	0(0)	—	—	—	—
Female *Hypanthidium*	6(4)*	1(1)*	1(1)	1(1)	1(1)	5(3)*	4(4)*	1(1)	0(0)
Female Euglossines	5(3)*	12(6)*	4(2)	0(0)	0(0)	8(6)*	9(7)*	2(2)	1(1)

*Used in statistical analyses.

pollination by bees. This suggests that neither birds nor butterflies have played any role in the repeated evolution of pink bracts or sepals.

I propose that most of the *variation* in floral colors (bracts, resin, sepals, stigmas) is nonadaptive in origin, although there may have been stronger selection for contrasting colors (from vegetation) in large-bee-pollinated species than small-bee-pollinated species. [This may reflect greater reliance on vision over longer approach distances (large bees) and greater reliance on scent over shorter distances (small bees).] The variation in colors of bracts, resin, sepals, and stigmas (other than green) probably reflects different responses to the same selective pressure (to be readily visible) as determined by the different genetic contexts in which the adaptation evolves. For example, pink bracts may evolve in respose to selection for showy blossoms in populations that already synthesize anthocyanin pigments for other purposes, whereas white bracts may evolve in response to the same selective pressure in populations lacking anthocyanin synthesis. This interpretation is supported by the observation that pigments responsible for variation in bract color are usually expressed in other organs. For example, species with pink to purple bracts, such as *D. dioscoreifolia, D. aristolochiifolia, D. schippii,* and *D. spathulata,* have reddish pigments in their young leaves and stems. Species with white or green bracts lack the reddish pigment in leaves and stems. This suggests that the reddish pigment (an anthocyanin) may have evolved as a protective pigment in leaves and stems of some species. Adaptation along "lines of least resistance" led to pink or purple bracts in those species already synthesizing anthocyanins and white or pale green bracts in those species not synthesizing anthocyanins. (Note that because the bracts are modified leaves, deep green is the ancestral bract color.)

This series of hypotheses can be tested by field experiments. Bract, resin, and floral colors can be manipulated in the field to determine whether pollinating bees are differentially responsive to different colors or combinations of colors. The underlying assumption that bees use bracts that differ in color from leaves to find the blossoms should also be tested in field experiments.

A final hypothesis deserves attention. Diverse blossom color may have evolved as adaptations reducing interspecific pollination (by increasing floral constancy in pollinating bees). This hypothesis deserves testing with field experiments and additional comparative data. Two observations are inconsistent with this hypothesis, however. No character displacement in bract and resin colors has been observed in species/ecotypes showing character displacement in other characters (Armbruster, 1985), although it would be expected if bract and resin color play a role in floral constancy. Second, inconstancy in floral visitation by female euglossines seems to be common, even when they are visiting *Dalechampia* species with very different bract and resin colors (Armbruster and Herzig, 1984). Neither observation falsifies the hypothesis, but they constitute evidence against it (by weak inference; Platt, 1964).

Adaptive Compromises in Style Form and Function

Sprengel (1793) explicitly addressed the problem of flower parts having multiple, sometimes conflicting functions in his discussion of the "design" of flowers and nectaries. Nectaries and associated floral structures must make nectar available to deserving pollinators, protect it, if possible, from nectar thieves (nonpollinating, nectar-feeding insects), and prevent it from being washed away or diluted by rain. Darwin (1877) also wrote of multiple functions of floral parts in orchids. Implicit in his writing was the idea that conflicting selective pressures sometimes generate adaptive compromises in floral design. Very few papers since Darwin's time have addressed empirically the issue of adaptive compromise in the evolution of floral structure and function (but see Galen, 1983; Primack, 1987; Galen and Newport, 1988; Campbell, 1991).

One selective conflict that Darwin (1871) discussed in great detail was the conflict between natural selection and sexual selection. Darwin restricted his examples to animals, but the concept of sexual selection has also been usefully applied to plants (reviewed in Willson and Burley, 1983; Willson, 1991). In plants, as in animals, sexual selection comprises those selective pressures that relate to access to mates: usually male-male competition (pollen competition in plants) and female choice (selective pollination, fertilization, and zygote abortion in plants). Because the flower is the subject of most sexual selection, as well as natural selection generated by pollinators and other factors, one could expect floral structures sometimes to be under conflicting selective pressures.

The distinction between sexual and natural selection is not always clear in plants. Although selection associated with pollen competition and female choice is purely sexual selection, that associated with pollinators and pollination is a mixture of natural selection (differential offspring production) and sexual selection (differential access to mates). The above caveat notwithstanding, I will consider selection associated with pollen competition and female choice as predominantly sexual selection and selection generated by pollinators as a concordant mix of natural and sexual selection.

Sprengel (1793) and Darwin (1877) both pointed out that the length of the style and location of the stigma must be compatible with pollinator length and behavior. At the same time, the style provides the arena for pollen competition and female choice. The stigma size and shape and style length may influence these processes, and hence themselves be influenced by sexual selection. One could expect that the style and stigmas of flowers have evolved by adaptive compromise to both sexual selection (pollen competition) and selection generated by pollinators.

The amount of the style surface that is stigmatic can affect both the pollination ecology and intensity of pollen competion/female choice. For example, having a stigmatic surface over much of the length of the style (as is observed in

Dalechampia; see below) may increase pollen arrival rates, but incurs a large potential cost in terms of pollen competition and female choice. The reason for this is as follows. The intensity and beneficial effects of pollen competition depend on the fact that the genetically influenced vigor of a pollen grain is a major determinant of the pollen's success as a father (Mulcahy, 1979). As random effects play a larger role in who fathers zygotes, the systematic beneficial effect of pollen competition is decreased. Variation in the length of the route that tubes must grow to reach the ovules may introduce a random component in which grains are successful. Increasing the stigmatic surface, especially along the style, will increase the variation in the length that pollen tubes must grow to reach the ovules. A grain that, by chance, lands on the stigmatic region at the base of the style has a short distance for its tube to grow to reach the ovules and will likely be successful, even if of an inferior genotype. A genetically superior pollen grain landing at the tip of the style will have a much greater distance to grow its tube, and will likely fail to fertilize an ovule if there are sufficient grains nearer the ovary.

Thus, selection to increase the stigmatic surface to improve the pollination process may be countered by selection to increase the intensity of pollen competition, if more pollen grains commonly land on the stigma than there are ovules to be fertilized and pollen competition results in superior offspring (see Mulcahy and Mulcahy, 1975, 1987; Mulcahy, 1979; Windsor et al., 1987; Snow and Spira, 1991*a, b*).

A possible example of how these selective conflicts are resolved through adaptive compromise is found in the evolution of *Dalechampia*-style shape and function. The length of the style affects the gland-stigma distance, which in turn affects which floral visitors will be effective pollinators. Gland-stigma distance, through its interaction with gland area, has significant influence on reproductive fitness (Armbruster, 1990). Thus, style length is presumably affected by selective pressures generated by interaction with pollinators. At the same time, style length affects the intensity of pollen competition and is potentially affected by sexual selection. Similarly, the area of the stigmatic surface also affects both pollination function and the intensity of pollen competition.

Methods

I mapped the pollination ecology and shape of styles and stigmas of extant neotropical *Dalechampia* onto the inferred phylogenetic tree of these species (see Armbruster, 1993). Using parsimony, I inferred the pollination ecology and style and stigma morphology of the ancestral taxa; this exercise was conducted and the results visualized and plotted using MacClade (Maddison and Maddison, 1992). From this reconstruction, I was able to estimate the history of evolution of pollination ecology and style and stigma shape. Using this historical reconstruction, I inferred the likely selective forces influencing the evolution of style and stigma shape.

Results and Discussion

Most *Dalechampia* species have enlarged stigmatic surfaces that cover most of the style (Fig. 9.2). The expanded stigma appears to be associated with fragrance secretion throughout the genus, which suggests that the large stigmatic surface evolved to promote fragrance production for attraction of pollinators. The large stigma may have also (or instead) promoted pollen receipt and/or self-pollination. An interesting possibility is that the lateral stigmas may allow for pollination by inefficient pollinators when pollinators of the most appropriate size and behavior are unavailable. This might reduce extinction rates and/or increase speciation rates and explain why lateral stigmas predominate in the genus. In any case, the large stigmatic surface appears to have evolved in response to selection generated by pollination function.

As was explained above, an enlarged stigma is potentially disadvantageous in that it increases the randomness in which pollen genotypes are successful in pollen competition. Pollen competition is likely to be important in *Dalechampia* only if more pollen lands on the stigmas in a fairly short period of time than there are ovules to be fertilized. *Dalechampia* species have three ovules per pistillate flower and nine ovules per blossom. The stigmas remain receptive

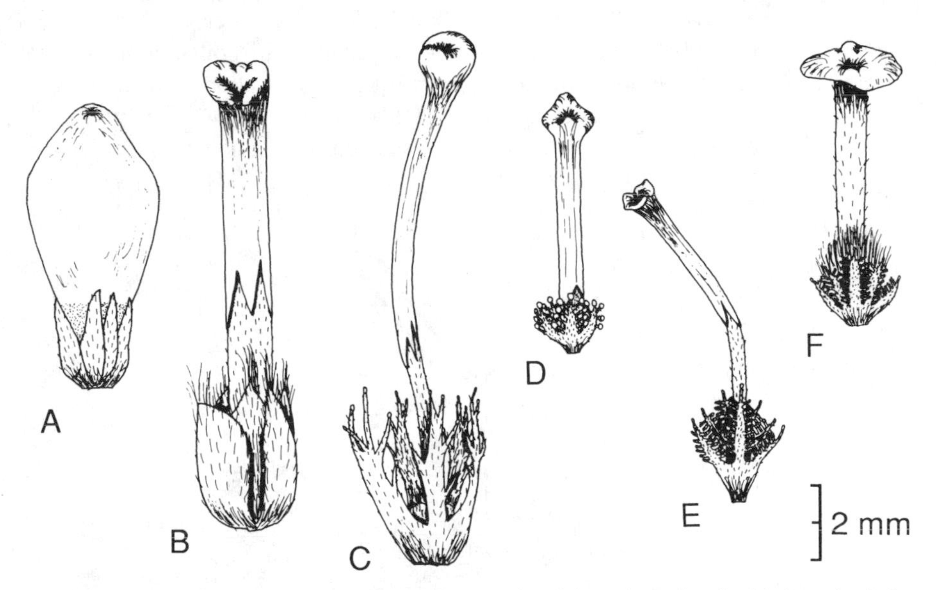

Figure 9.2. Styles of selected *Dalechampia* species. Stigmatic surface is distal unstippled area. (A) *Dalechampia fragrans*. (B) *D. pentaphylla*. (C) *D. spathulata*. (D) *D. scandens*. (E) *D. parvifolia*. (F) *D. pernambucensis*. Adapted from Armbruster et al. (1994).

throughout the 2-day pistillate phase and the 5–7 day bisexual phase of blossom development (e.g., Armbruster et al., 1992). With adequate pollination, usually all ovules produce seed; abortion is usually very limited under the usual high-light conditions in which most species grow (see Armbruster, 1982). In a survey of five species in the field (subject to natural insect pollination), all flowers accumulated between 5 and 100 times as much pollen on the stigmas as there were ovules, usually during the first few days of stigma receptivity (Armbruster et al., 1995). Pollen often arrives in clumps of 10–50 grains on a single visit by a bee (Armbruster, unpublished observations). Thus, pollen competition probably often occurs in most *Dalechampia* species.

Several evolutionary "options" exist for populations caught in the apparent adaptive dilemma of having large stigmas (for proper pollination) at the cost of increased randomness in who fathers zygotes. A population could evolve a longer style, with the stigma staying the same size but restricted to the distal region of the longer style. In this way, the proportional variation [i.e., coefficient of variation (CV); Sokal and Rohlf, 1981] in length of tube routes is reduced and thereby also the random component of who sires offspring. However, other pressures may oppose changes in style length: the optimal "fit" of the style and GSD to the principal pollinators (as well as the length that pollen tubes can grow). Thus, changes in style length driven by pollen competition and/or female choice may be opposed by selection generated by pollinators (Fig. 9.3).

Another "option" may be to expand the tip of the style laterally and confine the stigma to this region. In this way, stigmatic-surface area and fragrance secretion could be held nearly constant and the intensity of pollen competition increased, without disrupting blossom geometry and pollination. This apparent "solution" has evolved independently in at least two *Dalechampia* lineages, including that containing *D. dioscoreifolia* and that containing *D. pernambu-*

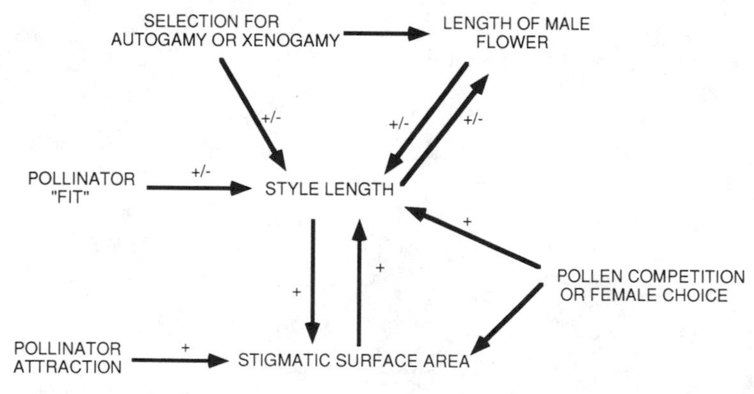

Figure 9.3. Hypothesized interactions between floral characters as generated by natural selection and sexual selection.

censis; both species have stigmas confined to the distal face of a platelike stylar tip (Figs. 9.2, and 9.4).

An alternative response to the conflicting pressures of pollination and pollen competition/female choice may have been taken by other *Dalechampia* species. Pollen grains that land on the lateral stigmatic surface of the style are not "permitted" to grow tubes directly toward the ovules. Instead, their tubes must grow to the stylar tip, where they reverse direction and grow back down the central core of the style (Armbruster et al., 1995). Seed set from these pollen grains is similar to that from grains at the style tip, and the tubes from both appear to grow at about the same speed (Armbruster et al., 1995). The unusual pollen-tube route may be a nonadaptive consequence of the evolutionary expansion of the stigmatic surface. Nevertheless, the result of this indirect pollen-tube

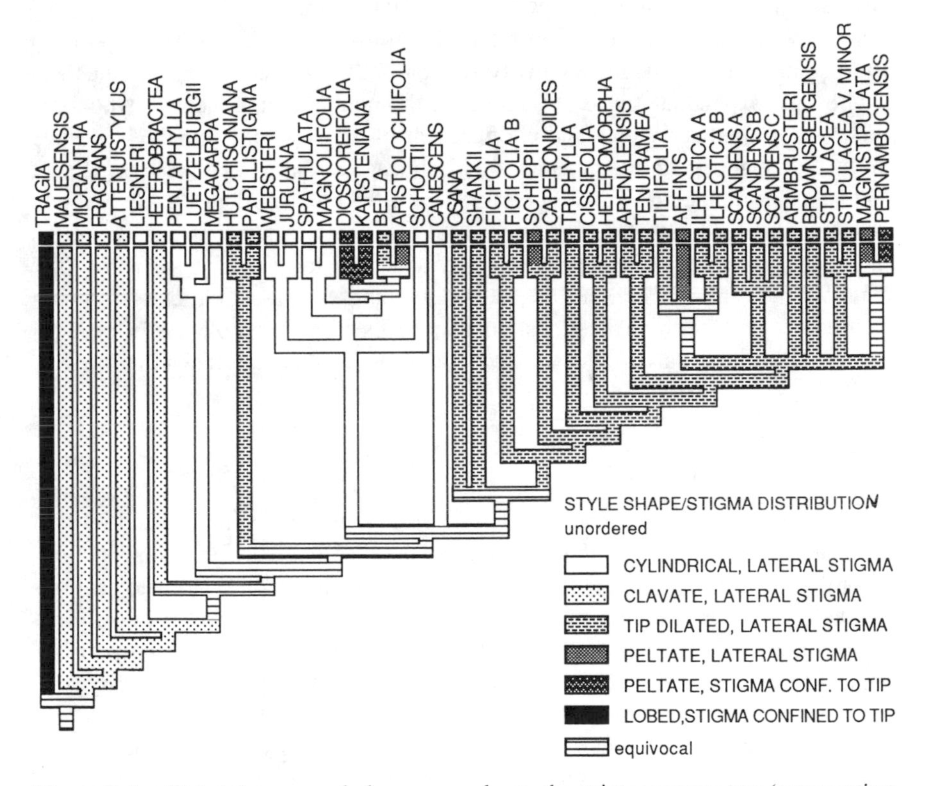

Figure 9.4. Style/stigma morphology mapped onto the strict consensus tree (representing the consensus of 552 most parsimonious estimated phylogenies) of *Dalechampia* species. The trees were based on parsimony analysis of 54 morphological characters and 46 taxa. Pattern of shading indicates inferred ancestral conditions and history of character transition. *Tragia* is sister genus. Additional details on the analysis are presented in Armbruster (1993).

growth is to nearly double the functional length of the style, while keeping the floral geometry the same and not disrupting pollen placement on, and pollen receipt from, pollinators. Hence, indirect pollen-tube growth appears to increase the intensity of pollen competition and/or female choice by reducing the proportional variance (CV) in the length of tube paths and thereby reducing the random component in who fathers offspring.

Mapping the stigma and style shape onto phylogeny allows inference of the history of the evolution of style and stigma morphology (Fig. 9.4). This analysis indicates that the expanded stigma with large lateral surfaces originated early in the genus and has persisted in many lineages. The enlarged stigma is by far the most common morphology in the genus today. It presumably evolved as a way to increase fragrance production and/or improve rates of pollen arrival on the stigmas. Reverse pollen-tube growth also apparently originated early in the evolution of the genus and is now represented in all major lineages. It is tempting to suggest that this unusual feature evolved to partially "correct" the problem generated by the expanded stigmas by reducing the random component in who sires offspring. Flared stylar tips and confinement of the stigma to near the tip have originated independently in at least two lineages. These may also represent "attempts" to reduce the random component in who sires offspring. Styles may also have evolved increased lengths to solve this problem, but evolution in this direction appears to have been limited by selection to maintain pollinator "fit" (Fig. 9.5). These explanations are ad hoc hypotheses that should be tested with experiments to determine the fitness consequences of indirect pollen-tube growth, expanded stigmas, flared stylar tips, and the other unusual features of this system.

Conclusions

The present review illustrates how diverse information is required to achieve an integrated understanding of the evolution of floral form and pollination function. Information on phylogeny, although rarely available in evolutionary and ecological studies, is of crucial importance if the morphology and pollination of different species are to be compared. Phylogenetically corrected statistical comparisons are also critical for hypothesis generation and testing. Fortunately, both phylogenetic and comparative statistical methods have improved significantly and become more generally accessible in recent years. Detailed ecological information is also a necessary component in such studies. However, this information is time-consuming to collect for one species; to collect it for multiple species may require many years of field work. (The ecological data for the present study represent nearly two decades of field work.) Patience and persistence are prerequisites for such endeavors. Nevertheless, the evolutionary insights gained from such broad comparative studies can be obtained in no other way.

It is apparent that the evolution of floral form and function is more complicated

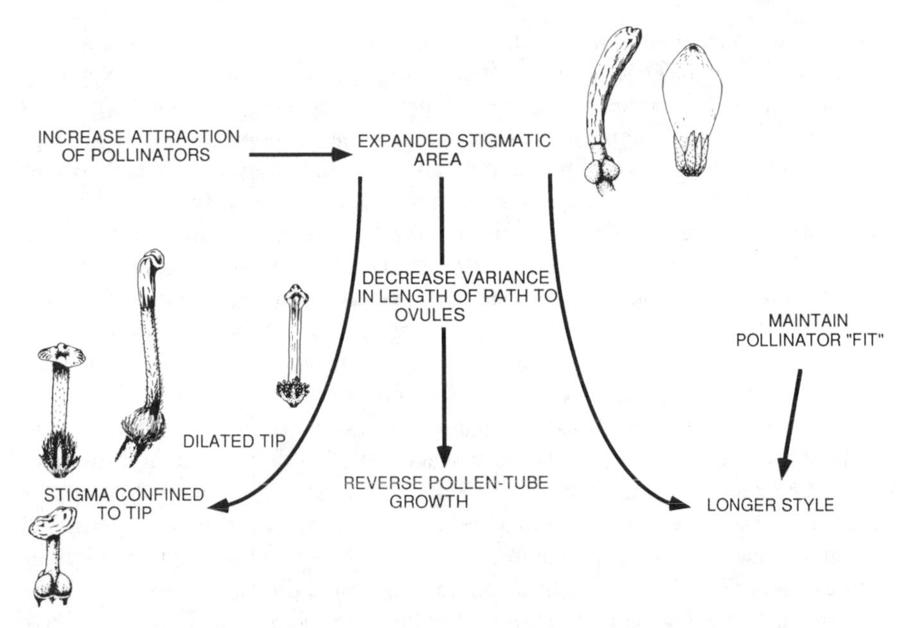

Figure 9.5. Hypothesized paths of evolution of *Dalechampia* style morphology and function.

than Darwin first thought. Several genetic and nonadaptive factors never imagined by Darwin (at least for plants) may often play important roles in floral evolution; these include sexual selection and genetic constraint. Nevertheless, the results of the present study are generally consistent with many of Sprengel's and Darwin's early insights into floral function (e.g., the role of pollinator-mediated selection and adaptive compromise).

It seems safe to conclude that, in *Dalechampia* at least, adaptation to pollinators has played a major role in floral evolution. Pollinators have generated strong selective pressures leading to the integration of floral characters into adaptive combinations. The selective covariance of ecologically important floral traits is expressed both among populations within species and among species. However, sexual selection associated with pollen competion may have generated conflicting evolutionary pressures acting in opposition to selection generated by pollinators. Hence, adaptive compromise appears to have been important in the evolution of blossom morphology.

Nonadaptive factors appear to have also played prominent roles in the evolution of blossoms. Genetic correlation appears to have constrained the evolution of involucral bracts. Selection for showy blossoms has operated in a diversity of genetic contexts. This has led to nonadaptive diversification of blossom colors; blossom colors are not obviously associated with pollination ecology.

Even this fairly comprehensive understanding of the evolution of the form and function of *Dalechampia* flowers falls short in many areas. We still lack a clear understanding of the genetic control of variation in floral morphology as expressed within and among populations. We do not yet know how important natural selection vs. genetic constraint is in creating among-population and among-species correlations of floral traits. We lack data from many neotropical species and most paleotropical species in the genus. There may be many new twists in the story as new species are added and new regions and biotic environments studied (see, e.g., Armbruster et al., 1993). There is need for integrative, multi-species comparative studies to be conducted on other plant groups. It is imposible to say whether the trends seen in *Dalechampia* will hold up elsewhere in the angiosperms. I cannot overemphasize the importance of conducting studies on a large sample of species. A study of the pollination ecology of one species provides insights into the ecology of that species. Knowledge increases additively with additional species studied. Only when we reach a large enough number does it become possible to pose and statistically test evolutionary hypotheses about patterns in the pollination data. When we consider data on floral form and function from a statistically meaningful number of species for which the phylogeny can be inferred, our insights include a historical dimension. With insights into the historical course of evolution of pollination ecology, it is possible to ask new kinds of questions: for example, questions about relative rates of evolutionary change in mutualistic relationships with pollinators, commonness and causes of homoplasy in pollination ecology and floral morphology, and effects of pollination system on speciation and extinction rates.

Future improvements in our understanding of the evolution of floral form and function are likely to come from four quarters. One will be in the area of improved methods of phylogeny reconstruction; especially promising is molecular (DNA) phylogenetics. A second will be in the development and application of new genetic methods for studying the genetics of population and species divergence. Quantitative genetic methods and molecular mapping of quantitative trait loci can be applied to the products of interpopulational crosses as well as traditional intrapopulation crosses. This will help elucidate the genetic basis of intrapopulation, interpopulation, and interspecific variation and covariation of traits (Conner and Via, 1993; Armbruster and Schwaegerle, 1995). The third source of improved understanding will come from studies of floral development, including heterochrony and the developmental basis of pleiotropy (e.g., Guerrant, 1982; Armbruster, 1991; see review in Diggle, 1992). The fourth source of advancement will be the synthesis of data from pollination ecology, phylogeny, statistical comparative methods, developmental biology, and quantitative and molecular genetics for obtaining an integrative perspective on floral evolution.

Acknowledgments

I thank David Lloyd for inviting me to contribute to this volume, numerous collaborators for help in the field and lab over the past two decades, Spencer

Barrett, Candace Galen, Susan Mazer, and an anonymous reviewer for comments on the manuscript, and the National Science Foundation (United States) for financial assistance (grants DEB-7824218, BSR-8509031, BSR-9006607, BSR-9020265).

References

Armbruster, W.S. 1982. Seed production and dispersal in *Dalechampia* (Euphorbiaceae): Divergent patterns and ecological consequences. *Am. J. Bot.,* 69: 1429–1440.

Armbruster, W.S. 1984. The role of resin in angiosperm pollination: Ecological and chemical considerations. *Am. J. Bot.,* 71: 1149–1160.

Armbruster, W.S. 1985. Patterns of character divergence and the evolution of reproductive ecotypes in *Dalechampia scandens* (Euphorbiaceae). *Evolution,* 39: 733–752.

Armbruster, W.S. 1988. Multilevel comparative analysis of the morphology, function, and evolution of *Dalechampia* blossoms. *Ecology,* 69: 1746–1761.

Armbruster, W.S. 1990. Estimating and testing the shapes of adaptive surfaces: The morphology and pollination of *Dalechampia* blossoms. *Am. Nat.,* 135: 14–31.

Armbruster, W.S. 1991. Multilevel analysis of morphometric data from natural plant populations: Insights into ontogenetic, genetic, and selective correlations in *Dalechampia scandens. Evolution,* 45: 1229–1244.

Armbruster, W.S. 1992. Phylogeny and the evolution of plant-animal interactions. *BioSci.,* 42: 12–20.

Armbruster, W.S. 1993. Evolution of plant pollination systems: Hypotheses and tests with the neotropical vine *Dalechampia. Evolution,* 47: 1480–1505.

Armbruster, W.S. and A.L. Herzig. 1984. Partitioning and sharing of pollinators by four sympatric species of *Dalechampia* (Euphorbiaceae) in Panama. *Ann. Missouri Bot. Gard.,* 71: 1–16.

Armbruster, W.S. and K.E. Schwaegerle. 1995. Causes of covariation of phenotypic traits among populations. *J. Evol. Biol.* (submitted).

Armbruster, W.S. and G.L. Webster. 1979. Pollination of two species of *Dalechampia* (Euphorbiaceae) in Mexico by euglossine bees. *Biotropica,* 11: 278–283.

Armbruster, W.S., M.E. Edwards, J.F. Hines, R.L.A. Mahunnah, and P. Munyenyembe. 1993. Evolution and pollination of Madagascan and African *Dalechampia* (Euphorbiaceae). *Res. Explor.,* 9: 430–444.

Armbruster, W.S., A.L. Herzig, and T.P. Clausen. 1992. Pollination of two sympatric species of *Dalechampia* (Euphorbiaceae) in Suriname by male euglossine bees. *Am. J. Bot.,* 79: 1374–1381.

Armbruster, W.S., C.S. Keller, M. Matsuki, and T.P. Clausen. 1989. Pollination of *Dalechampia magnoliifolia* (Euphorbiaceae) by male euglossine bees (Apidae: Euglossini). *Am. J. Bot.,* 76: 1279–1285.

Armbruster, W.S., P. Martin, J. Kidd, R. Stafford, and D.G. Rogers. 1995. Indirect pollen-tube growth in *Dalechampia* (Euphorbiaceae): Possible reproductive and evolutionary significance. *Am. J. Bot.* 82: 51–56.

Arnold, S.J. 1992. Constraints on phenotypic evolution. *Am. Nat.*, 140: S85–S107.

Bell, G. 1989. A comparative method. *Am. Nat.*, 133: 553–571.

Berg, R. 1960. The ecological significance of correlation pleiades. *Evolution*, 17: 171–180.

Brooks, D.R. and D.A. McLennan. 1991. *Phylogeny, Ecology, and Behavior*. Univ. Chicago Press, Chicago, IL.

Cammerloher, H. 1931. *Blutenbiologie I*. Gebruder Borntraeger, Berlin.

Campbell, D.R. 1989. Measurements of selection in a hemaphroditic plant: Variation in male and female pollination success. *Evolution*, 43: 318–334.

Campbell, D.R. 1991. Effects of floral traits on sequential components of fitness in *Ipomopsis aggregata*. *Am. Nat.*, 137: 713–737.

Campbell, D.R., N.M. Waser, and M.V. Price. 1994. Indirect selection of stigma position in *Ipomopsis aggregata* via a genetically correlated trait. *Evolution* 48: 55–68.

Campbell, D.R., N.M. Waser, M.V. Price, E.A. Lynch, and R.J. Mitchell. 1991. Components of phenotypic selection: Pollen export and flower corolla width in *Ipomopsis aggregata*. *Evolution*, 45: 1458–1467.

Chase, M.W. and H.G. Hills. 1992. Orchid phylogeny, flower sexuality, and fragrance-seeking bees. *BioSci.*, 42: 43–49.

Cheverud, J.M. 1982. Phenotypic, genetic, and environmental integration in the cranium. *Evolution*, 36: 499–516.

Cheverud, J.M. 1984. Quantitative genetics and developmental constraints on evolution by natural selection. *J. Theor. Biol.*, 110: 155–172.

Clarke, A.G. 1987. Genetic correlations: The quantitative genetics of evolutionary constraints. In *Genetic Constraints on Adaptive Evolution* (V. Loeschcke, ed.), Springer-Verlag, Berlin, pp. 25–45.

Conner, J. and S. Via. 1993. Patterns of phenotypic and genetic correlations among morphological and life-history traits in wild radish, *Raphanus raphanistrum*. *Evolution*, 47: 704–711.

Cox, P.A. 1990. Pollination and the evolution of breeding systems in Pandanaceae. *Ann. Missouri Bot. Gard.*, 77: 816–825.

Darwin, C. 1859. *The Origin of Species*. Republished 1958 New America Library, New York.

Darwin, C. 1871. *The Descent of Man and Selection in Relation to Sex*. J. Murray, London.

Darwin, C. 1877. *The Various Contrivances by Which Orchids Are Fertilised by Insect*. Republished 1984 Univ. Chicago Press, Chicago, IL.

Diggle, P.K. 1992. Development and the evolution of plant reproductive characters. In *Ecology and Evolution Plant Reproduction* (R. Wyatt, ed.), Chapman & Hall, New York, pp. 326–355.

Dobzhansky, T. 1959. Evolution of genes and genes in evolution. *Cold Spring Harbor Symp.*, 24: 15–30.

Dobzhansky, T. 1970. *Genetics of the Evolutionary Process*. Columbia Univ. Press, New York.

Donoghue, M.J. 1989. Phylogenies and the analysis of evolutionary sequences, with examples from seed plants. *Evolution,* 43: 1137–1156.

Epperson, B.K. and M.T. Clegg. 1987. First-pollination primacy and pollen selection in the morning glory, *Ipomoea purpurea. Heredity,* 58: 5–14.

Faegri, K. and L. van der Pijl. 1979. *The Principles of Pollination Ecology,* 3rd ed. Pergamon, Oxford.

Falconer, D.S. 1989. *Introduction to Quantitative Genetics,* 3rd ed. Longman, London.

Felsenstein, J. 1985. Phylogenies and the comparative method. *Am. Nat.,* 125: 1–15.

Galen, C. 1983. The effects of nectar thieving ants on seed set in floral scent morphs of *Polemonium viscosum. Oikos,* 41: 245–249.

Galen, C. 1985. Regulation of seed-set in *Polemonium viscosum*: Floral scents, pollination, and resources. *Ecology,* 66: 792–797.

Galen, C. 1989. Measuring pollinator-mediated selection on morphometric floral traits: Bumble bees and the alpine sky pilot, *Polemonium viscosum. Evolution,* 43: 882–890.

Galen, C. and M.E.A. Newport. 1988. Pollination quality, seed set and flower traits in *Polemonium viscosum:* Complementary effects of variation in flower scent and size. *Am. J. Bot.,* 75: 900–905.

Galen, C. and M.L. Stanton. 1989. Bumble bee pollination and floral morphology: Factors influencing pollen dispersal in the alpine sky pilot, *Polemonium viscosum* (Polemoniaceae). *Am. J. Bot.,* 76: 419–426.

Gould, S.J. and R.C. Lewontin. 1979. The spandrels of San Marco and the Panglossian paradigm: A critique of the adaptationist programme. *Proc. Roy. Soc. Lond., B,* 205: 581–598.

Gould, S.J. and E.S. Vrba. 1982. Exaptation—a missing term in the science of form. *Paleobiology,* 8: 4–15.

Grant, V.I. and K.A. Grant. 1965. *Flower Pollination in the Phlox Family*. Columbia Univ. Press, New York.

Gueinzius, W. 1858. On the habits of the hymenoptera of Natal. In *Proceedings of the Entomological Society of London,* pp. 9–11.

Guerrant, E.O., Jr. 1982. Neotenic evolution of *Delphinium nudicaule* (Ranunculaceae): A hummingbird-pollinated larkspur. *Evolution,* 36: 699–712.

Harvey, P.H. and M.D. Pagel. 1991. *The Comparative Method in Evolutionary Biology*. Oxford Univ. Press, Oxford, UK.

Heinrich, B. and P.H. Raven. 1972. Energetics and pollination ecology. *Science,* 176: 597–602.

Lande, R. and S.J. Arnold. 1983. The measurement of selection on correlated characters. *Evolution,* 37: 1210–1226.

McDade, L.A. 1992. Pollinator relationships, biogeography, and phylogenetics. *BioSci.,* 42: 21–26.

Maddison, W.P. and D. Maddison. 1992. *MacClade, Version 3. Analysis of Phylogeny and Character Evolution*. Sinauer, Sunderland, MA.

Miles, D.B. and A.E. Dunham. 1993. Historical perspectives in ecology and evolutionary biology: The use of phylogentic comparative analyses. *Ann. Rev. Ecol. Syst.*, 24: 587–619.

Mitchell, R.J. and R.G. Shaw. 1992. Heritability of floral traits for the perennial wildflower, *Penstemon centranthifolius. Heredity,* 7: 185–192.

Müller, H. 1879. Die Wechselbeziehungen zwischen den Blumen und den ihre Kreuzung vermittelden Insekten. In *Handbuch der Botanik I.* (Schenck, ed.), Eduard Trewent, Breslau, pp. 1–112.

Mulcahy, D.L. 1979. The rise of angiosperms: A genecological factor. *Science,* 206: 20–23.

Mulcahy, D.L. 1983. Models of pollen tube competition in *Geranium maculatum*. In *Pollination Biology,* (L. Real, ed.), Academic Press, New York. pp. 152–160.

Mulcahy, D.L. and G.B. Mulcahy. 1975. The influence of gametophytic competition on sporophytic quality in *Dianthus chinensis. Theor. Appl. Genet.,* 46: 277–280.

Mulcahy, D.L. and G.B. Mulcahy. 1987. The effects of pollen competition. *Am. Sci.,* 75: 44–50.

Pax, F. and K. Hoffmann. 1919. Euphorbiaceae-Dalechampieae. Das Pflanzenreich IV. 147. XII (Heft 68): 1–59.

Platt, J.R. 1964. Strong inference. *Science,* 146: 347–353.

Primack, R.B. 1987. Relationships among flowers, fruits, and seeds. *Ann. Rev. Ecol. Syst.,* 18: 409–430.

Ridley, M. 1983. *The Explanation of Organic Diversity*. Oxford Univ. Press, New York.

Ridley, M. 1992. Darwin sound on comparative method. *Trends Ecol. Evol.,* 7: 37–40.

Rieseberg, L.H., M.A. Hanson, and C.T. Philbrick. 1992. Androdioecy is derived from dioecy in Datiscaceae: Evidence from restriction site mapping of PCR-amplified chloroplast DNA fragments. *Syst. Bot.,* 17: 324–336.

Riska, B. 1986. Some models for development, growth, and morphometric correlation. *Evolution,* 40: 1303–1311.

Riska, B. 1989. Composite traits, selection response, and evolution. *Evolution,* 43: 1172–1191.

SAS Institute, Inc. 1985. *SAS User's Guide: Statistics. Version 5*. Cary, NC.

Sazima, M., I. Sazima, and R.M. Carvalho-Okano. 1985. Biologia floral de *Dalechampia stipulacea* (Euphorbiaceae) e sua pollinizacao por *Euglossa melanotricha* (Apidae). *Revista Brazileira de Biologia,* 45: 85–93.

Schemske, D.W. and C.C. Horvitz. 1984. Variation among floral visitors in pollination ability: A precondition for mutualism specialization. *Science* 225: 519–521.

Schemske, D.W. and C.C. Horvitz. 1988. Plant-animal interactions and fruit production in a neotropical herb: A path analysis. *Ecology,* 69: 1128–1137.

Schemske, D.W. and C.C. Horvitz. 1989. Temporal variation in selection on a floral character. *Evolution,* 43: 461–465.

Snow, A.A. and T.P. Spira. 1991*a*. Differential pollen-tube growth rates and nonrandom fertilization in *Hibiscus moscheutos* (Malvaceae). *Am. J. Bot.,* 78: 1419–1426.

Snow, A.A. and T.P. Spira. 1991*b*. Pollen vigor and the potential for sexual selection in plants. *Nature,* 352: 796–797.

Sokal, R.R. and F.J. Rohlf. 1981. *Biometry*. W.H. Freeman, San Francisco, CA.

Sprengel, C.K. 1793. *Das entdeckte Geheimnis der Natur im Bau und der Befruchtung der Blumen*. Viewig, Berlin.

Stanton, M.L. and R.E. Preston. 1988. A qualitative model for evaluating the effects of flower attractiveness on male and female fitness in plants. *Am. J. Bot.,* 75: 540–544.

Stanton, M.L., A.A. Snow, and S.N. Handel. 1986. Floral evolution: Attractiveness to pollinators influences male fitness. *Science,* 232: 1625–1627.

Stebbins, G.L. 1950. *Variation and Evolution in Plants*. Columbia Univ. Press, New York.

Stebbins, G.L. 1971. Adaptive radiation of reproductive characteristics in angiosperms. I. Pollination mechanisms. *Ann. Rev. Ecol. Syst.,* 1: 307–326.

Stebbins, G.L. 1974. *Flowering Plants. Evolution above the Species Level*. Harvard Univ. Press, Cambridge, MA.

Tedin, O. 1925. Vererbung, Variation, and Systematik der Gattung *Camelina*. *Hereditas,* 6: 275–386.

Thomson, J.D. and J. Brunet. 1990. Hypotheses for the evolution of dioecy in seed plants. *Trends Ecol. Evol.,* 5: 11–16.

Thomson, J.D. and D.A. Stratton. 1985. Floral morphology and cross pollination in *Erythronium grandiflorum* (Liliaceae). *Am. J. Bot.,* 72: 433–437.

Via, S. and R. Lande. 1985. Genotype-environment interaction and the evolution of phenotypic plasticity. *Evolution,* 39: 505–522.

Wagner, G.P. 1984. Coevolution of functionally constrained characters: Prerequisites for adaptive versatility. *Biol. Syst.,* 17: 51–55.

Wagner, W.L., S.G. Weller, and A.K. Sakai. 1994. Phylogeny, biogeography, and evolution in *Schiedea* and *Alsinidendron* (Caryophyllaceae). In *Hawaiian Biogeography. Evolution on a Hot-Spot Archipelago* (W.L. Wagner and V. Funk, eds.), Smithsonian Institution Press, Washington, DC.

Waser, N.M. 1983. The adaptive nature of floral traits: Ideas and evidence. In *Pollination Biology* (L. Real, ed), Academic Press, New York. pp. 242–277.

Waser, N.M. and M.V. Price. 1981. Pollinator choice and stabilizing selection for flower color in *Delphinium nelsonii*. *Evolution,* 35: 376–390.

Webster, G.L. and B. Webster. 1972. The morphology and relationships of *Dalechampia scandens* (Euphorbiaceae). *Am. J. Bot.,* 59: 573–586.

Weller, S.G., A.K. Sakai, W.L. Wagner, and D.R. Herbst. 1990. Evolution of dioecy in *Schiedea* (Caryophyllaceae: Alsinoideae) in the Hawaiian Islands: Biogeographical and ecological factors. *Syst. Bot.*, 15: 266–276.

Whitten, W.M., A. Young, and N. Williams. 1989. Function of glandular secretions in fragrance collection by male euglossine bees (Apidae: Euglossini). *J. Chem. Ecol.*, 15: 1285–1295.

Willson, M.F. 1991. Sexual selection, sexual dimorphism and plant phylogeny. *Evol. Ecol.*, 5: 69–87.

Willson, M.F. and N. Burley. 1983. *Mate Choice in Plants: Tactics, Mechanisms and Consequences*. Princeton Univ. Press, Princeton, NJ.

Windsor, J.A., L.E. Davis, and A.G. Stephenson. 1987. The relationship between pollen load and fruit maturation and the effect of pollen load on offspring vigor in *Cucurbita pepo*. *Am. Nat.*, 129: 643–656.

Wyatt, R. 1983. Pollinator-plant interactions and the evolution of breeding systems. In *Pollination Biology* (L. Real, ed.), Academic Press, New York. pp. 51–95.

Zeng, Z.-B. 1988. Long-term correlated response, interpopulation covariation, and interspecific allometry. *Evolution*, 42: 363–374.

10

The Evolution of Floral Form: Insights from an Alpine Wildflower, *Polemonium viscosum* (Polemoniaceae)

*Candace Galen**

Introduction

The morphological diversity of flowers has long fascinated evolutionary biologists. In his treatise on the origin of species, Darwin (1859) described a close correspondence between corolla length in clover and tongue length of its bumblebee pollinators to illustrate the concept of coevolution. Similarly, the elaborate structural mimicry of euglossine female bees by flowers of neotropical orchids and the unique inflorescence structure (synconia) of wasp-pollinated figs provide some of the most compelling evidence for evolutionary specialization in flowering plants (Feinsinger, 1983). Major phylogenetic shifts in floral morphology are presumed to reflect historical shifts in the availability or diversity of animal pollinators (Grant and Grant, 1965; Crepet, 1983). Modern comparative studies that map pollination syndromes onto statistically robust phylogenies afford promising tests of hypotheses on the adaptive radiation of flower form at a macroevolutionary scale (Donoghue, 1989; Systma, *et al.* 1991; Armbruster 1993 see also Chapter 9). However, to truly understand the process of floral evolution as it proceeds, we must turn our attention to the natural populations that comprise the arena for pollinator-mediated selection and gene flow. Recent advances in the analysis of heredity in morphometric traits and selection on them offer us a powerful set of tools with which to attempt this task (Lande and Arnold, 1983; Arnold and Wade, 1984; Mitchell-Olds and Shaw, 1987; Shaw, 1987). Here, I describe some of my own efforts in this regard, focusing on the evolution of clinal variation in flower size in the alpine wildflower, *Polemonium viscosum* (Polemoniaceae). This chapter considers the extent to which two basic requirements for the evolution of floral traits in response to pollinator-mediated selection

*Division of Biological Sciences, University of Missouri, Columbia, MO 65211.

(see Endler, 1986) are fulfilled in natural populations of this species, heritable variation in floral morphology and strong selection on flower morphology by pollinators. Whenever possible, I relate this work to similar studies on other taxa, highlighting those that involve plant populations in their natural environments. These studies, I believe, provide us with some of the sharpest insights about the potential for pollinator-mediated selection to drive the evolution of floral form.

Floral Biology of *Polemonium viscosum*

Plants of the alpine skypilot, *Polemonium viscosum,* are long-lived herbaceous perennials that branch rhizomatously at the base to form compact clusters of shoots. Plants are highly self-incompatible and depend on a diverse array of insect pollinators for cross-pollination and sexual reproduction (Galen and Kevan, 1980). In populations that we have sampled electrophoretically, multilocus outcrossing rates do not differ significantly from 1.0 (Galen et al., 1991).

Polemonium viscosum is an important element of the North American Rocky Mountain flora found commonly above treeline from New Mexico and Arizona to the Canadian Rockies (Davidson, 1950). Within the alpine, *P. viscosum* has a broad habitat affinity, occurring from treeline or krummholz to the exposed fellfields at the highest elevations. Pronounced clinal variation occurs across this elevational gradient in several floral traits of the skypilot, including fragrance, corolla color, and flower size (Galen and Kevan, 1980; Galen et al., 1987). For example, corolla flare, measured as the distance from tip to tip of opposite corolla lobes (Fig. 10.1), increases 12% on average ($P < 0.0001$) from krummholz to higher tundra habitats in five sites that we have surveyed (Galen et al., 1987). The trend of flowers to increase in size and showiness with elevation is not atypical of montane and alpine plant species, but is somewhat paradoxical (Bliss, 1985). From a growth economy perspective, one might expect the opposite under the austere climatic conditions and shortened growing season of the high alpine—increasing reliance on self-pollination with a concommitant reallocation of resources to maintenance and storage (Grime, 1979; Bliss, 1985).

For *P. viscosum,* production of large showy flowers incurs at least three kinds of resource costs: (1) increased use of structural carbohydrates since the dry weight of larger flowers exceeds that of small ones; (2) increased energy expenditure because larger flowers produce nectar sugar at a higher rate than small ones (Cresswell and Galen, 1991); and (3) increased expenditure of nitrogen and phosphorous, as larger flowers produce more pollen grains than small ones (Galen and Stanton, 1989). That flower size in the skypilot increases at high elevations suggests that plants having the highest investment in flower display occur under environmental conditions that afford the least opportunity for resource acquisition. To understand why this is so, I have been studying the ecological relation-

Figure 10.1. Polemonium viscosum. Brackets shown at (A) denote measurement of corolla length and those at (B) denote measurement of corolla flare.

ships and evolutionary responses of the alpine skypilot to its insect pollinators since 1978. Most of this work has been done at the center of the species' geographic range, on Pennsylvania Mountain in the Park Range of central Colorado (39°15′N; 106°07′W).

On Pennsylvania Mountain, *P. viscosum* is continuously distributed across an elevational gradient from 3500 m at treeline to 4000 m at the summit. As elsewhere in the Rocky Mountains (Macior, 1974), insect pollinator abundances vary with elevation at this site (Fig. 10.2; Galen, 1983*a*). Seventy-five % of the pollinator visits and more than 90% of the seeds set by plants of the skypilot can be attributed to bumble bee activity (principally queens of *Bombus kirbyellus* and *B. sylvicola*) at high tundra locations. Near treeline, however, bumble bees account for only about 5% of visits and about 50% of seeds set, whereas smaller flies and bees account for the remainder of pollination.

Corresponding changes in the importance of bumble bees as pollinators and in the size of flowers produced by plants of the skypilot suggest that bumble bees may play a paramount role in driving the evolution of flower size in this species. Such parallel shifts in floral attractants and pollinator activity have been cited frequently as evidence for the efficacy of pollinator-mediated selection on floral traits in other taxa (e.g., Miller, 1981; Armbruster, 1985). This interpretation assumes that genetic variation for flower size exists within natural populations of *P. viscosum* and bumble bees exert strong enough selection favoring plants

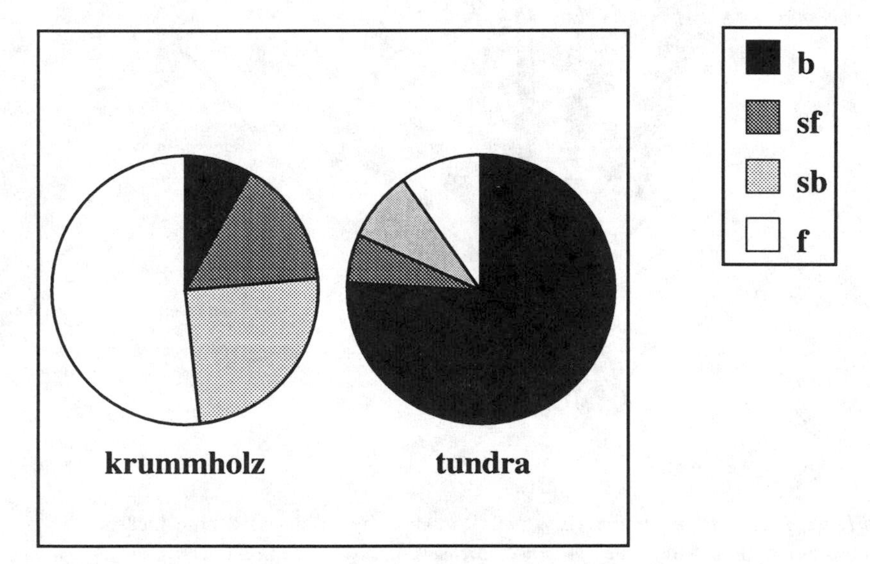

Figure 10.2. Pollinator pool for *Polemonium viscosum* (skypilot) on Pennsylvania Mountain, Colorado, showing relative visitation frequencies of insect visitors (b, bumble bees; sb, solitary bees; sf, syrphid flies; f, anthomyid and muscid flies) to flowers at 3530 m (krummholz) and 3640 m (tundra). From Galen (1983*a*).

with large flowers to bring about and maintain an increase of more than 10% in flower size over successive generations. In the pages that follow, I consider evidence for each of these assumptions, in turn.

Sources of Variation in Flower Form

Without doubt, the best evidence for heritable variation in flower form within natural plant populations comes from heterostylous species, where the genetic basis of morphological differences among floral variants is well understood (Darwin, 1877; Lewis and Jones, 1992). However, for taxa showing continuous variation in flower morphology, surprisingly little is known about sources of variation in floral traits within natural populations. Measurements of heritability on greenhouse grown plants suggest that genetic contributions to variation in flower morphology may be quite large (e.g., Rick et al., 1978; Schoen, 1982; Shore and Barrett, 1990; Holtsford and Ellstrand, 1992; Mitchell and Shaw, 1992; Anderson and Widén, 1993). But greenhouse derived estimates of heritability are probably inflated because of artificially reduced environmental heterogeneity. For example, even modest variation in temperature can strongly influence floral development and, consequently, flower morphology (Kinet et al., 1985; Holtsford and Ellstrand, 1992).

Our current lack of knowledge on the extent of heritable variation in floral traits of naturally occurring plant populations reflects the combined difficulties of performing quantitative genetic analyses in the field and on traits expressed late in the life cycle. I am aware of only three studies that have overcome these difficulties. These studies have special importance because the traits analyzed are known to have close relationships to components of reproductive success in natural populations. In field plantings of wild radish (*Raphanus sativus,* Brassicaceae) petal area, a correlate of pollen dispersal (Young and Stanton, 1990) showed little heritable variation under any of several density regimes (Mazer and Schick, 1991). However, other studies have found significant heritable variation in floral advertisements. Schwaegerle and Levin (1991) report heritabilities of about 0.15 for corolla tube length and petal width in *Phlox drummondii* (Polemoniaceae). In the montane wildflower, *Ipomopsis aggregata* (Polemoniaceae), variation in stigma-anther separation, a correlate of female reproductive success, has a strong genetic basis under natural conditions (Campbell, et al., 1994). Clearly, it is premature to draw any generalizations, but evidence exists that heritable variation in flower form may fuel floral evolution in at least some taxa.

In *P. viscosum,* I have obtained estimates for genetic and ontogenetic contributions to several aspects of floral display including annual flower number, scape height, corolla length and flare. The latter three traits show modest positive phenotypic correlations and together influence plant attractiveness to bumble bee pollinators (Galen, 1989). To estimate genetic contributions to floral traits, I measured maternal parent-offspring resemblance under natural environmental conditions. This approach is not without its limitations, as the heritability estimates obtained include possible maternal effects (Falconer, 1981). However, because maternal effects generally diminish for traits expressed late in an organism's development, the bias in floral traits may be less than that commonly seen at seed or seedling life-cycle stages (Roach and Wulff, 1987).

I measured floral traits of 53 offspring belonging to 40 maternal families that were either hand-outcrossed to randomly selected pollen donors in the surrounding krummholz population ($N = 19$) or were selected randomly from the krummholz population for pollination by bumble bees in a screen-house experiment ($N = 21$; Galen, 1989). Paternity for offspring of bumble bee-pollinated plants represents an unknown draw from a possible donor population of 94 individuals. However, as plants of the skypilot are highly self-incompatible, it can be assumed that all progeny used for genetic analyses were derived from outcross matings, regardless of pollination treatment. Maternal parents were pollinated in the summers of 1986 and 1987, seed stratification took place over the winter of each year, and seedlings germinating in the greenhouse were transplanted to the field after their first adult leaves developed the following spring. Offspring were planted in a randomized design along transects through the parental habitat (Galen, 1993). To avoid repeated sampling, floral traits measured at the oldest age of flowering were analyzed for heritability.

Age at flowering varied from 2–6 years and strongly influenced all four aspects of floral display (Table 10.1; Galen, in press). As plants mature and increase in size (Galen, 1993), their allocation to floral display increases accordingly. For example, a 1-year delay in reproduction is associated with a 10% increase in flower number and an 8% increase in corolla length.

Although strong stage or age-specific influences on components of reproductive effort are not unexpected (e.g., Schaffer and Schaffer, 1979), such effects complicate the genetic analysis of variation in floral characters. In essence, flower size or flower number at each age can be viewed as a unique age-specific trait just as survivorship or fecundity are age-specific population parameters. Such an approach, while desirable, requires that measurements of traits be made at the same ages on maternal parents and their offspring, and is impractical in plants that, like the skypilot, have a delayed onset of reproduction and flower intermittently thereafter. Instead, I used multiple regression analysis to factor out the effects of offspring age on floral traits before estimating heritabilities. For the flower size components, corolla length and flare, regression showed significant resemblance of maternal parent and offspring values with heritabilities close to 1.0 (Table 10.1). However, the resemblance between maternal parents and offspring in two other aspects of floral display, scape length or inflorescence height and annual flower number, was negligible. These findings indicate that interrelated aspects of floral attractiveness to pollinators vary widely in their evolutionary

Table 10.1. Genetic and ontogenetic effects on components of floral display in the alpine skypilot, Polemonium viscosum *(Galen, in press). Morphometric traits were log-transformed before regression.*

	Maternal	Offspring	Age			Seed Parent			Heritability
Trait	$\bar{x}\pm SD$	$\bar{x}\pm SD$	B_1*	t	$P<$	B_1	t	$P<$	$h^2\pm 2SE$
Corolla flare (mm)	14.2±1.6	14.4±2.0	0.029	2.233	0.03	0.51	2.554	0.015	1.02±0.80
Corolla length (mm)	23.6±2.1	22.7±2.6	0.032	2.999	0.01	0.51	3.160	0.003	1.02±0.64
Scape length (mm)	79.2±17.5	48.7±15.0	0.125	3.310	0.003	0.25	1.076	ns	
Flower number	15.3±10.4 (48)	10.0±4.1 (48)	1.123	2.764	0.01	−0.052	0.989	ns	

*Regression coefficient from multiple regression of F_1 trait value on age and maternal trait value.

potential (see also Mitchell and Shaw, 1992). Nonetheless, the high heritabilities observed for corolla length and flare are consistent with the view that the elevational cline in flower size of the skypilot represents an adaptive response to changes in its pollinator fauna (but see Houle, 1992). Occasional gene flow along the elevational cline probably also introduces alternative allele combinations into predominantly large-flowered or small-flowered subpopulations, contributing to the high heritability of flower size components (Galen et al., 1991; Galen, 1992).

Pollinator-Mediated Selection on Floral Form

Our present understanding of pollinator-mediated selection on flower form is much advanced relative to our knowledge of the underlying sources of variation in flower morphology. Several profitable approaches to studying selection on floral traits have been taken in recent years. These include observational studies addressing relationships between components of reproductive success and floral traits in natural populations (e.g., Johnston, 1991) and experimental manipulations of pollinator or trait distributions to explore mechanisms of selection on floral display (Stanton et al., 1991; Anderson and Widén, 1993). To truly pin down the role of pollinators in accounting for selection on floral form, experiments are essential. Only through experimental manipulation can the role of pollinators in accounting for correlations between morphometric floral traits and components of plant reproductive success be discerned from other underlying mechanisms (Stanton and Galloway, 1990; Young and Stanton, 1990; Armbruster, 1991; Mitchell and Shaw, 1992; Conner and Via, 1993). In *P. viscosum,* for example, plant age has correlated effects on flower number and flower size. Environmental heterogeneity in resource availability can act in a similar manner, such that correlations between fecundity and floral attributes in the field shed little light on pollinator-mediated selection (Mitchell-Olds and Shaw, 1987; Rausher, 1992).

A compelling demonstration of indirect selection on flower morphology due to genetic correlations with other aspects of floral display is provided by Campbell et al. (1994). In an elegant series of experiments, they tested whether selection by hummingbird pollinators on stigma exsertion in *Ipomopsis aggregata* resulted from direct positive effects of stigma exsertion on pollinator efficiency. Stigma exsertion in natural populations of this species is positively correlated with pollen receipt, but is also strongly related to the amount of time spent by a flower in the pistillate phase (Campbell, 1991). By independently manipulating stigma exsertion and the timing of stigma presentation, Campbell et al. (1994) showed that the increased pollen loads accruing on exserted stigmas reflected only the longer time available for pollen deposition. Genetic studies in the field demonstrated a heritable basis to the relationship between the two aspects of floral display. This landmark study highlights the importance of understanding the mechanisms, as well as the patterns of pollinator-mediated selection on floral traits.

Pollinator-mediated selection can operate directly on flower form to enhance the efficacy or frequency of pollination by influencing either male or female components of reproduction. Like Campbell et al.'s study, the most informative evidence of direct pollinator-mediated selection on flower form also comes from research incorporating both observational and experimental approaches. Nilsson (1988) tested Darwin's hypothesis that selection for effective pollination drives the evolution of corolla tube depth by exploring the mechanistic basis for strong positive correlations between corolla length and reproductive success in natural populations of orchids. Experimental constriction of corolla tubes at differing depths influenced the efficacy of pollen deposition and pollinia removal in three orchid species.

Our studies of pollinator-mediated selection on flower size in the skypilot have also drawn on experimental approaches to test the basis of relationships between phenotypic variation and fitness components observed in nature. In the field, bumble bees visit individuals with broadly-flared corollas more often than predicted from their encounter frequency with large-flowered plants. Bumble bee preference for large flowers yields a predicted 10% advantage in seed production, with an increase in corolla flare from 14–16 mm (Galen and Newport, 1987). Because plants with broadly-flared flowers export pollen more efficiently with successive bumble bee visits, the predicted advantage through pollen dispersal associated with broad corolla morphology is even greater than that achieved through female function (Galen and Stanton, 1989). For example, an increase in corolla flare from 14–16 mm should confer a 14% enhancement in pollen exported per bumble bee encounter.

These relationships are consistent with direct selection on flower morphology by bumble bee pollinators. But flower size in the skypilot shows a positive phenotypic correlation with nectar sugar production rate, as in many entomophilous species (e.g., Plowright, 1981; Harder et al., 1985; Stanton and Preston, 1988; Ashman and Stanton, 1991; Cresswell and Galen, 1991). Only experimental manipulation can dissect the separate effects of flower morphology and floral reward on the probability of bumble bee pollination and clarify whether our view of direct selection on flower size is warranted. By presenting plants with extreme combinations of corolla flare and nectar sugar content to bumble bees in a full factorial design, we showed that both aspects of floral display influence the behavior of bumble bee pollinators (Cresswell and Galen, 1991). Bumble bees entering experimental arrays of the skypilot visited plants having broadly-flared flowers first, regardless of nectar reward, and spent more time probing broadly-flared flowers than narrow ones. Probe time is correlated with pollen export in the skypilot (Galen and Stanton, 1989). However, once on a plant, the number of flowers visited by a bumble bee before departure was solely influenced by nectar reward level. It follows that insofar as bumble bee pollination enhances plant reproductive success, both flower size and nectar sugar production should experience direct positive selection. However, indirect selection could also operate on corolla flare through its relationship with nectar sugar production, if this phenotypic correlation has an underlying genetic basis (e.g., Mitchell and Shaw, 1992).

Depending on the strength of other deterministic and stochastic influences on seed production, advantages associated with enhanced pollination success might or might not ultimately affect plant fitness (Campbell, 1991). To test whether bumble bee preferences translate into enhanced reproduction through male or female function, I performed a series of experiments in natural populations and synthetic arrays of the skypilot. Synthetic populations are a convenient tool for evolutionary studies, because the intensity of selection can be manipulated by the researcher to match that seen in nature or explore effects of extreme bouts of selection on the evolution of fitness-related traits. In designing and conducting such an experiment, however, the investigator must have in mind which of these objectives is desired. If the decision is made to match natural levels of selection, then knowledge of the bounds of the selection factor (e.g., pollinator visitation rate) in natural populations is essential. Pollinator visitation and plant fecundity in selection experiments that I have conducted on *P. viscosum* are at the upper end, but within the range, of natural visitation rates and seed set for the Pennsylvania Mountain population (Galen, 1989).

For effects mediated through paternity, I asked whether traits influencing the number of flowers visited on a plant affect that plant's success in siring seeds on subsequently visited individuals. As corolla flare enhances a plant's probability of receiving a bumble bee visit when encountered and nectar sugar production influences bumble bee tenure once on an inflorescence, addressing this simple question sheds light on the potential for direct selection through paternity on both traits. I set up experimental arrays of 10–12 flowering skypilots with known genotypes at a polymorphic glutamine amino acid transferase (GOT) locus (Galen, 1992). Each array was comprised of a donor homozygous for a unique allele that was absent from all possible recipients. The distribution of the donor's allele in seedling progeny of the array provided a direct estimate of paternity. Additionally, because the self-incompatibility system in *P. viscosum* inhibits self pollen germination, a count of the donor's pollen donation to the first recipient visited could be obtained by simply scoring the number of pollen tubes entering stigmas (Galen, et al., 1989). Results from this experiment showed that plants receiving more bumble bee visits export more pollen and that greater pollen export enhances both pollen delivery to, and paternity on, the next plant visited. However, since either resources or ovule number (typically about 15) limit seed set at high values of pollen receipt, the relationship between pollen export (floral attractiveness) and paternity at this proximal scale is characterized by diminishing returns. Conversely, as pollen export increases, donors sire seeds in a greater proportion of available recipients. At the within patch scale then, a given level of attractiveness translates into a linear increase in gene flow. Because substantial predispersal risks associated with flower predation by nectar-thieving ants and seed predation by grazers (elk, deer) are spatially aggregated in populations of the skypilot (Galen, 1983*b*, 1990), pollen donors dispersing their genes widely among flowering conspecifics are likely to realize higher fitness than those reaching only a few recipients.

Annual fecundity in natural populations of the skypilot is pollination-limited, so probably traits increasing overall pollen receipt also enhance seed production (Galen, 1985). However, this general relationship does not directly address the role of any particular kind of pollinator in exerting selection on floral attractants through female function. To test specifically whether bumble bees exert selection on floral display through fecundity in the skypilot, I have measured the relationship between female components of reproductive success and flower size in the presence and absence of bumble bee pollinators. My experiments address potential selection during any one bout of reproduction, but it should be borne in mind that plants of the skypilot reproduce repeatedly during their lifetimes. To measure bumble bee-mediated selection through fecundity, I brought 94 randomly selected plants from the krummholz population into a flight cage in the field before their flowers opened. I then imported queens of the bumble bee species, *Bombus kirbyellus* and *B. sylvicola,* cleaned of pollen, into the flight cage for pollination. More than 40 bumble bees were allowed to visit the experimental array and their foraging sequences were recorded to discern the number of flowers frequented or the attractiveness of individual plants to bumble bee pollinators (Galen, 1989). From the relationship of seed production to floral traits, I calculated selection on corolla flare, corolla length and scape length as well as on suites of traits, combined, using phenotypic selection analysis (Lande and Arnold, 1983).

If resources limit both flower size and seed production, however, underlying heterogeneity in resource availability or plant size might give rise to relationships between seed set and floral display in the absence of pollinator-mediated selection (Rausher, 1992). Such relationships could obscure the detection of pollinator-mediated selection, even under the relatively controlled environment of the flight cage. Two controls were performed to check for possible relationships between flower size and seed set in the absence of bumble bee pollinators. In the first, 50 plants neighboring the original locations of bumble bee-pollinated individuals were screened from pollinators, assigned randomly to pollen donors from 1–50 m away, and outcrossed by hand. In the second control group, 53 nearby plants were surrounded by hardware cloth (0.5×0.5 cm^2 mesh) that excluded bumble bees, but allowed smaller pollen- and nectar-gathering flies and solitary bees to visit the flowers. Variation in flower size and seed set was comparable in the three treatment groups, although seed set was somewhat reduced in the absence of bumble bee or hand-pollination (Table 10.2; Galen, in press). The experiment showed that only with bumble bee-mediated pollination does variation in corolla flare influence seed production. In both control treatments, the relationship between corolla flare and seed set was negligible. Findings are similar for corolla length, although it showed a moderate correlation with the seed production of

Table 10.2. *Relationship of flower size components to annual fecundity under contrasting pollination regimes (Galen, 1989; Galen, in press). Morphometric traits were log-transformed and seed set was standardized relative to its maximum value in each population before regression.*

Treatment (n)	Corolla Length $\bar{x}\pm$SE	Corolla Flare $\bar{x}\pm$SE	Seed Set $\bar{x}\pm$SE	Corolla Length Regression				Corolla Flare Regression			
				$B_1\pm$SE	F	$P<$	R^2	$B_1\pm$SE	F	$P<$	R^2
Bumble bee-pollinated (94)	23.22±0.31	14.32±0.25	71.3±6.4	3.08±0.41	56.96	0.0001	37%	1.80±0.35	25.23	0.0015	11%
Randomly-pollinated (50)	24.55±0.74	15.34±0.19	71.7±6.6	2.93±1.04	7.94	0.01	15%	0.38±0.82	0.22	ns	0.5%
Bumble bees excluded (53)	24.29±0.32	13.67±0.28	49.2±6.7	0.08±0.94	0.01	ns	0.0%	0.49±0.45	1.18	ns	2.2%

randomly pollinated plants. Consequently, I cannot reject the possibility that relationships between corolla length and annual fecundity under bumble bee pollination reflect resource or size heterogeneity among plants as well as pollinator preference.

The strength of selection S on corolla flare and length exerted by bumble bees through female function depends on the stage at which fitness or offspring production is monitored for the maternal parent. For example, one could calculate selection differentials based on seed number, seedling number, the number of established progeny, or the number of flowering progeny. Each successive stage is more closely related to true fitness or the number of F2 seeds produced over a plant's lifetime. However, as more events intervene between pollination and the stage at which fitness is estimated, the relationship of traits acting at pollination to fitness may become more difficult to detect. Weakened correlations follow from the increasing influence of random sources of mortality (Cabana and Kramer, 1991). With the exception of studies that address the fitness consequences of mating-system correlates (e.g., Waser and Price, 1989; Holtsford and Ellstrand, 1990; Willis, 1993; Wolfe, 1993; Dole and Ritland, 1993), consideration of the fitness effects of plant reproductive traits seldom go beyond the stage of seed production, and consequently may misrepresent the relationships between reproductive traits and lifetime fitness (but see Stanton, et al., 1986). To address this point for the alpine skypilot, I calculated coefficients of bumble-bee-mediated selection on the two heritable aspects of floral display, corolla length and corolla flare, based on the production of seeds and establishment of juvenile offspring (age 6) in the field (Table 10.3; Galen, In press). Both traits undergo strong positive directional selection with bumble bee-mediated pollination, regardless of whether fitness is measured early or late in the progeny life cycle. As predicted theoretically (Cabana and Kramer, 1991), however, the correlations of the two traits with relative fitness decline markedly with time since pollination. Because my sample sizes are relatively large ($N \approx 100$), relationships between aspects of floral display and fitness remain statistically significant, but the confidence levels associated with these relationships drop by over an order of magnitude.

Is the strength of bumble bee-mediated selection sufficient to account for a 12% increase in corolla flare of the skypilot from krummholz to high tundra habitats? One way to explore this question is to ask how many generations of such selection would be necessary to evolve a tundra flower morphology from that found at lower elevations. The prediction equation of quantitative genetics provides a tool for such an exploration (Falconer, 1985). According to this equation, the response to selection R on a given trait can be predicted as the product of S and the trait heritability h^2. Maternal parent-offspring regression yielded a heritability estimate of 1.02 ± 0.80 (2 SE). Heritability cannot exceed 1.00 although sampling errors often yield greater estimates. Using S of 17% based on offspring establishment, we can bound the single-generation response of corolla flare to bumble bee-mediated selection at 4–17%. All else being equal,

Table 10.3. Standardized selection differential S *for flower size components based on relative fitness at early and late life-cycle stages (Galen, 1989; Galen, in press). Measurements are based on 94 selected parents. Traits were log-transformed before analysis.*

		Correlation with Relative Fitness		
Floral Trait	Fitness Measurement	*r**	*P*<	*S*
Corolla length	Seed set	0.41	0.0001	9%
	Established offspring†	0.20	0.04	18%
Corolla flare	Seed set	0.31	0.002	7%
	Established offspring†	0.24	0.01	17%

*Pearson's product moment correlation.

†Number of offspring to reach age 6 in the field.

selection over 1–3 generations could account for the evolution of the more expensive, broadly flared flowers seen in bumble bee-pollinated populations from the smaller flowers of krummholz *P. viscosum*.

Conclusions and Future Directions

Studies integrating genetic and ecological approaches are central to our understanding of pollinator-mediated selection. Such work gains much in value when done in the natural environments within which traits are expressed and pollinators choose among species and individuals. The main drawbacks are that (1) because of environmental heterogeneity, large sample sizes are needed to detect genetic effects, and (2) field studies of evolution in most naturally occurring organisms take a long time. My research on flower size evolution in the skypilot has taken up the better part of a decade, and I am still collecting data on the response of floral traits to pollinator-mediated selection. Funding sources, Ph.D. program schedules, and tenure decisions are not sensitive to these kinds of constraints. These caveats set forth, for those in the position to study the evolution of floral form, *in situ,* there are exciting questions to address. It should be painfully clear from this chapter that we know little about sources of phenotypic variation in floral display, although sex allocation models and our thoughts on the importance of pollinator-mediated selection in the adaptive radiation of flowering plant groups assume a genetic component. Moreover, it is becoming increasingly clear that accelerated habitat alteration (fragmentation, destruction) is drastically changing pollinator abundances (e.g., Jennersten, 1988). An understanding of the evolutionary and ecological risks involved mandates the integration of theory and experiment on plant/pollinator interactions. With the emergence of new and powerful tools for monitoring selection through paternal as well as maternal function (such as DNA fingerprinting, RAPD, etc.; Schaal et al, 1991; Newbury and Ford-Lloyd, 1993) and analytical methods for estimating quantitative genetic parameters [such as restricted maximum likelihood (REML) analysis of heritability; Shaw, 1987; Shaw and Waser, 1994], questions that have traditionally resisted rigorous empirical study can now be resolved, setting the stage for a renaissance in our understanding of the evolution of floral form.

Another neglected aspect of floral evolution concerns the physiological costs of floral advertisement or showiness. The evolution of floral traits is certainly influenced by the resource costs of pollinator attraction, yet we know little about nutrient or even carbon loss from floral structures. In part this reflects the focus of most plant ecophysiologists on leaf processes. However, flower morphology directly affects opportunities for fruit photosynthesis with pronounced consequences for seed size and number (Werk and Ehleringer, 1983; Williams et al., 1985; Galen et al., 1993). Moreover, petal tissue can incur carbon and nitrogen costs of investment that exceed those of female or male function (Ashman 1994; Galen et al., 1993). The linkage of physiological and evolutionary approaches (e.g., Geber and Dawson, 1990) holds much promise for evaluating such costs and incorporating them into models of floral trait evolution.

Acknowledgments

I thank J.E. Cresswell, L. Galloway, T. Gregory, B. Lubinski, D. Masters, K. Tack, M.L. Stanton, B. Steiner, G. Sutcliffe, H. West, K. Ziebold, and K. Zimmer-Doll for assistance in the field and greenhouse. R. Mitchell, S.C.H. Barrett, and an anonymous reviewer provided valuable suggestions on the manuscript. Discussions with M.E.A. Newport and M. Gromko were especially helpful in the formulative stages of this project. Access to the field site and facilities on Pennsylvania Mountain was granted by the University of Colorado at Colorado Springs. Research reported in this study was supported by NSF grants BSR-8604726, BSR-8915310, and BSR-9196058.

References

Anderson, S. and B. Widén. 1993. Pollinator-mediated selection on floral traits in a synthetic population of *Senecio integrifolius* (Asteraceae). *Oikos,* 66: 72–79.

Armbruster, W.S. 1985. Patterns of character divergence and the evolution of reproductive ecotypes of *Dalechampia scandens* (Euphorbiaceae). *Evolution,* 39: 733–752.

Armbruster, W.S. 1991. Multilevel analysis of morphometric data from natural plant populations: Insights into ontogenetic, genetic and selective correlations in *Dalechampia scandens*. *Evolution,* 45: 1229–1244.

Armbruster, W.S. 1993. Evolution of plant pollination systems: Ecophylogenetic hypotheses and tests with neotropical *Dalechampia*. *Evolution,* 47: 1480–1505.

Arnold, S.J. and M. Wade. 1984. On the measurement of natural and sexual selection: Theory. *Evolution,* 38: 709–719.

Ashman, T.-L. 1994. A dynamic perspective on the physiological cost of reproduction in plants. Am. Nat. 144: 300–316.

Ashman, T.-L. and M. Stanton. 1991. Seasonal variation in pollination dynamics of sexually dimorphic *Sidalcea oregana* spp. *spicata* (Malvaceae). *Ecology,* 72: 993–1003.

Bliss, L.C. 1985. Alpine. In *Physiological Ecology of North American Plant Communities* (B.F. Chabot and H.A. Mooney, eds.), Chapman & Hall, New York, pp. 41–65.

Cabana, G. and D.L. Kramer. 1991. Random offspring mortality and variation in parental fitness. *Evolution,* 45: 228–234.

Campbell, D.R. 1991. Effects of floral traits on sequential components of fitness in *Ipomopsis aggregata. Am. Nat.,* 137: 713–737.

Campbell, D.R., N.M. Waser, and M.V. Price. 1994. Indirect selection of stigma position in *Ipomopsis aggregata* via a genetically correlated trait. *Evolution.* 48: 55–68.

Conner, J. and S. Via. 1993. Patterns of phenotypic and genetic correlations among morphological and life-history traits in wild radish, *Raphanus raphanistrum. Evolution,* 47: 704–711.

Crepet, W.L. 1983. The role of insect pollination in the evolution of the angiosperms. In *Pollination Biology* (L. Real, ed.), Academic Press, Orlando, FL, pp. 31–50.

Cresswell, J.E. and C. Galen. 1991. Frequency-dependent selection and adaptive surfaces for floral character combinations: The pollination of *Polemoniun viscosum. Am. Nat.,* 138: 1342–1353.

Darwin, C. 1859. *On the Origin of Species by Means of Natural Selection.* John Murray, London.

Darwin, C. 1877. *The Different Forms of Flowers of the Same Species.* John Murray, London.

Davidson, J.R. 1950. The genus *Polemonium* (Tournefort) L. *Univ. Calif. Publ. Bot.,* 23: 209–282.

Dole, J. and K. Ritland. 1993. Inbreeding depression in two *Mimulus* taxa measured by multigenerational changes in the inbreeding coefficient. *Evolution,* 47: 361–373.

Donoghue, M.J. 1989. Phylogenies and the analysis of evolutionary sequences, with examples from seed plants. *Evolution,* 43: 1137–1156.

Endler, J.A. 1986. *Natural Selection in the Wild.* Princeton Univ. Press, Princeton, NJ.

Falconer, D.S. 1981. *Introduction to Quantitative Genetics.* Longman, New York.

Feinsinger, P. (1983). Coevolution and pollination. In *Coevolution* (D.J. Futuyma and M. Slatkin, eds.), Sinauer, Sunderland, MA, pp. 282–310.

Galen, C. 1983*a*. The Ecology of Floral Scent Variation in *Polemonium viscosum* Nutt. (Polemoniaceae). Ph.D. diss., Univ. Texas, Austin.

Galen, C. 1983*b*. The effects of nectar thieving ants on seedset in floral scent morphs of *Polemonium viscosum. Oikos,* 41: 245–249.

Galen, C. 1985. Regulation of seed-set in *Polemonium viscosum*: Floral scents pollination, and resources. *Ecology,* 66: 792–797.

Galen, C. 1989. Measuring pollinator-mediated selection on morphometric floral traits: Bumble bees and the alpine sky pilot, *Polemonium viscosum. Evolution,* 43: 882–890.

Galen, C. 1990. Limits to the distributions of alpine tundra plants: Herbivores and the alpine skypilot, *Polemonium viscosum. Oikos,* 59: 355–358.

Galen, C. 1992. Pollen dispersal dynamics in an alpine wildflower, *Polemonium viscosum. Evolution,* 46: 1043–1051.

Galen, C. 1993. Cost of reproduction in *Polemonium viscosum*: Phenotypic and genetic approaches. *Evolution,* 47: 1073–1079.

Galen, C. In press. Rates of floral evolution: adaptation to bumblebee pollination in an alpine wildflower, *Polemonium viscosum*. Evolution: In press.

Galen, C., T. Gregory, and L.F. Galloway 1989. Costs of self pollination in a self-incompatible plant, *Polemonium viscosum*. Am. J. Bot. 76: 1675–1680.

Galen, C. and P.G. Kevan. 1980. Scent and color, floral polymorphism and pollination biology in *Polemonium viscosum* Nutt. *Am. Midl. Nat.,* 104: 281–289.

Galen, C. and M.E.A. Newport. 1987. Bumblebee behavior and selection on flower size in the sky pilot, *Polemonium viscosum. Oecologia,* 74: 20–23.

Galen, C., T.E., Dawson, and M.L. Stanton. 1993. Carpels as leaves: Meeting the carbon cost of reproduction in an alpine buttercup. *Oecologia,* 95: 187–193.

Galen, C. and M.L. Stanton. 1989. Bumble bee pollination and floral morphology: Factors influencing pollen dispersal in an alpine wildflower, *Polemonium viscosum* (Polemoniaceae). *Am. J. Bot.,* 75: 419–426.

Galen, C., J.S. Shore, and H. DeYoe. 1991. Ecotypic divergence in alpine *Polemonium viscosum*: Genetic structure, quantitative variation, and local adaptation. *Evolution,* 45: 1218–1228.

Galen, C., K.A. Zimmer, and M.E.A. Newport. 1987. Pollination in floral scent morphs of *Polemonium viscosum*: A mechanism for disruptive selection on flower size. *Evolution,* 41: 599–606.

Geber, M.A. and T.E. Dawson. 1990. Genetic variation and covariation between leaf gas exchange, morphology and development in *Polygonum arenastrum,* an annual plant. *Oecologia,* 85: 153–158.

Grant, V. and K.A. Grant. 1965. *Flower Pollination in the Phlox Family*. Columbia Univ. Press, New York.

Grime, J.P. 1979. *Plant Strategies and Vegetation Process*. John Wiley & Sons New York.

Harder, L.D., J.D. Thomson, M.B. Cruzan, and R.S. Unnasch. 1985. Sexual reproduction and variation in floral morphology in an ephemeral lily, *Erythronium americanum. Oecologia,* 67: 286–291.

Holtsford, T.P. and N.C. Ellstrand. 1990. Inbreeding effects in *Clarkia tembloriensis* (Onagraceae) populations with different natural outcrossing rates. *Evolution,* 44: 2031–2046.

Holtsford, T.P. and N.C. Ellstrand. 1992. Genetic and environmental variation in floral characters which influence the outcrossing rate of *Clarkia tembloriensis*. *Evolution,* 46: 216–225.

Houle, D. 1992. Comparing evolvability and variability of quantitative traits. *Genetics,* 130: 195–204.

Jennersten, O. 1988. Pollination in *Dianthus deltoides* (Caryophyllaceae): Effects of habitat fragmentation on visitation and seed set. *Cons. Biol.,* 2: 359–366.

Johnston, M.O. 1991. Natural selection on floral traits in two species of *Lobelia* with different pollinators. *Evolution,* 45: 1468–1479.

Kinet, J.M., R.M. Sachs, and G. Bernier. 1985. *The Physiology of Flowering. III. The Development of Flowers.* CRC Press, Inc., Boca Raton, FL.

Lande, R. and S.J. Arnold. 1983. The measurement of selection on correlated characters. *Evolution,* 37: 1210–1226.

Lewis, D., and D.A. Jones. 1992. The Genetics of Heterostyly. *In* S.C.H. Barrett (ed.) *Evolution and Function of Heterostyly.* Springer-Verlag, Berlin. pp 129–148.

Macior, L.M. 1974. Pollination ecology of the Front Range of the Colorado Rocky Mountains. *Melanderia,* 15: 1–59.

Mazer, S.J. and C.T. Schick. 1991. Constancy of population parameters for life-history and floral traits in *Raphanus sativus* L. II. Effects of planting density on phenotype and heritability estimates. *Evolution,* 45: 1888–1907.

McDade, L. 1992. Pollinator relationships, biogeography, and phylogenetics. *BioScience,* 42: 21–26.

Miller, R.B. 1981. Hawkmoths and the geographic patterns of floral variation in *Aquilegia caerulea. Evolution,* 35: 763–774.

Mitchell, R.J. and R.G. Shaw. 1992. Heritability of floral traits for the perennial wildflower, *Penstemon centranthifolius. Heredity,* 71: 185–192.

Mitchell-Olds, T. and R.G. Shaw 1987. Regression analysis of natural selection: Statistical inference and biological interpretation. *Evolution,* 41: 1149–1161.

Newbury, H.J. and B.V. Ford-Lloyd 1993. The use of RAPD for assessing variation in plants. *Plant Growth Reg.,* 12: 43–51.

Nilsson, L.A. 1988. The evolution of flowers with deep corolla tubes. *Nature,* 334: 147–149.

Plowright, R.C. 1981. Nectar production in the boreal forest lily, *Clintonia borealis. Can. J. Bot.,* 59: 156–160.

Rausher, M.D. 1992. The measurement of selection on quantitative traits: Biases due to environmental covariances between traits and fitness. *Evolution,* 46: 616–626.

Rick, C.M., M. Holle, and R.W. Thorp. 1978. Rates of cross-pollination in *Lycopersicon pimpinellifolium*: Impact of genetic variation in floral characters. *Plant Syst. Evol.,* 129: 31–44.

Roach, D.A. and R.D. Wulff. 1987. Maternal effects in plants. *Ann. Rev. Ecol. Syst.,* 18: 209–235.

Schaal, B.A., S.L. O'Kane, Jr., and S.H. Rogstad. 1991. DNA variation in plant populations. *Trends Ecol. Evol.,* 6: 329–333.

Schaffer, W.M. and M.D. Schaffer. 1979. The adaptive significance of variations in reproductive habit in the Agavaceae. II. Pollinator foraging behaviour and selection for increased reproductive expenditure. *Ecology,* 60: 1051–1069.

Schoen, D.J. 1982. Genetic variation and the breeding system of *Gilia achilleifolia. Evolution,* 36: 352–360.

Schwaegerle, K.E. and D.A. Levin. 1991. Quantitative genetics of fitness traits in a wild population of *Phlox. Evolution,* 45: 169–177.

Shaw, R.G. 1987. Maximum likelihood approaches applied to quantitative genetics of natural populations. *Evolution,* 41: 812–826.

Shaw, R.G. and N.M. Waser. 1994. Quantitative genetic interpretations of postpollination reproductive traits in plants. *Am. Nat.,* 143: 617–635.

Shore, J.S. and S.C.H. Barrett. 1990. Quantitative genetics of floral characters in homostylous *Turnera ulmifolia* var. *angustifolia* Willd. (Turneraceae). *Heredity,* 64: 105–112.

Stanton, M.L., H.J. Young, N.C. Ellstrand, and J.M. Clegg. 1991. Consequences of floral variation for male and female reproduction in experimental populations of wild radish, *Raphanus sativus* L. *Evolution,* 45: 268–280.

Stanton, M.L. and L.F. Galloway. 1990. Natural selection and allocation to reproduction in flowering plants. In *Sex Allocation and Sex Change: Experiments and Models* (M. Mangel, ed.), American Mathematical Society, Providence, RI, pp. 1–50.

Stanton, M.L. and R.E. Preston. 1988. Ecological consequences and phenotypic correlates of petal size variation in wild radish, *Raphanus sativus* (Brassicaceae). *Am. J. Bot.,* 75: 528–539.

Stanton, M.L., A.A. Snow, and S.N. Handel. 1986. Floral evolution: Attractiveness to pollinators influences male fitness. *Science,* 232: 1625–1627.

Systma, K.J., J.F. Smith, and P.E. Berry 1991. The use of Chloroplast DNA to assess biogeography and evolution of morphology, breeding systems, and flavonoids in *Fuchsia* sect. Skinnera (Onagraceae). Syst. Bot. 16: 257–269.

Waser, N.M. and M.V. Price. 1989. Optimal outcrossing in *Ipomopsis aggregata*: Seed set and offspring fitness. *Evolution,* 43: 1097–1109.

Werk, K.S. and J.R. Ehleringer. 1983. Photosynthesis by flowers in *Encelia farinosa* and *Encelia californica* (Asteraceae). *Oecologia,* 57: 311–315.

Williams, K., Koch, G.W., and H.A. Mooney. 1985. The carbon balance of flowers of *Diplacus aurantiacus. Oecologia,* 66: 530–535.

Willis, J.H. 1993. Effects of different levels of inbreeding on fitness components in *Mimulus guttatus. Evolution,* 47: 864–876.

Wolfe, L.M. 1993. Inbreeding depression in *Hydrophyllum appendiculatum*: Role of maternal effects, crowding, and parental mating history. *Evolution,* 47: 374–386.

Young, H.J. and M.L. Stanton. 1990. Influences of floral variation on pollen removal and seed production in wild radish. *Ecology,* 71: 536–547.

11

Deceit Pollination in the Monoecious, Neotropical Herb *Begonia oaxacana* (Begoniaceae)

Douglas W. Schemske, Jon Ågren,† and Josiane Le Corff**

> And there at last, I finally discovered that these flowers are fertilized by certain flies which, deceived by their appearance, assume there is nectar in the funnel and therefore crawl into it.
>
> —C.K. Sprengel (1793)

Where flowers pollinated by an animal vector lack rewards, pollination occurs by "mistake" (Baker, 1976), and the attraction of pollinators to the rewardless flowers is by deceit (Willson and Ågren, 1989). Deceit pollination was first described by Christian Konrad Sprengel 200 years ago, following his observations of insect visitation to rewardless orchid flowers (Chapter 12) that resembled nectar-bearing flowers. This mode of pollination differs from the more typical system in which flowers attract pollinating animals by providing them with food resources such as nectar or pollen (Proctor and Yeo, 1973; Faegri and van der Pijl, 1979), nest-building materials (Chapter 9), or chemicals used in acquiring mates (Kimsey, 1980; Williams, 1983; Schemske and Lande, 1984). A variety of deceit pollination systems have been proposed (Vogel, 1978; Wiens, 1978; Little, 1983; Dafni, 1984, 1992), including (1) pseudocopulation, in which rewardless flowers resemble females of the pollinating insects, and are pollinated by males during attempted matings (Kullenberg, 1961; van der Pijl and Dodson, 1966); (2) interspecific mimicry in which the rewardless flowers of one species resemble the rewarding flowers of another species (Gentry, 1974; Dafni and Irvi, 1981; Nilsson, 1983), and (3) intersexual mimicry, found in some animal-pollinated species with unisexual flowers (e.g., monoecy, dioecy) where the rewardless female flowers resemble the rewarding male flowers of the same species (Baker, 1976; Bawa, 1977, 1980; Willson and Ågren, 1989; Ågren and Schemske, 1991).

*Department of Botany, University of Washington, Seattle, WA 98195.

†Department of Ecological Botany, University of Umeå, S-901 87, Umeå, Sweden.

*Send correspondence to D.W. Schemske (Tel. 206-543-9450; Fax 206-685-1728; E-mail: Schemske@botany.washington.edu).

In this paper, we focus on the ecology and evolution of deceit pollination in plant species with unisexual flowers. Baker (1976) observed a high incidence of this mode of pollination in predominately tropical families, and suggested that it was probably a widespread phenomenon. His hypothesis has recently been verified by Renner and Feil (1993), who in a survey of 29 genera and 21 families of tropical dioecious species, found that "about a third of the species offer no reward in the female morph, pollination by deceit apparently being common" (p. 1100).

Resemblance between rewarding male and nonrewarding female flowers has been observed for a number of species with unisexual flowers, and one of the most challenging questions concerning the evolution of deceit pollination in these species is:, "When does resemblance constitute mimicry?" (Willson and Ågren, 1989; p. 26). Although some resemblance may be due to developmental and genetic constraints resulting from shared ancestry, only when there is evidence of selection for similarity can it be concluded that resemblance is due to intersexual mimicry (Willson and Ågren, 1989). Selection for similarity is expected when the reproductive success of rewardless female flowers is a function of their degree of resemblance to rewarding male flowers (Ågren and Schemske, 1991), but we are unaware of any studies where it has been demonstrated. The most compelling evidence to date for intersexual mimicry is in those species where the similarity between male and female flowers is due to the evolution of nonhomologous floral parts (Willson and Ågren, 1989). For example, in the monoecious tropical tree *Jacaratia dolichaula* (Caricaceae), the rewardless female flowers lack corolla tubes, but possess petaloid stigmatic lobes that strongly resemble the white corolla lobes of the rewarding male flowers (Bawa, 1980).

If there is a cost to the pollinator of "mistake" pollination, natural selection should favor floral visitors that can identify and avoid rewardless flowers. There is evidence for such pollinator discrimination in several species with rewardless female flowers (Kaplan and Mulcahy, 1971; Ågren et al., 1986; Dukas, 1987; Ågren and Schemske, 1991; Charlesworth, 1993). This motivates the following question: What characterisitics of rewardless female flowers contribute to the deception of pollinators? The intersexual mimicry hypothesis predicts that female flowers will receive more pollinator visits, and therefore achieve higher reproductive success, when their floral display resembles that of the male (Ågren and Schemske, 1991). What is not clear is whether the optimal female display mimics the average male phenotype in the population, or instead, mimics the male phenotype with the most attractive flowers and highest reward level (Ågren and Schemske, 1991; Schemske and Ågren, 1995). On the one hand, rewardless female flowers may face a lower risk of discrimination by pollinators, and therefore a higher visitation rate, if they bear a close resemblance to the mean male phenotype. This would result in stabilizing selection on female floral characters, with an optimum female phenotype that is equal to the mean phenotypic value of male flowers (Schemske and Ågren, 1995). Alternatively, if pollinators display a strong preference for the most rewarding male flowers and reward levels are correlated with flower size or fragrance, plants bearing female flowers in which the attractive features are exaggerated may be favored. This would result in directional selection on the female floral characters that indicate reward levels expected by pollinators visiting male flowers. When this is the case, a trade-off between the investment in attractive structures by female flowers and the number of flowers produced could serve to limit the evolution of increasing attractiveness.

In insect-pollinated species of tropical, monoecious *Begonia* (Begoniaceae), only male flowers produce a reward to pollinators and female flowers may attract flower visitors by resembling male flowers (van der Pijl, 1978; Wiens, 1978; Ågren and Schemske, 1991; Schemske and Ågren, 1995). There is a positive correlation between the average size of male and female attractive structures (e.g., sepal length, sepal width) among different species of *Begonia* (Ågren and Schemske, 1995), but there are some species in which male flowers are larger than female flowers (*B. oaxacana, B. involucrata*), and others in which female flowers are larger (*B. cooperi, B. estrellensis;* Ågren and Schemske, 1995). Moreover, some insect-pollinated species of *Begonia* produce male and female flowers with different petal numbers (Schemske and Ågren, personal observation). These observations illustrate that female *Begonia* flowers do not always strongly resemble males, and motivate empirical investigation of the ecology and evolution of deceit pollination in this group.

Here we present the results of observations and experiments designed to examine the ecology and evolution of deceit pollination in the Neotropical perennial herb *B. oaxacana* A.D.C. We address the following questions: (1) What is the degree of resemblance in floral display between male and female flowers, and what characters of female flowers might contribute to the deception of pollinators? (2) Do pollinators discriminate between male and female flowers as measured by the visitation rate per flower and/or the time spent per flower? (3) Is seed production limited by the amount of pollen received? (4) What is the effect of temporal overlap in male and female flowering on female reproductive success? And (5) Is there directional selection on female flower size, and if so, is the evolution of larger female flowers constrained by a trade-off between female flower size and number?

Material and Methods

Natural History

Begonia oaxacana (section *Hexaptera:* Burt-Utley, 1985) is a 0.3–1.5 m tall terrestrial herb found in small openings, and along streams and trails in the understory of wet montane forests at 1500–3000 m from southern Mexico to Panama (Smith and Schubert, 1958, 1961). Flowering begins in the dry season

(January) and continues until early in the wet season (mid-April). The anthers of male flowers are dark yellow and the stigmas of female flowers are green when first open, becoming greenish yellow as the flower ages. Male and female flowers each have two white sepals (Fig. 11.1). In addition, males have two smaller, white petals arranged perpendicular to the sepals, whereas most female flowers have only a single petal (Fig. 11.1). Male flowers are more conspicuous than female flowers. The sepals of male flowers are quite flat and lie fully

Figure 11.1. Morphology and spectral qualities of male (left) and female (right) flowers of *Begonia oaxacana* viewed with UV wavelengths eliminated (top), and in UV light (bottom). Note that both the anthers of male flowers and stigmas of female flowers strongly absorb UV, whereas the sepals and petals of both sexes reflect UV.

perpendicular to the anthers, whereas the sepals of females are only partly open. In addition male flowers are displayed slightly outward after opening, whereas female flowers face downward. Neither male nor female flowers produce any noticeable scent.

Most plants have several inflorescences that bear both male and female flowers, often with one or more flowers of each sex open simultaneously on a single inflorescence. This pattern of flowering is in marked contrast to that of *B. involucrata* (Ågren and Schemske, 1991; Schemske and Ågren, 1995) and a number of other insect-pollinated *Begonia* species (Schemske et al., personal observation) that exhibit complete temporal separation of male and female flowers both within and between inflorescences of the same plant. Male and female flowers of *B. oaxacana* do not touch, so automatic self-pollination as observed in *B. hirsuta* and *B. semiovata* (Ågren and Schemske, 1993) is not possible, though there is an opportunity for geitonogamous selfing. The fruits of *B. oaxacana* do not dry on the plant and hence differ from most other Central American *Begonia* in that the seeds are not wind-dispersed. It is likely that mature fruits with seeds fall to the ground, where they may be consumed and dispersed by ground-foraging birds and/or mammals.

Study Site

This study was conducted in February 1992 and from January to May 1993 on the northwestern slope of Volcan Barva, Heredia Province, Costa Rica. The study population, composed of approximately 300 plants, was located along trails and in forest openings and forest understory near Laguna Barva, at an altitude of about 2800 m. Trailside plants and those in forest openings were densely clumped, whereas plants in the understory were less dense and more uniformly distributed. The vegetation of this area is classified by the system of Holdridge et al. (1971) as tropical montane rain forest (Hartshorn, 1983).

Floral Display

To compare the spectral properties of male and female flowers in ultraviolet light, photographs of flowers taken in visible light (wavelengths <400 nm eliminated; Schott filter GG-400) were compared to those taken in ultraviolet (UV) light (wavelengths >365 nm eliminated; Schott filter UG-1). All photographs were taken in natural daylight with a Nikon F camera, Kodak T-max film, ASA 3200.

To compare the size and shape of the floral display of male and female flowers, the number of sepals and petals, and the length and width of one sepal and one petal were recorded in 1992 for two flowers of each sex from each of 20 plants. Measurements were made to the nearest 0.1 mm with digital calipers. Floral parts were removed from the plants and then laid flat, to best estimate the true floral dimensions. For each flower, we calculated the mean sepal diameter d as the mean of the sepal length and width, and estimated sepal area SA by the formula $SA = \pi (d/2)^2$. Because sepals were essentially round, this method provided an adequate representation of sepal area. Floral characters were analyzed by a mixed-model ANOVA to determine the effects of sex (fixed effect), plant (random effect), and the sex $\times$ plant interaction (random effect). Sequential Bonferroni tests were used to determine if results were significant at a "table-wide" level of $\alpha = 0.05$ (for $k = 5$ comparisons of each source of variation) following the procedures described by Rice (1989).

In 1993, we quantified the relationship between the size of the reward-producing structures (the anthers) and sepal size in male flowers, and between stigma size and sepal size in female flowers. The length and width of one sepal, and the length and width of the area encompassed by the anthers or stigmas were determined for 40 male and 40 female flowers, each collected from a different plant. These measurements were made on intact flowers to best represent the display as seen by pollinators. Because the sepals of *B. oaxacana* are somewhat cup-shaped, the measurements of intact flowers will be smaller than those for detached floral parts. Sepal area was measured as discussed above, and the area encompassed by anthers and stigmas was calculated from the formula for an ellipse, $A = \pi ab$, where a is the radius along the principal axis, and b the radius along the secondary axis that is perpendicular to the primary axis. A linear regression of stigma and anther area on sepal area was conducted, and the slopes and intercepts of the regression for each sex were compared using the ANCOVA procedure of Superanova™ (Abacus Concepts, Inc., Berkeley, CA, 1989). All measurements were made to the nearest 0.1 mm with a pair of digital calipers.

To document inflorescence sex ratios, in 1992 and 1993 we recorded the number and gender of open flowers in all inflorescences in plots established for conducting pollinator observations (see below).

Flower Visitor Behavior

During peak flowering (February) in 1992 and 1993, we observed flower visitors to *B. oaxacana* at two sites approximately 10 m apart. Site 1 was exposed to the sun for several hours during midday, whereas site 2 was more shaded. Together, the two sites included >50 plants and >400 inflorescences. The sites were observed simultaneously 3 days each year for 3–3.5 h, yielding a total of 10 h of pollinator observation per site in 1992, and 9 h of observation per site in 1993. For each flower visitor entering the observation sites, we recorded the number and gender of flowers visited per inflorescence, and the time spent in each flower. Over the 2 years of observations, the bumblebee *Bombus (Pyrobombus) ephippiatus* Say made >97% of all flower visits. An unknown species of syrphid fly made all other visits, but is not included in the analyses below.

To determine if pollinators displayed a preference for male or female flowers, we used a *G*-test (Sokal and Rohlf, 1981) to compare the number of visits observed to flowers of each sex for each site and day to the number of visits expected, based on the number of open male and female flowers at each site. We recorded the gender of all open flowers at each of the two sites at the end of each day. The mean time per visit to male and female flowers in each year was compared by a Mann–Whitney *U*-test (Sokal and Rohlf, 1981), based on the combined pollinator observation data for all days in both sites.

Pollen Limitation of Seed Production

To determine if seed production was pollen-limited, in 1993 we marked two inflorescences on each of 41 plants, and two newly opened female flowers on each inflorescence. One flower on each inflorescence received supplemental pollen on three different days (from 2–3 different pollen parents each day), while the other served as a control. Both supplemental and control flowers were unbagged and, therefore open to pollinators. The pollinations were conducted in late February, and the marked flowers in both the supplemental and control treatments were censused until the seeds had matured in late June. To investigate the one-tailed hypothesis that flowers receiving supplemental pollen would produce more mature fruits and/or seeds, we compared the two treatments for the number of fruits produced by a paired sign test and the mean number of mature seeds per fruit by a paired *t*-test. Only fruits producing some seeds were included in the analysis of seed number.

The Effect of Overlap in Male and Female Flowering on Female Reproductive Success

Because male and female flowers of *B. oaxacana* are often open at the same time, geitonogamous selfing is possible if pollinators move from male to female flowers. This may increase pollination success if pollinators are limiting, but could reduce seed production if there is inbreeding depression for seed maturation. To examine the fitness consequences of temporal overlap in male and female flowering within inflorescences, in 1993 we created an experimental category of inflorescence in which all male buds were removed prior to the maturation of male or female flowers. Each "male removal" inflorescence was paired with a control inflorescence of similar age and size on the same plant. A total of 53 plants were marked in late February, and the number of mature fruits on control and treatment inflorescences was determined at the time of seed maturation in late June. The number of fruits per inflorescence, proportion of female flowers maturing fruit, and mean seed number per fruit (based on two fruits chosen randomly from each of 30 paired control and treatment inflorescences) were compared by paired *t*-tests to investigate the one-tailed hypothesis that the female reproductive success of controls should exceed that of the male removal treatment. Note that this experiment does not control for the possible effects of floral display on pollinator visitation. Reduced female reproductive success in treatment inflorescences could, therefore, indicate greater attractiveness of inflorescences with male flowers.

Inbreeding depression has been observed at the seed stage in a wide variety of plant species (Sorenson, 1969; Schemske, 1983; Husband and Schemske, in press), including the self-pollinating *B. semiovata* (Ågren and Schemske, 1993). To determine if geitonogamous selfing could reduce seed number in *B. oaxacana*, in 1993 we bagged two inflorescences on each of 25 plants with mesh material to exclude pollinators, and on two different days we pollinated two flowers on each inflorescence, one with self-pollen and one with outcross pollen (two to three different outcross pollen parents each day). Bags were removed after petals and sepals had dropped. We compared the self and outcross treatments for the proportion of flowers producing fruit and for the mean seed number per fruit (based on one fruit per inflorescence of each cross type taken from a subsample of plants) by a one-tailed paired *t*-test.

Selection on Female Flower Size

The intersexual mimicry hypothesis for species with unisexual flowers proposes that the reproductive success of females is a function of their similarity to males. Female flowers may receive more visits if they (1) mimic the average male in the population, or (2) mimic the most rewarding (e.g., largest) males. To determine the relationship between flower size and pollinator visitation, we constructed artificial female inflorescences of three different sizes: (1) the mean "male" phenotype, (2) small flowers (mean − 2 SD), and (3) large flowers (mean + 2 SD), and scored the attractiveness of each to pollinators in the field in 1993. The artificial female flowers possessed two sepals and one petal made from white, vinyl-coated window shades, and a stigmatic area represented by a single plastic map pin with a round, yellow head (Fig. 11.2). In visible light both the window shade material and map pins were strikingly similar to the colors of the female flower parts they were meant to mimic. In UV light the window shade material was somewhat more reflective than the actual sepals and petals, whereas the plastic heads of the map pins absorbed UV just as intensely as did the stigmas of actual female flowers (Fig. 11.2; see below).

The "mean" female flower was as similar in shape and size to the actual mean male flower as was possible (see Table 11.1). The "small" and "large" flowers were, respectively, two standard deviations smaller, or larger, than the mean flower size, based on the standard deviation of male sepal width. To construct flowers, the mean flower was first drawn on a computer using a graphics program, then the large and small flowers were made by reducing or enlarging the image to the correct size. The size of the small petal in the mean flower was equal to

Figure 11.2. Morphology and spectral qualities of artificial female flowers of *Begonia oaxacana* viewed with UV wavelengths eliminated (left), and in UV light (right). Note that the map pin used to represent the stigma strongly absorbs UV, whereas the material used to simulate the sepals and petal reflects UV. The flower shown here is equal in size to the mean male in the population.

its mean size in males, and was enlarged or reduced by the same amount as the sepals. Paper copies of a given flower size were stapled to the window shade material and cut out by hand. Throughout this paper the term flower size refers only to the size of the petaloid sepals; we did not vary the size of the artificial stigmas (= map pins).

Inflorescences were made from gutter screens that are commercially con-

Table 11.1. Size of petaloid sepals and petals (mm) in male and female flowers of Begonia oaxacana *at Volcan Barva, Costa Rica (*N = *40 flowers of each sex for all measurements). All characters exhibit significant differences between the sexes (Table 11.2).*

	Male	Female
Character	Mean±SD	Mean±SD
Sepal length	12.0±1.38	10.3±1.10
Sepal width	12.2±1.48	10.8±1.50
Petal length	10.0±1.25	7.8±1.07
Petal width	5.6±0.65	4.2±0.53
Sepal area	116.0±25.75	88.7±19.95

structed with flexible wires extending from a circular, metal center. Each "inflorescence" was painted light green to mimic the supporting structures of actual inflorescences. The tip of each wire was wrapped with several turns of green tape, and artificial flowers were secured to the ends of the wires by pushing the map pins through the flowers and into the tape. By twisting the wires, we could manipulate the positions of the artificial flowers to match the natural floral display. Inflorescences of *B. oaxacana* can be unisexual (all male or all female) or bisexual, with varying numbers of open flowers of either sex. We chose a single floral display of two equal-sized, female flowers per inflorescence for our experiment because this was the average number of flowers in all-female inflorescences in each of the two years of our study (see below). Manipulating both flower size and the number of male and female flowers per inflorescence would provide a more complete test of the mimicry hypothesis, but we found that this was not feasible due to the large number of replicates required.

The artificial inflorescences were placed at the height of natural inflorescences on green bamboo stakes and located at 0.25m intervals along three parallel transects in the vicinity of pollinator observation site 1 (see above). Each day we randomly positioned 15 inflorescences of each of the three flower sizes. To ensure that our artificial inflorescences did not increase the local density of flowering plants as perceived by pollinators, we bagged 45 female-phase inflorescences at the observation site. The inflorescences were observed simultaneously by two observers on 3 successive days (February 21, 22, and 23) for a total of 14 h. We recorded the number of approaches and visits to artificial flowers. An approach was followed by either a rejection, in which a pollinator flew to within 1 cm of a flower and then rejected it, or a flower visit, in which a pollinator landed on the flower. We determined whether the frequency of approaches (= rejections + visits) and visits was independent of flower size (*G*-test). We used linear regression to determine the form of the relationship between the total number of approaches and sepal area.

Relationship Between Female Flower Size and Number

Directional selection to increase female flower size could be constrained by an allocation trade-off between flower size and number, as has been observed in *B. involucrata* (Schemske and Ågren, 1995). To determine the relationship between female flower size and the number of female flowers per inflorescence, we measured the length and width of one sepal from each of two flowers per inflorescence for 47 inflorescences, and counted the total number of female flowers produced by each inflorescence. Only inflorescences with all female flowers open were included in this analysis. Female flower size was estimated as sepal area by the formula given above (see "Floral Display"), and the relationship between mean sepal area per inflorescence and the number of female flowers per inflorescence was determined by a Spearman rank correlation.

Results

Floral Display

Male and female flowers of *B. oaxacana* are very similar in their spectral qualities, but differ substantially in size and shape. Comparison of the UV properties of males and females indicated that both the anthers of male flowers and stigmas of female flowers strongly absorb UV, whereas the sepals and petals of flowers of both sexes reflect UV to a modest degree (Fig. 11.1). Thus, the anthers and stigmas appear as UV-absorbing targets to pollinators against the reflective background of the sepals and petals. All male and female flowers in our sample had two large sepals. All males also had two smaller petals, whereas 92% of the female flowers only had one petal (the remaining 8% had two petals; N = 40 flowers of each sex). Thus, on average, the female floral display is somewhat more asymmetric than that of the male (Fig. 11.1). All floral characters were significantly larger in male than female flowers (Tables 11.1 and 11.2). The petaloid sepals of male flowers were 17% longer and 12% wider than the sepals of female flowers, and the petals of male flowers were 28% longer and 33%

Table 11.2. Mixed model ANOVA of the effects of sex (fixed effect), plant (random effect), and the sex × plant interaction (random effect) on floral characters in a Costa Rica population of Begonia oaxacana. *Asterisks indicate significance at the "tablewide" level α=0.05, as determined by sequential Bonferroni tests. P values are the single-test significance levels.*

Character	Source of Variation	*df*	SS	*F*	*P*
Sepal length	Sex	1	55.3	50.8*	<0.0001
	Plant	19	44.3	1.6	0.0912
	Sex × plant	19	20.7	0.8	0.7243
	Residual	40	56.6		
Sepal width	Sex	1	38.0	19.5*	0.0003
	Plant	19	76.4	2.7*	0.0041
	Sex × plant	19	37.0	1.3	0.2338
	Residual	40	59.7		
Petal length	Sex	1	94.4	117.2*	<0.0001
	Plant	19	48.1	2.4*	0.0109
	Sex × plant	19	15.3	0.8	0.7429
	Residual	40	42.8		
Petal width	Sex	1	41.9	110.6*	<0.0001
	Plant	19	12.4	3.3*	0.0007
	Sex × plant	19	7.2	1.9	0.0424
	Residual	40	7.9		
Sepal area	Sex	1	14,936	31.9*	<0.0001
	Plant	19	15,014	1.8	0.0573
	Sex × plant	19	8884	1.1	0.4147
	Residual	40	17,496		

wider than those of female flowers. The sepal area of male flowers was 31% greater than that of females (Tables 11.1 and 11.2).

The regression of anther area on sepal area was positive and highly significant for male flowers, as was the regression of stigmatic area on the sepal area of female flowers (Fig. 11.3). The slopes of these regressions were not significantly different, but the intercept for males was significantly greater than that of females (Fig. 11.3). The mean area encompassed by the anthers of male flowers was significantly greater than the mean stigmatic area of females (18.4 mm^2; SD = 4.4 vs. 14.2 mm^2; SD = 4.0; t = 4.6, N = 80, $P < 0.0001$).

In 1992, 22% of the inflorescences scored (N = 424) were male-phase (with only male flowers open), 42% female-phase, and 36% mixed (Table 11.3). On average, male-phase inflorescences had 1.2 ± 0.44 (1–3) open flowers [mean

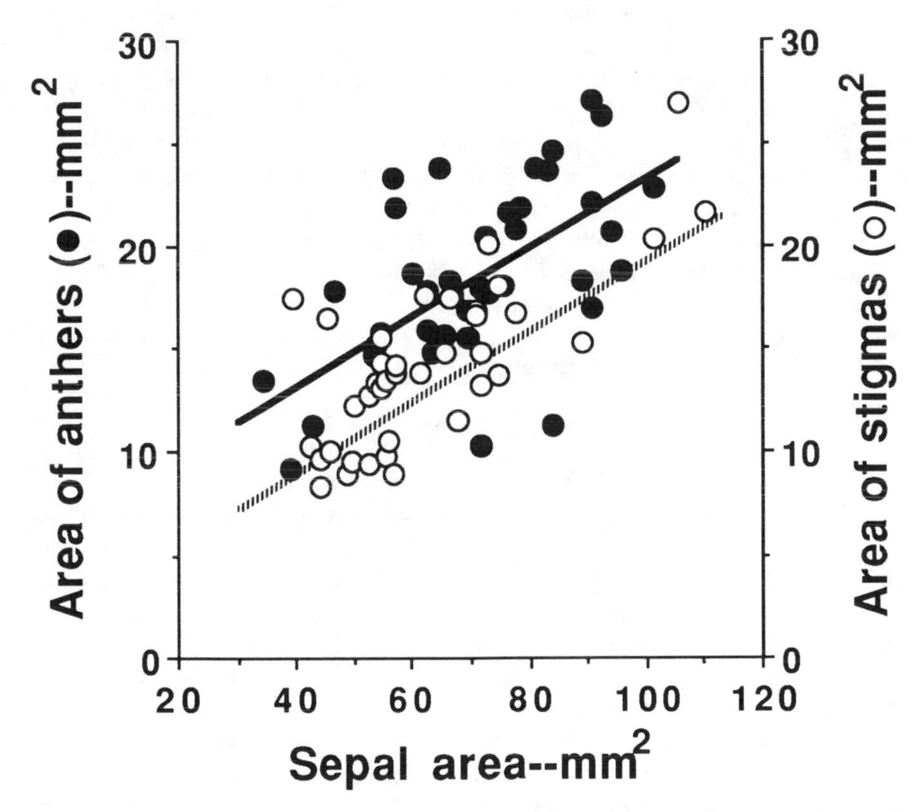

Figure 11.3. Regression of the area encompassed by anthers, or stigmas, on the sepal area of male, and female flowers. For male flowers: Area of anthers= 0.15 sepal area + 7.74; F=17.6, P<0.001, N=40. For female flowers: Area of stigmas= 0.17 sepal area + 3.15; F=44.5, P<0.0001, N=40. ANCOVA indicated that the slopes of the regressions for male and female flowers are not significantly different (F=2.24, df=1,76; NS), whereas the intercepts are significantly different (F=16.71, df=1,77; P<0.0001).

Table 11.3. Frequency (%) of inflorescences with different numbers of open male and female flowers in Begonia oaxacana. *Numbers above the slash are for 1992 (*N *= 424), numbers below the slash are for 1993 (*N *= 523). Data were taken on the first day of pollinator observations in both years.*

Number of Female Flowers	Number of Male Flowers			
	0	1	2	3
0	—	18.6/18.4	3.1/5.9	0.5/0.6
1	18.9/14.7	16.0/15.5	4.7/4.4	0/0.6
2	13.9/13.4	8.5/9.4	0.9/3.6	0/0
3	7.3/3.8	4.0/3.3	0.7/1.3	0/0
4	1.9/3.3	0.5/1.2	0/0	0/0
5	0/0.6	0/0	0/0	0/0
6	0/0	0/0.2	0/0	0/0

± SD (range); $N = 94$], female-phase inflorescences 1.8 ± 0.88 (1–4) open flowers ($N = 178$), and mixed inflorescences 2.8 ± 0.89 (2–5) open flowers ($N = 152$). In 1993, 25% of the inflorescences scored ($N = 523$) were male-phase, 36% female-phase, and 39% mixed (Table 11.3). On average, male-phase inflorescences had 1.3 ± 0.25 (1–3) open flowers [mean ± SD (range); $N = 130$], female-phase inflorescences 1.9 ± 0.99 (1–5) open flowers ($N = 187$), and mixed inflorescences 2.9 ± 0.88 (2–7) open flowers ($N = 206$).

In both 1992 and 1993, the floral sex ratio for the two pollinator observation sites combined was female-biased, averaging 66.7% females in 1992 and 67.8% in 1993 (average for 3 days of observation at each site each year; data from Table 11.4).

Flower Visitor Behavior

In 1992, during a total of 20 h of pollinator observation, 1358 flower visits by the bumble bee *Bombus* (*Pyrobombus*) *ephippiatus* were scored, and on each day at each site this species displayed a highly significant preference for male flowers (Table 11.4). The same strong preference for male flowers was found in 1993, during which 1832 visits were recorded in 18 h of observation (Table 11.4). In 1992, the daily visitation rate to male flowers was 4.8–6.3 times greater than that to female flowers (mean = 5.48, $N = 6$), and in 1993, the daily visitation rate to males was 4.8–9.5 greater than that to females (mean = 6.92, $N = 6$) (Table 11.4). In 1992, the mean visitation rate (visits per flower per hour) calculated for all days and sites combined was 0.417 to male flowers and 0.075 to female flowers; a 5.6-fold preference for males (Fig. 11.4). The corresponding values for 1993 were 0.377 for males and 0.057 for females; a 6.6-fold preference for males (Fig. 11.4).

Bumblebee visits to male flowers were significantly longer than those to female flowers in both years of observation (Mann–Whitney U-test, $P < 0.0001$; Fig.

Table 11.4. Pollinator visitation to male and female flowers of Begonia oaxacana *at Volcan Barva, Costa Rica. All visits were by* Bombus (Pyrobombus) ephippiatus. *Deviations from expected frequencies, based on the flower sex ratio, were tested by the G-statistic* ($P \leqslant 0.0001$ *in all cases).*

				Number of Flowers in Patch		Number of Flowers Visited					Relative Visitation
Year	Day	Site	No. Hours	Male	Female	Male	Female	*G*	Visits/Male/hr	Visits/Female/hr	Rate (M/F)
1992											
	1	1	3.0	78	152	147	48	140.0	0.628	0.105	6.0
	1	2	3.0	216	413	154	61	124.1	0.238	0.049	4.8
	2	1	3.5	102	207	187	67	175.1	0.524	0.092	5.7
	2	2	3.5	164	304	137	52	109.8	0.239	0.049	4.9
	3	1	3.5	100	214	195	80	176.0	0.557	0.107	5.2
	3	2	3.5	155	330	172	58	177.3	0.317	0.050	6.3
1993											
	1	1	3.0	325	383	297	69	192.9	0.305	0.060	5.1
	1	2	3.0	103	322	128	83	126.1	0.414	0.086	4.8
	2	1	3.0	249	378	258	43	273.2	0.345	0.038	9.1
	2	2	3.0	126	449	118	85	124.3	0.312	0.063	4.9
	3	1	3.0	280	410	564	87	595.9	0.671	0.071	9.5
	3	2	3.0	107	409	68	32	103.5	0.212	0.026	8.1

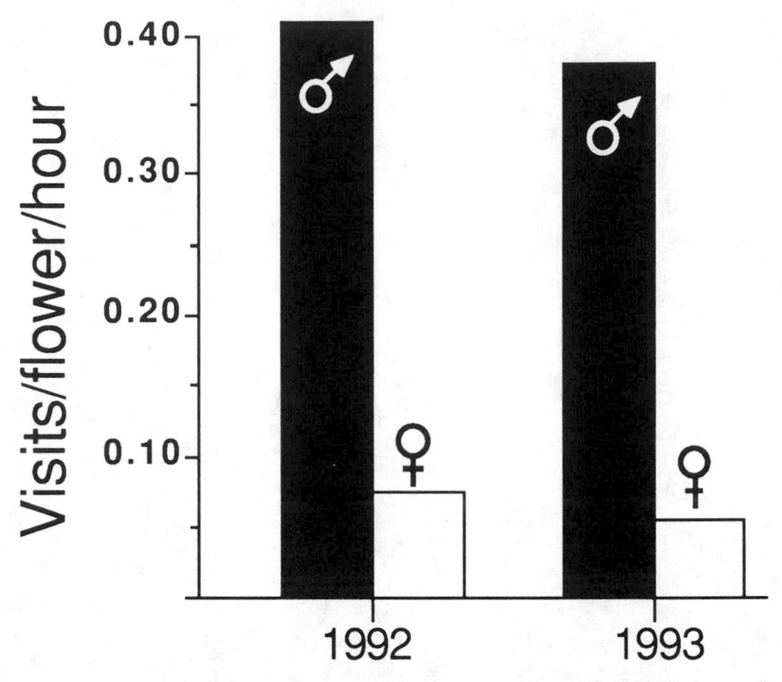

Figure 11.4. Mean visitation rate (visits/flower/hour) for all days and sites combined for male flowers (solid bars) and female flowers (open bars) of *Begonia oaxacana* in 1992 and 1993.

11.5). In 1992, the average time per visit to male flowers was 6.2 sec, as compared to 1.4 sec per visit to females, and in 1993 the average time per visit to males was 5.3 sec, as compared to 1.2 sec per visit to females (Fig. 11.5). Thus, for 1992 and 1993, respectively, the duration of visits to male flowers by bumblebees was 4.6 and 4.4 times longer than that to females.

In 1992, the bumblebee *B. ephippiatus* visited an average of 1.1 ± 0.4 (1–4) flowers per inflorescence [mean ± SD (range); $N = 1118$], and 88% of the inflorescence visits included only a single flower. The mean number of open flowers (male or female) per inflorescence for all days of pollinator observation taken together was 1.97 ($N = 6$; range of daily means 1.85–2.08). In 76 instances (6% of all inflorescence visits), both male and female flowers were visited on the same inflorescence. In 54% of these cases, the male flower was visited before the female, a distribution not significantly different from the 50:50 expectation of no preference ($G = 0.5$, $df = 1$, $P > 0.05$). In 1993, the bumblebee visited an average of 1.2 ± 0.2 (1–6) flowers per inflorescence [mean ± SD (range); $N = 1571$], and 86% of the inflorescence visits included only a single flower. The mean number of open flowers (male or female) per inflorescence for all days of pollinator observation taken together in 1993 was 2.26 ($N = 6$; range of daily means 1.91–2.57). During the observations conducted in this year, both

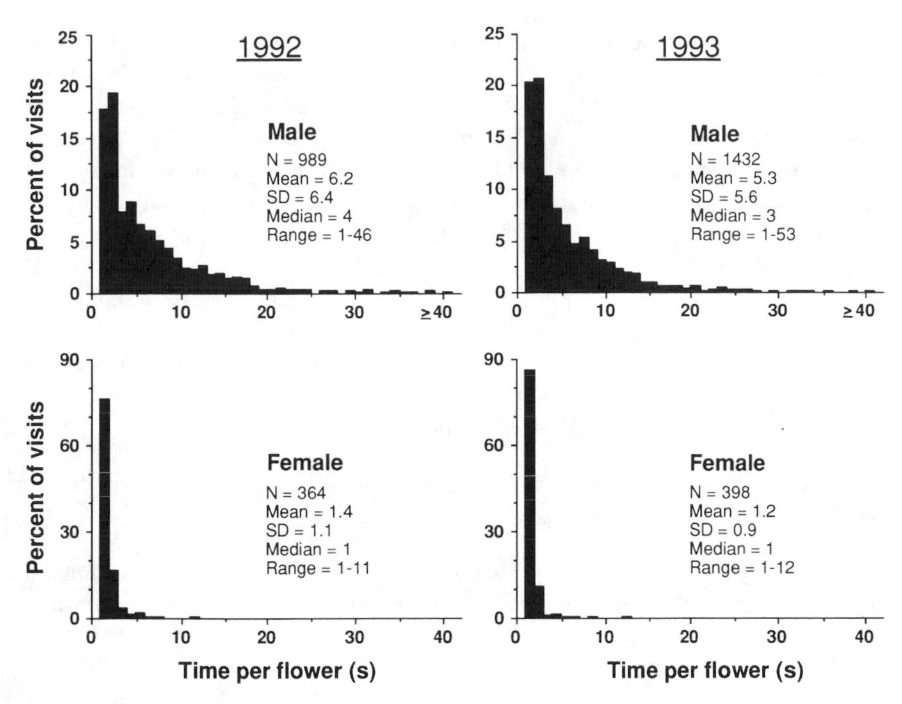

Figure 11.5. Time (in seconds) spent in male and female flowers of *Begonia oaxacana* by the bumblebee *Bombus (Pyrobombus) ephippiatus* Say. The time in male flowers was significantly greater than that in female flowers in both years (Mann–Whitney *U* test, $P<0.0001$).

male and female flowers were visited on the same inflorescence 138 times (9% of all inflorescence visits). In 81% of these cases, the male flower was visited before the female, a distribution significantly different from the 50:50 expectation ($G = 41.8$, $df = 1$, $P < 0.0001$). Although only a small fraction of the total number of inflorescence visits included visits to both male and female flowers, 41 of 366 visits to female flowers in 1992 (11%) were immediately preceded by a visit to a male flower in the same inflorescence, and in 1993 the corresponding figures were 112 of 399 (28%).

Pollen Limitation of Seed Production

The total number of fruits per plant present at the time of the final census did not differ significantly ($P = 0.25$, one-tailed paired sign test) between the supplemental and open pollination treatments (supplemental: mean = 0.88 fruits/plant, SD = 0.75, $N = 41$; open: mean = 0.98, SD = 0.79, $N = 41$). Some fruits probably matured and dropped to the ground before we conducted our final census; thus, the percentage of flowers producing mature fruit is probably higher than that reflected in our study (supplemental: 44%; open: 49%).

To compare mature seed number per fruit for fruits produced from supplemental and open pollinations, we counted the number of mature seeds in one fruit from each of the pollination treatments for the 27 inflorescences (22 plants) having at least one fruit produced from each treatment. This approach allowed us to statistically examine the effects of pollination treatment on seed number, while controlling for variation in seed number among plants. Because the analysis above indicated that there was no effect of treatment on the probability of fruit production, the subset of inflorescences producing fruits from both treatments should represent an unbiased sample for the purpose of comparing seed number per fruit. The number of seeds per fruit was significantly higher ($P = 0.059$; one-tailed paired t-test) for inflorescences in the supplemental treatment (mean = 1727, SD = 463, $N = 27$), as compared to those in the open pollination treatment (mean = 1556, SD = 489, $N = 27$) (Fig. 11.6).

The Effect of Overlap in Male and Female Flowering on Female Reproductive Success

Removal of male buds from experimental inflorescences did not affect female reproductive success. There was no significant difference (one-tailed paired t-

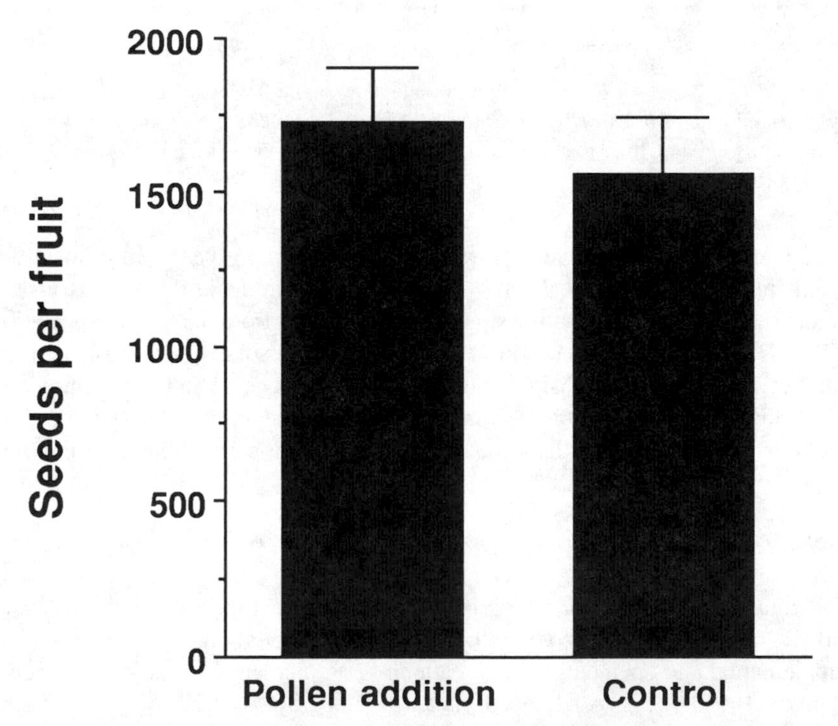

Figure 11.6. Mean seed production per fruit (± 2 SE) for flowers receiving supplemental pollination, and for control flowers receiving only natural pollination (t=1.62, N=28, P=0.059; one-tailed, paired t-test).

tests; $P > 0.30$ for all comparisons) between treatment (male buds removed) and control inflorescences for (mean ± SD) the number of mature fruits per inflorescence (treatment: 2.57 ± 1.36, $N = 51$; control: 2.48 ± 1.58, $N = 50$), the proportion of flowers producing mature fruit (treatment: 0.67 ± 0.34, $N = 51$; controls: 0.61 ± 0.36, $N = 50$), or the number of seeds per fruit (treatment: 1492 ± 357, $N = 30$; control: 1526 ± 404, $N = 30$). In addition, there was no significant difference (one-tailed paired *t*-tests; $P > 0.50$ for all comparisons) between self- and outcross pollination treatments for (mean ± SD) the proportion of flowers producing mature fruit (self: 0.44 ± 33, $N = 25$; outcross: 0.54 ± 0.38, $N = 25$), or the number of seeds per fruit (self: 1694 ± 333, $N = 14$; outcross: 1679 ± 382, $N = 14$).

Selection on Female Flower Size

Observations at Artificial Flowers

We observed a total of 575 approaches (= rejections + visits) to artificial flowers by the bumblebee *B. ephippiatus*. Most approaches ($N = 557$) ended with the bumblebee rejecting the flower just prior to landing, as is typical for bumblebees approaching natural female flowers of *B. oaxacana,* whereas the remainder ($N = 18$) were followed by a visit to a single artificial flower. For both approaches and visits, there was significant heterogeneity among flower sizes in the number of pollinator observations, and a striking preference for large flowers (Fig. 11.7). The number of approaches was greatest for large flowers ($N = 322$), intermediate for "mean" flowers ($N = 173$), and lowest for small flowers ($N = 80$). Large flowers also elicited the greatest number of visits ($N = 11$), as compared to the number of visits recorded at "mean" ($N = 5$), and small flowers ($N = 2$). In addition, we found a strong linear relationship between the number of pollinator observations and flower size for both kinds of approach behavior (number of approaches and rejections = 2.14 petal area − 67.28, $P = 0.03$, $R^2 = 0.998$; number of approaches and visits = 0.08 petal area −3.68, $P = 0.06$, $R^2 = 0.990$).

Relationship Between Female Flower Number and Flower Size

We observed a nonsignificant, negative correlation between the mean petal area of female flowers and total number of female flowers per inflorescence ($r_s = -0.16$, $N = 47$). The average number of female flowers per inflorescence was 4.81 (SD = 1.9, $N = 47$, range 2–8).

Discussion

Pollinator Behavior and Female Reproductive Success

Pollination of *B. oaxacana* occurs when pollen-collecting bumblebees mistake the rewardless female flowers for polleniferous male flowers. This deceit pollination

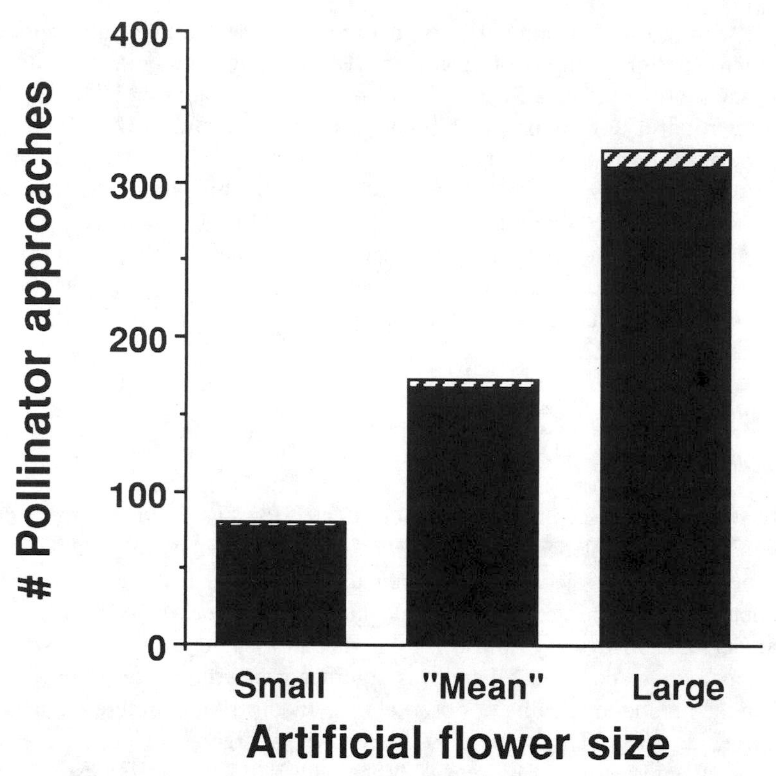

Figure 11.7. Pollinator approaches to artificial female flowers by the bumblebee *Bombus (Pyrobombus) ephippiatus* Say. The "mean" flower size is equal to the mean of male flowers, and the small and large flowers are, respectively, 2 SD smaller and 2 SD larger than the mean male (based on width of the petaloid sepal). The sepal areas for the three flower sizes are 67, 115, and 181 mm^2 for small, "mean," and large flowers, respectively. The filled bars indicate the number of approaches followed by rejection, and the hatched bars the number of approaches followed by a visit. For approaches (=rejections + visits) and visits, pollinator behavior departed significantly from the null expectation that flower size does not affect attractiveness (approaches: $G=158.9$, $df=2$, $P<0.0001$; visits: $G=7.1$, $df=2$, $P<0.05$).

system is typical of that found in other insect-pollinated *Begonia* we have studied in the Neotropics (Ågren and Schemske, 1991; Schemske and Ågren, personal observation). In both years of observation, the bumble bee *B. ephippiatus* was the primary pollinator of *B. oaxacana,* comprising >97% of all flower visits. The bumblebee displayed a striking preference for male flowers in both years, with a visitation rate to males more than five times that to females (Fig. 11.4).In addition, bumblebee visits to male flowers were more than five times longer than those to female flowers in both years (Fig. 11.5). Our results support the findings of Charlesworth (1993) that pollinators often discriminate against rewardless female flowers.

The strength of selection on pollinators to discriminate rewarding male from nonrewarding female flowers will depend on the cost of "mistake" visits to females. The cost to the bumblebee pollinators of *B. oaxacana* is probably small, given that male and female flowers are in close proximity, and pollinators quickly leave female flowers after landing (Fig. 11.5). In other deceit pollination systems where male and female flowers differ in rewards, the cost of mistake pollination can be much higher. For example, in some "dioecious" fig (*Ficus*) species, individual plants are functionally male, bearing only short-styled flowers, or functionally female, bearing only long-styled flowers (Kjellberg et al., 1987; Bronstein, 1988). Wasps pollinating female plants die without ovipositing and, therefore, experience strong selection to avoid female trees (Kjellberg et al., 1987).

Our experiments revealed that seed production in *B. oaxacana* was significantly pollen-limited (Fig. 11.6), yet the percentage increase in fecundity following supplemental pollination was smaller than might be expected (10%) given the low visitation rate to female flowers (mean for the 2 years of observation = 0.07 visits per flower per hour; Table 11.4). One factor that probably contributes to the surprisingly high reproductive success of female flowers is their longevity. Indirect evidence indicates that female flowers of *B. oaxacana* survive longer than males. Censuses of inflorescences in the bud stage give an estimate for the average proportion of male buds of 0.46 (Schemske and Ågren, unpublished), which can be compared to the average proportion of open male flowers found in this study of 0.32 (see "Floral Display" under "Results"). Thus, open males are less frequent than would be expected from the bud sex ratio, suggesting that female flowers live longer than males. Although we did not collect complete phenological information, our censuses of flowers in the hand-pollination experiments revealed that female flowers live a minimum of 10 days, and the population average is probably at least 14 days. If we assume that pollinators at Volcan Barva are active on *B. oaxacana* for 5 h per day, a figure consistent with our field observations, and each female flower lives on average 14 days, then the probability that a female flower will receive at least one visit in its lifetime given a visitation rate of 0.07 visits per flower per hour (Table 11.4) is 0.99 ($= 1 - .93^{70}$). Thus, the long lifespan of female flowers appears to compensate for low pollinator visitation rates, and is likely to have a strong selective value where pollinators discriminate against nonrewarding female flowers.

We conclude that deceit pollination in *B. oaxacana* is very effective. The only comparative data available are for *B. involucrata,* where the visitation rate by the primary pollinator *Trigona grandipennis* to male-phase inflorescences was more than seven times that to female-phase inflorescences, and visits to male flowers were more than 10 times longer than visits to female flowers (Ågren and Schemske, 1991). Seed production in this species was 40% higher after supplemental pollination (Ågren and Schemske, 1991), indicating a greater degree of pollinator limitation than in *B. oaxacana*. Female flowers of *B. involucrata* are also long-lived, but we do not have the data needed to estimate the proportion of its female flowers that may fail to receive a visit.

The proportion of *B. oaxacana* inflorescences with both male and female flowers open was 36% in 1992 and 39% in 1993. This is a higher frequency of mixed inflorescences than has been recorded for other perennial *Begonia* species [e.g., *B. cooperi* 0% (N = 99 inflorescences scored), *B. estrellensis* 7% (N = 68), *B. involucrata* 3% (N = 130), *B. multinervia* 0% (N = 61), and *B. urophylla* 0% (N = 245); Ågren and Schemske, unpublished] and approaches the proportion observed in the highly selfing (Ågren and Schemske, 1993) annual herb *B. semiovata* [48% (N = 73), Ågren and Schemske, unpublished]. The presentation of male and female flowers together could affect female reproductive success both because mixed inflorescences may be more attractive to pollinators than all-female inflorescences (they have more open flowers and include reward-producing male flowers), and because they facilitate geitonogamous pollination. In *B. oaxacana,* geitonogamous selfing may result from pollen transfer between male and female flowers of the same inflorescence, or between flowers on different inflorescences of the same plant. Fruit set and seed number per fruit were the same following hand-selfing and outcrossing, so increased pollen deposition through geitonogamy could increase female reproductive success if outcross pollen is limiting. However, the results of the male-removal experiment indicate that the reproductive success of female flowers in control inflorescences was not greater than that of females in experimental inflorescences from which all male flowers had been removed. Although this experiment was not designed to distinguish between the effects of floral display and geitonogamy on female reproductive success, the findings indicate that neither of these effects is very important for seed production in *B. oaxacana*.

Selection on Female Floral Traits

The female flowers of *B. oaxacana* possess a number of floral characters that may serve to increase their resemblance to male flowers. The stigmas are greenish yellow and the anthers are yellow, both stigmas and anthers strongly absorb UV (Fig. 11.1), and the flowers of both sexes possess two large sepals that appear white in visible light and moderately reflect UV wavelengths. However, male flowers have two petals, as compared to the single petal present in most female flowers, and the petals and sepals of male flowers are larger than those of females. This suggests that the attraction of pollinators to female flowers of *B. oaxacana* does not require close resemblance to conspecific male flowers.

Yellow, flavonoid-based pigmentation of pollen and anthers is found in both entomophilous and anemophilous plants, and may function as protection from UV radiation (Osche, 1983; Lunau, 1992). Thus, the coloration and UV properties of male *Begonia* flowers are not unique, although they may have influenced the evolution of attractive features of the rewardless female flowers. The yellow color and strong UV absorption of stigmas and anthers have been observed in all insect-pollinated species of *Begonia* examined to date (Schemske and Ågren, personal observation), suggesting that mimicry of anthers by the stigmas of female flowers may play a significant role in the deceit pollination of insect-pollinated *Begonia* species. The similarity of nonhomologous floral characters, such as the stigmas and anthers of *Begonia,* is consistent with the hypothesis for the evolution of resemblance by intersexual mimicry (Willson and Ågren, 1989). Because a phylogeny of *Begonia* is not available, we do not know if characters such as stigma color and UV absorption in female flowers represent the ancestral condition, or have evolved to increase the resemblance of female to male flowers. The intersexual mimicry hypothesis would predict a tendency toward increased resemblance in more advanced groups of insect-pollinated *Begonia.* In addition, the evolution of automatic selfing from primarily outcrossing ancestors should relax selection for similarity, and we would expect anthers and stigmas to show less resemblance in selfing species. This, however, is not supported by our observation that in the section Doratometra, the two highly selfing species *B. semiovata* and *B. hirsuta* (Ågren and Schemske, 1993) and the outcrossing species *B. tonduzii* all have yellow (or greenish yellow), UV-absorbing stigmas. Clearly, information on the phylogeny of *Begonia* would be very useful for determining if the evolution of male and female floral traits is consistent with the intersexual mimicry hypothesis.

Male flowers of *B. oaxacana* open more widely than female flowers and face outward after opening, whereas female flowers face downward. The display of female flowers is very similar to that of newly opened male flowers, which may make it more difficult for pollinators to distinguish females from young male flowers. Similarity in the display of female flowers and young male flowers is also found in *B. fischeri,* a species restricted to aquatic habitats (Schemske and Ågren, personal observation). Male flowers of *B. fischeri* open facing downward, and over the course of their lifetime (approximately 4 days) the flowers rotate until they eventually face outward. By contrast, the female flowers of *B. fischeri* open facing downward and do not change their position with age (Schemske and Ågren, personal observation). Because pollinators show a strong preference for male flowers facing downward (Schemske and Ågren, personal observation), females may achieve a higher pollinator visitation rate by remaining in this position.

Because female flowers of *B. oaxacana* never fully open, the small petals are not well exposed, which could explain why it is not critical for female flowers to bear the same number of petals as male flowers. Our finding that the size of the small inconspicuous petals differed more between male and female flowers than did the size of the showy petaloid sepals may indicate that the petals of female flowers do not play an important role in pollinator attraction. Different numbers of petals (or petaloid sepals or tepals) are also found in some highly selfing *Begonia* species (e.g., *B. hirsuta* and *B. semiovata;* Ågren and Schemske, 1993). These species do not require pollinators, which may explain their inconspicuous floral display and the striking dissimilarity of their male and female flowers.

If the reward levels of male flowers are correlated with attractive features such as size or fragrance, rewardless female flowers that resemble the most attractive males may have the highest pollination success. The significant positive correlation we observed between sepal area and anther area in male flowers (Fig. 11.3) suggests that the insect visitors to male flowers of *B. oaxacana* could use attractive structures as indicators of the size of the reward. This then provides the necessary conditions for directional selection on the attractive structures of the rewardless female flowers. Our observations of pollinator approaches to artificial female flowers of *B. oaxacana* indicated a strong preference for large flowers, with a nearly linear relationship between the number of pollinator approaches and sepal area (Fig. 11.7). Identical results were obtained in experiments conducted with artificial flowers of *B. involucrata* (Schemske and Ågren, 1995). Thus, female flowers that resemble the male floral display with the highest reward level may receive more visits than females that resemble the average male flower in the population. Other experiments with rewardless artificial flowers have indicated that pollinators often prefer large flowers (Kugler, 1943; Lunau, 1991; but see Knoll, 1922). These observations support the generalization made by Sprengel that pollinators are attracted to large flowers: "Because the final purpose of the corolla, which always applies, is that the flower catches the eyes of insects from afar, it must always be as large as possible" (Sprengel, 1793).

If pollinator preference was the only selective factor influencing female floral display in *B. oaxacana,* female flowers should evolve to increase in size. However, female flowers are smaller than males (Table 11.1), suggesting genetic and/or ecological constraints to the evolution of female flower size. Genetic constraints could include the lack of heritable variation for female flower size and genetic correlations between female flower size and other reproductive traits (Schemske and Ågren, 1995). The evolution of larger female flowers could be constrained if genes that suppress anther production in female flowers also have a negative effect on the size of attractive structures like sepals and petals. In many dioecious and gynodioecious species, female plants produce flowers with smaller corollas than male (or hermaphroditic) plants (Darwin, 1877; Baker, 1948; Lloyd and Webb, 1977; Eckhart, 1992; Ashman and Stanton, 1991; also Chapter 8; but see Meagher, 1992). We do not have the information needed to evaluate the role of genetic constraints for the evolution of female flower size in *B. oaxacana,* but our observation that female flowers are larger than males in some species of *Begonia* (e.g., *B. convallariodora, B. cooperi,* and *B. estrellensis*) indicates that female flower size can evolve independently of male flower size.

Female flower size could be constrained by a trade-off between flower size and flower number. In the gynodioecious herb *Sidalcea oregana,* females produce smaller flowers than hermaphrodites, but produce more open flowers per inflorescence (Ashman and Stanton, 1991), and in the dioecious herb *Silene latifolia,* females produce fewer, but larger, flowers than males (Meagher, 1992). In *B. involucrata* there is a highly significant, negative correlation between the number of female flowers per inflorescence and mean female flower size (Schemske and Ågren, 1995). In the present study of *B. oaxacana,* we found a nonsignificant, negative correlation between the number of female flowers per inflorescence and mean female flower size. The relatively small range in the number of female flowers per inflorescence in this species (2–8 flowers per inflorescence) as compared to that of *B. involucrata* (15–65 flowers per inflorescence) reduces the likelihood of detecting a correlation between flower size and number. Meagher (1992) attributed the weak relationship between female flower size and number in the dioecious herb *Silene latifolia* to a small range in flower number among female plants.

Summary and Future Directions

A suite of traits in female flowers contribute to the resemblance between male and female flowers in *B. oaxacana,* and despite the striking preference of pollinators for male flowers, females receive enough pollen to achieve high reproductive success. Our experiment to examine the relationship between female flower size and pollinator visitation represents a first step in identifying the selective forces operating in deceit pollination systems. Other experiments, such as altering the size, color, and spectral qualities of stigmas, as compared to the characteristics of anthers, would be useful for evaluating the potential for intersexual mimicry in *Begonia*. In addition, estimating the genetic variances and covariances for male and female flower traits would elucidate the importance of genetic constraints on floral evolution. Finally, further study of deceit pollination and intersexual mimicry would benefit from consideration of phylogenetic data that could be used to answer questions such as (1) Has the degree of floral resemblance evolved from the "ancestral state" in insect-pollinated *Begonia* due to changes in the mating system and/or the selective value of resemblance?, and 2) Does resemblance involve different floral characters in clades/species pollinated by different kinds of pollinators? We conclude that a synthetic study of deceit pollination and intersexual mimicry will require information on the form and magnitude of selection on male and female floral traits, the genetic architecture of floral traits, and phylogenetic relationships among species.

Acknowledgments

We are grateful to B. Best for assistance in the field, and S.C.H. Barrett, B. Best, and two anonymous reviewers for comments on the manuscript, R.W. Brooks for identification of bees, the Costa Rica National Park Service for permission to conduct this study, and the staff of Volcan Barva National Park for their assistance. This study was supported by a grant from the Mellon Foundation awarded through the Smithsonian Institution to D.W. Schemske, and by grants from the Swedish Natural Sciences Research Council and the Swedish Council for Forestry and Agricultural Research to J. Ågren.

References

Ågren, J. and D.W. Schemske. 1991. Pollination by deceit in a neotropical monoecious herb, *Begonia involucrata*. *Biotropica,* 23: 235–241.

Ågren, J. and D.W. Schemske. 1993. Outcrossing rate and inbreeding depression in two annual monoecious herbs, *Begonia hirsuta* and *B. semiovata*. *Evolution,* 47: 125–135.

Ågren, J. and D.W. Schemske. 1995. Sex allocation in the monoecious herb *Begonia semiovata*. *Evolution*, 49: 121–130.

Ågren, J., T. Elmqvist, and A. Tunlid. 1986. Pollination by deceit, floral sex ratios and seed set in dioecious *Rubus chamaemorus* L. *Oecologia* (*Berlin*), 70: 332–338.

Ashman, T.-L. and M. Stanton. 1991. Seasonal variation in pollination dynamics of sexually dimorphic *Sidalcea oregana* ssp. *spicata* (Malvaceae). *Ecology,* 72: 993–1003.

Baker, H.G. 1948. Corolla-size in gynodioecious and gynomonoecious species of flowering plants. *Proc. Leeds Phil. Soc.* 5: 136–139.

Baker, H.G. 1976. "Mistake" pollination as a reproductive system with special reference to the Caricaceae. In *Tropical Trees: Variation, Breeding and Conservation* (J. Burley and B.T. Styles, eds.), Academic Press, New York, pp. 161–169.

Bawa, K.S. 1977. The reproductive biology of *Cupania guatemalensis* Radlk. (Sapindaceae). *Evolution,* 31: 52–63.

Bawa, K.S. 1980. Mimicry of male by female flowers and intrasexual competition for pollinators in *Jacaratia dolichaula* (D. Smith) Woodson (Caricaceae). *Evolution,* 34: 467–474.

Bronstein, J.L. 1988. Mutualism, antagonism, and the fig-pollinator interaction. *Ecology,* 69: 1298–1302.

Burt-Utley, K. 1985. A revision of Central American species of *Begonia* section Gireoudia (Begoniaceae). *Tulane Studies Zool. Bot.,* 25: 1–131.

Charlesworth, D. 1993. Why are unisexual flowers associated with wind pollination and unspecialized pollinators? *Am. Nat.,* 141: 481–490.

Dafni, A. 1984. Mimcry and deception in pollination. *Ann. Rev. Ecol. System.,* 15: 259–278.

Dafni, A. 1992. *Pollination Ecology: A Practical Approach*. Oxford Univ. Press, Oxford.

Dafni, A. and Y. Irvi. 1981. Floral mimicry between *Orchis israelitica* Baumann and

Dafni (Orchidaceae) and *Bellevalia flexuosa* Boiss. (Liliaceae). *Oecologia,* 49: 229–232.

Darwin, C. 1877. *The Various Contrivances by Which Orchids are Fertilised by Insects,* 2nd ed. John Murray, London.

Dukas, R. 1987. Foraging behavior of three bee species in a natural mimicry system: Female flowers which mimic male flowers in *Ecballium elaterium* (Cucurbitaceae). *Oecologia (Berlin),* 74: 256–263.

Eckhart, V.M. 1992. The genetics of gender and the effects of gender on floral characters in gynodioecious *Phacelia linearis* (Hydrophyllaceae). *Am. J. Bot.,* 79: 792–800.

Faegri, K. and L. van der Pijl. 1979. *The Principles of Pollination Ecology.* Pergamon Press, Oxford.

Gentry, A.H. 1974. Flowering phenology and diversity in tropical Bignoniaceae. *Biotropica,* 6: 64–68.

Hartshorn, G.S. 1983. *Plants: Introduction.* In *Costa Rican Natural History* (D.H. Janzen, ed.), Univ. Chicago Press, Chicago, IL, pp. 118–157.

Holdridge, L.R., W.C. Grenke, W.H. Hatheway, T. Liang, and J.A. Tosi, Jr. 1971. *Forest Environments in Tropical Life Zones: A Pilot Study.* Pergamon Press, New York.

Husband, B.C. and D.W. Schemske. 1995. Evolution of the timing and magnitude of inbreeding depression in plants. *Evolution* (in press).

Kaplan, S.M. and D.L. Mulcahy. 1971. Mode of pollination and floral sexuality in *Thalictrum. Evolution,* 25: 659–668.

Kimsey, L.S. 1980. The behaviour of male orchid bees (Apidae, Hymenoptera, Insecta) and the question of leks. *Animal Behav.,* 28: 996–1004.

Kjellberg, F., P.H. Gouyon, M. Ibrahim, M. Raymond, and G. Valdeyron. 1987. The stability of the symbiosis between dioecious figs and their pollinators: A study of *Ficus carica* L. and *Blastophaga psenes* L. *Evolution,* 41: 693–704.

Knoll, F. 1922. Insekten und Blumen. Experimentelle Arbeiten zur Vertiefung unserer Kenntnisse über die Wechselbeziehungen zwischen Pflanzen und Tieren. III. Lichtsinn und Blumenbesuch des Falters von *Macroglossum stellatarum. Abhandlungen der Zoologisch-Botanischen Gesellschaft in Wien,* 12: 123–377.

Kugler, H. 1943. Hummeln als Blütenbesucher. Ein Beitrag zur experimentellen Blütenökologie. *Ergebnisse der Biologie,* 19: 143–323.

Kullenberg, B. 1961. Studies in *Ophrys* pollination. *Zool. Bidr. Upps.,* 34: 1–340.

Little, R.J. 1983. A review of floral food deception mimicries with comments on floral mutualism. In *Handbook of Experimental Pollination Biology* (C.E. Jones and R.J. Little, eds.), Van Nostrand Reinhold, New York, pp. 294–309.

Lloyd, D.G. and C.J. Webb. 1977. Secondary sex characters in plants. *Bot. Rev.,* 43: 177–216.

Lunau, K. 1991. Innate flower recognition in bumblebees (*Bombus terrestris, B. lucorum;* Apidae): Optical signals from stamens as landing reaction releasers. *Ethology,* 88: 203–214.

Lunau, K. 1992. A new interpretation of flower guide colouration: Absorption of ultraviolet light enhances colour saturation. *Plants System. Evol.,* 183: 51–65.

Meagher, T.R. 1992. The quantitative genetics of sexual dimporphism in *Silene latifolia* (Caryophyllaceae). I. Genetic variation. *Evolution,* 46: 445–457.

Nilsson, A. 1983. Mimesis of bellflower (*Campanula*) by the red helleborine orchid (*Cephalanthera rubra*). *Nature,* 305: 799–800.

Osche, G. 1983. Optische Signale in der Coevolution von Pflanze und Tier. *Berichte der Deutschen Botanischen Gesellschaft,* 96: 1–27.

Pijl, L. van der. 1978. Reproductive integration and sexual disharmony in floral functions. In *The Pollination of Flowers by Insects* (A.J. Richards, ed.), Academic Press, London, pp. 79–88.

Pijl, L. van der and C.H. Dodson. 1966. *Orchid Flowers: Their Pollination and Evolution.* Univ. Miami Press, Coral Gables, FL.

Proctor, M. and P. Yeo. 1973. *The Pollination of Flowers.* William Collins Sons and Company Ltd. London.

Renner, S.S. and J.P. Feil. 1993. Pollinators of tropical dioecious angiosperms. *Am. J. Bot.,* 80: 1100–1107.

Rice, W.R. 1989. Analyzing tables of statistical tests. *Evolution,* 43: 223–225.

Schemske, D.W. 1983. Breeding system and habitat effects on fitness components in three neotropical *Costus* (Zingiberaceae). *Evolution,* 37: 523–539.

Schemske, D.W. and J. Ågren. 1995. Deceit pollination and selection on female flower size in *Begonia involucrata*: An experimental approach. *Evolution*, 49: 207–214.

Schemske, D.W. and R. Lande. 1984. Fragrance collection and territiorial display by male orchid bees. *Anim. Behav.,* 32: 935–937.

Smith, L.B. and B.G. Schubert. 1958. Begoniaceae. Flora of Panama. *Ann. Missouri Bot. Gard.,* 45: 41–67.

Smith, L.B. and B.G. Schubert. 1961. Begoniaceae. Flora of Guatemala. *Fieldiana Bot.,* 24: 157–185.

Sokal, R.R. and F.J. Rohlf. 1981. *Biometry,* 2nd ed. W.H. Freeman, New York.

Sorenson, F.C. 1969. Embryonic genetic load in coastal Douglas-fir *Pseudotsuga menziesii* var. *menziesii. Am. Nat.,* 103: 389–398.

Sprengel, C.K. 1793. *Das Entdeckte Geheimniss der Natur im Bau und in der Befruchtung der Blumen.* Vieweg, Berlin.

Vogel, S. 1978. Evolutionary shifts from reward to deception in pollen flowers. In *The Pollination of Flowers by Insects* (A.J. Richards, ed.), Academic Press, London, pp. 89–96.

Wiens, D. 1978. Mimicry in plants. *Evol. Biol.,* 11: 365–403.

Williams, N.H. 1983. Floral fragrances as cues in animal behavior. In *Handbook of Experimental Pollination Biology* (C.E. Jones and R.J. Little, eds.), Van Nostrand Reinhold, New York, pp. 50–72.

Wilson, M.F. and J. Ågren. 1989. Differential floral rewards and pollination by deceit in unisexual flowers. *Oikos,* 55: 23–29.

12

Reproductive Success and Gender Variation in Deceit-Pollinated Orchids

*Anna-Lena Fritz and L. Anders Nilsson**

Introduction

In hermaphroditic plants differences in floral attractiveness to pollinators are expected to influence male and female reproductive success and thus functional gender (Nilsson, 1992). If reproductive success is limited by pollinator visitation, more attractive floral features are predicted to result in both more pollen exported and more seeds produced (Stanton and Preston, 1988). But what, more exactly, is the shape of the relationship between floral display and reproductive success in plants? This is clearly a critical issue that has to be clarified if floral evolution is to be understood. A greater individual display size in terms of number of flowers has been shown to result in an accelerating increase in the removal or receipt of pollen and thus probably male and/or female plant fitness via pollination (Schemske, 1980; Campbell, 1989*a;* see also Chapter 6). More commonly, however, a positive linear relationship between pollination success and floral display has been found (Willson and Price, 1977; Firmage and Cole, 1988; Campbell, 1989*a*; Broyles and Wyatt, 1990; Pleasants and Zimmerman, 1990). In some studies a decelerating increase in the visits by insects per flower, a measure that may be related to pollination success, has been reported (Galen and Newport, 1987; Andersson, 1988; Klinkhamer et al., 1989). An increase followed by a decrease in pollinator visitation with increasing inflorescence size has also been found (Bell, 1985; Pleasants and Zimmerman, 1990). Thus, the shape of animal-generated plant fitness curves in relation to floral display seems variable and deserves further study, especially because most studies so far have focussed on plants that are attractive through not only their floral display but also their pollinator-memory influencing floral rewards such as nectar.

*Department of Systematic Botany, Villavägen 6, S-752 36 Uppsala, Sweden.

Although hermaphrodite plants' female reproductive success can be estimated by convenient measures (e.g., fruit production), estimation of their male reproductive success constitutes a difficult methodological problem (see reviews by Bertin, 1988; Stanton et al., 1992; Snow and Lewis, 1993). However, male reproductive success can be estimated by the indirect method of assessing the amount of pollen removed from flowers by pollinators (Willson and Rathcke, 1974; Lloyd, 1980; Cruzan et al., 1988; Broyles and Wyatt, 1990). Although the relationship between removal and sequential delivery may be complex (Wilson and Thomson, 1991), a significant relationship has been found between pollinium removal and donation success or seeds sired in one asclepiad and one orchid species (Broyles and Wyatt, 1990; Nilsson et al., 1992). Such plants thus offer the possibility of using pollinium removal as a measure of male reproductive success, although because of pollinium waste by pollinators, this estimate has the disadvantage of generally overestimating the male function of individuals (Broyles and Wyatt, 1990). Pollinium waste is greatest (and thus the measure least efficient) in small populations, where the ratio ("equivalence") between fruits produced and pollinia removed is small (Fritz and Nilsson, 1994). Despite its drawbacks, however, the "pollinium flux" method (see, e.g., Willson and Rathcke, 1974; Bell, 1985) offers excellent possibilities for increasing our knowledge about ecologically and evolutionarily important reproductive patterns within plant populations.

The purpose of the present study was to investigate how male and female reproductive success and functional gender vary with size, initiation, and duration of the floral display and with the density of flowering individuals in natural populations of nectarless orchids. Such plants, accordingly termed "Scheinsaftblumen" by Sprengel (1793), act by deceit (see Chapter 11) and thus do not induce memory-based foraging behavior such as floral constancy in their pollinators (Heinrich, 1975, 1979). Therefore, the lack of a reward in these plants provides an opportunity to test the effect of display-related plant traits exclusively, without having to consider the often complex effects of reward variation.

Materials and Methods

During the summers of 1988–1993, populations of three species of orchids were studied on the island of Gotland in the Baltic Sea, Sweden. The orchids were *Orchis spitzelii* Sauter ex. Koch (10 populations), *O. palustris* Jacq. (8 populations), and *Anacamptis pyramidalis* (L.) Rich. (10 populations). Location and further information on the populations studied are presented elsewhere (Fritz and Nilsson, 1994). *Orchis spitzelii* flowers from mid–May to mid–June and grows in open, dry pine woods. It has a moderate number (mean = 10.7, SD = 4.1, N = 1381) of light purple flowers on a single, elongate spike. On Gotland this orchid is mainly pollinated by bumble bee queens of the species *Bombus pascu-*

orum (Scop.) (Fritz, 1990). *Orchis palustris* usually grows in sedge-dominated fens and flowers from mid–June to mid–July. It has a relatively low number (mean = 6.8, SD = 2.7, N = 1395) of purple flowers on a loose, elongate spike. It is pollinated on Gotland primarily by workers of *Bombus pascuorum* (Fritz, unpublished). *Anacamptis pyramidalis* grows mainly in grazed seashore and dry inland ("alvar") meadows, and flowers from mid–June to mid–July. It has numerous (mean = 27.2, SD = 11.2, N = 330) bright purple flowers on a pyramid-shaped spike. This species is pollinated on Gotland by a variety of butterflies; for example, *Melitaea cinxia* (L.) and *Mesoacidalia aglaja* (L.) (Nymphalidae), *Maniola jurtina* (L.) (Satyridae), and especially *Aporia crataegi* (L.) (Pieridae) (Fritz and Nilsson, 1994).

The three orchid species are outcrossing, self-compatible (self-pollen gives rise to embryo formation), nectarless deceivers with low levels of natural fruit set on Gotland, whereas hand-pollination routinely results in near 100% fruit set, indicating that within-season fruit production is not resource-limited (Fritz, unpublished). Flowering individuals were marked, and when all flowers on an individual's spike were open, we quantified the display by counting the flowers and measuring the length of the inflorescence (individuals of the three species rarely produced more than one flowering ramet, but in the few cases this occurred, ramets were treated as separate individuals). We also measured the distance from the plant to the nearest flowering conspecific individual and counted the number of flowering neighbors within a radius of 1 m in order to estimate density.

In order to evaluate the importance of initiation and duration of floral display on reproductive success, individuals were monitored for their exact start as well as termination of flowering in two different years in one large population of each of the two *Orchis* species. The population of *O. palustris* (located at UTM 63,11,6;3,72,0) was studied in 1988 and 1993, and that of *O. spitzelii* (at UTM 63,21,0;6,62,0) in 1988 and 1992 (in Table 12.1, these populations are denoted by the letter A).

Individual male reproductive success was estimated as the number of pollinia removed from flowers and female reproductive success as the number of fruits produced. Individuals were checked for pollinium removal immediately after termination of flowering, and fruit production was recorded approximately 2 weeks later when the ovaries of pollinated flowers were distinctly swollen. How well fruit production reflected actual pollen receipt was investigated by selecting one population per species in which naturally pollinated flowers were checked for fruit production. In such flowers, fruit set was 96.7% (N = 454 flowers) in *O. spitzelii*, 98.7% (N = 396) in *O. palustris*, and 100% (N = 84) in *A. pyramidalis*. Thus, fruit set was closely related to pollen deposition and a good estimate of female pollination success and also reflected the pollinator-mediated functional gender of plants.

Functional gender of an individual was measured as functional femaleness G_i according to the following equation (Lloyd, 1980):

*Table 12.1. The number of flowers, spike length, number of flowering conspecific neighbors within 1 m, and distance to nearest conspecific neighbor in populations of three nectarless orchids (*Anacamptis pyramidalis, Orchis palustris, *and* O. spitzelii*) in Gotland, Sweden.*

Population	Year	*N*	No. of Flowers Mean±SD	Spike Length (cm) Mean±SD	Neighbor Density Mean±SD	Neighbor Distance (cm) Mean±SD
A. pyramidalis						
A	1989	23	28.3±12.0	2.4±0.8	1.7±1.3	124±198
B	1989	27	19.7±7.7	2.1±0.8	1.0±0.9	80±94
C	1989	4	25.7±14.2	2.4±1.4	1.5±1.0	59±30
D	1989	18	21.4±8.1	2.0±0.7	0.6±0.8	243±294
	1990	32	25.7±10.6	2.0±0.6	1.2±1.2	131±143
E	1989	25	28.6±11.6	2.3±1.0	1.2±0.9	94±96
	1990	22	33.9±7.7	3.1±0.6	0.3±0.6	319±233
F	1989	33	38.0±12.8	3.7±1.1	3.5±2.9	48±55
	1990	28	30.5±12.8	2.8±0.6	4.5±2.0	26±26
G	1989	9	31.6±15.2	2.8±0.8	0.9±0.9	78±61
	1990	19	25.0±6.6	2.6±0.8	1.3±1.1	142±210
H	1989	18	24.0±7.6	2.5±0.9	0.3±0.5	212±194
	1990	24	22.5±6.7	2.2±0.6	1.1±0.8	174±286
I	1990	24	24.2±9.4	2.1±0.7	0.5±0.7	118±98
J	1990	24	25.2±9.3	2.2±0.7	1.8±1.6	110±119
O. palustris						
A	1988	350	6.2±2.6	4.3±1.8	0.6±0.8	176±178
	1989	252	7.2±2.9	4.2±1.7	0.7±0.9	127±109
	1990	76	7.9±2.8	4.1±1.4	0.6±0.7	190±227
	1993	198	7.2±2.3	4.4±1.5	0.3±0.6	198±157
B	1989	71	7.7±2.8	4.7±1.6	10.0±3.3	25±14
	1990	46	5.9±2.6	3.2±1.4	1.9±7.3	154±167
C	1988	44	7.5±3.1	No data	0.3±0.5	189±249
	1989	70	6.4±2.1	4.3±1.4	0.7±0.7	105±68
	1990	50	5.2±1.9	3.3±1.3	0.3±0.4	249±234
D	1989	60	8.6±2.5	4.8±1.4	0.4±0.6	140±91
	1990	49	5.9±2.7	3.6±1.6	0.8±1.1	155±134
E	1989	30	6.1±2.7	4.5±1.7	0.1±0.2	291±168
	1990	17	4.9±2.6	4.9±2.2	0	289±166
F	1989	32	7.9±2.3	5.3±1.6	0.6±0.9	243±234
G	1990	34	5.4±1.8	3.8±1.3	0.2±0.4	238±188
H	1990	20	5.8±1.8	4.3±1.2	0.2±0.4	466±324
O. spitzelii						
A	1988	296	10.8±4.0	4.8±1.7	4.7±4.6	46±68
	1989	319	11.4±4.0	4.9±1.8	5.9±5.2	33±58
	1990	236	10.2±3.6	5.0±1.8	3.8±3.3	42±55
	1992	193	9.6±3.9	4.5±1.9	8.0±7.1	61±40

Table 12.1. (continued)

Population	Year	*N*	No. of Flowers Mean±SD	Spike Length (cm) Mean±SD	Neighbor Density Mean±SD	Neighbor Distance (cm) Mean±SD
O. spitzelii (continued)						
B	1989	36	15.2±7.1	6.4±2.9	1.8±2.2	87±103
	1990	26	12.8±4.8	5.9±2.0	2.4±1.6	140±260
C	1989	34	10.0±3.7	4.3±1.9	2.2±2.2	123±218
	1990	12	11.1±3.8	5.3±1.7	0	500±260
D	1989	43	10.6±2.2	4.2±1.0	1.2±1.4	128±198
	1990	17	10.7±3.0	5.2±1.9	0.4±0.5	324±367
E	1989	36	11.3±4.3	4.7±1.9	1.2±1.4	98±103
	1990	15	11.9±3.9	6.0±1.6	0.4±0.8	326±360
F	1989	26	8.3±2.7	4.2±2.0	0.9±1.2	112±106
	1990	16	7.9±3.4	3.7±1.3	0.8±0.8	133±148
G	1989	24	10.4±2.7	5.2±1.3	1.1±0.8	141±273
	1990	3	11.0±1.7	5.8±1.1	0.7±0.6	345±566
H	1989	6	7.3±2.1	3.3±1.2	2.0±1.5	211±390
	1990	18	8.3±2.7	4.8±1.6	3.0±2.6	111±235
I	1989	10	10.7±2.2	3.7±1.2	0.9±0.9	152±165
J	1989	10	13.2±3.8	5.3±2.0	1.1±0.6	61±35

$$G_i = \frac{f_i}{f_i + m_i E} \tag{1}$$

where f_i is the number of fruits produced and m_i the number of pollinia removed from its flowers. E is an equivalence factor measured for each population as the sum of all fruits divided by the sum of all pollinia removed. With this measure, a plant with a pure male gender will have the value 0 and a plant with a pure female gender the value 1. The analysis assumes that all removed pollinia have equal probabilities of contributing to male reproductive success. The relationship between reproductive success as well as functional gender and each studied plant character was evaluated. We performed simple and quadratic regressions for this relationship in each population. Very small populations (with $\leq$ 10 individuals) were not included in the regression analyses. Standardized regression coefficients were determined by performing analysis of covariance (ANCOVA) on reproductive success and gender, with year and population nested within year as class variables and the plant character as a covariate (see Campbell, 1989*a*). Variables of reproductive success per flower were subject to arcsine square root transformation before analysis. Significance levels for ANCOVAs were based on type IV sums of squares. ANCOVAs and regressions utilized the GLM procedure of SAS (SAS Institute, 1985).

Results

Floral Display Size

General statistics on number of flowers, spike length, conspecific plant density, and nearest-neighbor distance are given in Table 12.1, whereas number of flowering plants, pollinium removal, and fruit set for populations are reported elsewhere (Fritz and Nilsson, 1994). Correlations between plant characters are presented in Table 12.2. Size and flowering characters were significantly intercorrelated in all combinations. Spike length and number of flowers were positively correlated ($r = 0.82$) in each of the three orchid species. The larger the display, the earlier the anthesis initiation and the longer the flowering period. Pollinium removal as well as fruit production had a higher correlation with spike length than flower number in the two *Orchis* species, whereas in *Anacamptis pyramidalis* these two variables of reproductive success were only slightly more correlated with the number of flowers (Table 12.2). We therefore decided to use spike length rather than the number of flowers as a measure of inflorescence size in the following analyses.

In each of the three species, individuals with larger spikes had more pollinia removed and fruits produced (within-population regression coefficients in ANCOVA, $P < 0.0001$, Table 12.3). There were significant differences in pollinium removal with the interaction between spike length and year (ANCOVA, $P < 0.05$, Table 12.4), so data on male reproductive success were analyzed separately for each year (Fig. 12.1A). Both *Anacamptis pyramidalis* and *O. palustris* had their highest overall rates of pollinium removal in 1990, whereas for *O. spitzelii* it was highest in 1989. The relation between fruit set and spike length varied significantly between years in the two *Orchis* species (ANCOVA, $P < 0.01$; Fig. 12.1B), but not in *A. pyramidalis* (ANCOVA, $P < 0.90$, Table 12.4).

An overall ANCOVA of relations between reproductive success per flower with spike length showed a highly significant increase in pollinia removed for all species and in fruit set for two species, *A. pyramidalis* and *O. spitzelii* (Table 12.3). There were highly significant differences among years in pollinia removed per flower in both *Orchis* species, and in *O. palustris* also a significant interaction between spike length and year for pollinia removed per flower (Table 12.4). *Orchis palustris* also showed differences among years in fruits per flower in relation to spike lengh. No strong among-year differences were present in *A. pyramidalis*.

Regression analyses of reproductive success per flower on spike length or number of flowers detected many cases with populations in each species where there was a significant linear increase in male or female total reproductive success with display size (Table 12.5). In all, 53 such significant positive relationships were found; only four were negative. Adding a quadratic term to the regression model indicated a significant accelerating increase in four (three with female and

Table 12.2. Pearson correlation coefficient matrices for plant characters in Anacamptis pyramidalis *(N=330),* Orchis palustris, *and* O. spitzelii. *Sample sizes in the latter two species are given below the coefficient values.*

A. pyramidalis	No. of Flowers	Spike Length	Pollinia Removed	Fruit Set	Neighbor Distance
No. of flowers	1				
Spike length	0.82***	1			
Pollinia removed	0.84***	0.73***	1		
Fruit set	0.53***	0.51***	0.70***	1	
Neighbor distance	0.08 ns	0.02 ns	0.14**	0.17**	1
Neighbor density	0.11 ns	0.16**	0.09 ns	−0.06 ns	−0.47***

O. palustris	No. of Flowers	Spike Length	Pollinia Removed	Fruit Set	Start of Flowering	Flow-ering Span	Neighbor Distance
No. of flowers	1						
	1395						
Spike length	0.82***	1					
	1351	1351					
Pollinia removed	0.46***	0.53***	1				
	1395	1351	1395				
Fruit set	0.36***	0.43***	0.68***	1			
	1395	1351	1395	1395			
Start of flowering	−0.42***	−0.42***	−0.31***	0.34***	1		
	546	546	546	546	546		
Flowering span	0.59***	0.51***	0.36***	0.28***	−0.29***	1	
	546	546	546	546	546	546	
Neighbor distance	−0.09 ns	−0.06*	0.09 ns	−0.02 ns	0.14***	−0.05 ns	1
	1395	1351	1395	1395	546	546	1395
Neighbor density	0.07**	0.05 ns	−0.05 ns	−0.03 ns	−0.13**	−0.02 ns	−0.27***
	1395	1351	1395	1395	546	546	1395

continued

Table 12.2. (continued)

O. spitzelii	No. of Flowers	Spike Length	Pollinia Removed	Fruit Set	Neighbor Distance		
	1						
No. of flowers	1381						
	0.82***	1					
Spike length	1363	1363					
	0.44***	0.47***	1				
Pollinia removed	1366	1361	1366				
	0.36***	0.40***	0.70***	1			
Fruit set	1381	1363	1366	1381			
	−0.26***	−0.21***	−0.35***	−0.29***	1		
Start of flowering	476	476	476	476	476		
	0.44***	0.42***	0.15***	0.15***	0.16***	1	
Flowering span	476	476	476	476	469	469	
	0.08 ns	0.06*	0.20***	0.04 ns	−0.04 ns	0.00 ns	1
Neighbor distance	1365	1355	1358	1365	469	469	1365
	0.11 ns	−0.04 ns	−0.25***	−0.10***	0.00 ns	−0.05 ns	−0.33***
Neighbor density	1366	1356	1359	1366	469	469	1365

*$P<0.05$.

**$P<0.01$.

***$P<0.001$.

ns = Not significant.

Table 12.3. Results of ANCOVAs for male and female reproductive success (pollinium removal and fruit set, respectively) and functional femaleness in Anacamptis pyramidalis, Orchis palustris, *and* O. spitzelii *(for years and populations included, see Table 12.1). The following model statement was used in the GLM procedure (SAS Institute, 1985): Dependent variable = Year population (year) spike length spike length * year, with year and population as class variables.* B = *within-population regression coefficient and SE = standard error of estimate.*

Species	Dependent Variable	$B \pm SE$	$P(B=0)$	Model R^2
A. pyramidalis	Pollinia removed	8.20±0.61	0.0001	0.66
	Pollinia removed/flower	0.06±0.02	0.002	0.43
	Fruits produced	3.14±0.36	0.0001	0.48
	Fruits produced/flower	0.07±0.02	0.0003	0.34
	Functional femaleness	0.08±0.02	0.0001	0.14
O. palustris	Pollinia removed	0.75±0.05	0.0001	0.46
	Pollinia removed/flower	−0.002±0.01	0.85	0.29
	Fruits produced	0.51±0.04	0.0001	0.32
	Fruits produced/flower	0.03±0.009	0.0007	0.17
	Functional femaleness	0.02±0.008	0.005	0.04
O. spitzelii	Pollinia removed	0.94±0.05	0.0001	0.45
	Pollinia removed/flower	0.04±0.006	0.0001	0.27
	Fruits produced	0.70±0.05	0.0001	0.28
	Fruits produced/flower	0.04±0.007	0.0001	0.14
	Functional femaleness	0.01±0.006	0.06	0.08

one with male reproductive success) and a decelerating relationship in eight cases, thus in a minor fraction of all population cases.

In all six *A. pyramidalis* populations, female reproductive success always exceeded male reproductive success at large spike sizes, a relationship resulting in a higher average functional femaleness for plants with larger floral display. *Orchis palustris* and *O. spitzelli* showed three and two populations, respectively, with significant although small increases in functional femaleness with spike length (Table 12.5). In *O. spitzelii,* population D in 1989 and G in 1989 had a higher female than male reproductive success with increasing spike length, and functional femaleness was near a significant increase ($P < 0.07$). In *O. palustris,* three populations (A in 1989 and 1990, and B in 1993), and in *O. spitzelii,* two populations (A in 1989 and 1992, and C in 1989) showed a very similar increase in both male and female reproductive success per flower with spike length, a pattern that consequently did not result in any increase in functional femaleness.

Analyses including all populations of each species showed a highly significant positive relationship in the three orchids for functional femaleness with spike length (ANCOVA, $P < 0.01$; Table 12.6). In *O. spitzelii,* there were also differences in functional femaleness across years as well as among populations within years. In no species was there a significant interaction between spike length and year for functional femaleness ($P > 0.22$). Mean functional femaleness

Table 12.4. ANCOVAs for male and female reproductive success, as well as functional femaleness on spike length in Anacamptis pyramidalis, Orchis palustris, *and* O. spitzelii.

Dependent Variable and Source of Variation	*A. pyramidalis*				*O. palustris*				*O. spitzelii*			
	df	Sum of Squares	*F*	*P*<	*df*	Sum of Squares	*F*	*P*<	*df*	Sum of Squares	*F*	*P*<
Pollinia removed												
Year	1	49.19	1.02	0.31	3	36.79	4.48	0.004	3	42.58	3.02	0.03
Population (year)	13	582.51	8.61	0.0001	11	759.82	26.43	0.0001	16	1880.95	24.99	0.0001
Spike length	1	15668.09	325.66	0.0001	1	1585.87	579.34	0.0001	1	1726.55	367.07	0.0001
Spike length * year	1	186.81	3.88	0.05	3	53.30	6.49	0.0002	3	175.31	12.42	0.0001
Residual	313	15058.88			1332	3649.79			1337	6288.62		
Pollinia removed/flower												
Year	1	0.01	0.15	0.70	3	7.59	20.05	0.0001	3	2.60	10.73	0.0001
Population (year)	13	8.27	13.42	0.0001	11	51.56	37.16	0.0001	16	28.76	22.23	0.0001
Spike length	1	1.01	21.32	0.0001	1	4.47	35.42	0.0001	1	4.39	54.32	0.0001
Spike length * year	1	0.03	0.56	0.46	3	2.44	6.46	0.0002	3	0.43	1.77	0.16
Residual	313	14.80			1332	168.04			1336	108.02		

Fruits produced												
Year	1	0.74	0.04	0.84	3	9.44	1.52	0.21	3	2.22	0.17	0.92
Population (year)	13	2232.42	9.96	0.0001	11	423.50	18.64	0.0001	16	495.57	7.18	0.0001
Spike length	1	1865.22	108.21	0.0001	1	616.03	298.19	0.0001	1	1093.55	253.42	0.0001
Spike length * year	1	0.27	0.02	0.90	3	24.78	4.00	0.008	3	117.35	9.07	0.0001
Residual	313	5395.39			1332	2751.79			1339	5777.91		
Fruits produced/flower												
Year	1	0.02	0.36	0.55	3	1.00	3.07	0.03	3	0.42	1.67	0.18
Population (year)	13	5.38	9.11	0.0001	11	20.56	17.20	0.0001	16	10.60	7.99	0.0001
Spike length	1	1.33	29.21	0.0001	1	5.70	52.48	0.0001	1	4.67	56.26	0.0001
Spike length * year	1	0.04	0.96	0.33	3	0.25	0.77	0.52	3	0.24	0.95	0.42
Residual	313	14.23			1332	144.75			1338	110.97		
Functional gender												
Year	1	0.01	0.13	0.72	3	0.04	0.16	0.92	3	0.80	3.73	0.01
Population (year)	13	0.73	1.10	0.36	11	1.14	1.39	0.17	16	5.05	4.42	0.0001
Spike length	1	1.54	30.21	0.0001	1	1.31	17.54	0.0001	1	0.51	7.11	0.008
Spike length * year	1	0.02	0.36	0.55	3	0.04	0.17	0.91	3	0.31	1.46	0.22
Residual	313	15.92			1160	86.85			1117	79.79		

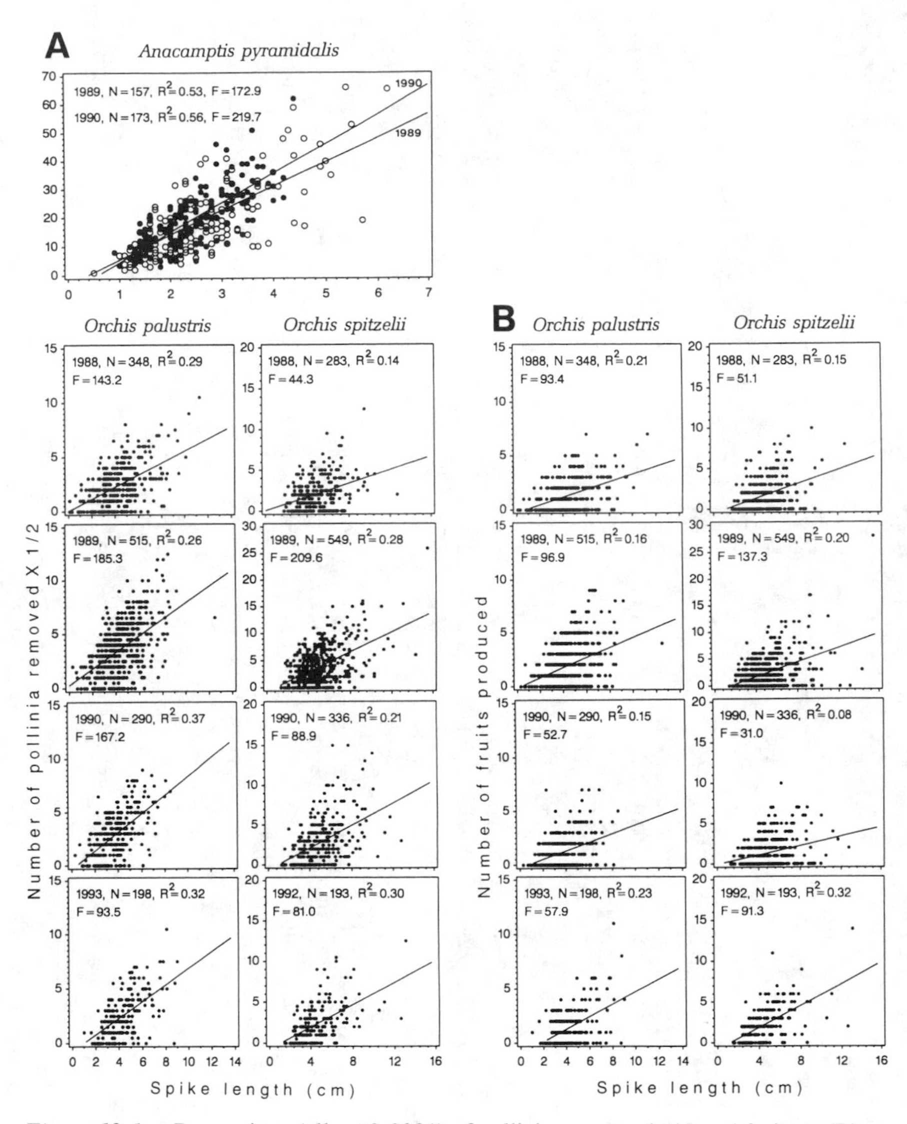

Figure 12.1. Regressions (all $p<0.0001$) of pollinium removal (A) and fruit set (B) on floral display size (spike length) in different years for three species of orchids (*Anacamptis pyramidalis, Orchis palustris,* and *O. spitzelii*).

approached 0.5 for larger spike sizes (Fig. 12.2), and there was a significant negative relationship in all three species between the variance in functional femaleness within spike-size class and spike size (Spearman correlation coefficients; *A. pyramidalis*, $r = -0.52$, $P < 0.001$, $N = 35$; *O. palustris*, $r = -0.59$, $P < 0.0001$, $N = 78$; *O. spitzelii*, $r = -0.26$, $P < 0.05$, $N = 80$ classes; the class width was 1 mm in these analyses). It should be noted that

Table 12.5. Summary results of simple regressions of male (M) and female (F) reproductive success (pollinium removal and fruit set, respectively) per flower as well as functional femaleness (G) *on traits of floral display in populations of three species of nectarless orchids (*Anacamptis pyramidalis, Orchis palustris, *and* O. spitzelii*). A value followed by + or − indicates number of significant (*$P<0.05$*) positive or negative regressions, respectively. Number of flowers was log-transformed before analysis. For populations (>10 flowering individuals) and years included, see Table 12.1.*

Species	Trait	M	F	G
A. pyramidalis	Spike length	5+	6+	6+
	Number of flowers	3+	3+	3+
O. palustris	Spike length	6+ 1−	6+	3+
	Number of flowers	4+ 3−	2+	2+
O. spitzelii	Spike length	5+	8+	2+
	Number of flowers	2+	3+	1+

this relationship for *O. spitzelii* first became significant when relatively many individuals at the right-hand end of the distribution were grouped together (see Fig. 12.2, but note that in this figure class width is 1 cm).

Floral Display Initiation and Duration

In *O. palustris,* individuals that started to flower early in the season had a significantly higher functional femaleness than late flowering ones (ANCOVA, $B \pm SE = -0.016 \pm 0.005$, $P < 0.001$). Such a relationship was not present

*Table 12.6. ANCOVAs of functional femaleness in three nectarless orchid species (*Anacamptis pyramidalis, Orchis palustris, *and* O. spitzelii*).*

Species	Source	*df*	SS	*F*	*P*
A. pyramidalis	Year	1	0.006	0.13	0.72
	Population (year)	13	0.730	1.10	0.36
	Spike length	1	1.536	30.21	0.0001
	Spike length*year	1	0.018	0.36	0.55
	Residual	313	15.918		
O. palustris	Year	3	0.036	0.16	0.92
	Population (year)	11	1.145	1.39	0.17
	Spike length	1	1.314	17.54	0.0001
	Spike length*year	3	0.039	0.17	0.91
	Residual	1160	86.854		
O. spitzelii	Year	3	0.798	3.73	0.01
	Population (year)	16	5.047	4.42	0.0001
	Spike length	1	0.508	7.11	0.008
	Spike length*year	3	0.313	1.46	0.22
	Residual	1117	79.786		

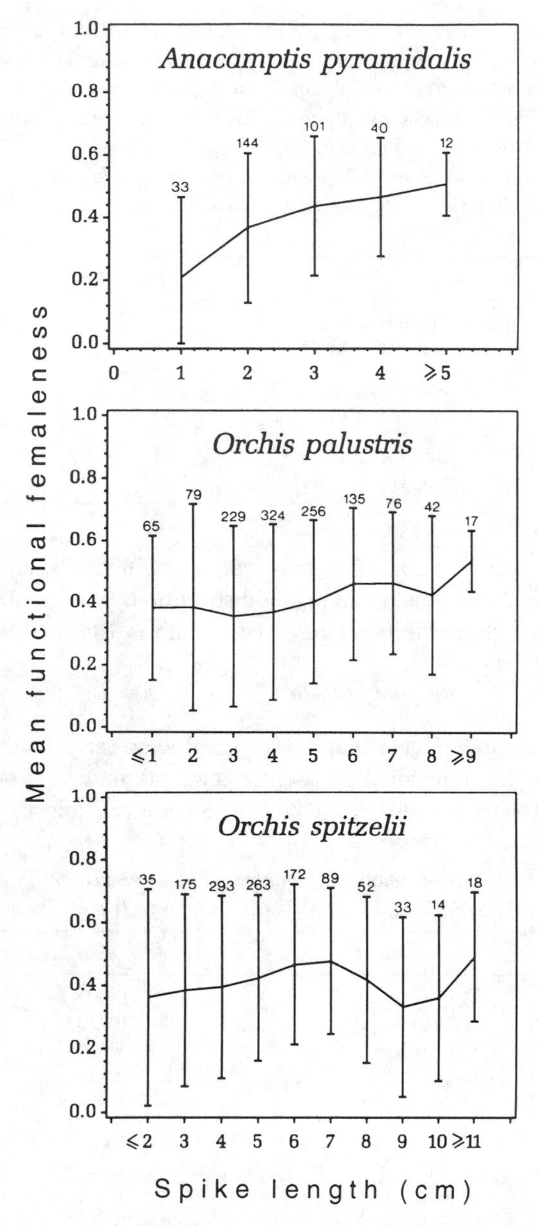

Figure 12.2. Graphs of mean functional femaleness as a function of floral display size (spike length) in three species of orchids (*Anacamptis pyramidalis, Orchis palustris*, and *O. spitzelii*). Spike length has been divided into size classes of 1 cm. Vertical bars represent one standard deviation and the number above bars indicate the number of observations.

in *O. spitzelii* ($B \pm SE = 0.005 \pm 0.005$, $P > 0.33$). Individual flowering time-span varied significantly between years in *O. palustris* [in 1988, 18 ± 4 (mean ± SD) days; in 1993, 25 ± 4 days], so data from each year were analyzed separately in this orchid. Results showed a significant positive relation between functional femaleness and flowering time-span in 1988 ($B \pm SE = 0.013 \pm 0.004$, $P < 0.004$), but not in 1993 ($B \pm SE = -0.004 \pm 0.005$, $P > 0.39$). In *O. spitzelii,* there was such a positive relation ($B \pm SE = 0.005 \pm 0.002$, $P < 0.03$), and there was no difference between the years 1988 and 1992 (ANCOVA, $P > 0.99$).

A multiple regression analysis including functional femaleness as the dependent variable and the independent variables spike length, start of flowering, and duration of flowering in *O. palustris* showed significant effects of spike size, flowering initiation, and duration of flowering in 1993. Neither of these three traits had any significant effect on functional femaleness in 1988. In *O. spitzelii,* spike length showed a significant effect in 1988, but none of the three traits had any significant effect on functional femaleness in 1992.

Floral Display Density

The density of individuals (Table 12.1) was highest in *O. spitzelii,* with a mean distance of 66 ± 138 cm between flowering individuals and 4 ± 5 neighbors within 1 m ($N = 1365$), compared to 128 ± 178 cm and 2 ± 3 neighbors, and 171 ± 175 cm and 1 ± 3 neighbor in *A. pyramidalis* ($N = 330$) and *O. palustris* ($N = 1395$), respectively. Pollinium removal increased significantly with the distance to the nearest flowering conspecific individual (*O. spitzelii,* $B \pm SE = 0.002 \pm 0.0009$, $P < 0.05$; *O. palustris,* $B \pm SE = 0.002 \pm 0.0008$, $P < 0.005$; and *A. pyramidalis,* $B \pm SE = 0.02 \pm 0.005$, $P < 0.003$), and in *O. spitzelii* and *O. palustris* also decreased with the number of neighboring individuals ($B \pm SE = -0.11 \pm 0.03$, $P < 0.0001$ and $B \pm SE = -0.15 \pm 0.06$, $P < 0.02$, respectively). The corresponding ANCOVA models were significant for the number of fruits set and the distance between flowering individuals and number of neighboring individuals in only *O. spitzelii* ($P < 0.002$). There was also in the two *Orchis* species a negative correlation between mean distance between flowering individuals and population size [Spearman rank correlation coefficients; *O. palustris,* $r = -0.90$, $P < 0.001$, $N = 15$; *O. spitzelii,* $r = -0.80$, $P < 0.001$, $N = 19$, but not in *A. pyramidalis* ($r = -0.20$, $P > 0.51$, $N = 15$); population sizes are given in Fritz and Nilsson, 1994].

Discussion

In principle, the study plants function according to a reproductive mode that approaches sequential hermaphroditism: Individuals tend to be male first and then hermaphrodite by adding a female function. The change in functional gender with size is directly related to a change in pollinator visitation that, in turn, is an indirect response to variation in resources invested in floral display (Nilsson, 1992). Data from the related orchid *Orchis morio* indicate that phenotypic (structural) gender also changes with size, small plants being more male and large plants more female in terms of relative pollen and ovule production, that is, the P/O ratio decreases with size (Nilsson and Pettersson, unpublished). There is nothing to suggest that the present study species do not show similar specialization in phenotypic gender change as *O. morio*. An increase in phenotypic femaleness with size has been documented in other species as well, and is probably a widespread reproductive strategy in nutritionally resource-limited plants (see Klinkhamer and de Jong, 1993; Kudo, 1993 and refs. cited there). Clearly, structural and functional gender covary in hermaphroditic plants, femaleness increasing with size through both types of gender. When all the flowers on *O. morio* individuals are hand-pollinated, flowering frequency in successive years is negatively related to plant size in the year of hand-pollination (Nilsson, unpublished). This suggests, as predicted also by the size-related variation in P/O ratio, that resource limitation is negatively correlated with size in nectarless orchids. An increase in phenotypic femaleness toward greater hermaphroditism at larger inflorescence sizes meets the same predictions as the so-called "size advantage hypothesis," where selection will favor an individual that reproduces as a male when small and changes structurally to a female when large, if female reproductive success increases with size relatively more than male reproductive success (Charnov, 1982).

Female and male fitness gains with size are to be expected in plant species that are strongly pollinator-limited (Stanton and Preston, 1988), for example, such as nectarless plants that act by deceit to attract pollinators. In a fairly large number of the populations studied in each of the three species, both male and female reproductive success per flower increased linearly with floral display size. In some cases, the increase in reproductive success was accelerating. Sexual selection (*sensu* Willson, 1990) may therefore operate through both sex functions. "Mutual sexual selection" with the two sex functions exerting selection on each other may be a relevant term for the process responsible for floral-display evolution in pollinator-limited plants (Nilsson, 1992).

Individuals with larger floral displays not only showed distinctly greater values, but also less variance in functional femaleness (most clear in *A. pyramidalis*) that approached 0.5 for larger spikes (Fig. 12.2). The large variance in smaller size classes was caused in part by a relatively high number of purely male- or female-functioning individuals (with pollinium removal or receipt only). This means that large individuals (probably old and/or in resource acquisition, exceptionally successful plants) have a more stabilized hermaphrodite gender than small plants and a greater probability of achieving true pollinator-generated hermaphroditism with a function half as male and half as female.

Generally, animal pollinators will tend to give small-display plants a functionally male role (or plants will stay sexually inactive neuters when not visited), compared to large-display plants that will be given a hermaphrodite role (Nilsson, 1992). However, time-related flowering traits may also influence functional gender especially as in the present cases, when plants are pollinator-limited. The higher functional femaleness for early flowering individuals in *O. palustris* was probably in part caused by strong intercorrelations: individuals with larger spikes started to flower earlier and flowered longer than those with smaller spikes (Table 12.2). Earliness and duration of flowering were also correlated with spike length in *O. spitzelii,* but functional femaleness was only significantly related to the duration of flowering in this species. Campbell (1989*b*), in contrast, found a positive correlation for functional femaleness (measured as pollen-analogous dye donated and received) with date of anthesis initiation for protandrous *Ipomopsis aggregata*. Campbell explained this by differences among populations in anthesis since an analysis of covariance did not indicate any significant within-population effects of anthesis date on functional femaleness. In the two *Orchis* species, a positive relationship between functional femaleness and flowering duration seems plausible because pollinium removal and receipt will tend to reach their respective fixed maxima (set by the number of flowers produced) with time as long as there are pollinators available, eventually yielding functionally perfect hermaphrodites (as defined here). A positive relationship between flowering duration and functional hermaphroditism may therefore be the rule in pollinator-limited plants.

Both density of flowering individuals and pollinator abundance in a plant's populations vary throughout a flowering season, and such factors may affect competition for pollen removal and receipt among plants over time (Chaplin and Walker, 1982; Campbell and Waser, 1989). For example, a high plant density can cause intense intraspecific competition for pollinators, leaving small unattractive plants unvisited (Zimmerman, 1980). As reported above, the average distance between flowering individuals was negatively correlated with population size in the two *Orchis* species. The positive relationship between pollinium removal and distance to the nearest flowering neighbor as revealed by ANCOVA can be explained by a higher degree of pollinator visit-limitation in large populations. This resulted in a decreased proportion of pollinia removed (as well as in a higher proportion of nonvisited individuals) in large populations compared to small ones; in *A. pyramidalis,* however, there were no nonvisited individuals and thus probably less competition for pollinators (Fritz and Nilsson, 1994).

The increase in pollinium removal with interplant distance might also be an effect of within-population pollinator behavior and movement in response to deceit. To the best of our knowledge, there are no detailed reports on observation of pollinator movements in nectarless plant populations of different densities. However, it is known that bumble bees have a tendency to fly longer distances while moving through areas with depleted resources or nectarless, deceitful orchids (Heinrich, 1975, 1979). Such behavior thus could have contributed to the density- and distance-dependent pollinium removal recorded in the present study. Interestingly, in A. pyramidalis and in O. palustris, there was no density or distance effect on fruit set as revealed by ANCOVA. At least in the *Orchis* species, this probably resulted from pollinium waste being relatively greater in small (i.e., sparse) rather than large populations (Fritz and Nilsson, 1994). Clearly, the density of pollinators and their food plants will also influence pollination success (Jennersten, 1988; Laverty, 1992), and thus also probably mating structure and functional gender of plant populations. However, pollinator density and abundance of food plants were factors not investigated in the present study. Whether variation in pollinator density and the co-occurring flora affects functional gender within plant populations is an issue that deserves to be examined in the future.

Conclusions

The present study, using natural populations of three hermaphroditic, deceit-pollinated (nectarless) orchids as model systems, documents that when pollination is limited, pollinators influence the pattern of reproductive success as well as functional gender of plants. This occurs in response to variation in the size, timing, and duration of the floral display and the density of flowering individuals. In the study populations, male and female reproductive success, in terms of the number of pollinia removed from flowers by insects and the number of fruits set per flower, respectively, most often exhibited a linear increase although sometimes showed an accelerating increase with floral display size. These patterns of reproductive success suggest the presence of mutual intersexual selection in plants via differential pollinator attraction. Functional gender (femaleness) increases and approaches 0.5 and becomes less variable for plants with larger displays and may also be promoted by long flowering periods. In each of the three species, the number of pollinia removed, and in one species the number of fruit set, were found to increase with interplant distance, reflecting relaxed competition for pollinators. Clearly, in plant populations animal pollinators make small-display plants reproduce relatively more as males and large-display plants have the greatest chances of acting as functional hermaphrodites. This pattern is probably universal in populations of pollinator-limited angiosperms.

Acknowledgments

We are indebted to Spencer Barrett, Jon Ågren, and anonymous reviewers for comments on the manuscript. This work was supported by the Swedish World Wide Fund for Nature to A.-L. Fritz and the Swedish Natural Science Research Council to L. Anders Nilsson.

References

Andersson, S. 1988. Size-dependent pollination efficiency in *Anchusa officinalis* (Boraginaceae): Causes and consequences. *Oecologia*, 76: 125–130.

Bell, G. 1985. On the function of flowers. *Proc. R. Soc. Lond.*, B 224: 223–265.

Bertin, R.I. 1988. Paternity in plants. In *Plant Reproductive Ecology: Patterns and Strategies* (J. Lovett Doust and L. Lovett Doust, eds.), Oxford Univ. Press, New York, pp. 30–59.

Broyles, S.B. and R. Wyatt. 1990. Paternity analysis in a natural population of *Asclepias exaltata*: Multiple paternity, functional gender, and the "pollen-donation hypothesis." *Evolution*, 44: 1454–1468.

Campbell, D.R. 1989*a*. Inflorescence size: Test of the male function hypothesis. *Am. J. Bot.*, 76: 730–738.

Campbell, D.R. 1989*b*. Measurements of selection in a hermaphroditic plant: Variation in male and female pollination success. *Evolution*, 43: 318–334.

Campbell, D.R. and N.M. Waser. 1989. Variation in pollen flow within and among populations of *Ipomopsis aggregata*. *Evolution*, 43: 1444–1455.

Chaplin, S.J. and J.L. Walker. 1982. Energetic constraints and the adaptive significance of floral display of a forest milkweed. *Ecology*, 63: 1857–1870.

Charnov, E.L. 1982. *The Theory of Sex Allocation*. Princeton Univ. Press, Princeton, NJ.

Cruzan, M.B., P.R. Neal, and M.F. Willson. 1988. Floral display in *Phyla incisa*. Consequences for male and female reproductive success. *Evolution*, 42: 505–515.

Firmage, D.H. and F.R. Cole. 1988. Reproductive success and inflorescence size in *Calopogon tuberosus* (Orchidaceae). *Am. J. Bot.*, 75: 1371–1377.

Fritz, A.-L. 1990. Deceit pollination of *Orchis spitzelii* (Orchidaceae) on the island of Gotland in the Baltic: A suboptimal system. *Nord. J. Bot.*, 9: 577–587.

Fritz, A.-L. amd L.A. Nilsson. 1994. How pollinator-mediated mating varies with population size in plants. *Oecologia,* 100: 451–462.

Galen, C. and M.E.A. Newport. 1987. Bumblebee behavior and selection on flower size in the sky pilot, *Polemonium viscosum*. *Oecologia*, 74: 20–23.

Heinrich, B. 1975. Bee flowers: A hypothesis on flower variety and blooming times. *Evolution*, 29: 325–334.

Heinrich, B. 1979. Resource heterogeneity and patterns of movement in foraging bumblebees. *Oecologia*, 40: 235–245.

Jennersten, O. 1988. Pollination in *Dianthus deltoides* (Caryophyllaceae): Effects of habitat fragmentation on visitation and seed set. *Conserv. Biol.*, 2: 359–366.

Klinkhamer, P.G.L. and T.J. de Jong. 1993. Phenotypic gender in plants: Effects of plant size and environment on allocation to seeds and flowers in *Cynoglossum officinale*. *Oikos*, 67: 81–86.

Klinkhamer, P.G.L., T.J. de Jong, and G.J. de Bruyn. 1989. Plant size and pollinator visitation in *Cynoglossum officinale*. *Oikos*, 54: 201–204.

Kudo, G. 1993. Size-dependent resource allocation pattern and gender variation of *Anemone debilis* Fisch. *Plant Species Biol.*, 8: 29–34.

Laverty, T.M. 1992. Plant interactions for pollinator visits: A test of the magnet species effect. *Oecologia*, 89: 502–508.

Lloyd, D.G. 1980. Sexual strategies in plants. III. A quantitative method for describing the gender of plants. *New Zeal. J. Bot.*, 18: 103–108.

Nilsson, L.A. 1992. Animal pollinators adjust plant gender in relation to floral display: Evidence from *Orchis morio* (Orchidaceae). *Evol. Trends Plants*, 6: 33–40.

Nilsson, L.A., E. Rabakonandrianina, and B. Pettersson. 1992. Exact tracking of pollen transfer and mating in plants. *Nature*, 360: 666–668.

Pleasants, J.M. and M. Zimmerman. 1990. The effect of inflorescence size on pollinator visitation of *Delphinium nelsonii* and *Aconitum columbianum*. *Collect. Bot.*, 19: 21–39.

SAS Institute. 1985. *SAS User's Guide: Statistics Version*, 5th ed. SAS Institute, Cary, NC.

Schemske, D.W. 1980. Evolution of floral display in the orchid *Brassavola nodosa*. *Evolution*, 34: 489–493.

Snow, A.A. and P.O. Lewis. 1993. Reproductive traits and male fertility in plants: Empirical approaches. *Ann. Rev. Ecol. Syst.*, 24: 331–351.

Sprengel, C.K. 1793. *Das entdeckte Geheimnis der Natur in Bau und in der Befruchtung der Blumen*. Vieweg, Berlin.

Stanton, M.L. and R.E. Preston. 1988. A qualitative model for evaluating the effects of flower attractiveness on male and female fitness in plants. *Am. J. Bot.*, 75: 540–544.

Stanton, M.L., T.-L. Ashman, L.F. Galloway, and H.J. Young. 1992. Estimating male fitness of plants in natural populations. In *Ecology and Evolution of Plant Reproduction* (R. Wyatt, ed.), Chapman & Hall, New York, pp. 62–90.

Willson, M.F. 1990. Sexual selection in plants and animals. *Trends in Ecol. Evol.*, 5: 210–214.

Willson, M.F. and P.W. Price. 1977. The evolution of inflorescence size in *Asclepias* (Asclepiadaceae). *Evolution*, 31: 495–511.

Willson, M.F. and B.J. Rathcke. 1974. Adaptive design of the floral display in *Asclepias syriaca* L. *Am. Midl. Nat.*, 92: 47–57.

Wilson, P. and J.D. Thomson. 1991. Heterogenity among floral visitors leads to discordance between removal and deposition of pollen. *Ecology*, 72: 1503–1507.

Zimmerman, M. 1980. Reproduction in *Polemonium*: Competition for pollinators. *Ecology*, 61: 497–501.

13

Stylar Polymorphisms and the Evolution of Heterostyly in *Narcissus* (Amaryllidaceae)

Spencer C.H. Barrett, David G. Lloyd,† and Juan Arroyo‡*

Introduction

In outcrossing hermaphrodite plants, the separate functions of pollen dispersal and pollen receipt may interfere with one another so that fitness as a paternal or maternal parent is compromised (van der Pijl, 1978; Bawa and Opler, 1975; Lloyd and Yates, 1982; Lloyd and Webb, 1986; Webb and Lloyd, 1986; Bertin and Newman, 1993; Harder and Barrett, 1995). This is particularly likely in flowers in which the sex organs are close together and mature at the same time. Interference can potentially take several forms, including the obstruction by pistils of efficient pollen dispatch by pollinators, stamens restricting access by pollinators to stigmas, thus reducing pollen deposition, and the deleterious effects of self-pollination on maternal function due to stigmatic, stylar, or ovular clogging. Although there is some experimental evidence for self-pollen interference (Shore and Barrett, 1984; Barrett and Glover, 1985; Bertin and Sullivan, 1988; Palmer et al., 1989; Waser and Price, 1991; Scribailo and Barrett, 1994), the other two forms of pollen-pistil interference have seldom been investigated (see, however, Barrett and Glover, 1985; Kohn and Barrett, 1992*a*).

Selection to reduce interference by segregating sex organs in time or space has been proposed as a major influence on floral evolution, resulting in a variety of conditions including dichogamy, herkogamy, and various forms of gender expression such as monoecy (Lloyd and Webb, 1986; Webb and Lloyd, 1986; Bertin, 1993; Bertin and Newman, 1993; Harder and Barrett, 1995; also see

*Department of Botany, University of Toronto, Toronto, Ontario, Canada M5S 3B2.

†Department of Plant & Microbial Sciences, University of Canterbury, Christchurch 1, New Zealand.

‡Departamento de Biología Vegetal y Ecología, Universidad de Sevilla, 41080 Sevilla, Spain.

Chapter 6). Traditionally, these floral traits have been interpreted as antiselfing mechanisms, which reduce the harmful effects of inbreeding. However, their frequent occurrence in self-incompatible taxa casts doubt on this as a general explanation for their adaptive significance. Although selection to reduce pollen-pistil interference is likely to be of evolutionary significance in outcrossing plants, few studies in floral biology have examined this problem compared with the large literature on antiselfing mechanisms and inbreeding depression (reviewed in Richards, 1986; Charlesworth and Charlesworth, 1987; Jarne and Charlesworth, 1993; Thornhill, 1993).

In animal-pollinated plants, the spatial separation of female and male reproductive organs within flowers (*herkogamy*) may be constrained by the requirements of proficient cross-pollination. If stigmas and anthers are too widely separated, pollinators may contact only one set of sex organs or touch them with different body parts while visiting flowers. The conflict between the avoidance of self-interference and precision in pollination is reduced in heterostylous plants because of the occurrence within populations of two or three usually self-incompatible morphs that differ reciprocally in the placement of female and male reproductive organs within flowers. Plants of each morph are strongly herkogamous, thus avoiding self-interference; however, the placement of female and male sex organs in a reciprocal and complementary manner among the morphs promotes more precise cross-pollen transfer. Although, heterostyly has been studied intensively since Darwin's (1877) early work, there is still debate concerning its evolution and adaptive significance (reviewed in Barrett, 1992*a*). Models of the evolution of heterostyly differ in the pathways by which the polymorphism is thought to originate and in the emphasis placed on proficient cross-pollination, inbreeding avoidance and pollen-stigma interference as selective forces responsible for evolution of the principal features of the polymorphism (Charlesworth and Charlesworth, 1979*b;* Charlesworth, 1979; Ganders, 1979; Richards, 1986; Barrett, 1990; Lloyd and Webb, 1992*a,b*). Unfortunately, a major limitation in testing these models is the paucity of plant groups in which putative stages in the evolutionary assembly of the polymorphism can be identified unequivocally.

In this chapter, we investigate the patterns of sex-organ variation that occur in populations of the perennial geophyte *Narcissus* (Amaryllidaceae) and discuss the selective mechanisms responsible for the evolution and maintenance of this variation. Our interest in *Narcissus* was initially sparked by a historical debate concerning the nature of sexual polymorphisms in various species and whether "true" heterostyly occurs in the genus. Several early workers reported the occurrence of heterostyly in species of *Narcissus* (Henriques, 1887, 1888; Fernandes, 1935, 1964, 1965, also see Crié, 1884; Wolley-Dod, 1886; Des Abbayes, 1935). However, its occurrence in the genus was later disputed by Bateman (1952*a*) and most modern workers interested in the evolution of heterostyly. Skepticism over the presence of heterostyly in *Narcissus* was based on the concern that only a limited sampling of variation had been involved in early reports and continuous variation in sex-organ position appeared to exist in natural populations (Bateman, 1952a; also see Baker, 1964). In addition, the finding that several *Narcissus* species possess an incompatibility system with many mating types, rather than a system with only two (distyly) or three (tristyly) (Bateman, 1954; Dulberger, 1964), probably biased modern workers against accepting the earlier reports. For many workers the association of diallelic (intramorph) incompatibility with the style-stamen polymorphism that characterizes heterostylous plants is a *sine qua non* for the occurrence of "true" heterostyly. Recently, however, Lloyd et al. (1990) reconsidered the evidence presented in earlier papers and also examined sex-organ variation in horticultural cultivars of *N. triandrus*. They concluded that both distyly and tristyly occur in the genus, but suggested that field studies were necessary to confirm the true nature of the polymorphisms.

To clarify these issues, we have examined the patterns of sex-organ variation in natural populations of selected species of *Narcissus* from the Iberian peninsula. Of particular importance in these studies was to establish whether the relative positions of female and male sex organs vary continuously, or discrete morphs differing in sex-organ position occur. Our surveys reveal that three distinct types of sexual variation occur in *Narcissus,* with populations either monomorphic, dimorphic, or trimorphic for style length. These three fundamental conditions are illustrated in Figure 13.1. Here we document this variation in detail and address the following general questions: (1) What are the patterns of variation in stigma and anther position within and between species and do any species exhibit heterostyly? (2) What selective forces maintain style-length variation and what are the evolutionary relationships among different stylar conditions? (3) What floral traits are associated with the three types of sexual variation and which aspects of the pollination biology of species have influenced the evolution of stigma-height polymorphisms? Our results confirm that heterostyly occurs in *Narcissus,* but its expression differs in important ways from the widely held view of the polymorphism. These differences are of special significance because they not only provide clues as to how the polymorphism may have evolved in *Narcissus,* but they also give more general insights into the selective mechanisms responsible for the evolution of heterostyly in other groups.

Floral Biology of *Narcissus*

Before discussing the evolution of sexual polymorphisms in *Narcissus,* we briefly review the systematics and natural history of the genus, focusing in particular on floral biology. A remarkable feature of *Narcissus* is that although it is one of the most important perennials used in horticulture, with more than 20,000 registered names and over a century of breeding effort, the floral biology and pollination systems of most *Narcissus* species are unknown. Moreover, little is known of the phylogenetic relationships of taxa, and species circumscriptions

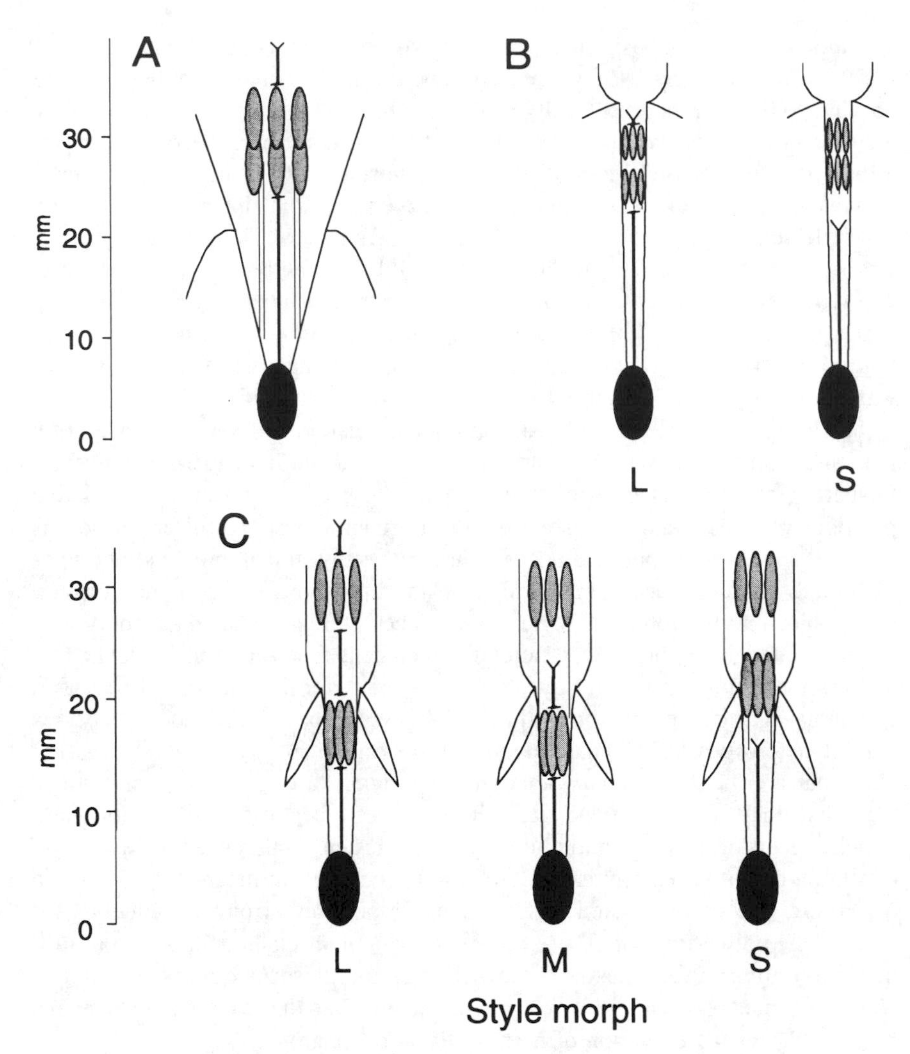

Figure 13.1. The three fundamental stylar conditions in *Narcissus*. Schematic illustrations of the flowers of *Narcissus* species monomorphic, dimorphic, and trimorphic for stigma height. All measurements are based on 50 flowers per morph sampled from a single population of each species (A) *N. bulbocodium*. Populations of this species are monomorphic for style length and possess approach herkogamous flowers with two anther levels. (B) *N. assoanus*. Two morphs that differ in stigma height are present in populations of this species. In the long-styled morph (L), stigmas are positioned above the two anther levels (approach herkogamy), whereas in the short-styled morph (S), stigmas are below the two anther levels (reverse herkogamy). (C) *N. triandrus*. Three morphs (L, M, and S) differing in style length occur in most populations of this species. Anther heights are similar in the long- and mid-styled morph, but differ in the short-styled morph. See text for further details.

vary widely, so there is little agreement on the number of species in the genus. Estimates have ranged from 16–150, with most authorities accepting between 35–70 species, divided among two subgenera and 10 sections (for further details, see Fernandes, 1951, 1967, 1975; Meyer, 1961; Webb, 1980; Valdés, 1987; Blanchard, 1990; Jefferson-Brown, 1991).

Floral Morphology

Members of the genus are perennial bulbs of rocky hills and montane areas of Europe and North Africa, particularly the Mediterranean region. The center of diversity of the genus is Spain and Portugal with a high concentration of species also occurring in Morocco. All species are winter-growing and summer-dormant with most blooming in early spring and a few (*N. cavanillesii, N. elegans, N. serotinus, N. broussonetii, and N. viridiflorus*) flowering in the autumn. Flowers vary in diameter from as little as 12 mm in forms of *N. bulbocodium* to over 12.5 cm in *N. nobilis* and, with the exception of *N. viridiflorus,* flowers usually possess white, yellow, or orange perianth parts. The most prominent features of *Narcissus* flowers are the corona, which forms a cylindrical cone extending beyond the tepals, and the fused perianth giving rise to a prominent floral tube. Corona morphology is quite variable, ranging from a tiny disk in *N. serotinus* to a long trumpetlike structure in *N. pseudonarcissus*. Floral tubes range from long and narrow in species of sections Jonquillae and Apodanthae to virtually absent in *N. cavanillesii*. Flower number per stem ranges from 1 in many species to 15–16 in *N. tazetta,* and flower orientation varies from ascending in *N. rupicola* to deflexed in *N. triandrus*.

Pollinator Visitation

Suprisingly little information has been collected on the pollination biology of *Narcissus* species, if we consider that many possess prominent showy flowers and some are highly scented. Our observations of *N. bulbocodium, N. papyraceus,* and *N. triandrus* from the Iberian peninsula indicate that flowers are visited primarily by bees and also butterflies and flies. The only night-flowering species in the genus, the strongly scented *N. viridiflorus,* is believed to be pollinated by small moths (Vogel and Müller-Doblies, 1975). One noteworthy aspect of our observations of insect visitors to *Narcissus* flowers is the very low frequency of visitation in populations, irrespective of the size of the floral display. For example, despite many hours observing populations of *N. assoanus, N. calcicola, N. gaditanus,* and *N. rupicola,* few pollinators were observed. Fernandes (1965) remarked on the curious lack of pollinators visiting *N. triandrus* and raised the possibility that the species may be pollinated by nocturnal insects. However, he doubted this was likely because of the cool night temperatures that occur during flowering in many of the montane sites that the species inhabits in central Portugal and Spain. Experimental studies of *N. longispathus* by Herrera (1995) revealed that fruit set in this species was pollen-limited. Because many *Narcissus* species flower during early spring when pollinator activity is low, pollen limitation may often occur.

Narcissus flower longevities range from 5–20 days, depending on the species and environmental conditions during flowering. The adaptive significance of this variation has not been investigated, but is likely to be associated with the levels of pollinator activity experienced by individual species (see Chapter 5). There is some evidence to support this view for *N. tazetta,* one of the few *Narcissus* species in which the floral biology of natural populations has been studied. Populations in Israel differ in habitat, flowering time, floral morphology, style-morph ratios, and the types of pollinators that visit populations (Dulberger, 1967 and personal communication; Arroyo and Dafni, 1995). Populations occurring in lowland marshes have floral longevities of 14–20 days and are serviced by the sphingid *Macroglossum stellarum,* which visits flowers quite infrequently. In contrast, floral longevities of hill populations range from 5–7 days, and flowers are visited regularly by the syrphid *Erystalis tenax* and solitary bees, including *Anthophora* sp. and *Proxylocopa olivieri*. Differences in the rates of fitness accrual as a result of contrasting levels of pollinator activity (see Chapter 5) may account for the differentiation of floral longevities in the two ecological races of *N. tazetta*. Herrera (1995) has recently conducted a detailed study of the floral biology of the early-flowering trumpet daffodil *N. longispathus* in southern Spain. Flowers of this species last for 17 days on average and are pollinated primarily by the andrenid bee *Andrena bicolor*. Herrera found that bees foraged only on sunny days when the temperature was above 12–17°C. Visitation to flowers was positively correlated to the average temperature inside coronas, which could be up to 8°C above ambient temperature.

Sexual Systems

Although detailed studies of dichogamy have not been conducted, our preliminary observations indicate that most *Narcissus* species are protandrous to varying degrees. Bateman (1954), however, reported strong protogyny in *N. bulbocodium* with anthers dehiscing "a day or more after the stigma becomes receptive." In species with two distinct stamen levels (e.g., *N. bulbocodium* and *N. triandrus*), anthers of the upper stamens frequently dehisce before anthers of the lower stamens. This pattern has also been reported in tristylous *Pontederia cordata* (Harder and Barrett, 1993). In some species, dichogamy is associated with the prolonged growth of styles. In members of sections Tazettae, Jonquillae, and Apodanthae, style growth in long-styled plants often proceeds during anther dehiscence so that stigmas are at the same level as the anthers during early male function, but project above them during stigma receptivity. This pattern of development can be interpreted functionally as a mechanism to reduce pollen-stigma interference. A similar phenomenom has been reported in distylous *Quinchamalium chilense* (Santalaceae) by Riveros et al. (1987).

Despite extensive breeding of *Narcissus* for ornamental purposes, there are few published studies on the compatibility or mating systems of wild or cultivated species. The most detailed study is that of Bateman (1954) published in the *The Daffodil and Tulip Year Book*. Based on controlled self- and cross-pollinations and observations of seed set, he reported the occurrence of self-sterility in *N. asturiensis, N. bulbocodium, N. calcicola, N. cyclamineus, N. poeticus, N. tazetta,* and *N. triandrus*. The strength of self-sterility varied in a few species (e.g., *N. bulbocodium* varieties and *N. triandrus*), with some plants setting few seeds upon selfing. This variation may explain the report of self-compatibility in *Narcissus triandrus* by Richards (1986, p. 241). Based on our own pollination studies, the following additional species can be added to the list of self-sterile taxa: *N. assoanus, N. elegans, N. jonquilla, N. papyraceus,* and *N. viridiflorus*. We have also confirmed the presence of self-sterility in *N. bulbocodium* and *N. triandrus*. In addition, we have shown that *N. serotinus* is thoroughly self-compatible, setting similar numbers of seeds following self- or cross-pollination. *Narcissus longispathus* is also highly self-compatible and flowers occasionally set fruit by autonomous self-pollination (Herrera, 1995).

Of relevance to the issue of heterostyly in *Narcissus* are the patterns of cross-compatibility found within and among members of different style-length morphs. In the four polymorphic species of *Narcissus* studied experimentally, individuals are self-sterile but cross-compatible with other members of the same style-length group. Hence, intramorph as well as intermorph crosses are fully interfertile, a pattern unexpected in heterostylous species that usually exhibit strong intramorph incompatibility. This situation was first found in trimorphic *N. triandrus* by Bateman (1952*a*) and has also been demonstrated in dimorphic *N. tazetta* by Dulberger (1964). We have confirmed the occurrence of intramorph compatibility in *N. triandrus,* as well as dimorphic *N. assoanus and N. papyraceus*. The absence of an association between floral heteromorphism and the incompatibility system responsible for self-sterility also occurs in distylous *Anchusa* of the Boraginaceae (Dulberger, 1970; Philipp and Schou, 1981; Schou and Philipp, 1983, 1984) and has implications for the evolution of heterostyly (see "Implications for Models of the Evolution of Heterostyly").

An interesting aspect of self-sterility in *Narcissus* concerns the underlying mechanism(s) responsible for reduced seed set upon self-pollination. Bateman (1954, p. 24) reported that "in *Narcissus* pollen tubes grow quite as fast on selfing as on crossing, for the full length of the style. The incompatibility must be late-acting, somewhere in the ovary, and perhaps even after fertilization." Dulberger (1964) found that self-sterility in *N. tazetta* was caused by a breakdown in ovule development after fertilization or at least after pollen tubes had grown through the micropyle and penetrated the embryo sac. Similar late-acting ovarian phenomena occur in *N. triandrus,* although in this species self-rejection apparently precedes fertilization (Sage et al., unpublished). These findings raise the issue of whether self-sterility in *Narcissus* results from a true late-acting self-incompatibility system (Seavey and Bawa, 1986; Waser and Price, 1991; Gibbs and Bianchi, 1993) or self-rejection is a manifestation of early-acting embryo abortion owing to inbreeding depression (Sorensen, 1969; Wiens et al., 1989; Krebs and Hancock, 1991; Husband and Schemske, 1995). Of course, both mechanisms could also influence patterns of seed set following self- and cross-pollination. Careful histological studies will be required to assess these possibilities (see Sage et al., 1994).

Although the mechanism(s) of self-sterility in *Narcissus* is not known, the late-acting rejection may bear important reproductive costs. Because pollen-tube growth rates of self- and outcross pollen are similar, seed set can be reduced significantly if self-pollen is deposited on stigmas before outcross pollen. This effect has been demonstrated experimentally in *N. tazetta* (Dulberger, 1964) and *N. triandrus* (Barrett et al., unpublished). In the latter species, application of self-pollen to stigmas 24 h prior to cross-pollen reduces seed set by 75% compared with outcrossed controls. As discussed below, the wastage of ovules that may arise because of prior self-pollination in *Narcissus* could have important ecological and evolutionary consequences, particularly with regard to the evolution of sexual polymorphisms in the genus.

Patterns of Style-Length Variation in Natural Populations of *Narcissus*

During 1990 and 1991, we collected population samples in Spain and Portugal of 11 *Narcissus* species distributed among seven sections of the genus to investigate the patterns of sex-organ variation within and between populations. In addition, we surveyed herbarium specimens, performed a literature search, and grew additional *Narcissus* taxa under glasshouse conditions in an effort to determine whether species of *Narcissus* not encountered in our field survey possessed sexual polymorphisms involving stigma and anther heights. Field studies involved random collection of a large sample (when possible) of flowering stems from each population. Detailed localities of all populations reported in this study are available from the senior author on request. Population samples were used for two purposes. First, measurements of a range of floral traits including style length and anther height were made on a subsample of flowers. Measurements were either conducted in the field with digital calipers, or the laboratory on preserved material fixed in 70% alcohol using a binocular microscope and digitizing tablet. When sexual polymorphisms were evident, the complete sample of stems was used to estimate floral morph frequencies in each population.

Based on our studies, sexual polymorphisms involving stigma height occur in four of the 10 sections of *Narcissus*. In three sections (Tazettae, Jonquillae, and Apodanthae), both monomorphic and dimorphic species occur. In *N. trian-*

drus, the sole member of section Ganymedes, most populations are trimorphic, but dimorphic populations also occur in some parts of the range. In the remaining six sections of *Narcissus,* no evidence of sexual polymorphisms was found. Although our surveys are not exhaustive and we have not observed all species in the field, we are reasonably confident that sexual polymorphisms are unlikely to be discovered in the monospecific sections (Tapeinanthus, Serotini, and Aurelia) or small sections Narcissus and Bulbocodium. Further work on the larger section Pseudonarcissus is desirable, particularly as the rare *N. cyclamineus* was reported by Henriques (1887) as possessing homostylous and herkogamous morphs. However, for reasons discussed more fully below, we believe the floral morphology of species in section Pseudonarcissus is unlikely to be associated with the evolution of sexual polymorphisms. It would therefore not be suprising to us if our findings on the distribution of polymorphisms for stigma and anther height among the four sections of *Narcissus* remain unchanged despite further exploration.

Monomorphic Populations

Figure 13.2 illustrates continuous variation of style length in four species of *Narcissus,* each from a different section of the genus. In these species and all others we have observed that are monomorphic for style length, stigmas are positioned at or above the two anther levels. Flowers in which stigmas are regularly held above the anthers are described as exhibiting *approach herkogamy* (Webb and Lloyd, 1986). In some species of *Narcissus,* stigma-anther separation is quite substantial (e.g., *N. bulbocodium,* 5–10 mm), in others (e.g., section Pseudonarcissus), it is quite small (<5 mm), and in yet others (e.g., *N. serotinus*), there is no separation between the sex organs and stigmas are positioned at the same height as the two anther levels. In most monomorphic species that we have observed, the two anther levels are positioned close together within the flower (e.g., *N. viridiflorus*) or are not distinguished by height (e.g., section Pseudonarcissus). Figure 13.3 illustrates the range of variation in sex-organ position in a population of *N. bulbocodium*. There is continuous variation in both stigma height and the degree of herkogamy. *Narcissus bulbocodium* is atypical for most species without sex-organ polymorphisms in possessing two anther levels that are differentiated by height, particularly during early anthesis.

Dimorphic Populations

Figure 13.4 illustrates patterns of style-length variation in four species of *Narcissus* from three sections of the genus. The observed bimodality of style length demonstrates that each species possesses a population-level polymorphism involving two morphs that differ in style length. In the long-styled morph, anthers are positioned below the stigma, whereas in the short-styled morph, they are above

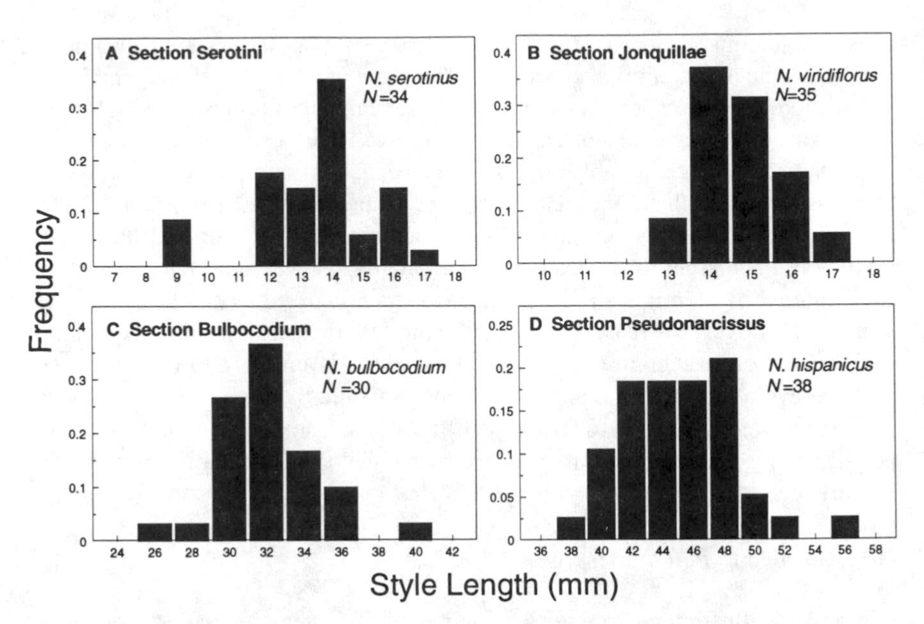

Figure 13.2. The distribution of style length in populations of four *Narcissus* species from different sections of the genus. Each species is monomorphic for style length. (A) *N. serotinus*. (B) *N. viridiflorus*. (C) *N. bulbocodium*. (D) *N. hispanicus*. *N* refers to the number of flowers sampled. In each population, a single flower was measured from plants sampled at random.

the stigma. The latter condition is referred to as *reverse herkogamy* (Webb and Lloyd, 1986). The occurrence of these two morphs within a population is characteristic of distylous species. Lloyd and Webb (1992*a*, p. 152) defined heterostyly in morphological terms as a "genetically determined polymorphism in which the morphs differ in the sequence of heights at which the anthers and stigmas are presented within their flowers." This description would appear to fit the dimorphic *Narcissus* populations described above. However, in elaborating on this definition, Lloyd and Webb indicated that for a species to be considered truly heterostylous, both anther and stigma heights should differ in a reciprocal manner between the morphs and they introduced the term *reciprocal herkogamy* to describe this condition.

Although the floral variation in dimorphic *Narcissus* species resembles heterostyly, in our opinion it does not fully meet the criteria required to define that particular floral polymorphism. This is because anther heights are not sufficiently differentiated in the two floral morphs and there is little correspondence in height between the lower stamens of the long-styled morph and stigmas of the short-styled morph (see Fig. 13.1B). Measurements of sex-organ position in six dimorphic populations are presented in Table 13.1. Although mean stigma heights are markedly differentiated in each morph, the positions of upper and lower stamens

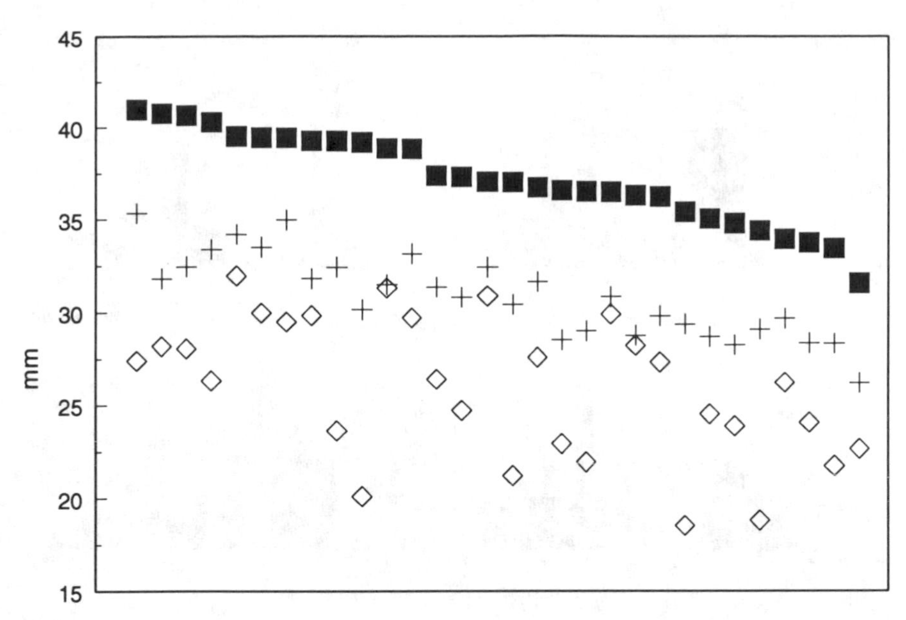

Figure 13.3. Variation in style length (■) and anther height (upper anthers +, lower anthers ◇) among individuals of *Narcissus bulbocodium* within a single population. Flowers are ranked by style length and each anther height measurement is the mean of the three anthers per level.

differ to only a limited degree. For example, in four of the six species examined, upper stamens in the two morphs are not significantly different in height. For lower stamens there are significant differences in all species, but the lower anthers of the short-styled morph are only a few millimetres higher in the floral tube than the corresponding anther level of the long-styled morph. Although these differences may be functionally important (see below), the clear lack of reciprocity in organ position probably disqualifies the dimorphic species of *Narcissus* from being considered truly distylous, particularly when a strictly morphological definition of the polymorphism is applied. Instead, we prefer to refer to the condition found in these species as involving a simple stigma-height dimorphism (also see Dulberger, 1964).

Considerable variation in the positions of stigmas and anthers occurs among individuals within dimorphic populations of *Narcissus* species. This is evident in Figure 13.5, which illustrates the range of variation in sex-organ position within a population of *N. assoanus*. Part of this variation is developmental in origin or associated with flower size differences among individuals; however, the basic dimorphism in stigma position is not obscured by these sources of variability. Among long-styled individuals, there is considerable variation in the degree of herkogamy. In most plants, the stigma is positioned above the two

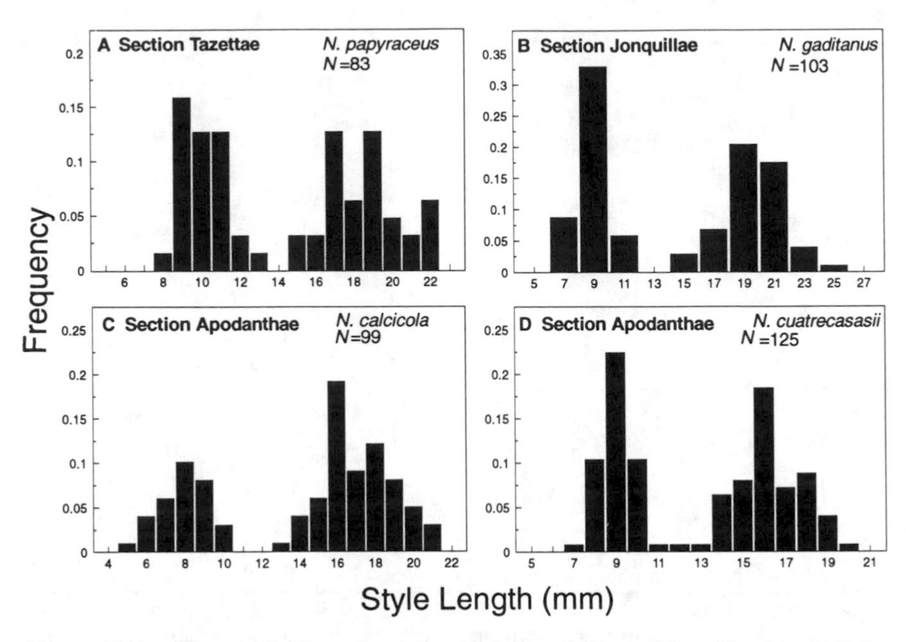

Figure 13.4. The distribution of style length in populations of four *Narcissus* species from three sections of the genus. Each population possesses a stigma-height dimorphism. (A) *N. papyraceus.* (B) *N. gaditanus.* (C) *N. calcicola.* (D) *N. cuatrecasasii. N* refers to the number of flowers sampled. In each population, a single flower was measured from plants sampled at random.

stamen levels; however, in some individuals the upper stamens are above the stigma and the lower stamens below it. This pattern was also evident in several other dimorphic species that were studied in detail (*N. calcicola, N. gaditanus, N. papyraceus,* and *N. rupicola*). Of particular functional significance was the finding that in all the dimorphic species studied, the average distance separating stigmas and anthers was considerably greater in the short-styled morph than the long-styled morph (Table 13.1). Averaged over the six species, the mean stigma-anther separation in the short-styled morph was approximately four times greater than in the long-styled morph. This consistent difference in the degree of herkogamy is likely to play an important role in influencing levels of self-pollination in the two sexual morphs. Elsewhere, there is good evidence that the distance separating stigmas and anthers has a strong effect on levels of self-pollination (e.g., Breese, 1959; Ennos, 1981; Thomson and Stratton, 1985; Barrett and Shore, 1987; Barrett and Husband, 1990; Murcia, 1990; Robertson and Lloyd, 1991).

Henriques (1887) reported long-, mid-, and short-styled morphs in *N. tazetta* of section Tazettae, and Fernandes (1964) "true trimorphic heterostyly" in several species in sections Jonquillae (*N. fernandesii*) and Apodanthae (*N. calcicola,*

Table 13.1. The positions of sexual organs in six species of Narcissus, *each displaying genetic polymorphism for stigma height. Values given are the mean and standard error (in parentheses) of stigma height, upper and lower stamen heights, and stigma-anther separation in a single population of each species. Sample sizes for the long- and short-styled morphs (L, S morph, respectively) were* N. papyraceus*: 33, 30;* N. assoanus*: 46, 51;* N. gaditanus*: 52, 51;* N. calcicola*: 81, 80;* N. cuatrecasasii*: 68, 54;* N. rupicola*: 25, 75. Stigma and anther heights were measured from the base of the style and stigma-anther separation was based on the closest anther to the stigma. Significant differences were assessed using t-tests.*

Species	Stigma Height			Upper Anther Height			Lower Anther Height			Stigma-Anther Separation		
Style Morph	L		S	L		S	L		S	L		S
Narcissus papyraceus	18.06	***	9.61	20.39	ns	20.47	15.78	**	16.82	0.03	***	3.24
	(0.279)		(0.293)	(0.297)		(0.311)	(0.259)		(0.272)	(0.185)		(0.194)
Narcissus assoanus	24.53	***	13.15	22.63	ns	23.15	18.06	***	19.71	0.84	***	4.91
	(0.322)		(0.305)	(0.345)		(0.327)	(0.307)		(0.292)	(0.237)		(0.225)
Narcissus gaditanus	18.62	***	8.10	15.95	ns	16.24	10.89	***	12.69	1.64	***	3.16
	(0.219)		(0.221)	(0.209)		(0.211)	(0.192)		(0.194)	(0.165)		(0.167)
Narcissus calcicola	17.84	***	8.00	15.11	**	15.97	10.43	***	13.17	1.20	***	3.47
	(0.157)		(0.158)	(0.195)		(0.196)	(0.18)		(0.182)	(0.144)		(0.145)
Narcissus cuatrecasasii	15.67	***	8.47	16.39	***	17.26	11.90	***	15.08	0.25	***	2.64
	(0.155)		(0.174)	(0.131)		(0.147)	(0.144)		(0.162)	(0.088)		(0.098)
Narcissus rupicola	15.91	***	9.21	20.63	ns	20.1	14.90	**	15.68	1.51	***	5.22
	(0.241)		(0.139)	(0.355)		(0.205)	(0.302)		(0.175)	(0.254)		(0.147)

ns = not significant.

**$p<0.01$.

***$p<0.001$.

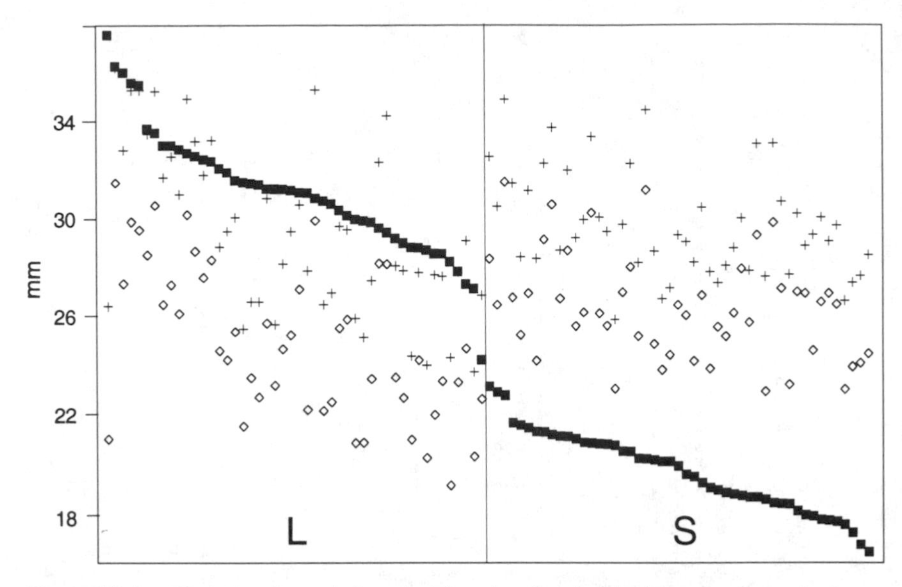

Figure 13.5. Variation in style length (■) and anther height (upper anthers +, lower anthers ◇) among individuals of *Narcissus assoanus* within a single population. Flowers are ranked by style length and each anther height measurement is the mean of the three anthers per level. L and S refer to the long- and short-styled morphs, respectively.

N. scaberulus, and *N. rupicola*). We suspect that this conclusion was based on observations of variation patterns similar to that illustrated in Figure 13.5. Individuals with stigmas positioned between the two stamen levels are likely to have been classified as mid-styled and the polymorphism interpreted as trimorphism. We believe this interpretation is erroneous and that only two sexual morphs are likely to occur in members of these three sections. The range of variation in stigma-anther position within the long-styled morph is continuous and the distance separating stamen levels does not warrant separation of a third class of plants classified as the mid-styled morph. Similarly, we see no merit in classifying individual plants with anthers and stigmas at similar heights as homostylous when they constitute part of the continuous variation displayed within one of the style-length morphs, as has been done for *N. tazetta* (see Yeo, 1975) and other polymorphic *Narcissus* species (Fernandes, 1965). Although this variation in sex-organ position may have important functional consequences for pollination, the use of these categories and the connotations from the heterostyly literature that they imply tend to obscure the fundamental dimorphism that occurs in populations of these species.

Floral Trimorphism

Narcissus triandrus appears to be the only species in the genus that displays a genuine genetic polymorphism for style length involving three discrete morphs

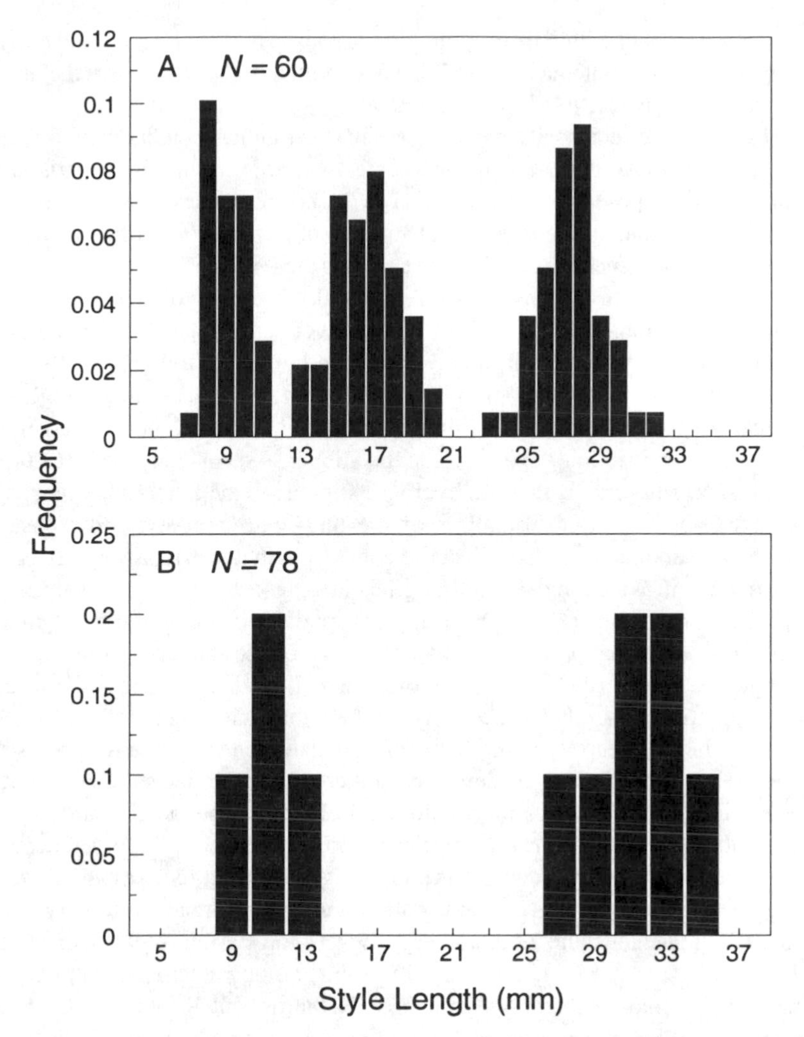

Figure 13.6. The distribution of style length in two populations of *Narcissus triandrus*. (A) A trimorphic population with long-, mid-, and short-styled plants (L, M, and S, respectively). (B) A dimorphic population with long- and short-styled plants (L and S, respectively). *N* refers to the number of flowers sampled. In each population, a single flower was measured from plants sampled at random.

(Fig. 13.6A). Most *N. triandrus* populations are comprised of long-, mid-, and short-styled individuals; however, dimorphic populations composed of long- and short-styled morphs (Fig. 13.6B) also occur in a restricted part of the geographical range of the species. In contrast to the dimorphic *Narcissus* species, in which stamen height differences between the morphs are slight, in *N. triandrus* three distinct stamen levels, which correspond to the positions of the three stigma heights, are evident within trimorphic populations (Figs. 13.1C and 13.7A). This finding contradicts Bateman's (1952*a*) assertion that in *N. triandrus* "the anthers are at the same two levels in all individuals."

Because of the reciprocity between stigma and anther heights in populations of *N. triandrus,* we believe that the species warrants recognition as tristylous, as originally proposed by Henriques (1887) and Fernandes (1935). However, the anther positions in the three floral morphs of *N. triandrus* differ from those of any tristylous species that has been described in the literature. Although mid- and short-styled morphs exhibit reciprocal positioning of anther and stigma heights, in the long-styled morph the "mid-level" anthers are positioned above stigmas of the mid-styled morph and correspond in height to long-level stamens of the mid- and short-styled morphs. As a result, average stamen heights are identical in the long- and mid-styled morphs and only the short-styled morph possesses a distinct mid-level stamen position. A second atypical feature of tristyly in *N. triandrus* is that, on average, stigmas in the long-styled morph are positioned above long-level anthers of the mid- and short-styled morphs. We believe these anomalies do not affect the conclusion that *N. triandrus* is tristylous; the sequence in which anthers and stigmas are presented within flowers is preserved, the sex organs occupy three distinct spatial location within each morph, and stigma-anther reciprocity is evident for two of the three organ levels.

Lloyd and Webb (1992*a*) suggested that the reciprocal herkogamy of heterostylous populations does not necessarily require that the heights of anthers and stigmas coincide precisely in the floral morphs, and indeed in many species they do not. However, these considerations indicate that the decision to classify a species like *N. triandrus* as morphologically heterostylous but exclude others, such as the dimorphic *Narcissus* species, is to some extent arbitrary, because it rests on judgments on the degree of reciprocity displayed by particular species. Although numerical indices can be used to quantify sex-organ reciprocity (Lloyd et al., 1990; Richards and Koptur, 1993; Eckert and Barrett, 1994), survey data indicate that wide variation in reciprocity occurs among heterostylous plants and those with stigma-height polymorphisms (Richards and Koptur, 1993; Eckert and Barrett, 1994; Richards et al., unpublished). This problem is reminiscent of early attempts to classify the quantitative variation in gender that exists in many flowering plants into discrete categories (see Lloyd, 1980; Barrett, 1992*b*).

In all dimorphic populations of *N. triandrus* that we have observed, the mid-styled morph is absent (Fig.13.7B). A similar pattern was also reported by Fernandes (1965). In some dimorphic populations of *N. triandrus,* mid-level stamens in the short-styled morph are positioned above the corresponding stamen level of this morph from trimorphic populations. This modification is similar to that described in populations of tristylous *Oxalis alpina* missing the mid-styled morph (Weller, 1992) and is probably a consequence of selection for efficient pollen transfer to stigmas of the long-styled morph. Differences between mid-level stamen position in the long-styled morph from trimorphic and dimorphic

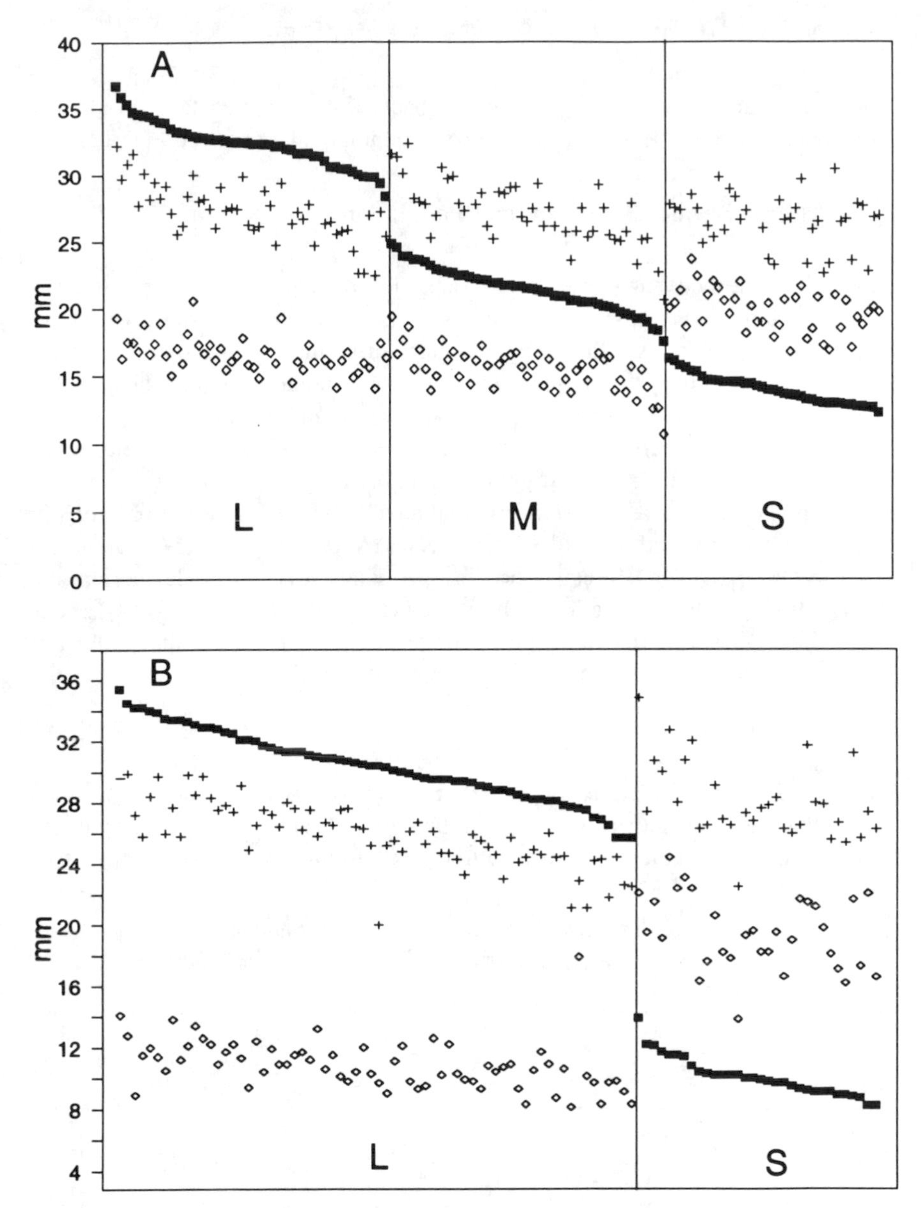

Figure 13.7. Variation in style length (■) and anther height (upper anthers +, lower anthers ◇) among individuals of *Narcissus triandrus* sampled from a trimorphic and dimorphic population. Flowers are ranked by style length and each anther height measurement is the mean of the three anthers per level. (A) In the trimorphic population, three distinct morphs are evident that differ in the sequence in which stigmas and anthers are presented within a flower. Note that unlike most tristylous species, anther heights are similar in the long-styled (L) and mid-styled (M) morphs. The short-styled (S) morph possesses distinct mid-level anthers corresponding in height to stigmas of the M morph. (B) In the dimorphic population, two distinct morphs (L and S) occur and the M morph is absent.

populations are less evident, presumably because this stamen level is already positioned above mid-level stigmas in trimorphic populations (see Fig. 13.1C).

Style-Morph Frequencies in Natural Populations

A considerable body of survey data exists on the frequencies of style morphs in populations of heterostylous plants (see Ganders, 1979; Barrett, 1993). In equilibrium populations with legitimate (intermorph) mating and no fitness differences among morphs, the frequencies of morphs should be equal for both the distylous and tristylous condition (Charlesworth and Charlesworth, 1979*b;* Heuch, 1979). This expectation is often observed, particularly in distylous populations. Because of the rarity of stigma-height polymorphisms in flowering plants, no comprehensive surveys have been conducted. To investigate morph ratios in *Narcissus,* we surveyed two widespread species (*N. papyraceus* and *N. triandrus*) extensively and several rarer species in less detail. Because style morphs are clearly distinguished morphologically, there is no ambiguity in classifying plants into different style-morph categories. Our results indicate an unusual pattern of style-morph structure that has not been reported elsewhere.

Dimorphic Populations

Surveys of nine populations of five dimorphic *Narcissus* species in sections Jonquillae and Apodanthae indicate a similar pattern (Table 13.2). In each population, the long-styled morph is most common with frequencies ranging from 0.61–

Table 13.2. Style-morph frequencies in nine dimorphic populations of five species of Narcissus *from Spain and Portugal. L. and S refer to the long- and short-styled morphs, respectively.*

Section Species		Morph Frequency L	Morph Frequency S	L/S Ratio	Number of Plants Sampled
Jonquillae					
N. assoanus	Pop. 1	0.76	0.24	3.17	287
	Pop. 2	0.62	0.38	1.63	145
N. gaditanus	Pop. 1	0.76	0.24	3.17	255
	Pop. 2	0.94	0.06	15.7	55
N. jonquilla	Pop. 1	0.85	0.15	5.67	26
Apodanthae					
N. calcicola	Pop. 1	0.68	0.32	2.13	355
	Pop. 2	0.61	0.39	1.56	307
N. cuatrecasasii	Pop. 1	0.77	0.23	3.35	248
	Pop. 2	0.72	0.28	2.57	202

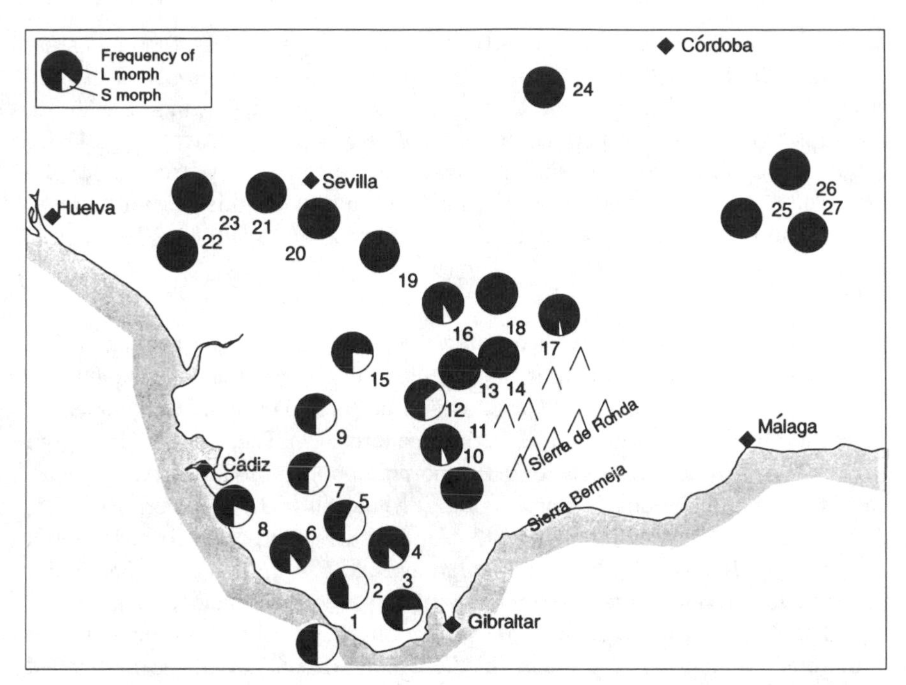

Figure 13.8. The frequencies of the long-styled (L) and short-styled (S) morphs of *Narcissus papyraceus* in a survey of 27 populations from southwest Spain. Populations occurring in the southern coastal portion of the region surveyed were dimorphic, whereas inland populations, particularly in mountainous areas, were more often monomorphic.

0.94. This pattern was also evident in the detailed survey of *N. papyraceus* populations from southern Spain (Fig. 13.8). Of 27 populations examined, 15 were dimorphic with the long-styled morph comprising 75% of all plants sampled (N = 1305 plants). All remaining populations contained long-styled plants only (N = 913 plants).

Monomorphic and dimorphic populations of *N. papyraceus* were nonrandomly distributed over the region surveyed. In the southern coastal area, populations were dimorphic with variable frequencies of the short-styled morph ranging from a slight excess in one population to deficiencies in most others. Populations further north, many of which occurred in more mountainous regions, were uniformly monomorphic. These patterns suggest that under certain ecological conditions the short-styled morph may be at a selective disadvantage. Elsewhere, Dulberger (1967) and Arroyo and Dafni (1995) found striking differences in style-morph ratios among dimorphic populations of the related *N. tazetta* from hill and marsh (lowland) habitats of Israel. In Dulberger's survey, morph ratios exhibited a slight excess of the short-styled morph in lowland populations growing on alluvial soils (mean frequencies: L morph = 0.45, S morph = 0.55; N = 6 populations, and N = 1814 plants). In contrast, hill populations occurring on rocky slopes displayed a predominance of the L morph (mean frequencies: L morph = 0.86, S morph = 0.14; N = 6 populations, and N = 711 plants). Arroyo and Dafni (1995) have suggested that pollinator-mediated selection may account for the observed patterns. It would seem worthwhile to investigate this possibility in *N. tazetta* and in *N. papyraceus* populations from southern Spain.

Trimorphic Populations

Narcissus triandrus is widely distributed throughout the Iberian peninsula. We surveyed 80 populations from four major geographical regions including southern, central, and Atlantic Spain and central and northern Portugal. Our survey of style-morph ratios revealed three distinct patterns: (1) The long-styled morph predominated in both trimorphic and dimorphic populations (mean frequencies in trimorphic populations: L morph = 0.57, M morph = 0.22, S morph = 0.21; N = 68 populations, and N = 14,515 plants; dimorphic populations: L morph = 0.71, S morph = 0.29; N = 12 populations, and N = 1535 plants). Fernandes (1935, 1965) found a similar pattern involving the predominance of the long-styled morph in a small sample of populations from central Portugal. (2) In trimorphic populations, a decrease in the average frequency of the mid-styled morph among the different regions was associated with an increase in the long-styled morph. In contrast, the average frequency of the short-styled morph was remarkably similar throughout the range (Table 13.3). (3) All dimorphic populations were concentrated in central and northern Portugal and Atlantic Spain. In these regions, the average frequency of the mid-styled morph in trimorphic populations was low in comparison with its frequency in central and southern Spain. Hence, as in dimorphic species of *Narcissus,* the long-styled morph in *N. triandrus* has a fitness advantage over the short-styled morph. In addition, in the northern part of the range of *N. triandrus,* the ecological conditions are apparently often unsuitable for the maintenance of floral trimorphism. When this

Table 13.3. Style-morph frequencies (standard deviation in parentheses) in 68 trimorphic populations of Narcissus triandrus *sampled from different regions of Spain and Portugal. L, M, and S refer to the long-, mid-, and short-styled morphs, respectively.*

Region	Morph Frequency			Number of Populations Sampled	Number of Plants Sampled
	L	M	S		
Southern Spain	0.45 (0.10)	0.35 (0.09)	0.20 (0.04)	10	2962
Central Spain	0.46 (0.13)	0.32 (0.11)	0.22 (0.05)	18	5536
Central Portugal	0.57 (0.11)	0.21 (0.10)	0.22 (0.06)	17	3512
Northern Spain and Portugal	0.70 (0.08)	0.09 (0.07)	0.22 (0.06)	23	2505

occurs, it is always the mid-styled morph that is lower in frequency or missing from populations.

Field studies of the floral biology of the morphs are required to determine the selective mechanisms responsible for the particular morph ratios found in *N. triandrus* and other species of *Narcissus*. Floral morphology and incompatibility in *Narcissus* species are not associated, hence, it is not unexpected that equilibrium morph ratios are quite different from those found in heterostylous species with diallelic incompatibility. Because all cross-pollinations in *Narcissus* are compatible, morph ratios are likely to be governed by the genetic system controlling floral polymorphism and the relative fitness of the morphs as male and female parents. Interestingly, in dimorphic *Anchusa officinalis* in which style length and incompatibility are apparently also uncoupled, population surveys indicate that the frequency of the long-styled morph also greatly exceeds that of the short-styled morph (Philipp and Schou, 1981).

Bateman (1968) proposed pollinator-mediated selection as a mechanism to account for variation in the frequencies of style morphs among populations of *N. triandrus*. He suggested that the style morphs may be adapted to different pollinators that vary in their abundance among populations. He also suggested that monomorphic species such as *N. bulbocodium* and *N. pseudonarcissus* were adapted to more "regular" pollinators and styles in these species had therefore been selected for an optimum length. Although it seems improbable that the style morphs represent "different ecotypes adapted to different pollination conditions" as Bateman (1968, p. 646) suggested, we believe that information on the interactions between pollinators and the floral morphs is likely to be the key to understanding morph ratios in *Narcissus* populations. In this regard, it is worth noting that our preliminary observations of pollinators visiting *N. triandrus* in different regions of the Iberian peninsula revealed a striking pattern associated with the distribution of trimorphic vs. dimorphic populations. Throughout much of the central and southern range of the species occupied exclusively by trimorphic populations, the most common floral visitor was the bee *Anthophora pilipes* Fabr. (=*A. acervorum* L.). In contrast, in the cooler Atlantic zone where dimorphic populations are primarily located, this bee was not observed and, instead, bumble bees (*Bombus* spp.) were the main floral visitors. It is possible that species of *Bombus* are less effective at mediating segregated pollen transfer among the three morphs and this may account for the apparent selective disadvantage to the mid-styled morph at the northern edge of the range.

Evolutionary Considerations

The inter- and intraspecific patterns of style-morph variation in *Narcissus* described above raise many questions concerning the evolutionary origins and maintenance of the polymorphisms. For example, what are the evolutionary relationships between stylar monomorphism, dimorphism, and trimorphism and do these represent stages in the assembly of tristyly in the genus? What selective forces are responsible for the evolution of stigma-height polymorphisms and why does the long-styled morph predominate in most populations? What evolutionary constraints might account for the rarity of heterostyly in *Narcissus* compared with the more common occurrence of stigma-height dimorphism in the genus? To clarify some of these issue, we next develop several hypotheses and a quantitative selection model of the evolution of sexual polymorphisms in *Narcissus* to guide future studies aimed at answering these questions.

Evolutionary Relationships Among Stylar Conditions

The occurrence of stylar monomorphism, dimorphism, and trimorphism among taxa of *Narcissus* is markedly uneven. Stylar monomorphism represents the most widespread condition, occurring in most species and sections, dimorphism is restricted to four sections, and trimorphism is apparently limited to the single species (*N. triandrus*) of section Ganymedes. Although phylogenetic analysis is required to determine the evolutionary relationships of taxa and polarity of stylar conditions, we assume as a working hypothesis that monomorphism represents the ancestral stylar condition in the genus and stylar polymorphisms have evolved from this state. Whether stylar trimorphism is derived from stylar dimorphism or directly from a monomorphic condition is not known. However, we consider the former to be more likely. We take this position because we find it difficult to imagine that a complex polymorphism like tristyly would not evolve via an intermediate condition involving two morphs. This was also the view taken by Charlesworth (1979) in her theoretical model of the evolution of tristyly.

The occurrence of stigma-height polymorphisms in a genus that contains heterostyly is relevant to recent models of the evolution of heterostyly developed by Lloyd and Webb (1992 *a, b*). In their models, they considered establishment of a stigma-height polymorphism to be the first stage in the evolution of reciprocal herkogamy. Stylar dimorphisms in *Narcissus* therefore take on added significance because they may represent an intermediate stage in the evolution of tristyly in section Ganymedes. It is not clear how similar the stylar dimorphisms that occur in extant members of the genus are to putative transitional stages that gave rise to tristyly in *N. triandrus*. Furthermore, the postulated sequence from monomorphism via dimorphism to trimorphism does not preclude the possibility that secondary reversions from dimorphism to monomorphism or from trimorphism to dimorphism have occurred through the loss of style morphs from populations. Indeed, it seems likely that reversions of this type might account for some of the monomorphic and dimorphic populations that occur in *N. papyraceus* and *N. triandrus,* respectively.

If we assume that stylar monomorphism is the ancestral state in *Narcissus,* it is of interest to consider what other floral characters occurred in the ancestors of species that now possess stylar polymorphisms. We consider three conditions particularly likely: wide perianths, approach herkogamy, and self-sterility.

Most *Narcissus* species with stylar monomorphism are characterized by either bowl-shaped or broadly tubular flowers. Pollinators visiting these species frequently enter the flower and contact anthers in an imprecise manner, rather than probing from the mouth of the tube. It seems likely that flowers of this type would not effectively promote reciprocal pollen transfer based on segregated contact of stigmas and anthers (see Lloyd and Webb, 1992*b,* Fig. 13.1) Accordingly, stylar polymorphisms in *Narcissus* are probably not associated with these types of floral shapes. In contrast, all species with stylar polymorphisms possess well-developed floral tubes that force pollinators into stereotypical positions while probing for nectar. This raises the possibility that a change from a relatively unspecialized pollination syndrome to one with greater precision, involving constricted floral tubes, long-tongued pollinators, and segregated contact of reproductive organs, may have provided the stimulus for the evolution of sexual polymorphisms in *Narcissus*. In this context, it is worth noting that among heterostylous families the types of flowers that have become heterostylous are predominantly those that possess prominent floral tubes. This is presumably because tube formation is advantageous in positioning insect mouth parts for segregated contact and efficient cross-pollination (Ganders, 1979; Lloyd and Webb, 1992*a*).

Both approach herkogamy and self-sterility are common in *Narcissus* species and occur in most of the sections for which information is available. It would seem unlikely given the rarity of their alternate states, namely, self-compatibility and reverse herkogamy, that the evolution of stylar polymorphisms involved species with these latter features. These arguments lead to the hypothesis that the reproductive systems of the ancestors of *Narcissus* taxa that currently possess stylar polymorphisms were largely outcrossing through self-sterility and possessed a floral morphology involving approach herkogamy and two stamen levels. This morphological phenotype was used as the ancestral state in Charlesworth's (1979) model of the evolution of tristyly. Lloyd and Webb (1992*a*) also concluded, based on comparative evidence, that the ancestors of heterostylous species were most likely to possess approach herkogamy. Following this view, the first step in the evolution of sexual polymorphisms was the invasion of an approach herkogamous population by a reverse herkogamous variant. Since the population may have been largely outcrossing, it seems likely that the selective forces responsible for the evolution of stylar polymorphisms in *Narcissus* did not involve inbreeding avoidance, but rather concerned aspects of pollen dispersal and possibly also pollen-stigma interference. To explore whether these features of floral biology may have been important, we next develop quantitative phenotypic selection models of this invasion process, in an effort to account for the evolution of stable stigma-height polymorphisms in *Narcissus*.

Selection of a Stigma-Height Dimorphism

As part of their theoretical work on the evolution of distyly, Lloyd and Webb (1992*b*) developed a phenotypic selection model for the evolution of a stable stigma-height polymorphism. Here we modify this model to take into account particular features of the floral biology of *Narcissus*. Specifically, we investigate to what extent the morph-specific differences in floral morphology, interacting with the particular type of late-acting ovarian self-sterility that occurs in *Narcissus,* may assist the invasion of herkogamous variants into populations monomorphic for style length. Recall that prior self-pollination of outcrossed *Narcissus* flowers can result in a significant reduction of seed set. This arises because the growth rates of self- and outcross pollen tubes are similar, and early arrival of self-pollen renders ovules that would otherwise be fertilized by outcross pollen tubes nonfunctional in some way. Hence, self-pollination can have significant reproductive consequences, depending on the arrival schedule of self- and outcross pollen. We refer to this loss of female fertility as *ovule discounting*. This differs from *seed discounting* (Lloyd, 1992), in which an increase in the amount of self-fertilized seed is accompanied by a concomitant decrease in some or all of the cross-fertilized seed. Because stigma-anther separation is much greater in the reverse herkogamous morph compared with the approach herkogamous morph (Table 13.1), we explore whether differences in ovule discounting between the morphs resulting from differential self-pollination favor the initial invasion of a reverse herkogamous morph. In addition, we examine whether features of the floral biology of *Narcissus* could result in the observed predominance of the long-styled morph (hereafter approach herkogamous morph) relative to the short-styled morph (hereafter reverse herkogamous morph).

The model describes a dimorphic population containing a fraction a of approach herkogamous plants and a complementary fraction, $r = 1 - a$, of reverse herkogamous plants. Both morphs are assumed to be identical, except in style length, with each plant producing f flowers with o ovules each. We additionaly assume that pollinators visit morphs randomly and pollen is transported only to the next plant visited by a pollinator. The average number of pollen grains transported between two plants (per flower) is q_{ij}, where i indicates the style morph of the donor plant and j identifies the morph of the recipient plant (e.g., q_{aa} . . . , whereas q_{ra} . . .) Given these morph frequencies and pollen-transfer proficiencies, the average approach herkogamous plant imports $f(aq_{aa} + rq_{ra})$ pollen grains and exports $f(aq_{aa} + rq_{ar})$ grains. Correspondingly, the average reverse herkogamous plant imports $f(aq_{ar} + rq_{rr})$ pollen grains and exports $f(aq_{ra} + rq_{rr})$ grains. For both morphs, we assume that the intensity of pollen dispersal is not sufficient to fertilize all available ovules.

Postpollination processes now act to determine seed production. Due to various causes (e.g., inviability and germination failure) only a fraction k of outcross pollen grains produce pollen tubes that successfully enter the ovary. We assume that a pollen tube growing toward a particular ovule cannot alter its course if another pollen tube has already entered the micropyle. Consequently, not all the pollen received by a plant fertilizes ovules, even though seed production is pollen-limited. Of particular interest is the reduced ovule availablity that results when a self-pollen tube disables an ovule. Because of ovule discounting, the average proportion of ovules available after self-pollination is v_a for approach plants and v_r for reverse plants.

The average fitness of plants of a given morph depends on the sum of female and male fertilities. For the approach herkogamous morph, female fertility is the number of ovules available per plant (fov_a) times the average pollen receipt ($aq_{aa} + rq_{ra}$) times the probability that a pollen grain received fertilizes an ovule (k), or $fov_a k(aq_{aa} + rq_{ra})$. The corresponding male fertility is the sum of siring success on other approach herkogamous flowers ($afov_a kq_{aa}$) and reverse herkogamous flowers ($rfov_r kq_{ar}$), or $fok(av_a q_{aa} + rv_r q_{ar})$. Hence, the total fitness of approach herkogamous plants is therefore:

$$w_a = fok[v_a(aq_{aa} + rq_{ra}) + av_a q_{aa} + rv_r q_{ar}]$$

Similarly, the total fitness for reverse herkogamous plants is:

$$w_r = fok[v_r(aq_{ar} + rq_{rr}) + av_a q_{ra} + rv_r q_{rr}]$$

The fitness advantage of reverse herkogamous plants is thus

$$w_r - w_a = fok\{a[q_{ar}v_r + v_a(q_{ra} - 2q_{aa})] - r[q_{ra}v_a + v_r(q_{ar} - 2q_{rr})]\}$$

Now consider whether the reverse herkogamous morph could invade a population of approach herkogamous plants (i.e., when $a \cong 1$ and $r \cong 0$). Invasion occurs when $w_r - w_a > 0$, or $q_{ar}v_r + v_a(q_{ra} - 2q_{aa}) > 0$, or alternatively,

Condition 1

$$\frac{q_{ar}\left(\frac{v_r}{v_a}\right) + q_{ra}}{2} > q_{aa}$$

If relative ovule availability is equal for the two morphs (i.e., $v_a = v_r$), this condition specifies that the average proficiency of pollen transfer between morphs must exceed the transfer proficiency among approach herkogamous plants (Eq. 5a of Lloyd and Webb, 1992*b*). Between-morph differences in ovule discounting modify this condition, so that invasion by the reverse herkogamous morph is more likely if approach herkogamous flowers lose relatively more ovules as a result of self-pollination (i.e., $v_a < v_r$). The approach herkogamous morph can invade a population of reverse herkogamous plants (i.e., when $a \cong 0$ and $r \cong 1$) if $w_r - w_a < 0$, or $-q_{ra}v_a - v_r(q_{ar} - 2q_{rr}) < 0$, or alternatively,

Condition 2

$$\frac{q_{ar} + q_{ra}\left(\frac{v_a}{v_r}\right)}{2} > q_{rr}$$

As with condition 1, the weighted average proficiency of pollen transfer between morphs must exceed transfer among reverse herkogamous plants; however, in this case greater ovule discounting by the approach herkogamous morph renders invasion less likely. Once one morph invades a population of the other morph, a stable stigma-height dimorphism persists only if both conditions 1 and 2 are satisfied.

Two questions now arise:

1. Are between-morph differences in ovule discounting sufficient to allow persistence of a dimorphic population? If stigma position affects only ovule discounting and not pollen dispersal, then $q_{ar} = q_{ra} = q_{aa} = q_{rr}$. Now if, for example, approach herkogamous flowers suffer more ovule discounting than reverse herkogamous flowers (i.e., $v_a < v_r$) so that $v_r/v_a > 1$, condition 1 holds, but condition 2 does not. As a result, the reverse herkogamous morph could not only invade a population monomorphic for the approach herkogamous condition, but it should spread to fixation. Hence, based on this model, a balanced dimorphism *cannot* be maintained solely by between-morph differences in ovule discounting, whereas differences in pollen-transfer probabilities are both necessary and sufficient.

2. If a balanced dimorphism develops, what is the equilibrium morph ratio (a^*/r^*)? At equilibrium $w_r - w_a = 0$ and the expected morph ratio is therefore

$$\frac{a^*}{r^*} = \frac{q_{ra}v_a + v_r(q_{ar} - 2q_{rr})}{q_{ar}v_r + v_a(q_{ra} - 2q_{aa})}$$

For *Narcissus,* we expect the proficiency of pollen transfer onto reverse herkogamous flowers to be independent of the donor morph (i.e., $q_{ar} = q_{rr} = q_{\cdot r}$). Given this simplification, the equilibrium frequency of approach herkogamous plants is

$$a^* = \frac{q_{ra}v_a - q_{\cdot r}v_r}{2v_a(q_{ra} - q_{aa})}$$

According to this expression, approach herkogamous plants will tend to predominate at equilibrium if (1) pollen transfer from reverse to approach herkogamous plants is more proficient than transfer among approach herkogamous plants (i.e., $q_{ra} > q_{aa}$ or $q_{ra}/q_{aa} > 1$). (2) Pollen transfer onto reverse herkogamous plants is less proficient than among approach herkogamous plants (i.e., $q_{aa} > q_{\cdot r}$ or $q_r/q_{aa} < 1$). (3) The ratio of ovule availability on reverse herkogamous plants relative to approach herkogamous plants is less than the ratio of pollen transfer among approach herkogamous plants to transfer onto reverse herkogamous plants (i.e., $v_r/v_a < q_{aa}/q_{\cdot r}$).

The first of these conditions could occur in *Narcissus* because of two floral mechanisms. In the reverse herkogamous morph, both absence of stylar interference with pollen dispatch by pollinators or the greater average height of anthers in this morph (Table 13.1) may result in more proficient cross-pollination to the approach herkogamous morph. The second condition involves pollen dispersal onto reverse herkogamous plants. Because of the lack of reciprocity between stigmas of reverse herkogamous plants and anthers in either morph, it seems likely that pollen transfer onto reverse herkogamous plants may be less proficient than that onto approach herkogamous plants. This suggestion was first proposed for *Narcissus* by Yeo (1975). In contrast to the first two conditions, the third requirement may not apply for *Narcissus* because the pattern of ovule discounting, if indeed it occurs, will likely be greater on approach herkogamous flowers and hence v_r/v_a will be >1, whereas if condition 1 is true, $q_{aa}/q_{\cdot r}$ will be <1.

The effects of different combinations of pollen transfer probabilities and ovule discounting on equilibrium morph ratios are illustrated in Figure 13.9. Over a broad range of conditions, the approach herkogamous morph predominates, particularly when pollen transfer between approach herkogamous plants is more proficient than that onto reverse herkogamous plants. A simple numerical example serves to illustrate these effects. If pollen transfer from reverse to approach herkogamous flowers is 50% more proficient than among approach flowers (i.e., $q_{ra} = 1.5q_{aa}$) and transfer onto reverse herkogamous flowers is 40% less proficient (i.e., $q_{\cdot r} = 0.6q_{aa}$), then the ratio of approach to reverse herkogamous morphs would be 0.90:0.10, in the absence of ovule discounting (i.e., $v_a = v_r = 1$). However, if the approach herkogamous morph has 20% fewer ovules available for outcrossing owing to ovule discounting (i.e., $v_a = 0.8v_r$), then the same transfer probabilities result in an equilibrium morph ratio of 0.75:0.25. Thus, although the maintenance of a stable stigma-height polymorphism requires frequency-dependent selection resulting from the promotion of intermorph pollen transfer, the potential occurrence of less ovule discounting in the reverse herkogamous morph of *Narcissus* removes the necessity for both morphs to have a promotion advantage. In addition, it reduces the minimum size of such an advantage, especially during the early stages of the establishment of the polymorphism.

Constraints on the Evolution of Heterostyly

The first step in Lloyd and Webb's (1992*b*) model for the evolution of distyly is the establishment of a stigma-height polymorphism. This type of polymorphism

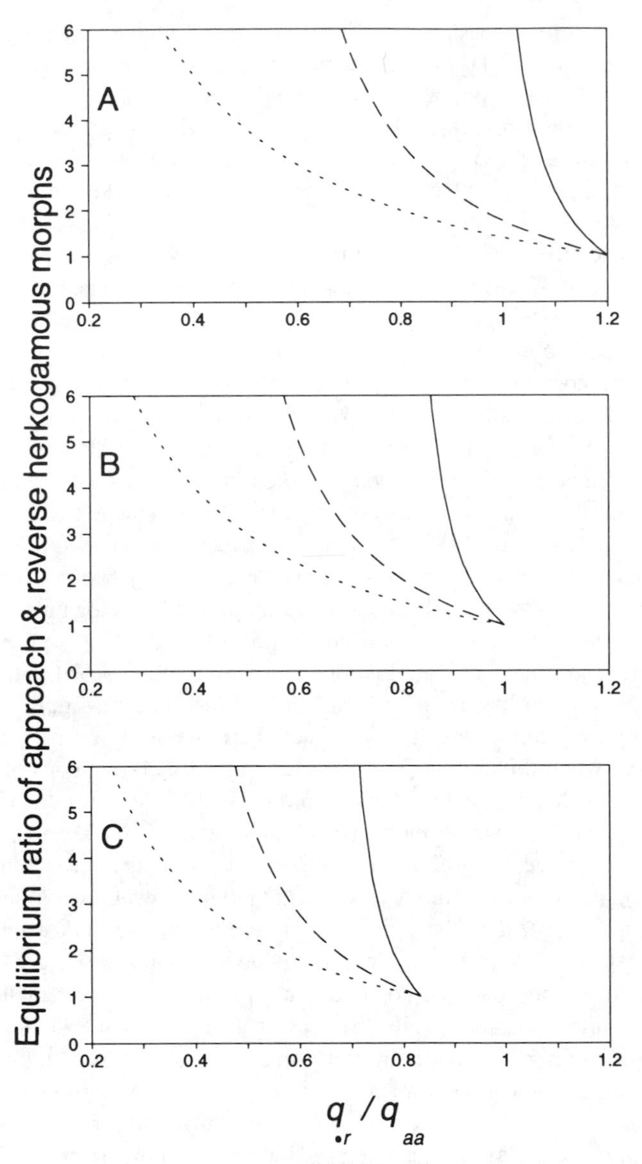

Figure 13.9. Examples of combinations of pollen-transfer probabilities and ovule discounting that result in an excess of approach herkogamous plants (L morph) compared with reverse herkogamous plants (S morph) at equilibrium in populations with a stigma-height dimorphism. The three panels illustrate the effects of different ratios of ovule availability on reverse vs. approach herkogamous plants: (A) $v_r/v_a=0.833$; (B) $v_r/v_a=1$; (C) $v_r/v_a=1.2$. Within each panel, the three curves illustrate equilibria for different values of q_{ra}/q_{aa} (solid curve, $q_{ra}/q_{aa}=1.2$; dashed curve, $q_{ra}/q_{aa}=1.6$; dotted curve, $q_{ra}/q_{aa}=2$).

is reported from only a handful of angiosperm species (e.g., *Chlorogalum angustifolium:* Jernstedt, 1982; *Epacris impressa:* O'Brien and Calder, 1989), and there is no evidence that this condition occurs commonly in heterostylous families. The rarity of stigma-height polymorphisms in heterostylous families led Lloyd and Webb (1992*a*) to suggest that this stage may be passed through quite rapidly during the evolution of heterostyly. In contrast, in *Narcissus* stigma-height polymorphisms are well established, occurring in at least a dozen or so species in four sections of the genus. On the other hand, heterostyly is very rare in *Narcissus*. If we assume that the postulated stages in the Lloyd and Webb model are correct, then the patterns in *Narcissus* indicate that there must be severe constraints preventing the transition from a stigma-height polymorphism to reciprocal herkogamy in the genus. It is not clear what constraints are involved, but they are likely to be associated with the morphology and pollination biology of *Narcissus* flowers.

The evolution of reciprocal herkogamy in groups with a stigma-height polymorphism requires selective modification of anther heights. Such modifications can be brought about by changes in flower size, alterations of filament attachment sites on the perianth, or elongation of the filaments themselves. Examples of such changes to anther position in heterostylous species are discussed in Richards and Barrett (1992) and Richards and Koptur (1993). Selection reducing anther height in the approach herkogamous morph of *Narcissus* may be developmentally constrained because of either packing restrictions associated with the very narrow diameter of the perianth tube in dimorphic species, or difficulties in the alteration of filament attachment sites. Shortening of filaments is unlikely because in dimorphic species these are already of limited length. Selection on the position of upper-level anthers may also be restricted because of the small coronas typical of species from dimorphic sections. However, because long-level anthers and stigmas of the long-styled morph are at similar heights, this alteration is likely to be of less selective importance. It is perhaps no coincidence that the floral morphology of tristylous *N. triandrus* is quite distinct from species in sections of the genus with stigma-height dimorphisms. Flowers of this species exhibit large tubular coronas and wider perianth tubes that permit a longer vertical area for the positioning of long-, mid-, and short-level stamens. The stigma-height polymorphisms in *Narcissus* may be evolutionarily stable because developmental aspects of floral architecture constrain selection of the distinct anther-height polymorphism that typifies most heterostylous plants.

In dimorphic *Narcissus* populations, there may be a limited zone within flowers for deployment of the two anther positions for optimal pollen dispatch by pollinators. A lowering of anther height in the approach herkogamous morph increasing reciprocity with short-level stigmas might impair effective pollen dispersal from this morph. Data collected by Harder and Barrett (1993) on pollen removal by long-tongued bees from the three anther levels in tristylous *Pontederia cordata* are relevant to these ideas. They found that removal of pollen from short-level anthers was highly unpredictable in comparison with removal from mid- and long-level anthers and depended on the extent to which the deeply recessed anthers contacted insect mouth parts. Anthers positioned within the opening of the perianth tube were considered to be in the most beneficial location for pollen dispersal, in part because anthers in this position contact pollinators more consistently than more inserted or exserted anthers. According to this hypothesis, anther position should be selected to maximize pollen dispersal and the position(s) occupied then determines the location(s) of the stigma to reduce self-pollination and self-interference. The occurrence of stigma-height dimorphism is an interesting floral condition in this regard. The presence of two stigma heights but only one anther height implies that the benefits associated with the avoidance of self-pollination in the reverse herkogamous morph owing to greater herkogamy may outweigh any losses in fitness incurred from a lack of precision in pollen dispersal to short-level stigmas.

Implications for Models of the Evolution of Heterostyly

Despite the isolated occurrence of heterostyly in *Narcissus,* its presence in the genus has an important bearing on models for the evolution of heterostyly. Until recently, most workers considered that reciprocal herkogamy in heterostylous plants evolved *after* the origin of diallelic incompatibility (Bateman, 1952*b*; Baker, 1966; Yeo, 1975; Ganders, 1979; Lewis and Jones, 1992). This was based on the idea that the style-stamen polymorphism primarily reduces the large wastage of incompatible pollen that would occur in a monomorphic population with only two or three incompatibility groups. Hence, diallelic incompatibility is generally considered a prerequisite for the evolution of reciprocal herkogamy, and indeed quantitative models of the evolution of distyly by Charlesworth and Charlesworth (1979*b*) require its establishment for the evolution of stamen and style polymorphisms. In these models, the primary selective force responsible for the evolution of diallelic incompatibility in heterostylous populations is inbreeding depression.

Particular anomalies for this interpretation of the evolution of heterostyly are taxa such as *Narcissus* and *Anchusa* in which heterostyly is associated with a self-sterility system in which all cross-pollinations are fertile. In these taxa, the Darwinian concepts of legitimate and illegitimate pollination do not apply, and selection of reciprocal herkogamy requires alternative explanations. These difficulties are to some extent reconciled by the Lloyd and Webb models (1992*a*, *b*), which put greater emphasis on the pollination biology of the morphs than the genetic consequences of particular mating patterns. They propose that reciprocal herkogamy is selected because it promotes efficient cross-pollination among plants rather than mating types. The selective forces in these models are independent of the type of compatibility system present in the ancestral population because inbreeding avoidance is considered unimportant for the evolution of reciprocal herkogamy. Hence, reciprocal herkogamy could evolve prior to the establishment of the incompatibility system in groups with diallelic incompatibility. This perspective differs from earlier interpretations that largely viewed heterostyly as an antiselfing mechanism, rather than a mechanical device that increases male fitness through active promotion of proficient cross-pollination (but see Kohn and Barrett, 1992*b;* Stone and Thomson, 1994). Although it seems likely that cross-promotion played the principal role in the evolution and maintenance of reciprocal herkogamy in *N. triandrus,* the reproductive costs associated with ovule discounting owing to self-pollination may have also been important, particularly during establishment of stigma-height polymorphisms in the genus.

Future Studies

Our preliminary survey of the floral biology of *Narcissus* species has exposed considerable diversity in floral traits associated with their pollination biology and sexual systems. Our studies confirm earlier reports of the occurrence of tristyly in *Narcissus* and clarify the nature of other stylar polymorphisms in the genus. Three fundamentally distinct patterns of stylar variation involving monomorphism, dimorphism, and trimorphism occur. This variation offers outstanding opportunities to investigate the adaptive significance of reproductive characters, particularly in species such as *N. papyraceus* and *N. triandrus,* where intraspecific variation is geographically structured and possibly correlated with ecological factors that could influence the pollination biology of populations. Field studies are required to determine whether pollinator-mediated selection on pollen dispersal can account for the transitions from one stylar condition to another in *Narcissus*.

The models we developed to account for the evolution of stylar polymorphisms in *Narcissus* incorporated various features of floral biology that require experimental verification. Our arguments concerning the likelihood that particular pollen transfers are more proficient than others (e.g., $q_{ra} > q_{aa} > q_{ar} = q_{rr}$) were based in part on evidence from other heterostylous plants. In taxa with conspicuous pollen-size heteromorphism, studies of pollen loads of naturally pollinated stigmas have demonstrated that the transport of pollen between anthers and stigmas of similar heights is considerably more proficient than between those at different heights (Ganders, 1974; Barrett and Glover, 1985; Piper and Charlesworth, 1986; Lloyd and Webb, 1992*b*). In light of these findings, we hypothesized that the transfer of pollen onto approach herkogamous flowers may be more proficient than that onto reverse herkogamous flowers, because anthers correspond more closely in height to the stigmas of the approach herkogamous morph. Unfortunately, polymorphic *Narcissus* species lack pollen-size heteromorphism (Barrett, unpublished), precluding opportunities to use these techniques to investigate the dynamics of pollen transfer. Alternative experimental approaches will therefore be required to verify the specific pollen-transfer relationships proposed for *Narcissus*.

The proposition that the pollen transfer $q_{ra} > q_{aa}$ was based, in part, on the assumption that stylar interference reduces the male fertility of approach herkogamous plants because of the close proximity of sex organs in the constricted floral tubes of dimorphic species (see Fig. 13.1B). According to this hypothesis, the placement of styles well below the anthers in reverse herkogamous flowers relieves this constraint and pollen dispatch by pollinators is therefore more proficient. There are virtually no data on this form of pollen-stigma interference to evaluate this possibility. In the only experimental study that has investigated whether stylar interference reduces male fertility, Kohn and Barrett (1992*a*) found no significant differences in the siring ability of long-styled plants of tristylous *Eichhornia paniculata,* with intact vs. excised styles. However, because of the close proximity of anthers and styles in dimorphic *Narcissus* species, we believe that stylar interference is likely to be considerably more important in comparison to *E. paniculata,* which possesses strongly herkogamous flowers and a flaring perianth tube. Experimental studies involving floral manipulations and estimates of male reproductive success are needed to determine the evolutionary significance of stylar interference.

The proximate mechanisms and ecological consequences of late-acting self-sterility in *Narcissus* species merit detailed consideration. Late-acting self-sterility systems appear to be much more widespread than previously appreciated (reviewed in Seavey and Bawa, 1986; Sage et al., 1994), but as yet there has been little discussion of their evolution and functional significance. To explore the implications of the ovule wastage that is frequently associated with late-acting systems (e.g., Cope, 1962; Dulberger, 1964; Crowe, 1971; Yeo, 1975; Waser and Price, 1991; Broyles and Wyatt, 1993), we introduced the concept of ovule discounting to describe the situation where ovules are excluded from cross-fertilization because they are rendered nonfunctional by self-pollen tubes as a consequence of prior self-pollination. Like pollen discounting (see Holsinger et al., 1984; Ritland, 1991; Kohn and Barrett, 1994; also see Chapter 6), ovule discounting has the potential to exert significant reproductive costs, particularly for species in which flowers receive abundant self-pollen.

In *Narcissus* the approximately 4-fold difference between the approach and reverse herkogamous morphs in the distance separating stigmas and anthers is likely to strongly affect levels of self-pollination through either autonomous or pollinator-mediated influences (see Dulberger, 1964; Yeo, 1975). Indeed, field observations of dimorphic species frequently revealed considerable self-pollen on stigmas of long-styled plants because of the close proximity of reproductive organs in this morph. Such effects are likely to be particularly pronounced in polymorphic species with horizontal or pendulous flowers, such as *N. tazetta* and *N. triandrus,* respectively. Although protandry could potentially reduce self-pollen inteference, populations frequently experience low pollinator visitation rates, resulting in delays in the removal of self-pollen. Under these circumstances, ovule discounting could be ecologically important. As demonstrated in our models, differences in ovule discounting between the morphs in conjunction with asymmetrical pollen transfer could help explain the predominance of the long-styled morph that characterizes many sexually polymorphic populations of *Narcissus* species. Investigations of the schedule and amount of self- and outcross pollen deposition on stigmas of the morphs and measures of ovule discounting are required to assess whether self-pollen interference plays a significant role in the floral biology of *Narcissus*.

Acknowledgments

We thank several colleagues especially J. Fernández Casas for help in locating populations of *Narcissus;* Fanny Strumas for carrying out floral measurements; William Cole for assistance in the field, data analysis, and preparation of the figures; Lawrence Harder for help with development of the quantitative models of selection for stigma-height polymorphisms; Carlos Herrera for providing an unpublished manuscript; Tammy Sage for advice on late-acting incompatibility; and Rivka Dulberger and Amots Dafni for encouragement to study *Narcissus*. The manuscript benefited from the constructive comments of Deborah Charlesworth, Rivka Dulberger, Sean Graham, Lawrence Harder, and Jennifer Richards. Research was funded by grants from the Natural Sciences and Engineering Research Council of Canada to SCHB and grant DGICYT PB91-0894 to JA. This chapter is dedicated to the memory of the late Professor Abílio Fernandes (University of Coimbra, Portugal) for his life-long contributions to our knowledge of the genus *Narcissus*.

References

Arroyo, J. and A. Dafni. 1995. Variations in habitat, season, flower traits and pollinators in dimorphic *Narcissus tazetta* L. (Amaryllidaceae) in Israel. *New Phytologist* 129: 135–146.

Baker, H.G. 1964. Variation in style length in relation to outbreeding in *Mirabilis* (Nyctaginaceae). *Evolution*, 18: 507–512.

Baker, H.G. 1966. The evolution, functioning and breakdown of heteromorphic incompatibility systems. I. The Plumbaginaceae. *Evolution*, 20: 349–368.

Barrett, S.C.H. 1990. The origin and adaptive significance of heterostyly. *Trends Ecol. Evol.*, 5: 144–148.

Barrett, S.C.H. (ed.) 1992*a*. *Evolution and Function of Heterostyly*. Springer-Verlag, Berlin.

Barrett, S.C.H. 1992*b*. Gender variation and the evolution of dioecy in *Wurmbea dioica* (Liliaceae). *J. Evol. Biol.*, 5: 423–444.

Barrett, S.C.H. 1993. The evolutionary biology of tristyly. In *Oxford Surveys in Evolutionary Biology*, Vol. 9 (D. Futuyma and J. Antonovics, eds.), Oxford Univ. Press, Oxford, pp. 283–326.

Barrett, S.C.H. and D.E. Glover. 1985. On the Darwinian hypothesis of the adaptive significance of tristyly. *Evolution*, 39: 766–774.

Barrett, S.C.H. and B.C. Husband. 1990. Variation in outcrossing rates in *Eichhornia paniculata*: The role of demographic and reproductive factors. *Pl. Sp. Biol.*, 5: 41–55.

Barrett, S.C.H. and J.S. Shore. 1987. Variation and evolution of breeding systems in the *Turnera ulmifolia* L. complex (Turneraceae). *Evolution*, 41: 340–354.

Bateman, A.J. 1952*a*. Trimorphism and self-incompatibility in *Narcissus*. *Nature*, 170: 496–497.

Bateman, A.J. 1952*b*. Self-incompatibility in angiosperms. I. Theory. *Heredity*, 6: 285–310.

Bateman, A.J. 1954. The genetics of *Narcissus*. I. Sterility. *Daffodil and Tulip Year Book* 19: 23–29. Royal Horticultural Society, London.

Bateman, A.J. 1968. The role of heterostyly in *Narcissus* and *Mirabilis*. *Evolution*, 22: 645–646.

Bawa, K.S. and P.A. Opler. 1975. Dioecism in tropical forest trees. *Evolution*, 29: 167–179.

Bertin, R.I. 1993. Incidence of monoecy and dichogamy in relation to self-fertilization. *Am. J. Bot.*, 80: 557–583.

Bertin, R.I. and C.M. Newman. 1993. Dichogamy in angiosperms. *Bot. Rev.*, 59: 112–152.

Bertin, R.I. and M. Sullivan. 1988. Pollen interference and cryptic self-fertility in *Campsis radicans*. *Am. J. Bot.*, 75: 1140–1147.

Blanchard, J.W. 1990. *Narcissus: A Guide to Wild Daffodils*. Alpine Garden Society, Surrey UK.

Breese, E.L. 1959. Selection for different degrees of out-breeding in *Nicotiana rustica*. *Ann. Bot.*, 23: 331–344.

Broyles, S.B. and R. Wyatt. 1993. The consequences of self-pollination in *Asclepias exaltata*, a self-incompatible milkweed. *Am. J. Bot.*, 80: 41–44.

Charlesworth, B. and D. Charlesworth. 1979*a*. The maintenance and breakdown of distyly. *Am. Nat.*, 114: 499–513.

Charlesworth, D. 1979. The evolution and breakdown of tristyly. *Evolution*, 33: 486–498.

Charlesworth, D. and B. Charlesworth. 1979*b*. A model for the evolution of distyly. *Am. Nat.*, 114: 467–498.

Charlesworth, D. and B. Charlesworth. 1987. Inbreeding depression and its evolutionary significance. *Ann. Rev. Ecol. Syst.*, 18: 237–268.

Cope, F.W. 1962. The mechanism of pollen incompatibility in *Theobroma cacao* L. *Heredity*, 17: 157–182.

Crié, L. 1884. Sur le polymorphisme floral du Narcisse des îles Glénans (Finistère). *Comptes Rendus de l'Académie des Sciences*, 98: 1600–1601.

Crowe, L.K. 1971. The polygenic control of outbreeding in *Borago officinalis*. *Heredity*, 27: 111–118.

Darwin, C. 1877. *The Different Forms of Flowers on Plants of the Same Species*. John Murray, London, UK.

Des Abbayes, H. 1935. Contribution a l'étude du Narcisse des îles Glenans (Finistère). *Bull. Soc. Sci. Bretagne*, 12: 1–9.

Dulberger, R. 1964. Flower dimorphism and self-incompatibility in *Narcissus tazetta* L. *Evolution*, 18: 361–363.

Dulberger, R. 1967. Pollination Systems in Plants in Israel. Ph.D. thesis, Hebrew Univ., Jerusalem.

Dulberger, R. 1970. Floral dimorphism in *Anchusa hybrida* Ten. *Isr. J. Bot.*, 19: 37–41.

Eckert, C.G. and S.C.H. Barrett. 1994. Tristyly, self-compatibility and floral variation in *Decodon verticillatus* (Lythraceae). *Biol. J. Linn. Soc.*, 53: 1–30.

Ennos, R.A. 1981. Quantitative studies of the mating system in two sympatric species of *Ipomoea* (Convolvulaceae). *Genetica*, 57: 93–98.

Fernandes, A. 1935. Remarque sur l'hétérostylie de *Narcissus triandrus* et de *N. reflexus* Brot. *Bol. Soc. Brot.*, Sér. 2, 10: 5–15.

Fernandes, A. 1951. Sur la phylogénie des espèces du genre *Narcissus* L. *Bol. Soc. Brot.*, Sér. 2, 25: 113–190.

Fernandes, A. 1964. Contribution à la connaissance de la génétique de l'hétérostylie chez le genre *Narcissus* L. I. Résultats de quelques croisements. *Bol. Soc. Brot.*, Sér. 2, 38: 81–96.

Fernandes, A. 1965. Contribution à la connaissance de la génétique de l'hétérostylie chez le genre *Narcissus* L. II. L'hétérostylie chez quelques populations de *N. triandrus* var. *cernuus* et *N. triandrus* var. *concolor*. *Genét. Ibérica*, 17: 215–239.

Fernandes, A. 1967. Contribution à la connaissance de la biosystématique de quelques espèces du genre *Narcissus* L. *Portugaliae Acta Biologica* (B), 9: 1–44.

Fernandes, A. 1975. L'évolution chez le genre *Narcissus* L. *Anal. Inst. Bot. Cavanilles*, 32 (2): 843–872.

Ganders, F.R. 1974. Disassortative pollination in the distylous plant *Jepsonia heterandra*. *Can. J. Bot.*, 52: 2401–2406.

Ganders, F.R. 1979. The biology of heterostyly. *N. Zeal. J. Bot.*, 17: 607–635.

Gibbs, P.E. and M. Bianchi. 1993. Post-pollination events in species of *Chorisia* (Bombacaceae) and *Tabebuia* (Bignoniaceae) with late-acting self-incompatibility. *Bot. Acta*, 106: 64–71.

Harder, L.D. and S.C.H. Barrett. 1993. Pollen removal from tristylous *Pontederia cordata*: Effects of anther position and pollinator specialization. *Ecology*, 74: 1059–1072.

Harder, L.D. and S.C.H. Barrett. 1995. Mating cost of large floral displays in hermaphrodite plants. *Nature* 373: 512–515.

Henriques, J.A. 1887. Observações sobre algumas especies de *Narcissus*, encontrados em Portugal. *Bol. Soc. Brot.*, 5: 168–174.

Henriques, J.A. 1888. Additamento ao catalogo das Amaryllideas de Portugal. *Bol. Soc. Brot.*, 6: 45–47.

Herrera, C.M. 1995. Floral biology, microclimate, and pollination by ectothermic bees in an early blooming herb. *Ecology*, 76: 218–228.

Heuch, I. 1979. Equilibrium populations of heterostylous plants. *Theor. Popul. Biol.*, 15: 43–57.

Holsinger, K.E., M.W. Feldman, and F.B. Christiansen. 1984. The evolution of self-fertilization in plants: A population genetic model. *Am. Nat.*, 124: 446–453.

Husband, B.C. and D.W. Schemske. 1995. Evolution of the magnitude and timing of inbreeding depression in plants. *Evolution* (in press).

Jarne, P. and D. Charlesworth. 1993. The evolution of the selfing rate in functionally hermaphrodite plants and animals. *Ann. Rev. Ecol. Syst.*, 24: 441–466.

Jefferson-Brown, M. 1991. *Narcissus*. Timber Press, Inc., Portland, OR.

Jernstedt, J.A. 1982. Floral variation in *Chlorogalum angustifolium* (Liliaceae). *Madroño*, 29: 87–94.

Kohn, J.R. and S.C.H. Barrett. 1992*a*. Floral manipulations reveal the cause of male fertility variation in experimental populations of tristylous *Eichhornia paniculata* (Pontederiaceae). *Func. Ecol.*, 6: 590–595.

Kohn, J.R. and S.C.H. Barrett. 1992*b*. Experimental studies on the functional significance of heterostyly. *Evolution*, 46: 43–55.

Kohn, J.R. and S.C.H. Barrett. 1994. Pollen discounting and the spread of a selfing variant in tristylous *Eichhornia paniculata*: Evidence from experimental populations. *Evolution* 48: 1576–1594.

Krebs, S.L. and J.F. Hancock. 1991. Embryonic genetic load in the highbush blueberry, *Vaccinium corymbosum* (Ericaceae). *Am. J. Bot.*, 78:1427–1437.

Lewis, D. and D.A. Jones. 1992. The genetics of heterostyly. In *Evolution and Function of Heterostyly* (S.C.H. Barrett, ed.), Springer-Verlag, Berlin, pp. 129–150.

Lloyd, D.G. 1980. Sexual strategies in plants. III. A quantitative method for describing the gender of plants. *N. Zeal. J. Bot.*, 18: 103–108.

Lloyd, D.G. 1992. Self- and cross-fertilization in plants. II. Selection of self-fertilization. *Int. J. Plant Sci.*, 153: 370–382.

Lloyd, D.G. and C.J. Webb. 1986. The avoidance of interference between the presentation of pollen and stigmas in angiosperms. I. Dichogamy. *N. Zeal. J. Bot.*, 24: 135–162.

Lloyd, D.G. and C.J. Webb. 1992*a*. The evolution of heterostyly. In *Evolution and Function of Heterostyly* (S.C.H. Barrett, ed.), Springer-Verlag, Berlin, pp. 151–178.

Lloyd, D.G. and C.J. Webb. 1992*b*. The selection of heterostyly. In *Evolution and Function of Heterostyly* (S.C.H. Barrett, ed.), Springer-Verlag, Berlin, pp. 179–208.

Lloyd, D.G. and J.M.A. Yates. 1982. Intrasexual selection and the segregation of pollen and stigmas in hermaphrodite plants, exemplified by *Wahlenbergia albomarginata* (Campanulaceae). *Evolution*, 36: 903–913.

Lloyd, D.G., C.J. Webb, and R. Dulberger. 1990. Heterostyly in species of *Narcissus* (Amaryllidaceae) and *Hugonia* (Linaceae) and other disputed cases. *Plant Syst. Evol.*, 172: 215–27.

Meyer, F.G. 1961. Exploring for wild *Narcissus*. *Am. Hort. Mag.*, 40: 212–220.

Murcia, C. 1990. Effect of floral morphology and temperature on pollen receipt and removal in *Ipomoea trichocarpa*. *Ecology*, 71: 1098–1109.

O'Brien, S.P. and D.M. Calder. 1989. The breeding biology of *Epacris impressa*. Is this species heterostylous? *Austr. J. Bot.*, 37: 43–54.

Palmer, M., J. Travis, and J. Antonovics. 1989. Temporal mechanisms influencing gender expression and pollen flow within a self-incompatible perennial, *Amianthium muscaetoxicum* (Liliaceae). *Oecologia*, 78: 231–236.

Philipp, M. and O. Schou. 1981. An unusual heteromorphic incompatibility system: Distyly, self-incompatibility, pollen load and fecundity in *Anchusa officinalis* (Boraginaceae). *New Phytol.*, 89: 693–703.

Pijl, L. van der. 1978. Reproductive integration and sexual disharmony in floral functions. In *The Pollination of Flowers by Insects* (A.J. Richards, ed.), Linn. Soc. Symp. Ser. No. 6, London, UK. pp. 79–88.

Piper, J. and B. Charlesworth. 1986. The evolution of distyly in *Primula vulgaris*. *Biol. J. Linn. Soc.*, 29: 123–137.

Richards, A.J. 1986. *Plant Breeding Systems*. G. Allen and Unwin, London.

Richards, J.H. and S.C.H. Barrett. 1992. The development of heterostyly. In *Evolution and Function of Heterostyly* (S.C.H. Barrett, ed.), Springer-Verlag, Berlin, pp. 85–127.

Richards, J.H. and S. Koptur. 1993. Floral variation and distyly in *Guettarda scabra* (Rubiaceae). *Am. J. Bot.*, 80: 31–40.

Ritland, K. 1991. A genetic approach to measuring pollen discounting in natural plant populations. *Am. Nat.*, 138: 1049–1057.

Riveros, M., M.T. Kalin Arroyo, and A.M. Humaña. 1987. An unusual kind of distyly in *Quinchamalium chilense* (Santalaceae) on Volcan Casablanca, Southern Chile. *Am. J. Bot.*, 74: 313–320.

Robertson, A.W. and D.G. Lloyd. 1991. Herkogamy, dichogamy and self-pollination in six species of *Myosotis* (Boraginaceae). *Evol. Trends Plants*, 5: 53–63.

Sage, T.L., R.I. Bertin, and E.G. Williams. 1994. Ovarian and other late-acting self-incompatibility systems. In *Genetic Control of Self-Incompatibility and Reproductive Development in Flowering Plants* (E.G. Williams, R.B. Knox, A.E. Clarke, eds.), Kluwer Academic Publishers, Dordrecht, The Netherlands, pp. 116–140.

Schou, O. and M. Philipp. 1983. An unusual heteromorphic incompatibility system II. Pollen tube growth and seed sets following compatible and incompatible crossings within *Anchusa officinalis* L. (Boraginaceae). In *Pollen: Biology and Implications for Plant Breeding*. D.L. Mulcany ed. Elsevier, New York pp. 219–227.

Schou, O. and M. Philipp. 1984. An unusual heteromorphic incompatibility system. III. On the genetic control of distyly and self-incompatibility in *Anchusa officinalis* L. (Boraginaceae). *Theor. Appl. Genet.*, 68: 139–144.

Scribailo, R.W. and S.C.H. Barrett. 1994. Effects of prior self-pollination on outcrossed seed set in tristylous *Pontederia sagittata* (Pontederiaceae). *Sex. Plant Reprod.* 7: 273–281.

Seavey, S.R. and K.S. Bawa. 1986. Late-acting self-incompatibility in angiosperms. *Bot. Rev.*, 52: 195–219.

Shore, J.S. and S.C.H. Barrett. 1984. The effect of pollination intensity and incompatible pollen on seed set in *Turnera ulmifolia* (Turneraceae). *Can. J. Bot.*, 62: 1298–1303.

Sorensen, F. 1969. Embryonic genetic load in Coastal Douglas-Fir *Pseudotsuga menziesii* var. *menziesii*. *Am. Nat.*, 103: 389–398.

Stone, J.L. and J.D. Thomson. 1994. The evolution of distyly: Pollen transfer in artificial flowers. *Evolution* 48: 1595–1606.

Thomson, J.D. and D.A. Stratton. 1985. Floral morphology and cross-pollination in *Erythronium grandiflorum* (Liliaceae). *Am. J. Bot.*, 72: 433–437.

Thornhill, N.W. (ed.) 1993. *The Natural History of Inbreeding and Outbreeding*. Univ. Chicago Press, Chicago, IL.

Valdés, B. 1987. *Narcissus*. In *Flora Vascular de Andalucia Occidental*, (B. Valdes, S. Talavera, and E. Fernández-Galiano, eds.), Vol. 3 Ketres, Barcelona, Spain, pp. 463–474.

Vogel, S. and D. Müller-Doblies. 1975. Eine nachtblütige Herbst-Narzisse Zwiebelbau und Blütenokologie von *Narcissus viridiflorus* Schousboe. *Bot. Jahrb. Syst.*, 96: 427–447.

Waser, N.M. and M.V. Price. 1991. Reproductive costs of self-pollination in *Ipomopsis aggregata* (Polemoniaceae). *Am. J. Bot.*, 78: 1036–1043.

Webb, D.A. 1980. *Narcissus*. In *Flora Europaea* 5 (T.G. Tutin, V.H. Heywood, N.A. Burges, D.M. Moore, D.H. Valentine, S.M. Walters, and D.A. Webb, eds.), Cambridge Univ. Press, Cambridge, pp. 78–84.

Webb, C.J. and D.G. Lloyd. 1986. The avoidance of interference between the presentation of pollen and stigmas in angiosperms. II. Herkogamy. *N. Zeal. J. Bot.*, 24: 163–178.

Weller, S.G. 1992. Evolutionary modifications to tristylous breeding systems. In *Evolution and Function of Heterostyly* (S.C.H. Barrett, ed.), Springer-Verlag, Berlin, pp. 247–272.

Wiens, D., D.L. Nickrent, C.I. Davern, C.L. Calvin, and N.J. Vivrette. 1989. Developmental failure and loss of reproductive capacity in the rare palaeoendemic shrub *Dedeckera eurekensis*. *Nature*, 338: 65–66.

Wolley-Dod, C. 1886. Polymorphism of organs in *Narcissus triandrus*. *Gard. Chronicle*, April: 468.

Yeo, P.F. 1975. Some aspects of heterostyly. *New Phytol.*, 75: 147–153.

14

Evolution of *Campanula* Flowers in Relation to Insect Pollinators on Islands

Ken Inoue, Masayuki Maki,† and Michiko Masuda†*

Introduction

Since Darwin's observations on the biota of the Galápagos, the evolution of island plants has attracted the attention of many investigators (Carlquist, 1974). Adaptive radiation of organisms on oceanic islands continues to be one of the most interesting topics in evolutionary biology. Opportunities for studying evolution are excellent on islands because of the high levels of variation. The development of the theory of island biogeography has offered a basis for the study of the ecology and evolution of island biotas in relation to neighboring continents (MacArthur and Wilson, 1967; MacArthur, 1972). The theory predicts that the number of species on an island will decrease with distance from the mainland and that species with greater colonizing ability are more likely to establish on a given island. Plant species on islands may evolve through interactions with flora and fauna that are different from those on the nearby mainland.

Plant breeding systems have received a great deal of attention because of the variation they exhibit in association with the colonization of islands. The most well-known example is the prevalence of self-compatibility in island plants (Baker, 1955, 1967; Stebbins, 1957; Cox, 1989). Self-incompatible species have a disadvantage when colonizing islands, especially after long dispersal, since at least two individuals must colonize an area for successful establishment to occur. On the other hand, a self-compatible species can establish from a single propagule. Paucity of certain types of pollinators on islands may also make self-compatibility more advantageous (Stebbins, 1957). In several species, the breakdown of self-

*Biological Institute and Herbarium, Faculty of Liberal Arts, Shinshu University, 3-1-1 Asahi, Matsumoto 390, Japan.

†Department of Biology, College of Arts and Science, University of Tokyo, Tokyo, Japan.

incompatibility is reported in island populations (Strid, 1970; McMullen, 1987). In particular, the breakdown of heterostyly is often observed in island populations, providing an opportunity to investigate the ecological factors responsible for the evolution of self-fertilization. This situation has been intensively investigated by Barrett and co-workers (Barrett and Shore, 1987; Barrett et al., 1989; Barrett and Husband, 1990).

A high proportion of dioecious plants occur on islands, especially in the Hawaiian Islands (Carlquist, 1974; Bawa, 1982; Baker and Cox, 1984), in comparison with most mainland floras. In addition, the incidence of cryptic dioecy was recently reported from the Ogasawara Islands (Japan) and Hawaiian Islands (Kawakubo, 1990; Mayer, 1990; see Mayer and Charlesworth, 1991). Thus, island floras offer excellent opportunities for studying the evolution of dioecy and other gender polymorphisms, as evidenced by recent research on *Schiedea* by Weller and Sakai (1990), *Wikstroemia* by Mayer and Charlesworth (1992), and *Bidens* by Sun and Ganders (1988). The prevalence of gender polymorphisms on islands is, however, seemingly inconsistent with the reproductive assurance argument. The incidence of dioecism on islands has been interpreted as one evolutionary response to avoid the deleterious effects of inbreeding depression, which often accompanies the loss of self-incompatibility (Baker, 1967). Pollinator availability may be critical in determining whether outcrossing evolves after the breakdown of self-incompatibility. Even if outcrossed progeny have higher fitness, outcrossing may be difficult to evolve when pollinator availability is low. Thus, it is expected that variation in mating systems may occur if pollinator faunas vary among different islands.

As expected, island populations often show wide variation in outcrossing rates (Glover and Barrett, 1986; Barrett and Husband, 1990; Inoue and Kawahara, 1990) and therefore represent model systems to investigate the evolutionary maintenance of mixed mating systems in plants. Theoretical models that incorporate inbreeding depression and the cost of outcrossing predict that obligate outbreeding or obligate inbreeding are stable endpoints of mating-system evolution (Fisher, 1941; Lande and Schemske, 1985) and mixed mating cannot be evolutionarily stable. However, many insect-pollinated plants show mixed mating and this has led to a debate on whether mixed mating can be evolutionarily stable under certain conditions (Aide, 1986; Schemske and Lande, 1986, 1987; Waller, 1986). Several models explain the occurrence of stable mixed mating (Uyenoyama, 1986; Iwasa, 1990; Yahara, 1992), and the validity of these models can be tested by measuring parameters in natural populations of island plants.

Considerable evidence indicates that pollinators have an important influence on the evolution of floral traits (Feinsinger, 1983; Waser, 1983; see also Chapters 9 and 10). Several studies have experimentally investigated the effects of pollinators on floral traits and demonstrated that pollinators indeed act as selective agents (Campbell et al., 1991; Stanton et al., 1991; Anderson and Widén, 1993). However, different pollinators may constitute different selective forces on floral traits and their interactions may be complicated (see Chapters 3 and 4). To understand the direct impact of pollinators on the evolution of floral traits, it is necessary to examine the evolution of floral characters resulting from a change in pollinators, but few studies have been conducted on this problem (but see Johnston, 1991). Regarding this point, islands provide excellent opportunities to carry out such studies, since plants on islands often show differentiation of floral traits from their mainland progenitors. It is possible to investigate the selective forces that promote such differentiation by comparing floral traits of mainland and island populations, as well as the pollinators that visit them. Such comparisons may enable inferences to be made on the different selective forces operating on populations.

In this study, we investigate the evolution of corolla size in island populations. Although the evolution of corolla size has recently received attention as one characteristic affected by the preference of pollinators (Charlesworth and Charlesworth, 1987*a;* Cohen and Shmida, 1993), there are few empirical studies on the topic (but see Chapters 10 and 11). In addition, we pay particular attention to variation in dichogamy and sex allocation to floral organs, since these aspects of floral biology are closely related to the evolution of mating systems (Charnov, 1982).

In this chapter, we present the results of our studies on floral differentiation of plants in the Izu Islands, a chain of small islands near mainland Japan (Honshu). In the Izu Islands several plant species exist that are differentiated from related species from mainland Honshu. These are often treated as varieties or sister species of the mainland taxa. *Campanula punctata* Lam. and *C. microdonta* Koidz. (Campanulaceae) are such a species pair and are the focus of the present study. In this chapter, we address the following specific questions: (1) Is observed floral differentiation the result of selection by pollinating insects? (2) If so, does selection operate directly through the behavior or morphology of pollinating insects or indirectly through a change in pollinator availability? (3) Can the observed change in breeding systems be explained by the present data and theoretical models?

Natural History of Campanula punctata *and* C. microdonta

Campanula punctata is a perennial herb that occurs in Honshu, mainland Japan. It over-winters as a rosette in lowland Japan and reproduces vegetatively with subterranean rhizomes. Vegetative shoots are produced more frequently before flowering. Flowering shoots disconnect from vegetative shoots during flowering and usually senesce after producing seeds. Flowers of *C. punctata* in lowland Japan are 5 cm long, campanulate, pendulous, white to pink, and open mainly in mid- to late June (Fig. 14.1, T).

Plants on the Izu Islands were formerly treated as a variety of *C. punctata* (*C. punctata* var. *microdonta*), but we consider it more appropriate to treat island

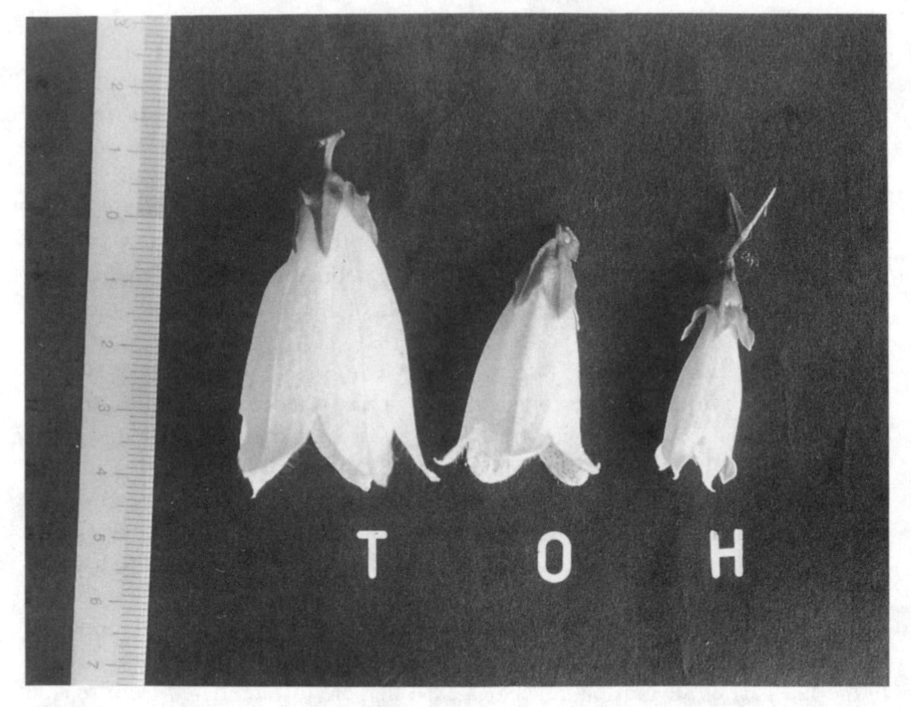

Figure 14.1. Flowers of *Campanula punctata* and *C. microdonta* showing variation in flower size. T, a flower from mainland, Honshu; O, from Oshima Island; H, from Hachijo Island.

populations as an independent species (*C. microdonta*) as several characteristics are different between mainland and island plants. Compared to mainland plants (*C. punctata*), island plants (*C. microdonta*) have several distinguishing features including smaller flowers (Fig. 14.1, H), larger number of flowers per plant, glabrous and broader leaves, smaller seeds, and later flowering (mainly mid–July). These characteristics are consistent among plants from the southern Izu Islands such as Hachijo Island and Miyake Island, but change somewhat in plants from the northern islands. For example, plants from Oshima Island have larger flowers than those from other islands (Fig. 14.1, O).

Description of the Izu Islands

The Izu Islands are a chain of islands extending from north to south near mainland Honshu (Fig. 14.2). We performed field work and sampled populations from all islands except Mikura Island (from north to south Oshima Island, Toshima Island, Niijima Island, Kozu Island, Miyake Island, and Hachijo Island). Oshima Island is the largest island (approx. 91 km^2) and is nearest to mainland Honshu, an

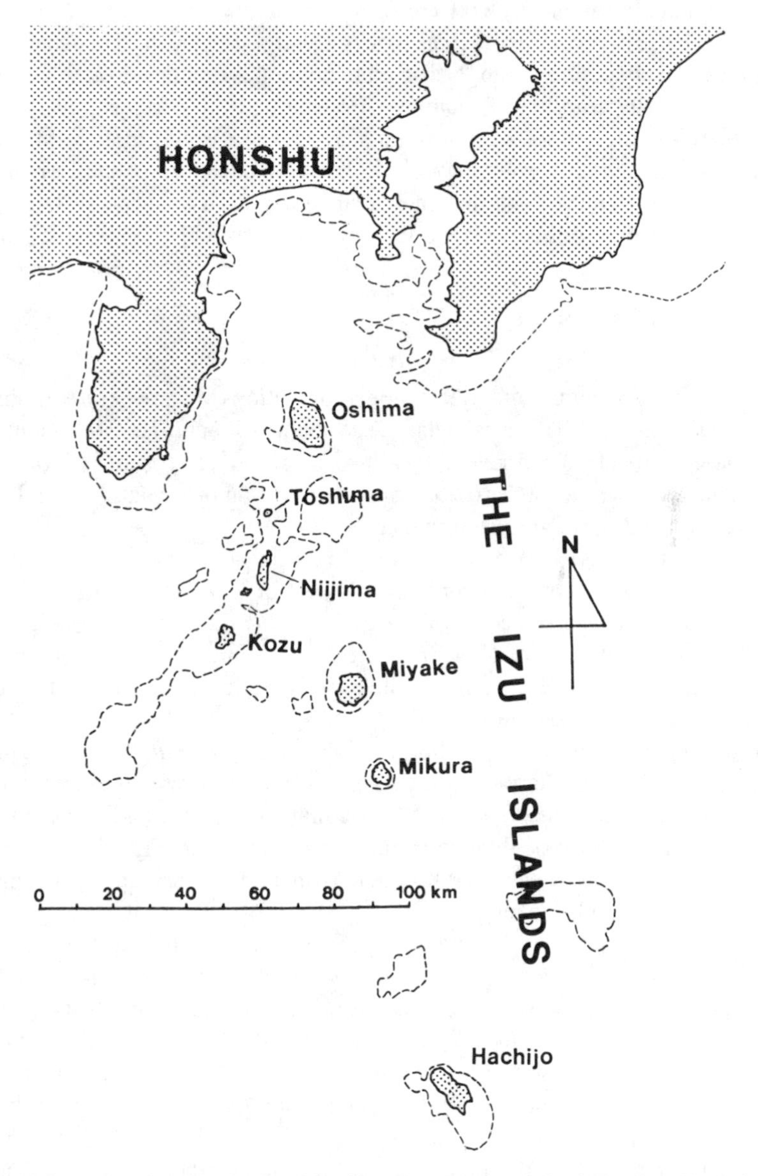

Figure 14.2. Map showing the Izu Islands and a part of mainland Honshu. Shaded areas represent present land surface. The broken line is the contour line at the 200-m sea depth.

approximate 20-km distance. The southernmost Hachijo Island is the next largest (68 km^2) and approximately 200 km from the mainland.

According to a recent geological theory (Karig, 1975), the Izu Islands were produced by subduction of the Pacific plate under the Philippine plate. Sea depth between the Izu peninsula of mainland Honshu and the northern Izu Islands is more than 400 m, and as the Würmian sea recession is 140 m at most, the Izu Islands and mainland Honshu have never been connected since the formation of the Izu Islands. This means that the organisms now living on the Izu Islands have established themselves on the islands after dispersal over the sea.

Variation in Floral Traits

We investigated several floral traits among populations of *C. punctata* (mainland) and *C. microdonta* (the Izu Islands). Data from mainland Honshu populations were mainly collected from several localities near Mount Kiyosumi, Chiba Prefecture. There is significant variation in such traits among mainland and island populations, and these are listed in Table 14.1.

Flowers of *C. punctata* from mainland Honshu are on average about 5 cm long and 3 cm wide. On the other hand, flowers of *C. microdonta* from Hachijo Island are nearly one-half the size of those of *C. punctata,* about 3 cm long and less than 2 cm wide. Flowers of *C. microdonta* from Oshima Island are intermediate between Honshu and Hachijo, and those of Toshima Island are intermediate between Oshima and Hachijo.

In common with other *Campanula* species, both *C. punctata* and *C. microdonta* exhibit protandry. The duration of the staminate phase varies considerably among populations, although the duration of the pistillate phase varies little. The duration of the staminate phase of flowers from Honshu and Oshima and from Hachijo and Miyake is approximately 50 h; whereas that of flowers from Toshima and Niijima is about 20 h longer.

We investigated the pollen/ovule (P/O) ratio of populations from different islands. P/O ratios decreased from north to south, with the distance between mainland and any given island ranging from 3220 in mainland plants to 580 on Hachijo Island. According to Cruden (1978), the values from Honshu and Oshima indicate xenogamy, those from Toshima and Niijima facultative xenogamy, and that from Hachijo indicates an intermediate mating system between facultative xenogamy and autogamy.

We also investigated the breeding systems of populations by bagging and self-pollination of flowers in the nursery and by bagging of flowers in natural populations. Most plants in Honshu and Oshima Island set few fruits and seeds after self-pollination, indicating strong self-incompatibility. Since a few plants set a small number of seeds, they may have a weak or leaky self-incompatibility system. Almost all plants on the other islands set large amounts of seed upon

Table 14.1. Variation in flower size, duration of staminate phase, pollen/ovule (P/O) ratios, and % fruit set in bagging among mainland Campanula punctata *and island* C. microdonta *populations (Inoue, 1988, 1990b; Masuda et al., unpublished). Flower size was measured in natural populations by collecting one flower from each plant. Data on duration of staminate phase were collected from plants grown from seed. Number of pollen grains and ovules were counted from buds collected from natural populations. Bagging experiments were conducted in natural populations.*

Population	Flower Length (mm) (SD, *N*)	Flower Width (mm) (SD, *N*)	Duration of Staminate Phase (hrs) (SD, *N*)	P/O Ratios (SD, *N*)	% Fruit Set in Bagging at Bud Stage (*N*)
Mainland (Honshu)	54.09 (3.78, 37)	24.45 (1.81, 37)	53.6 (16.1, 16)	3220 (270, 10)	5.5 (73)
Oshima Island	45.76 (5.20, 40)	19.45 (1.40, 40)	52.2 (12.9, 15)	2320 (390, 10)	11.3 (80)
Toshima Island	40.39 (4.15, 35)	18.22 (1.87, 35)	71.5 (12.9, 15)	1570 (200, 10)	100.0 (48)
Niijima Island	31.20 (2.40, 28)	16.02 (1.08, 28)	78.7 (16.0, 28)	1540 (210, 10)	100.0 (136)
Kozu Island	31.23 (3.14, 31)	13.92 (1.61, 31)	—	1130 (240, 10)	97.8 (46)
Miyake Island	30.90 (3.05, 30)	13.30 (1.56, 30)	59.3 (21.3, 20)	830 (80, 10)	90.6 (139)
Hachijo Island	30.59 (2.97, 32)	12.66 (1.52, 32)	48.4 (16.4, 18)	580 (80, 10)	97.0 (99)

N = Number of flowers.

self-pollination, indicating strong self-compatibility (Inoue and Amano, 1986; Inoue, 1988). This pattern involving a change from self-incompatibility to self-compatibility has also been observed in *Rhododendron kaempferi* at the Izu Islands (Inoue, 1993).

Outcrossing Rates

We estimated outcrossing rates t in island and mainland populations with two methods using allozyme markers. One method was based on Wright's fixation index (the deviation of heterozygotes from Hardy–Weinberg equilibria) and the other from open-pollinated progeny analysis (Table 14.2). For the former, we collected mature leaves from each population, assayed more than 30 plants electrophoretically, and calculated outcrossing rates from the fixation index f by the formula (Falconer, 1989)

$$t = \frac{(1 - f)}{(1 + f)}$$

This method assumes that outcrossing rates are constant among generations, inbreeding depression is absent among inbred plants (Ritland, 1989), and Wahlund effects are minimal. We also estimated multilocus outcrossing rates using progeny arrays. We collected families from populations, assayed seedlings grown in the laboratory, and estimated outcrossing rates by the method of Ritland and Jain (1981). The values obtained from the two methods corresponded roughly with each other (Spearman's correlation coefficient, $r_s = 0.943$, $P < 0.05$) (Table 14.2). Outcrossing rates using the multilocus method in the Honshu and

*Table 14.2. Estimated outcrossing rates of mainland (*Campanula punctata*) and island (*C. microdonta*) populations from multilocus estimates of progeny array (t; Maki et al., unpublished) and the deviations from Hardy–Weinberg equilibria (t_2; Inoue and Kawahara, 1990). Different letters after t indicate statistically significant differences (5% level) by Student's t-test modified by sequential Bonferonni (Rice, 1989). $Feq = (1-t)/(1+t)$ is the level of inbreeding expected in a population with a given outcrossing rate; f is Wright's fixation index; $\Delta F = f - Feq$.*

Population	t (SE) (No. of families, array, loci used)	Feq	t_2	f	ΔF
Mainland (Chiba)	0.851a (0.087) (19, 131, 7)	0.080	0.742	0.148	0.068
Oshima	0.931a (0.144) (31, 149, 5)	0.036	0.691	0.232	0.196
Toshima	0.333bc (0.082) (31, 277, 4)	0.500	0.535	0.303	−0.197
Niijima	0.557ab (0.173) (28, 161, 3)	0.285	0.566	0.277	−0.008
Kozu	— — —	0.366	0.464	—	—
Miyake	0.087c (0.077) (27, 159, 2)	0.840	0.157	0.729	−0.111
Hachijo	0.163bc (0.193) (28, 161, 1)	0.720	0.236	0.618	−0.101

Oshima populations were high, 85 and 93%, respectively. Outcrossing rates were considerably lower among the other island populations. At the northern Izu Island populations such as Toshima and Niijima values were 33 and 56%, respectively, indicating mixed mating system, whereas those at the southern Izu Islands of Hachijo and Miyake were 16 and 8%, respectively, indicating predominant self-fertilization. Estimated outcrossing rates decreased according to the distance from the mainland.

Table 14.2 indicates a tendency for heterozygote deficiency in the more outcrossing populations and heterozygote excess in populations with low outcrossing, although the sample size is small. This pattern has been reported in other studies (Brown, 1979; Barrett and Husband, 1990) and may be caused by the effects of inbreeding depression in facultative inbreeding populations and Wahlund effects in highly outcrossing ones.

Changes in Pollinator Fauna

It is generally known that islands have a smaller number of species than the corresponding mainland and animals or plants with specialized adaptations often find it difficult to establish on islands (MacArthur, 1972). This seems true for insect pollinators on the Izu Islands. Pollinator species decrease in number, and long-tongued pollinators such as bumble bees or swallowtails are absent or scarce. Elsewhere, the absence or scarcity of long-tongued pollinators has also been reported for other island systems (Carlquist, 1974; Feinsinger et al., 1982; Barrett et al., 1989). It may be difficult for social bees or hummingbirds with life spans exceeding the length of the flowering period to establish themselves on small islands. Pollinators on small islands, such as solitary bees or syrphid flies, have generally short adult life spans and/or small energy demands. Thus, mainland plants that are strongly dependent on long-tongued pollinators may find it particularly difficult to establish on small islands (Inoue, 1993).

Bombus diversus, the predominant pollinator of *C. punctata,* does not occur on the Izu Islands. Instead, another bumblebee species (*Bombus ardens*), occurring only on the Oshima Island, is one of the predominant pollinators of *C. microdonta*. On the other islands, no bumblebees have been observed and instead *C. microdonta* is pollinated mainly by halictid bees. At Toshima Island, one megachilid bee species also visits *C. microdonta* along with halictid bees (Inoue and Amano, 1986; Inoue, 1988). In mainland Honshu, megachilid bees are usually minor pollinators when *Bombus diversus* is present, but sometimes play an important role as pollinators when *B. diversus* is absent (Inoue, personal observation). On the mainland, bumblebees visiting *C. punctata* flowers will forage on almost all flowers in a patch, and individual flower are visited several times a day. The absence or scarcity of bumblebees is associated with lower pollinator availability on the islands. In Table 14.3, visitation frequencies of pollinating bees per hour at each population are shown (Inoue, 1990*a*). Visitation

Table 14.3. Pollinator availability in mainland Campanula punctata *and island* C. microdonta *populations, represented by pollinator visitation frequencies per flower per hour at peak flowering (modified from Inoue, 1990*a*).*

	Population						
Pollinating Bees	Mainland (Chiba)	Oshima	Toshima	Niijima	Kozu	Miyake	Hachijo
Bumblebees	0.93	0.56	0	0	0	0	0
Megachilid bees	0.18	0	0.29	0	0	0	0
Halictid bees	0.41	0.19	0.46	0.35	0.35	0.26	0.22
Anthophorid bees	0	0	0	0.09	0	0	0
Total	1.52	0.75	0.75	0.44	0.35	0.26	0.22
Observation time (h)	4.5	4.6	3.0	3.5	1.0	3.0	6.0

frequencies are lower in the Izu Islands compared with mainland Honshu and, in addition, decrease from north to south among the islands. These patterns suggest that the variation in *Campanula* flowers among mainland and island populations may result from the contrasting pollinator faunas present in the different areas.

Evolution of Flower Size

In the previous section, we described large variation in flower size among populations of *Campanula microdonta* and *C. punctata*. This variation raises the question: What factors determine the differences in flower size among *Campanula* populations? Two main hypotheses on the evolution of flower size have been proposed: the pollinator attraction hypothesis (see Willson, 1983; Bell, 1985; refer also to Cohen and Shmida, 1993) and the pollinator-matching hypothesis (Stebbins, 1971). The pollinator attraction hypothesis predicts that larger flowers attract pollinators more frequently than smaller flowers, thus increasing fitness. However, the costs of producing large flowers may also increase. Flower size is therefore determined by the balance between the benefits of increased reproductive success and the costs of producing larger flowers. The pollinator-matching hypothesis assumes that the size and form of flowers are determined by the size and shape of pollinators, and natural selection optimizes pollination efficiency per pollinator visit. Extremely large (or small) flowers may decrease the probability of pollen transfer and may be selected against.

Although these hypotheses are not mutually exclusive, each predicts relationships between pollinator responses or pollination efficiency and flower-size variation. The pollinator attraction hypothesis predicts that visitation frequencies of pollinating insects will increase with flower size as a decreasing function regardless of pollinator groups. However, the curves of visitation frequency against flower size may differ among different pollinating insects. If pollinator attraction is responsible for flower-size variation in *Campanula,* selection should favor the flower size when fitness measures such as visitation frequency/flower size are maximized. Pollinator attraction predicts that outcrossing rates may correlate positively with flower size. The pollinator matching hypothesis predicts a positive correlation between flower size and insect size, and that flower size may be almost a linear function of the pollinating insect size. Next, we examine correlates of flower size with other factors to discriminate between the two hypotheses.

Several Correlations

Figure 14.3 shows the relationship between outcrossing rate and mean flower width among the *Campanula* populations. Although there is a weak correlation between outcrossing rate and flower size (Pearson's correlation coefficient r = 0.834, $P < 0.05$; however, Kendall $\tau = 0.600$, NS), there are several inconsistencies. Flowers from the Toshima population are a little larger than those at Niijima, but the outcrossing rate of the former population is smaller. Further, flowers from the populations of Niijima, Miyake, and Hachijo are of similar size, but they exhibit a large variation in outcrossing rates. This suggests that outcrossing is not the only factor affecting selection on flower size in *Campanula*.

Figure 14.4 shows the relationship between mean pollinator body width and

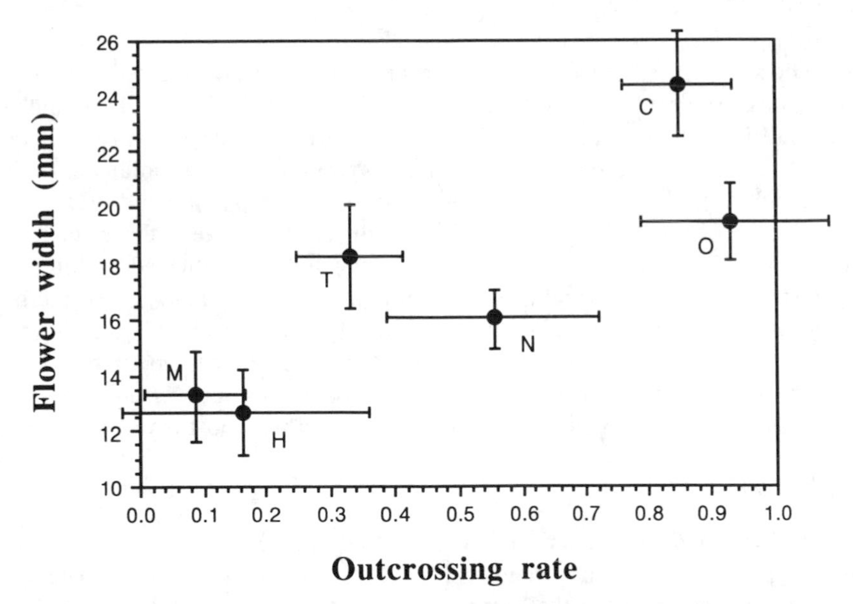

Figure 14.3. Relationship between outcrossing rate (with SE) and mean flower size (with SD) in island *Campanula microdonta* and mainland *C. punctata* populations. C, mainland (Chiba); H, Hachijo Island; and M, Miyake Island; N, Niijima Island; O, Oshima Island; T, Toshima Island.

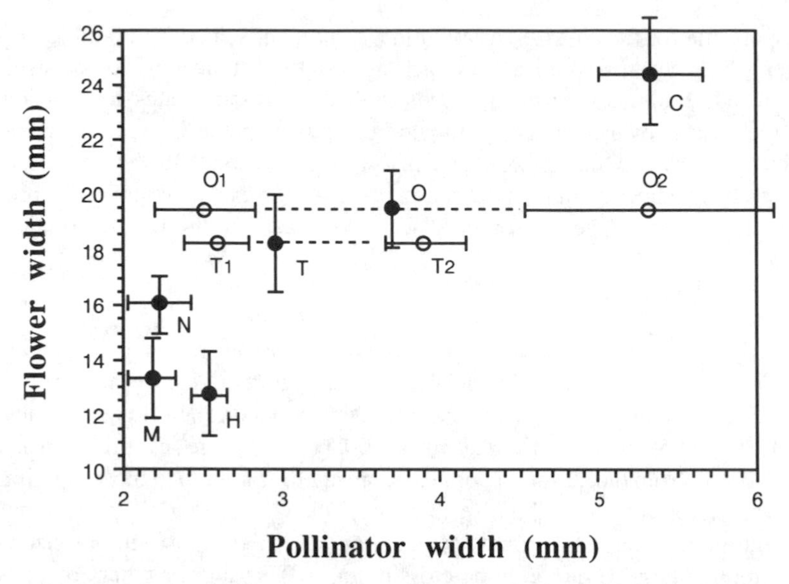

Figure 14.4. Relationship between mean pollinator body size (with SD) and mean flower size (with SD) in island *Campanula microdonta* and mainland *C. punctata* populations. C, mainland (Chiba); H, Hachijo Island; M, Miyake Island; N, Niijima Island; O, Oshima Island; T, Toshima Island.

mean flower width. The values of Oshima and Toshima were obtained by simply calculating the means of the two main pollinators [sample sizes: $N = 11$ (bumblebees) and 10 (halictid bees) for Oshima; $N = 5$ (megachilid bees) and 16 (halictid bees) for Toshima]. There is a significant correlation between pollinator body width and mean flower width (Pearson's correlation coefficient $r = 0.931$, $P < 0.01$; Kendall $'\tau = 0.733$, $P < 0.05$). If optimal flower size is determined by pollinator body size and optimal flower size in populations pollinated by different pollinators becomes intermediate between the sizes of small and large pollinators, intermediate flower size may be selected. This may have occurred in the Oshima and Toshima populations and supports the pollinator-matching hypothesis. However, these data do not directly reject other possibilities such as pollinator attraction and it is therefore necessary to test these other hypotheses by experiments.

Experimental Populations

In 1992, we conducted experiments to investigate the response of different bees to a variation in flower size. Plants collected from several islands and mainland Honshu were cultivated at the Botanical Gardens, University of Tokyo. In July when flowers were open, experimental populations were set up at Tokyo, and Nikko, Tochigi Prefecture containing considerable flower-size variation. Before bee visits occurred, each flower was measured for size, checked for sexual stage, and tagged for monitoring. Records on pollinator were then made including information on the species of bee, its visitation frequency, and residence time on each flower visited.

Figures 14.5a and 14.5b show the results of the experiment at Tokyo. Halictid bees and megachilid bees were major visitors to *Campanula* flowers. Preference for flower size between megachilid and halictid bees was different. Megachilid bees selectively foraged on larger flower (Jonckeere test, J = 493, $P < 0.001$) as predicted by the pollinator-attraction hypothesis, whereas halictid bees showed no significant preference for flower size (Jonckeere test, J = 20, NS).

Figures 14.5c and 14.5d show the results of the experiment at Nikko. Bumblebees, megachilid bees, and halictid bees were observed visiting *Campanula* flowers. Data for megachilid bees are not presented here because of small sample size; however, megachilid bees at Nikko showed the same tendency as observed at Tokyo. Halictid bees at Nikko showed no significant preference for flower size, indicating the same tendency as halictids at Tokyo (Jonckeere test, J =

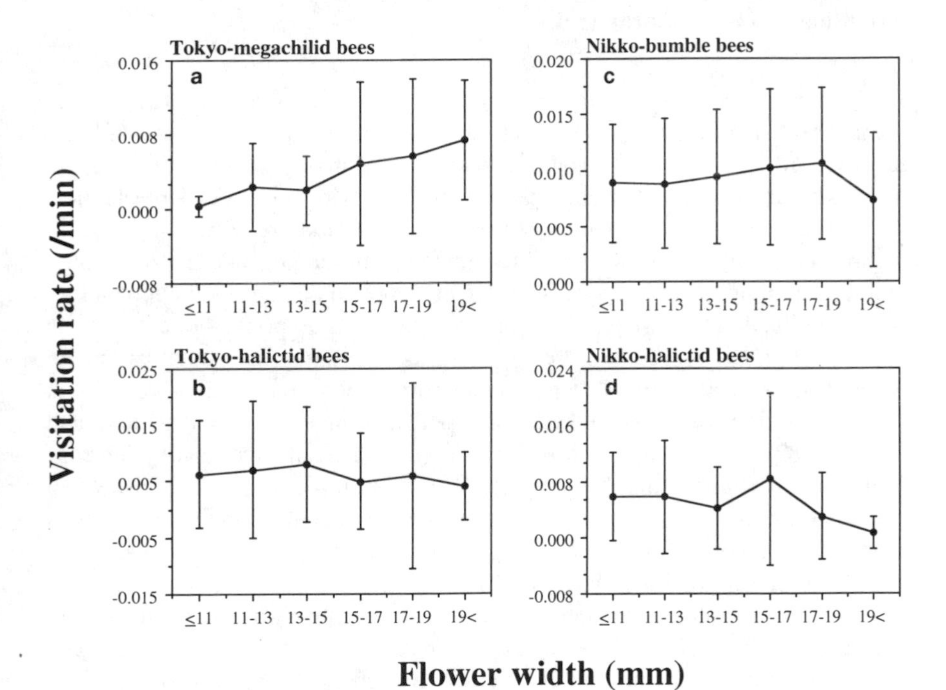

Figure 14.5. Mean visitation rates (with SD) of bee groups to experimental populations of *Campanula* showing flower-size variation in Tokyo and Nikko. Visitation rates of megachilid bees at Tokyo (a) and bumblebees at Nikko (c) were significantly different among flower sizes (Jonckeere test, $P<0.001$), whereas those of halictid bees at Tokyo (b) and Nikko (d) did not differ significantly. Observation time was 1300 min at Tokyo and 445 min at Nikko.

32, NS). Bumblebees at Nikko showed a preference for large flower size (Jonckeere test, $J = 149$, $P < 0.001$). Recall that plants with small flowers were collected from islands other than Oshima, where halictid bees are the main pollinators. The above results suggest that small flower size in southern island populations is maintained by selection against larger flower size. The preference of megachilids and bumblebees for large flower size may operate as one of several selective forces for large flower size in the Honshu, Oshima, and Toshima populations. It is notable that the above results do not contradict the size-matching hypothesis.

We assume in the above discussion that the visitation of pollinating insects is positively correlated with fitness. In some cases, this assumption may be incorrect (see Stanton et al., 1991; Wilson and Thomson, 1991). In the future it will be necessary to investigate the pollination efficiency of different pollinating insects and its relation to fitness.

Evolution of Other Floral Traits

Evolution of other floral traits such as duration of protandry, P/O ratio, and compatibility relationship may be associated with the evolution of diverse pollination systems in *Campanula*. Halictid bees with relatively small bodies carry a much smaller amount of pollen than bumblebees with relatively large bodies. To assure outcrossing by halictids with low visitation frequency, relative to bumblebees, it may be necessary to lengthen the duration of the staminate phase and rely on multiple visits by bees for effective pollen dispersal. The probability of self-pollination may increase if pollen grains adhere to pollen-collecting hairs after the stigmatic lobes open. Lengthening the duration of the staminate phase in protandrous self-compatible populations may therefore be interpreted as a mechanism to reduce self-pollination and pollen-stigma interference (Lloyd and Yates, 1982). Lengthening the duration of the staminate phase may increase energetic costs, and it may be appropriate to shorten the staminate phase when little outcross pollen is expected to be available in selfing populations such as in the Hachijo populations. Shortening of flower duration in selfing populations has also been reported by Aizen (1993).

Allocation to male reproductive structures in hermaphroditic plants should decrease with increasing self-fertilization rate (Charnov, 1982; Lloyd, 1987) and there is empirical support for this prediction (Schoen, 1982; McKone, 1987; Morgan and Barrett, 1989). A similar pattern was also observed in our study. P/O ratios are correlated with outcrossing rates, primarily as a result of reduced pollen production in selfing populations (Pearson's correlation coefficient, $r = 0.900$, $P < 0.02$; Fig. 14.6).

Plants of mainland and Oshima Island exhibit self-incompatibility, and plants on other island possess self-compatibility. It is noteworthy that the distribution

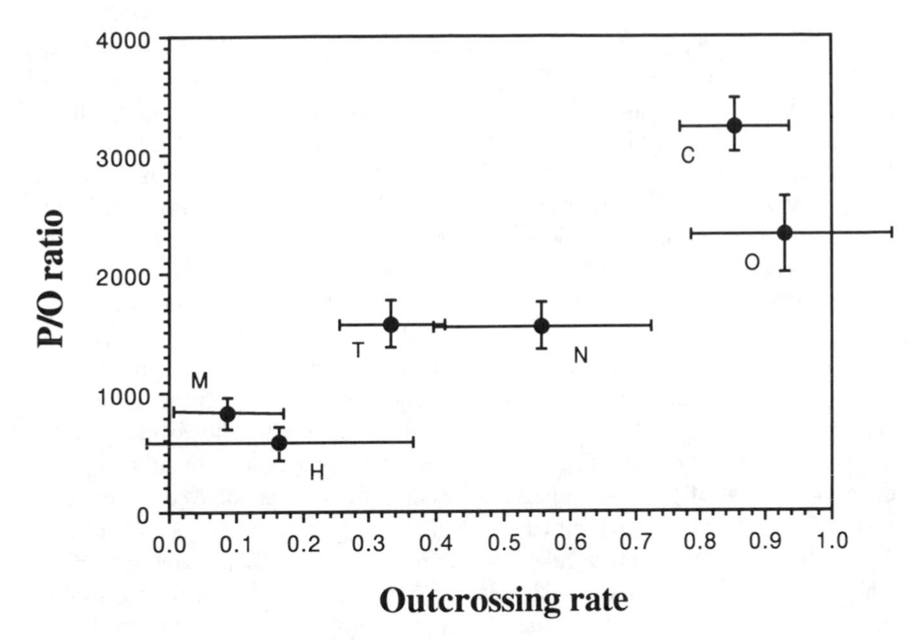

Figure 14.6. Relation of estimated outcrossing rates (with SE) and mean pollen-ovule ratios (with SD) in island *Campanula microdonta* and mainland *C. punctata* populations. C, mainland (Chiba); H, Hachijo Island; M, Miyake Island; N, Niijima Island; O, Oshima Island; T, Toshima Island.

of bumblebees and self-incompatible populations corresponds with each other. Regular pollinator service by bumblebees may aid in strengthening the self-incompatibility mechanism and promoting outbreeding. The population size of *Campanula* may also be associated with the maintenance of self-incompatibility. It may be difficult to maintain *S* alleles at the sporophytic self-incompatibility locus in the small populations that occur on islands other than Oshima (see Byers and Meagher, 1992). Oshima is the largest island in the Izu Islands, and population sizes are much larger than those on the smaller islands such as Toshima and Niijima. However, population size cannot explain the occurrence of self-compatibility in populations from Hachijo Island, as sizes there are larger than those from Oshima Island (Inoue, personal observation).

Evolution of Mating Systems

Inbreeding Depression

Inbreeding depression is thought to be one of the crucial factors in the evolution of mating systems. We therefore investigated inbreeding depression in the self-compatible populations of *C. microdonta* by collecting open-pollinated and self-pollinated seeds from Toshima, Niijima, and Miyake populations. Self-pollinated seeds were obtained by bagging flowers in the bud stage in natural habitats. Self-pollinated and open-pollinated seeds were obtained from more than 25 plants at each population. Seeds in each treatment were pooled, sown in 15 flats in the laboratory at Tokyo University, and the survival rates of each flat were counted. The performances of progeny were compared 10 months after sowing. Estimates of apparent inbreeding depression δ_o and estimated inbreeding depression δ are listed in Table 14.4. Apparent inbreeding depression is the difference in mean survival rates of open-pollinated seeds and self-pollinated seeds 10 months after sowing. Inbreeding depression δ is calculated under the formula $\delta = \delta_o/(t + \delta_o - t\delta_o)$, where t is the outcrossing rate (Yahara, unpublished; see Appendix 1). As data were obtained only until the seedling stage, the values for inbreeding depression are likely to be underestimates. It is noteworthy, however, that the largest values of inbreeding depression were found in the predominantly inbreeding population at Miyake. Most previous models predict that values of inbreeding depression will be small in inbreeding populations; however, our results indicate that inbreeding depression is relatively large in *Campanula* populations. Elsewhere, considerable inbreeding depression has been detected in highly selfing populations of *Clarkia temblorienesis* (Holtsford and Ellstrand, 1990) and *Begonia* spp. (Ågren and Schemske, 1993). In contrast, Barrett and Charlesworth (1991) reported no inbreeding depression in a selfing population of *Eichhornia paniculata* from the island of Jamaica. Although the reasons for these differences are obscure, high inbreeding depression may accumulate even in predominantly selfing populations if mutation rates for deleterious genes are high and a large number of loci are responsible for inbreeding depression (Charlesworth and Charlesworth, 1987*b*). It may be necessary to consider additional factors as well as inbreeding depression to explain mixed mating in plants. We next consider two hypotheses that address these issues.

Table 14.4. Inbreeding depression δ of Campanula microdonta *populations (Yahara et al., unpublished). Apparent inbreeding depression δ_o is the difference in mean survival rates between plants from open-pollinated seeds and those from self-pollinated seeds after 10 months' growth. Survival rates were the average of 15 flats from each population. Data on δ_o on Niijima Island were obtained from three replicates (mean±SD). Inbreeding depression δ is calculated under the formula $\delta = \delta_o/(t + \delta_o - t\delta_o)$, where t is the outcrossing rate.*

Population	t (SE)	δ_o	δ
Toshima	0.333 (0.082)	0.13	0.31
Niijima	0.557 (0.173)	0.12±0.04	0.20
Miyake	0.087 (0.077)	0.06	0.42

Two Hypotheses for Mixed Mating

As we have observed, pollinator availability or pollinator efficiency on islands is reduced in comparison with the mainland and, in addition, it decreases with the distance between the mainland and any given island. As a result, plants on isolated islands may suffer from pollinator limitation. If pollinator limitation is severe, selfing may be favored. On the other hand, if pollinator limitation is not severe, outbreeding may be advantageous. If pollinator efficiency changes over time, an intermediate selfing rate may become evolutionarily stable in an annual or a monocarpic perennial plant under certain conditions (Inoue, 1990*a;* Kakehashi and Inoue, unpublished; see Appendix 2; Fig. 14-7). On islands with low pollinator density, yearly fluctuations of pollinators may cause a large variation in pollinator efficiency. On the other hand, pollinator populations during flowering of mainland *C. punctata* populations are dense throughout the flowering season (Inoue, 1990*a*). Where yearly fluctuations in mainland pollinator populations occur, they may cause little variation in pollinator efficiency (Inoue, 1990*a*). Hence, mixed mating may evolve more easily in habitats with low pollinator availability such as islands.

The next hypothesis is concerned with resource allocation and reproductive assurance. *Campanula punctata* and *C. microdonta* are perennial; however,

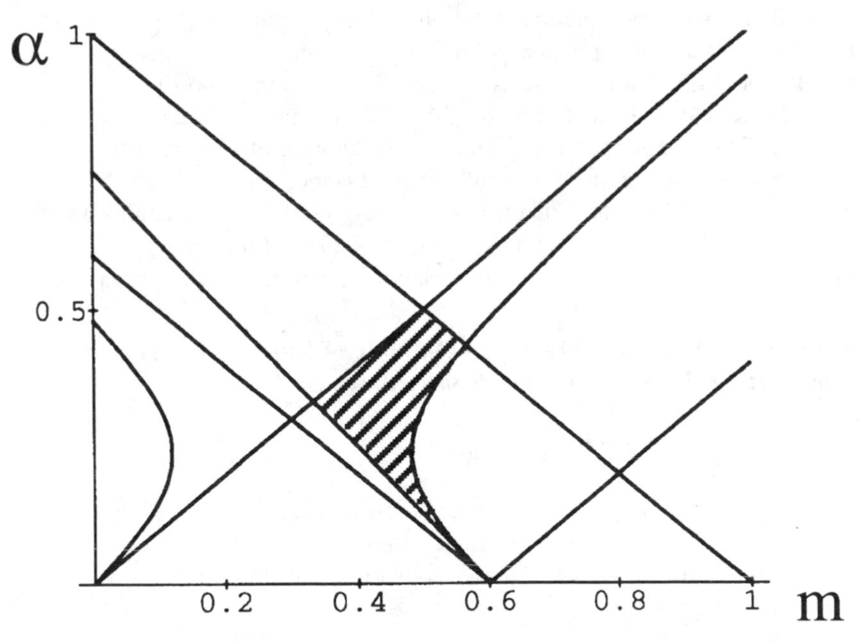

Figure 14.7. One example of the region where intermediate selfing rates are an ESS, where parameters $\delta=0.7$ and $r_1=0.1$, following the model of mixed mating with temporal variation in pollination efficiency (see Appendix 2).

shoots disconnect from one another after bolting and hence cannot recruit resources. If plants are self-incompatible and pollinators are scarce, fruit and seed set may be reduced. Actually, almost half the flowers in self-incompatible populations at Oshima set no fruit (Inoue, personal observation). Hence, there may be costs to maintaining self-incompatibility and outcrossing. On the islands other than Oshima, the costs to maintaining self-incompatibility may be too large because of low pollinator availability resulting in a shift to self-compatibility. In self-compatible campanulas, some degree of outcrossing occurs through protandry, and delayed self-pollination occurs if outcrossing fails (Inoue, personal observation; see Lloyd, 1980). When resources cannot be recruited in the following season, producing inbred seeds may be a better strategy than producing no seeds at all. In such situations, populations may inevitably show a mixed mating system.

Conclusions

Island populations of *Campanula microdonta* have floral characteristics different from mainland populations of *C. punctata*. These include smaller flowers, self-compatible breeding systems, smaller pollen/ovule (P/O) ratios, and a longer duration in the staminate stage. We found several correlations with these changes in floral characteristics and the change in pollinator fauna among mainland and island populations. Predominant pollinators change from bumblebees in mainland populations to halictid bees in island populations. Small flowers in island *C. microdonta* may be maintained by the behavior and morphology of halictid bees. Self-compatibility, low P/O ratio, and duration of staminate stage can be interpreted as adaptations to low pollinator efficiency on the islands. The mixed mating observed may be explained by the low pollinator availability in island populations. As gene flow among islands are usually limited, coevolution with local pollinators may be promoted in island populations of *C. microdonta*.

Appendix 1: Method to Obtain an Estimate of Inbreeding Depression from Open-Pollinated and Self-Pollinated Seeds*

Here we denote the following abbreviations:

w_o = mean fitness of outcrossed seeds
w_s = mean fitness of selfed seeds
w_f = mean fitness of open-pollinated seeds

We have

*T. Yahara (unpublished).

$$w_f = tw_o + (1 - t)w_s \tag{1}$$

where t is the outcrossing rate. Apparent inbreeding depression is

$$\delta_o = \frac{1 - w_s}{w_f} \tag{2}$$

Inbreeding depression is

$$\delta = \frac{1 - w_s}{w_o} \tag{3}$$

By eliminating w_o, w_s, and w_f from Equations (1–3), we can obtain the formula

$$\delta = \frac{\delta_o}{(t + \delta_o - t\delta_o)}$$

Appendix 2: A Model of Mixed Mating with Temporal Variation in Pollinator Efficiency†

We analyzed the effect of temporal variation in pollinator efficiency k, where k is defined as the proportion of ovules fertilized with activities of pollinators among nonselfed ovules ($0 < k \leq 1$). We assume that there are two types of years, $i = 1$ and $i = 2$, that take place with probability r_1 and r_2, respectively ($r_1 + r_2 = 1$). The pollinator efficiency is denoted by k_i in year i. Then the fitness in each year can be expressed as follows depending on the type of the years i:

$$\frac{\partial \phi(s|s^*)}{\partial s} = \phi_1^{r_1 - 1} \phi_2^{r_2 - 1} \left\{ r_1 \phi_2 \frac{\partial \phi_1}{\partial s} + r_2 \phi_1 \frac{\partial \phi_2}{\partial s} \right\} \tag{4}$$

In this equation, A and B represent strategies of plants. The quantity $\phi(B \mid A)$ represents the fitness of strategist B invading a population consisting of strategists A. The parameter b_x represents reproductive success via female function (seeds) and the parameter w_x via male function (pollen) of individuals with strategy X. Another parameter, s_x, is the selfing rate of strategists X. As for the other parameters, δ represents inbreeding depression.

From Equation (4), the average fitness in each year under temporal variation is given as follows calculating the geometric mean:

†M. Kakehashi and K. Inoue (unpublished).

$$\phi(B \mid A) = \phi_1(B \mid A)^{r_1}\, \phi_2(B \mid A)^{r_2} \tag{5}$$

By investigating the partial derivative of ϕ, we obtain the conditions that can result in an evolutionarily stable selfing rate, s^*. The intermediate selfing rate is the ESS when the following conditions are satisfied simultaneously:

$$\begin{gathered} 0 < k_1 \leq k_2 \leq 1 \\ \frac{1}{2}k_1 < 1 - \delta < \frac{1}{2}k_2 \\ \left\{r_1k_2\left(-\frac{k_1}{2} + 1 - \delta\right) + r_2k_1\left(-\frac{k_2}{2} + 1 - \delta\right)\right\}\left\{r_1\left(-\frac{k_1}{2} + 1 - \delta\right)\right. \\ \left. + r_2\left(-\frac{k_2}{2} + 1 - \delta\right)\right\} < 0 \end{gathered} \tag{6}$$

If we postulate $k_1 = m - \alpha$ and $k_2 = m + \alpha$, the above conditions are rewritten as follows:

$$\begin{gathered} 0 \leq \alpha \leq m \quad \text{and} \quad m + \alpha \leq 1 \\ \frac{m - \alpha}{2} < 1 - \delta < \frac{m + \alpha}{2} \\ \left\{r_1(m + \alpha)\left(-\frac{m - \alpha}{2} + 1 - \delta\right) + r_2(m - \alpha)\left(-\frac{m + \alpha}{2} + 1 - \delta\right)\right\} \\ \times \left\{r_1\left(-\frac{m - \alpha}{2} + 1 - \delta\right) + r_2\left(-\frac{m + \alpha}{2} + 1 - \delta\right)\right\} < 0 \end{gathered} \tag{7}$$

One example of the parameter space where intermediate selfing rates are the ESS is represented in Figure 14.7.

Acknowledgments

We thank Spencer Barrett and two anonymous reviewers for providing comments and editorial suggestions that improved an earlier draft of the paper. We also thank Tetsu Yahara and Masayuki Kakehashi for providing unpublished data and manuscripts, and Eiiti Kasuya for comments on statistical analysis. This work was partly supported by a Japanese Ministry of Education, Grant in Aid for Scientific Research on Priority Area (No. 319).

References

Ågren, J. and D.W. Schemske. 1993. Outcrossing rate and inbreeding depression in two annual monoecious herbs, *Begonia hirsuta* and *B. semiovata*. *Evolution,* 47: 125–135.

Aide, T.M. 1986. The influence of wind and animal pollination on variation in outcrossing rates. *Evolution,* 40: 434–435.

Aizen, M.A. 1993. Self-pollination shortens flower lifespan in *Portulaca umbraticola* H.B.K. (Portulacaceae). *Int. J. Pl. Sci.,* 154: 412–415.

Anderson, S. and B. Widén. 1993. Pollinator-mediated selection on floral traits in a synthetic population of *Senecio integrifolius* (Asteraceae). *Oikos,* 66: 72–79.

Baker, H.G. 1955. Self-compatibility and establishment after "long-distance" dispersal. *Evolution,* 9: 347–349.

Baker, H.G. 1967. Support for Baker's Law as a rule. *Evolution,* 21: 853–856.

Baker, H.G. and P.A. Cox. 1984. Further thoughts on dioecism and islands. *Ann. Missouri Bot. Gard.,* 71:244–253.

Barrett, S.C.H. and D. Charlesworth. 1991. Effects of a change in the level of inbreeding on the genetic load. *Nature,* 352: 522–524.

Barrett, S.C.H. and B.C. Husband. 1990. Variation in outcrossing rates in *Eichhornia paniculata*: The role of demographic and reproductive factors. *Plant Spec. Biol.,* 5: 41–55.

Barrett, S.C.H., M.T. Morgan, and B.C. Husband. 1989. The dissolution of a complex genetic polymorphism: The evolution of self-fertilization in tristylous *Eichhornia paniculata* (Pontederiaceae). *Evolution,* 43: 1398–1416.

Barrett, S.C.H. and J.S. Shore. 1987. Variation and evolution of breeding systems in the *Turnera ulmifolia* L. complex (Turneraceae). *Evolution,* 41: 340–354.

Bawa, K.S. 1982. Outcrossing and the incidence of dioecism in island floras. *Am. Nat.,* 119: 866–871.

Bell, G. 1985. On the function of flowers. *Proc. Roy. Soc. Lond. B,* 224: 223–265.

Brown, A.H.D. 1979. Enzyme polymorphism in plant populations. *Theor. Pop. Biol.,* 15: 1–42.

Byers, D.L. and T.R. Meagher. 1992. Mate availability in small populations of plant species with homomorphic sporophytic self-incompatibility. *Heredity,* 68: 353–359.

Campbell, D.R., N.M. Waser, M.V. Price, E.A. Lynch, and R.J. Mitchell. 1991. Components of phenotypic selection: Pollen export and flower corolla width in *Ipomopsis aggregata*. *Evolution,* 45: 1458–1467.

Carlquist, S. 1974. *Island Biology*. Columbia Univ. Press, New York.

Charlesworth, D. and B. Charlesworth. 1987*a*. The effect of investment in attractive structures on allocation to male and female functions in plants. *Evolution,* 41: 948–968.

Charlesworth, D. and B. Charlesworth. 1987*b*. Inbreeding depression and its evolutionary consequences. *Ann. Rev. Ecol. Syst.,* 18: 237–268.

Charnov, E. 1982. *The Theory of Sex Allocation*. Princeton Univ. Press, Princeton, NJ.

Cohen, D. and A. Shmida. 1993. The evolution of flower display and reward. *Evol. Biol.,* 27: 197–243.

Cox, P.A. 1989. Baker's law, plant breeding systems and island colonization. In *The*

Evolutionary Ecology of Plants (J.H. Bock and Y.B. Linhart, eds.), Westview Press, Boulder, CO, pp. 209–224.

Cruden, R.W. 1978. Pollen-ovule ratios: A conservative indicator of breeding systems in flowering plants. *Evolution,* 31: 32–46.

Falconer, D.S. 1989. *Introduction to Quantitative Genetics,* 3rd ed. Longman, Essex.

Feinsinger, P. 1983. Coevolution and pollination. In *Coevolution* (D.J. Futuyma and M. Slatkin, eds.), Sinauer, Sunderland, MA, pp. 282–310.

Feinsinger, P., J.A. Wolfe, and L.A. Swarm. 1982. Island ecology: Reduced hummingbird diversity and the pollination biology of plants, Trinidad and Tobago, West Indies. *Ecology,* 63: 494–506.

Fisher, R.A. 1941. Average excess and average effect of a gene substitution. *Ann. Eugen.,* 11: 53–63.

Glover, D.E. and S.C.H. Barrett. 1986. Variation in the mating system of *Eichhornia paniculata* (Spreng.) Solms. (Pontederiaceae). *Evolution,* 40: 1122–1131.

Holtsford, T.P. and N.C. Ellstrand. 1990. Inbreeding effects in *Clarkia tembloriensis* (Onagraceae) populations with different natural outcrossing rates. *Evolution,* 44: 2031–2046.

Inoue, K. 1988. Patterns of breeding system change in the Izu Islands in *Campanula punctata*: Bumblebee-absence hypothesis. *Plant Spec. Biol.,* 3: 125–128.

Inoue, K. 1990*a*. Evolution of mating systems in island populations of *Campanula microdonta*: Pollinator availability hypothesis. *Plant Spec. Biol.,* 5: 57–64.

Inoue, K. 1990*b*. Dichogamy, sex allocation, and mating system of *Campanula microdonta* and *C. punctata*. *Plant Spec. Biol.,* 5: 197–203.

Inoue, K. 1993. Evolution of mutualism in plant-pollinator interactions on islands. *Bioscience,* 18: 525–536.

Inoue, K. and M. Amano. 1986. Evolution of *Campanula punctata* Lam. in the Izu Islands: Changes of pollinators and evolution of breeding systems. *Plant Spec. Biol.,* 1: 89–97.

Inoue, K. and T. Kawahara. 1990. Allozyme differentiation and genetic structure in the island and mainland Japanese populations of *Campanula punctata* (Campanulaceae). *Am. J. Bot.,* 77: 1440–1448

Iwasa, Y. 1990. Evolution of the selfing rate and resource allocation models. *Plant Spec. Biol.,* 5: 19–30.

Johnston, M.O. 1991. Natural selection on floral traits in two species of *Lobelia* with different pollinators. *Evolution,* 45: 1468–1479.

Karig, D.E. 1975. Basin genesis in the Philippine Sea. *Init. Repts. Deep Sea Drilling Proj.,* 31: 857–879.

Kawakubo, N. 1990. Dioecism of the genus *Callicarpa* (Verbenaceae) in the Bonin (Ogasawara) Islands. *Bot. Mag. Tokyo,* 103: 57–66.

Lande, R. and D.W. Schemske. 1985. The evolution of self-fertilization and inbreeding depression. I. Genetic models. *Evolution,* 39: 24–40.

Lloyd, D.G. 1980. Demographic factors and mating patterns in angiosperms. In *Demography and Evolution in Plant Populations* (O.T. Solbrig, ed.), Blackwell, Oxford, pp. 67–88.

Lloyd, D.G. 1987. Allocations to pollen, seeds and pollination mechanisms in self-fertilizing plants. *Funct. Ecol.,* 1: 83–89.

Lloyd, D.G. and J.M.A. Yates. 1982. Intrasexual selection and the segregation of pollen and stigmas in hermaphrodite plants, exemplified by *Wahlenbergia albomarginata* (Campanulaceae). *Evolution,* 36: 903–913.

MacArthur, R.H. 1972. *Geographical Ecology,* Princeton Univ. Press, Princeton, NJ.

MacArthur, R.H. and E.O. Wilson 1967. *The Theory of Island Biogeography*. Princeton Univ. Press, Princeton, NJ.

Mayer, S.S. 1990. The origin of dioecy in Hawaiian *Wikstroemia* (Thymelaeaceae). *Mem. NY. Bot. Gard.,* 55: 76–82.

Mayer, S.S. and D. Charlesworth. 1991. Cryptic dioecy in flowering plants. *Trends in Ecol. Evol.,* 6: 320–325.

Mayer, S.S. and D. Charlesworth. 1992. Genetic evidence for multiple origins of dioecy in the Hawaiian shrub *Wikstroemia* (Thymelaeaceae). *Evolution,* 46: 207–215.

McKone, M.J. 1987. Sex allocation and outcrossing rate: A test of theoretical predictions using bromegrasses (*Bromus*). *Evolution,* 41: 591–598.

McMullen, C.K. 1987. Breeding systems of selected Galápagos Islands angiosperms. *Am. J. Bot.,* 74: 1694–1705.

Morgan, M.T. and S.C.H. Barrett. 1989. Reproductive correlates of mating system variation in *Eichhornia paniculata* (Spreng.) Solms (Pontederiaceae). *J. Evol. Biol.,* 2: 183–203.

Rice, W.R. 1989. Analyzing tables of statistical tests. *Evolution,* 43: 223–225.

Ritland, K. 1989. Gene identity and the genetic demography of plant populations. In *Plant Population Genetics, Breeding, and Genetic Resources* (A.H.D. Brown, M.T. Clegg, A.L. Kahler, and B.S. Weir, eds.), Sinauer Associates, Sunderland, MA, pp. 181–199.

Ritland, K. and S.K. Jain. 1981. A model for the estimation of outcrossing rate and gene frequencies using *n* independent loci. *Heredity,* 47: 35–52.

Schemske, D.W. and R. Lande. 1986. Mode of pollination and selection on mating system: A comment to Aide's paper. *Evolution,* 40: 436.

Schemske, D.W. and R. Lande. 1987. On the evolution of plant mating systems: A reply to Waller. *Am. Nat.,* 130: 804–806.

Schoen, D.J. 1982. Male reproductive effort and breeding system in an hermaphroditic plant. *Oecologia,* 53: 255–257.

Stanton, M., H.J. Young, N.C. Ellstrand, and J.M. Clegg. 1991. Consequences of floral variation for male and female reproduction in experimental populations of wild radish, *Raphanus sativus* L. *Evolution,* 45: 268–280.

Stebbins, G.L. 1957. Self fertilization and population variability in the higher plants. *Am. Nat.,* 91: 337–354.

Stebbins, G.L. 1971. Adaptive radiation of reproductive characteristics in angiosperms, I: Pollination mechanisms. *Ann. Rev. Ecol. Syst.*, 1: 307–326.

Strid, A. 1970. Studies in the Aegean flora. XVI. Biosystematics of the *Nigella arvensis* complex. *Opera Botanica,* 28: 1–169.

Sun, M. and F.R. Ganders. 1988. Mixed mating systems in Hawaiian *Bidens* (Asteraceae). *Evolution,* 42: 516–527.

Uyenoyama, M.K. 1986. Inbreeding and the cost of meiosis: The evolution of selfing in populations practicing biparental inbreeding. *Evolution,* 40: 388–404.

Waller, D.M. 1986. Is there disruptive selection for self-fertilization? *Am. Nat.*, 128: 421–426.

Waser, N.M. 1983. The adaptive nature of floral traits: Ideas and evidence. In *Pollination Biology* (L. Real, ed.), Academic Press, Orlando, FL, pp. 241–285.

Weller, S.G. and A.K. Sakai. 1990. The evolution of dicliny in *Schiedea* (Caryophyllaceae), an endemic Hawaiian genus. *Plant Spec. Biol.,* 5: 83–95.

Willson, M.F. 1983. *Plant Reproductive Ecology*. John Wiley & Sons, New York.

Wilson, P. and J.D. Thomson. 1991. Heterogeneity among floral visitors leads to discordance between removal and deposition of pollen. *Ecology,* 72: 1503–1507.

Yahara, T. 1992. Graphical analysis of mating system evolution in plants. *Evolution,* 46: 557–561.

Subject Index

Taxonomic Index